Fiber Diffraction Methods

Alfred D. French, Editor
Southern Regional Research Center

KennCorwin H. Gardner, Editor
E. I. Du Pont de Nemours & Company

Based on a symposium
sponsored by the
Cellulose, Paper and Textile Division
at the 178th Meeting of the
American Chemical Society,
Washington, D.C.,
September 10–14, 1979.

ACS SYMPOSIUM SERIES 141

AMERICAN CHEMICAL SOCIETY
WASHINGTON, D. C. 1980

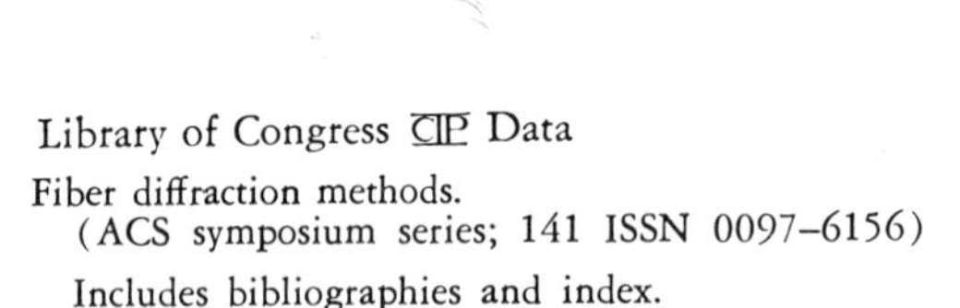

Library of Congress CIP Data

Fiber diffraction methods.
(ACS symposium series; 141 ISSN 0097-6156)

Includes bibliographies and index.

1. Polymers and polymerization—Analysis—Congresses. 2. Textile fibers, Synthetic—Analysis—Congresses. 3. X-rays—Diffraction—Congresses.
I. French, Alfred D., 1943– . II. Gardner, KennCorwin H., 1947– . III. American Chemical Society. Cellulose, Paper and Textile Division. IV. Series: American Chemical Society. ACS symposium series; 141.

QD380.F5 547.8′4046 80-21566
ISBN 0-8412-0589-2 ASCMC 8 141 1–518 1980

PRINTED IN THE UNITED STATES OF AMERICA

ACS Symposium Series

FOREWORD

The ACS SYMPOSIUM SERIES was founded in 1974 to provide a medium for publishing symposia quickly in book form. The format of the Series parallels that of the continuing ADVANCES IN CHEMISTRY SERIES except that in order to save time the papers are not typeset but are reproduced as they are submitted by the authors in camera-ready form. Papers are reviewed under the supervision of the Editors with the assistance of the Series Advisory Board and are selected to maintain the integrity of the symposia; however, verbatim reproductions of previously published papers are not accepted. Both reviews and reports of research are acceptable since symposia may embrace both types of presentation.

CONTENTS

PREFACE

This collection of papers was part of a unique symposium held during the 178th Meeting of the American Chemical Society. The symposium, Diffraction Methods for Structural Determination of Fibrous Polymers, had a pronounced international character, with scientists from 12 different countries. The speakers represented both the synthetic polymer and biopolymer fields, with contributions in each of the three classes of natural polymers: nucleic acids, proteins, and polysaccharides. Most important, the symposium centered on methods and techniques for studying fibrous polymers, methods that are usually taken for granted despite their inadequacies.

In this volume, along with "method" papers, are contributions describing new structures that illustrate the methods and assumptions needed to determine the structure of a new polymer. Also included are reviews of classes of polymers for which investigation and methods development have coincided.

The participants generally view fiber diffraction as the most useful method for determining the molecular arrangement of a polymer in the solid state, if the polymer is in the form of crystallites randomly ordered about a single axis. Other methods, such as IR spectroscopy, can provide information for evaluating a proposed structure. However, they are not usually as definitive as determining diffraction intensities, constructing a computer model of the polymer, and fitting the computer model to the diffraction data.

Electron diffraction patterns often can supplement fiber diffraction patterns by providing information such as accurate cell dimensions and a confirmation of the space group.

The sophistication of fiber diffraction has grown along with the development of digital computers. These techniques started with the calculation of diffraction intensities for a few proposed models for comparison with the diffraction pattern. At present, parameters of the models can be varied to produce the minimal variance for the observed and calculated diffraction intensities and simultaneously the minimal stereochemical or packing energy.

Progress continues to depend on applying computers to several outstanding problems. Several chapters deal with automated data collection and reduction. Better computer models and more efficient computer programs are being developed to determine the ranges of stereochemically feasible models to be considered. Another application reverses the usual procedures of structure determination. Instead of essentially correcting the observed data for disorder and amorphous scattering, a pattern that includes effects of these conditions is calculated. In this way, the conditions become parameters of the structure determination. Although not usually considered as "structural" information, the kind of disorder and its magnitude often have physical consequences.

One stumbling block is the limitation of our techniques. For example, are Hamilton's tests appropriate for current fiber diffraction studies? If applicable, these tests allow the calculation of the significance of a difference between *R* factors for two competing models. The tests are derived from analysis of variance, and the usual cautions for those analyses apply. But there are often large differences when different laboratories obtain data for the same substance. *R* factors between data sets range from 20 to 50% even though structures were refined for each set, giving *R* values (between observed data from one source and the model fitted to those data) of approximately 20%. Two factors, the standardization and distribution of refinement programs and the continuing effort to develop interactive graphics techniques to obtain and correct diffraction data, should soon bring added confidence to the fiber diffraction field.

Also, what is the best means of reporting final results when the positions of all the nonhydrogen atoms can be determined directly? Is there a legitimate role for modeling methods if individual atomic positions can be determined? Surely we know bond lengths and valence angles from model compounds more accurately than we could calculate them from the atomic positions derived in a fiber study. To the end of accurately knowing such intramolecular features, it would be an unusual situation indeed that would justify reporting those parameters derived from fiber data. To understand intermolecular interactions, however, the derived atomic positions, with their standard deviations, might be more useful. Calculation of intermolecular effects based on a modeling technique might introduce cumulative errors.

Future work should emphasize resolution of the above questions and continue the current strong emphasis on data collection and reduction.

The editors wish to thank the authors who participated in the symposium. In particular, we are grateful to Struther Arnott for his thorough treatment of fiber diffraction given in the first chapter.

Southern Regional Research Center
USDA
P.O. Box 19687
New Orleans, LA 70179

ALFRED D. FRENCH

Central Research & Development Department
Experimental Station
E. I. Du Pont de Nemours & Company
Wilmington, DE 19898

KENNCORWIN H. GARDNER

May 21, 1980

1

Twenty Years Hard Labor as a Fiber Diffractionist

STRUTHER ARNOTT

Department of Biological Sciences, Purdue University, West Lafayette, IN 47907

X-ray diffraction can be used to help determine the molecular geometry of polymers that prefer to be long helices rather than more complexly folded structures. It is usually possible to prepare specimens in which such helical molecules are aligned with their long axes parallel. Often further lateral organization occurs, but rarely to the degree of a three-dimensionally ordered single crystal. *Potentially* this is an advantage, since there is more information (about the Fourier transform (1, 2, 3) of a molecular structure) in the continuous intensity distribution in the diffraction pattern from a less well-ordered system than there is the "sampled" distribution characteristic of a single crystal. But, since "sampling" also implies local amplification of the molecular transform (at reciprocal lattice points), its absence results in much weaker diffraction signals and the theoretical advantage of knowing the continuous intensity variation is offset by the experimental difficulty of recording it accurately. A further complication is that there are a great many *kinds* of partially-ordered systems of helical molecules, each giving rise to different types of diffraction pattern in which both continuous intensity and Bragg maxima occur. If we wish quantitatively to analyze a diffraction pattern we, of course, have to succeed in modelling not only the molecular structure but also the molecular packing. This is true

0-8412-0589-2/80/47-141-001$07.50/0

for *any* diffraction pattern, but for fiber diffraction patterns there is additional complexity because the modes of packing are more varied and complex than in single crystals.

Fibrous biopolymers are afflicted also by the problem of phase determination common to all X-ray analyses of structure, and by the same limitations of resolution that affect diffraction analyses of most macromolecules even when (like globular enzymes) they are organized in single crystals. The ways in which these problems have been overcome for fibrous systems are quite commonplace, although the emphasis may be unfamiliar. These biopolymers do not usually have covalently-bound atoms heavy enough for facile phase determination. Nor, because of their high symmetry and tendency to disorder, is it easy to obtain isomorphous heavy-atom derivatives without multiple site occupancy. Therefore, many of the well-trodden paths that lead from sets of diffraction intensities to a unique solution of molecular structure are not available. More usually one builds a stereochemically plausible model of a residue that fits into a helix which has the dimensions and symmetry characteristics determined from the layer-line spacings and from the systematic absences and general intensity distribution in the diffraction pattern. Thereafter the problem is one of refinement. If fundamentally different initial models are conceived of, there is then the additional task of refining each possibility and adjudicating among optimized models of each kind by appropriate tests (4).

As I see it, structural biochemists and polymer scientists using fiber diffraction data should strive to mimic classical crystallographic studies so as to arrive at similarly credible solutions of structures by similarly noncontroversial methods of procedure that are similarly reproducible in other laboratories. We are obviously on the threshold of greatly improving the accuracy of intensity measurement but I will leave discussion of this to others and discuss in turn the different kinds of packing arrangements available to fibrous molecules, a general scheme for determining their structures and packings, and examples of arbitration among competing models.

Types of Disorder and Consequent Diffraction Effects

Although somewhat idealized, the following general model will serve to indicate the variety of packing modes that may be encountered. The model has parallel arrays of helical molecules with their long axes intersecting a plane perpendicular to them at points forming a (regular rhombic) net. In the present discussion we will ignore the fact that these nets are not infinite, although finite net area has the important consequence of broadening diffraction signals, thereby aggravating problems of intensity measurement. It should also be recognized that fibers typically consist of many small domains like our model and that these are parallel in respect of the helix axes' direction but no other. This means (for example) that when the domains are fully crystalline the diffraction from the fiber is like that from a rotated single crystal, with the penalty of overlapping diffraction signals for reciprocal lattice points with the same reciprocal space cylindrical polar radius (R in Fig. 1). However, for the moment we will discuss only the consequences of different types of disordering of molecular packing *within* each small domain.

An isolated helical molecule is in essence a "one-dimensional crystal" because of its axial periodicity. Its Fourier transform (1, 2, 3) is therefore confined to layer lines and on each layer line it is a continous function proportional to T_W where

$$T = \sum_{nj}\sum f_j\, J_n(2\pi R r_j)\, \exp[i\{n(\psi-\phi_j+\pi/2)+2\pi \ell z_j\}] \qquad \text{(i)}$$

If

$$G_n = \sum_j f_j\, J_n(2\pi R r_j)\, \exp[i(2\pi \ell z_j - n\phi_j)], \qquad \text{(ii)}$$

and

$$t_n = G_n \exp[in(\psi+\pi/2)], \qquad \text{(iii)}$$

then

$$T = \sum_n t_n \qquad \text{(iv)}$$

is an abbreviated form of (i) that will be found useful below. Symbols and definitions: c = axial repeat along length helix; $\underline{S}$ is the reciprocal (i.e. diffraction) space vector that has cartesian components (ξ,η,ζ); $\zeta = \ell/c$, ℓ (the layer line number) is an integer; $\underline{S} = \underline{R} + \underline{\zeta}$ vectorially; (R,ψ,ζ) are the cylindrical polar coordinates of point that has cartesian coordinates (ξ,η,ζ); (r_j,ϕ_j,Z_j) are the cylindrical polar coordinates of the jth atom of one residue of the helical molecule; $z_j = Z_j/c$; f_j is the scattering factor of the jth atom; $J_n(Y)$ = Bessel function of the first kind of order n and argument Y; $\Delta\phi_p$ the relative orientation of the pth helical molecule (as a fraction of 2π); Δz_p the axial displacement (as a fraction of c) of the pth helical molecule; $\underline{A}_p$ is a vector in the net on which the helices are arrayed; for an N-fold integral helix, n is an integer determined by the selection rule $n = \ell - Nm$, where $m = 0, \pm 1, \pm 2$, etc.

That T is a series of Bessel rather than trigonometric functions is merely a consequence of using cylindrical polar coordinates (r_j, ϕ_j, cz_j) for atoms in real space and $(R, \psi, \ell/c)$ for points in reciprocal space. Not only is this a convenient framework for describing a helical molecule, but it can lead to economies in computing T. For helices, only Bessel terms with

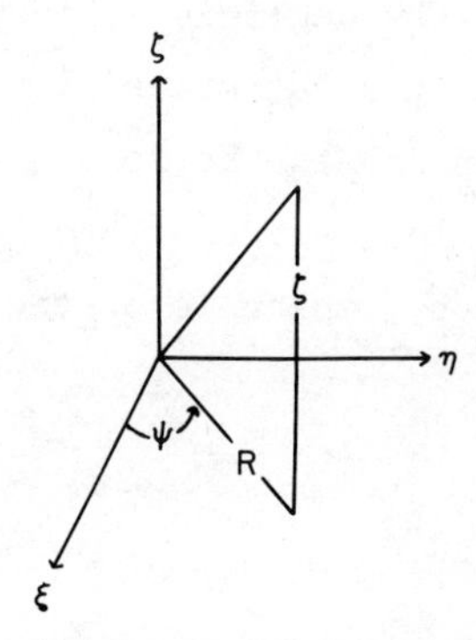

Figure 1. Reciprocal space coordinates: (R, ψ, ζ) are the cylindrical polar coordinates

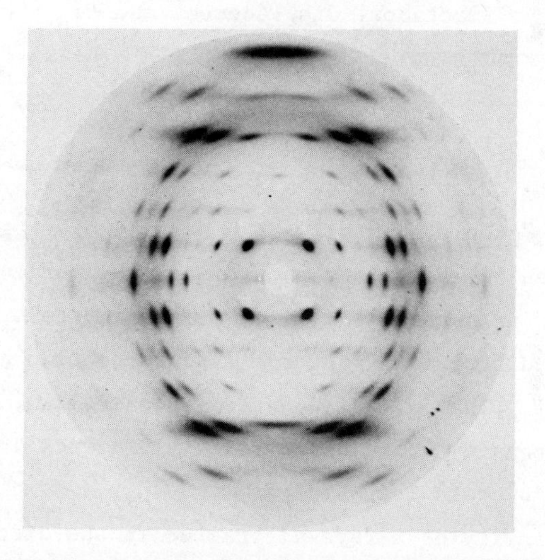

Figure 2. Fiber diffraction from the sodium salt of polycytidylic acid recorded on a flat film perpendicular to the indicident x-ray beam. The fiber was at a 75° angle to the beam to permit the meridional diffraction on layer line 6 to be recorded. The lattice is trigonal with a = b = *2.32 nm;* c = *1.86 nm. The molecules are 6_1 helices.*

$$n = \ell - Nm \quad (m = 0, \pm 1, \pm 2 \ . \ . \ . \ .) \tag{v}$$

are non-zero. (N is the number of residues in the repeat length of the helix). Moreover, for these values of the argument Y which we need to consider, $J_n(Y)$ is usually negligible for $n>25$. Thus, when N = 12 say, T can be approximated satisfactorily by no more than the first two or three terms of the series.

Let us now consider arrays of parallel helices. From eqns. (i), (iv), the Fourier transform of the pth molecule of the array is

$$T_p = \sum_n t_n \exp[2\pi i(\ell \Delta z_p - n\Delta\phi_p)] \exp[2\pi i \underline{A}_p . \underline{R}] \tag{vi}$$

where Δz_p is the fractional displacement and $\Delta\phi_p$ the change of orientation (as a fraction of 2π) of the pth helix.

<u>Oriented, polycrystalline specimens</u>. If, for example, all $\Delta z = \Delta\phi = 0$ the array would be a three-dimensionally ordered domain i.e. a "single crystal". The most highly organized fibers are ones in which many such domains are oriented with only the helix axes parallel. As I indicated above, such fibers provide X-ray diffraction patterns like those from a rotated single crystal. The not-quite-perfect parallelism of the domains causes the intensity to be distributed along arcs instead of concentrated in spots (see Fig. 2).

The intensity of X-rays diffracted in the direction corresponding to $\underline{S} = \underline{R} + \underline{\zeta}$ is

$$I = \sum_{pq}\sum T_p T_q^* \tag{vii}$$

where T_q^* is the complex conjugate of T_q. For all $\Delta z = \Delta\phi = 0$, all $T_p = T \exp[2\pi i \underline{A}_p \cdot \underline{R}]$, and therefore

$$I = TT^* \sum_{pq}\sum \exp[2\pi i(\underline{A}_p - \underline{A}_q)\cdot\underline{R}] \qquad \text{(viii)}$$

which corresponds to a "spotty" intensity distribution such as one obtains from a three-dimensionally-ordered crystal.

<u>"Statistical" crystallinity.</u> In slightly less-ordered specimens, molecules, at random, have displacements $(\Delta z_s, \Delta\phi_s)$ where s takes only a few values. The resulting diffraction effects are of the same general kind that are found when polymers have chain sense and when, as is often the case, a net is populated randomly with equal numbers of up-pointing and down-pointing molecules. Chain sense is exhibited by any helical molecule that does not possess a center of symmetry or a diad perpendicular to its screw axis. α-Poly-L-alanine is a classic example of a helical molecule for which the essential molecular geometry had been determined (5) but confirmation and refinement (6,7) had to await recognition that the molecular packing was of the "statistical" crystal type (8).

For a random distribution of molecules with displacements $(\Delta z_s, \Delta\phi_s)$, where a large fraction of the molecules have one of the *few* different $(\Delta z_s, \Delta\phi_s)$ values, we can define

$$\bar{T} = \langle \sum_n t_n \exp[2\pi i(\ell\Delta z_p - n\Delta\phi_p)] \rangle, \qquad \text{(ix)}$$

and

$$T_p = (\bar{T} + \Delta T_p)\exp 2\pi i \underline{A}_p \cdot \underline{R}, \qquad \text{(x)}$$

where

$$\sum_p \Delta T_p = 0. \qquad \text{(xi)}$$

(For "up-" and "down-pointing" disorder, $\bar{T}$ would be the average of the transforms of the (reference) "up-pointing" molecule and the "down-pointing" molecule.) In any case

$$I = \sum_{pq}\sum (\bar{T}+\Delta T_p)(\bar{T}^*+\Delta T_q^*) \exp[2\pi i(\underline{A}_p-\underline{A}_q)\cdot\underline{R}]$$

$$= \overline{TT^*} \sum_{pq}\sum \exp[2\pi i(\underline{A}_p-\underline{A}_q)\cdot\underline{R}] + \bar{T}^* \sum_{pq}\sum \Delta T_p \exp[2\pi i(\underline{A}_p-\underline{A}_q)\cdot\underline{R}]$$

$$+ \bar{T} \sum_{pq}\sum \Delta T_q^* \exp[2\pi i(\underline{A}_p-\underline{A}_q)\cdot\underline{R}] + \sum_{pq}\sum \Delta T_p \Delta T_q^* \exp[2\pi i(\underline{A}_p-\underline{A}_q)\cdot\underline{R}] \quad \text{(xii)}$$

The second and third summations in (xii) will be zero, the fourth would also be zero but for terms where $q = p$. This is because for non-integral values of α, the average value of $(\exp 2\pi i\alpha)$ is zero so that summations over a sufficiently large and varied number of complex exponential terms like this are also negligible. When α is integral the exponential has value unity. In particular, when $\alpha = \underline{A}_p\cdot\underline{R}$ or $\underline{A}_q\cdot\underline{R}$ the non-integral condition exists *between* reciprocal net points and the integral case *at* net points, so that $\sum_{pq}\sum \exp 2\pi i(\underline{A}_p-\underline{A}_q)\cdot\underline{R}$ has significant values only at points corresponding to these reciprocal net points. The vanishing of certain terms in equation (xii) depends also on $\sum_p \Delta T_p = 0$. Later in equations (xvi), (xix), (xxi), the vanishing terms are a result of the generally non-integral values of Δz_p and $\Delta\phi_p$ and the functions of them that are the arguments of the complex exponential terms. For the present case we note that

$$I = \overline{TT^*} \sum_{pq}\sum \exp[2\pi i(\underline{A}_p-\underline{A})\cdot\underline{R}] + \sum_p \Delta T_p \Delta T_p^* \quad \text{(xiii)}$$

The first summation in (xiii) is the same as the intensity from a three dimensionally ordered crystal in which every unit cell would contribute the same amplitude proportional to $\bar{T}$, and indicates that there will be a Bragg diffraction pattern corresponding to this "average" or "statistical" crystal. The second summation in (xiii) contains the squared amplitudes of the differences between the transform of each molecule and the average transform (see ix) and is a continuous function on all layer lines. Therefore, the Bragg pattern will, in the general case, be overlaid everywhere by continuous intensity. Fig. 3 provides an example of such a pattern.

"Screw disorder" or "semicrystallinity". Another common type of partially ordered array for helices occurs when the Δz_p have random values for different molecules but each $\Delta\phi_p = \Delta z_p$. The helical molecules are therefore like threaded bolts which have been screwed into the net plane by randomly varying amounts. In this case it is important to recall from (iv) and (v) that

$$T = t_\ell + \sum_m t_{\ell-Nm}, \quad m = \pm 1, \pm 2, \text{ etc.}, \tag{xiv}$$

and from (vi) that

$$T_p = \{t_\ell \exp[2\pi i(\Delta z_p - \Delta\phi_p)] + \sum_m t_{\ell-Nm} \exp[2\pi i\ell\Delta z_p] \exp[2\pi i(nm-\ell)\Delta\phi_p]\} \exp[2\pi i\underline{A}_p \cdot \underline{R}]. \tag{xv}$$

Consequently

$$I = \sum_{pq}\sum t_\ell t^*_\ell + \sum_{m'} t_\ell t^*_{\ell-Nm'} \exp[-2\pi i m' N\Delta\phi_q] + \sum_m t_{\ell-Nm} t^*_\ell \exp[2\pi i m N\Delta\phi_p] + \sum_{mm'}\sum t_{\ell-Nm} t^*_{\ell-Nm'} \exp[2\pi i N(m'\Delta\phi_q - m\Delta\phi_p)] \exp[2\pi i(\underline{A}_p - \underline{A}_q) \cdot \underline{R}]. \tag{xvi}$$

In (xvii) the summations over m and m' are zero, in general. But in the double summation where p = q and m' = m, this is not the case, with the result that

$$I = t_\ell t^*_\ell \sum_{pq}\sum \exp[2\pi i(\underline{A}_p - \underline{A}_q) \cdot \underline{R}] + \sum_{mp}\sum t_{\ell-Nm} t^*_{-Nm'} \tag{xvii}$$

What (xvii) shows is that the J_ℓ Bessel function terms provide single-crystal-like diffraction and all other Bessel terms continuous intensity. Although this kind of array is not the only one for which Bragg and continuous intensity both occur in the same pattern, it is mainly for this that the unfortunate term "semi-crystalline" has sometimes been used. Examples are *C*-DNA (9,10), ribosomal RNA fragments (11). Fig. 4 shows an example of the diffraction from such an array of molecules. The helices in this case are 12-fold therefore J_ℓ has large values only for small values of ℓ. It follows that the Bragg diffraction is confined to the center of the pattern.

"Oriented" specimens. The least ordered situations we will consider occur when the helical molecules have axes parallel and either all $\Delta z_p = 0$ and all $\Delta\phi_p$ have different, unrelated values, or all $(\Delta z_p, \Delta\phi_p)$ are unrelated to each other.

To explore the case when all $\Delta z_p = 0$ but every $\Delta\phi_p$ has a different value we use (xv) to obtain,

$$T_p = \{t_\ell \exp[-2\pi i\ell\Delta\phi_p] + \sum_m t_{\ell-Nm} \exp[-2\pi i(\ell-Nm)\Delta\phi_p]\}$$

$$\times \exp[2\pi i A_p \cdot \rho], \qquad \text{(xviii)}$$

and therefore

$$I = \sum\sum_{pq} \{t_\ell t^*_\ell \exp[2\pi i\ell(\Delta\phi_q - \Delta\phi_p)] + (t_\ell \exp[2\pi i\ell(\Delta\phi_q - \Delta\phi_p)]$$

$$\times \sum_{m'} t^*_{\ell-Nm'} \exp[-2\pi i Nm'\Delta\phi q)] + (t^* \exp[2\pi i\ell(\Delta\phi_q - \Delta\phi p)]$$

$$\times \sum_m t_{\ell-Nm} \exp[2\pi i Nm\Delta\phi_p)] + \sum\sum_{mm'} t^*_{\ell-Nm} t_{\ell-Nm'}$$

$$\times \exp[2\pi i N(m\Delta\phi_p - m'\Delta\phi_q)] \exp(2\pi i\ell(\Delta\phi_q - \Delta\phi_p)]$$

$$\times \exp(2\pi i(\underline{A}_p - \underline{A}_q) \cdot \underline{R}]. \qquad \text{(xix)}$$

Similarly for the case where all $(\Delta z_p, \Delta\phi_p)$ are unrelated

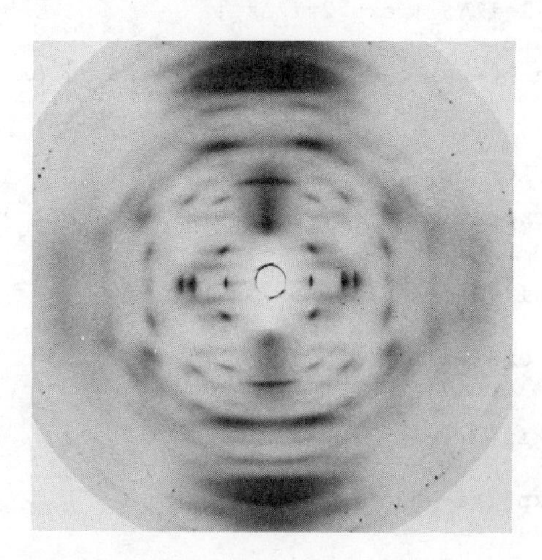

Figure 3. Fiber diffraction from a "statistically disordered" fiber of the sodium salt of poly d(GC) · poly d(GC). The molecules in this structure form an unusual left-handed DNA duplex in which the dinucleotide pCpG is the molecular asymmetric unit. The unit cell is trigonal with a = b = *1.91 nm;* c = *4.35 nm. The space group is probably $P226_5$. The molecular symmetry is itself 226_5, and the statistical structure arises from a random choice of a molecular diad to a point along a particular direction.*

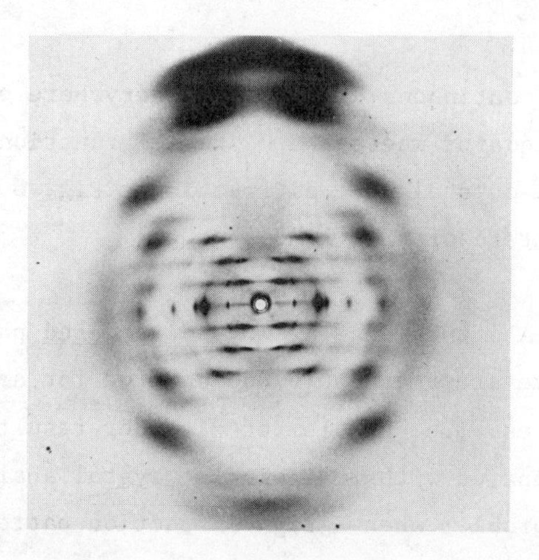

Figure 4. Diffraction from a "screw-disordered" arrangement of poly U · poly A · poly U molecules. The Bragg pattern from the fiber containing these triple-stranded polynucleotide molecules corresponds to a trigonal cell with a = b = *2.71 nm;* c = *3.65 nm. The molecules are 12_1 helices.*

$$T_p = \{t_\ell \exp[-2\pi i\ell\Delta\phi_p] \exp[2\pi i\ell\Delta z_p]$$
$$+ \sum_m t_{\ell-Nm} \exp[-2\pi i(\ell-Nm)\Delta\phi_p] \exp[2\pi i\ell\Delta z_p]\} \exp[2\pi i\underline{A}_p\cdot\underline{R}], \qquad (xx)$$

and

$$I = \sum_{pq}\sum t_\ell t_\ell^* \exp[2\pi i\ell(\Delta\phi_q-\Delta\phi_p)] \exp[2\pi i\ell(\Delta z_p-\Delta z_q)]$$
$$+ (t_\ell \exp[2\pi i\ell(\Delta\phi_q-\Delta\phi p)] \exp[2\pi i\ell(\Delta z_p-\Delta z_q)]$$
$$\times \sum_{m'} t_{\ell-Nm'}^* \exp[-2\pi iNM'\Delta\phi_q])$$
$$+ (t_\ell^* \exp[2\pi i\ell(\Delta\phi_q-\Delta\phi_p)] \exp[2\pi i\ell(\Delta z_p-\Delta z_q)]$$
$$\times \sum_m t_{\ell-Nm} \exp[2\pi i\, Nm\Delta\phi_p])$$
$$+ \sum_{mm'}\sum t_{\ell-Nm}t_{\ell-Nm'}^* \exp[2\pi iN(m\Delta\phi_p-m'\Delta\phi_q)] \exp[2\pi i\ell(\Delta\phi_q-\Delta\phi_p)]$$
$$\times \exp[2\pi i\ell(\Delta z_p-\Delta z_q)]\} \exp[2\pi i(\underline{A}_p-\underline{A}_q)\cdot\underline{R}]. \qquad (xxi)$$

Therefore,

$$I = t_o t^* \sum_{pq}\sum \exp[2\pi i(\underline{A}_p-\underline{A}_q)\cdot\underline{R}] + \sum_{mp}\sum \{t_\ell t_\ell^* + t_{\ell-Nm}t_{\ell-Nm}^*.\} \qquad (xxii)$$

This indicates continuous diffraction everywhere except in that region of the equator where the J_o Bessel function component provides single-crystal like diffraction. Fig. 5 shows a fiber diffraction pattern of this kind.

Comment. All the less- or more-disordered packing modes introduced above are frequently encountered for arrays of helical molecules. The problem of disorder results in an additional (compared with most single crystal analyses) deconvolution problem when X-ray diffraction patterns of such systems are being interpreted. Although complications from disorder effects are not unique to fibrous systems, they are more frequently encountered there. I suspect that this has been a major disincentive to routine, quantitative determinations of molecular structures that occur only in fibrous form. However the deconvolution problem is not insoluble since rotating-anode X-ray generators, and toroidal focusing devises (12) and the like now combine to give better quality diffraction patterns in reasonable time, while film-scanners and computers make it possible more conveniently to extract information from complex patterns (13,14). Both theory and experiment are therefore evolving to a point where thorough diffraction analyses of less-than-fully-crystalline specimens need not be viewed with dismay.

Other Problems and Solutions in the Diffraction Analysis of Fibrous Biopolymers.

The problem of low resolution and paucity of data. In addition to having a greater variety of packing modes than are usually seen with single crystals, fibrous structures typically provide diffraction data only of lowish resolution. Relatively few diffracted X-ray intensities from fibers correspond to periodicities less than 0.25nm and one has therefore to forego all thought of independently determining the cartesian coordinates of each atom with a precision likely to be useful. (For purposes of identifying possible hydrogen-bonds or important non-bonded contacts a precision of 0.01-0.02nm in atomic positions is sufficient for most molecular biological applications.)

Being able to record only low resolution data also implies that these are relatively few in number. Here some relief is provided by the high symmetry of regular helical molecules. Many helical polyncleotides, for example, provide 100-200 independent X-ray reflections that can be used to determine the molecular geometry of the one nucleotide residue from which all the others can be generated by symmetry operations. A comparable data set for a transfer RNA single crystal with, say, 80 nucleotides in the crystal asymmetric unit would be 8,000-16,000 (i.e. an order of magnitude greater than what are, ordinarily available).

The linked-atom least-squares solution. A convenient way of overcoming many of the limitations of the X-ray diffraction data from fibers is to supplement them with other information such as expected bondlengths, bond-angles and conformation-angles. The expected values come from surveys of accurate single-crystal analyses of relevant small molecules. Surveys of single crystal analyses of monomers suitable for polysaccharide and polynucleotide work are available (15,16). Typically, bond-lengths are distributed about the mean with a standard deviation of 0.002nm and are, therefore, in polymer analyses with the precision we aim for, kept fixed at the mean values observed in monomers. Bond-angles show standard deviations from the mean of about 2.5°. Whether they are kept fixed at the mean values in monomers, or are allowed to vary from them in a "stiffly elastic" fashion, depends on how many further degrees of freedom the molecule is judged to require. Ultimately this is determined by the quantity and quality of data available in a particular case. Conformation-angles have generally less predictable values and are usually the main free variables. In addition, the conformation-angles in any residue are subject to the constraints of being compatible with a helical structure for which the axial translation and rotation per residue are known. In the case, for example, of the nucleotide residue in Fig. 6 the usual structure variables would be α, β, γ, δ, ε, χ and ζ would be fixed if the furanose ring was considered to have a fixed, standard shape. The translation and rotation per residue provide two relationships among the variables, leaving only four degrees of structural freedom. Optimum values of these can be determined by constrained linked-atom least-squares procedures. The

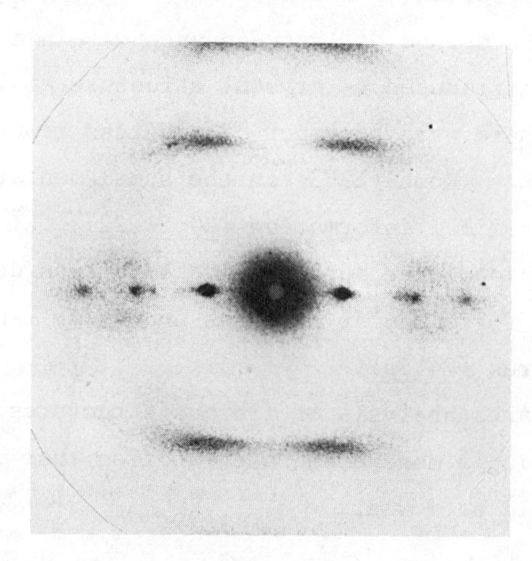

Figure 5. Diffraction from an uniaxially oriented specimen of stretched calf thymus DNA (courtesy M. H. F. Wilkins). The molecules are approximately 14_{13} helices with c = *10.6 nm. The Bragg reflections on the equator can be indexed on the basis of a rhombic net with* a = b = *2.30 nm.*

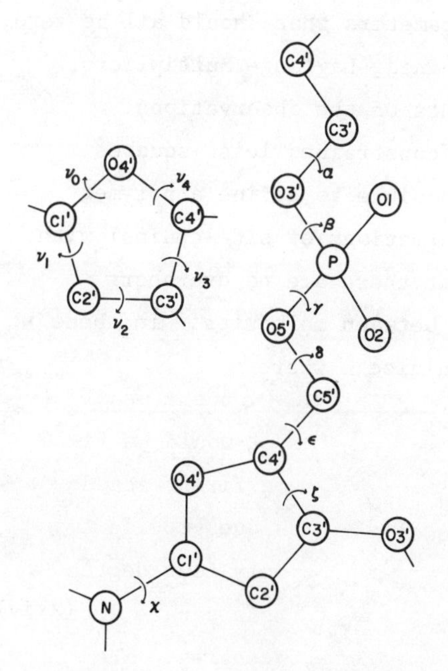

Figure 6. Conformational variables of a nucleotide residue

linked-atom description of a molecular chain was introduced by Eyring (17). The method of least-squares was invented by Gauss (18) and introduced to crystal structure analysis by Hughes (19). There were a number of progress reports at the I.U.Cr. Conference, Rome (1963) on the supplementation of X-ray data by stereochemical information (20,21,22). Since then the strategies for biopolymer applications have been developed mainly by Diamond (23) for globular proteins and by my laboratory for fibrous structures (24,25).

Least-squares analysis of fibrous structures may therefore involve minimizing a more elaborate function that is commonly used. This could be Φ where

$$\Phi = \sum_m (w_m F_m^2) + \sum_h \lambda_h G_h \qquad \text{(xxiii)}$$

In (xxiii) the ΔF_m are the differences between model and experimental X-ray structure amplitudes, the G_h are a set of exact relationships among the parameters that should all be zero, the λ_h are the initially undetermined, Lagrange Multipliers, and the w_m are the relative weights of the observations.

Even with the linked-atom, constrained least-squares strategy (24) it is not always possible to define a polymer structure (particularly the conformations of side-chains) with sufficient accuracy to ensure that there are no overshort non-bounded distances within and between molecules. In these circumstances it is useful to minimize Ω where

$$\Omega = \sum_m w_m \Delta F_m + S\sum_j \varepsilon_j + \sum_h \lambda_h G_h,$$

$$\varepsilon_j = k_j({}_o d_j - d_j)\ ,\ d_j < {}_o d_j,$$

$$\varepsilon_j = 0,\ d_j \geq {}_o d_j. \qquad \text{(xxiv)}$$

The second term in (xxiv) is used to ensure the stereochemical acceptability of the model and includes all non-bonded interatomic distances, d_j, less than some specified minimum distance, ${}_od_j$. S, an overall weight is determined empirically, to ensure that $< w_m \Delta F_m > = < S_j \varepsilon_j >$. The values of the constants k_j and ${}_od_j$ have been derived for each type of interatomic interaction from Buckingham energy functions of the form

$$E = -Bd^{-6} + A\exp(-\mu d) \qquad \text{(xxv)}$$

using published values of B, A, μ. Each ${}_od_j$ is chosen to be 0.02nm greater than the value of d corresponding to minimum E to ensure that all short contacts are driven to larger values. The values of k_j are chosen so that $\partial[k_j({}_od_j - d_j)^2]/\partial d$ closely approximates $\partial E/\partial d$ in the range $[E_{min} - 0.05] < d > [E_{min}]$.

It is possible also to use eqn. (xxiv) to take account of *attractive* interatomic forces such as arise from hydrogen bonds or the interactions between a cation and atoms of its coordination shell, and to maintain the distances between the atoms involved near prescribed standards. This is achieved by arranging that when the distance, d_j, between two such atoms falls within the range ${}_Ad_j$ to ${}_Bd_j$, $({}_Ad_j \leq d_j \leq {}_Bd_j \leq {}_od_j)$, then the appropriate contribution, ε_j, to the second term in eqn. (xxiv) becomes

$$\varepsilon_j = k_j\,({}_od_j - d_j)^2,\; 0 < d_j < {}_Ad_j \text{ or } {}_Bd_j > d_j > {}_od_j,$$
$$\varepsilon_j = {}_Ik_j\,({}_Id_j - d_j),\; {}_Ad_j \leq d_j \leq {}_Bd_j,$$
$$\varepsilon_j = 0,\; d_j \geq {}_od_j. \qquad \text{(xxvi)}$$

The quantities ${}_Id_j$ and ${}_Ik_j$ are the "ideal" distances and corresponding weights.

If one ignores the X-ray intensities and sets all w_m to zero in eqn. (xxiv) one, in effect, builds a model to certain sterochemical specifications. This is indeed how initial models for subsequent X-ray refinement can most conveniently be created.

Linked-atom least-squares refinements of fibrous structures were introduced by Arnott and Wonacott (24) in 1966 and since then have been used to refine a wide range of polymers including polyesters (24,26,27) polypeptides (28,29,30,31,32), polynucleotides (33), and polysaccharides (34-45). Earlier analyses involved mainly eqn. (xxiii) but since 1975 eqn. (xxiv) has been preferred (25). Fig. 7 illustrates a scheme for systematic structure analyses based on fiber diffraction data.

The phase problem and the problem of arbitration. Fibrous structures are usually made up of linear polymers with helical conformations. Direct or experimental solution of the X-ray phase problem is not usually possible. However, the extensive symmetry of helical molecules means that the molecular asymmetric unit is commonly a relatively small chemical unit such as one nucleotide. It is therefore not difficult to fabricate a preliminary model (which incidently provides an approximate solution to the phase problem) and then to refine this model to provide a "best" solution. This process, however, provides no assurance that the solution is unique. Other stereochemically plausible models may have to be considered. Fortunately, the linked-atom least-squares approach provides a very good framework for objective arbitration: independent refinements of competing models can provide the best models of each kind; the final values of Ω or its components (eqn. xxiv) provide measures of the acceptability of various models; these measures of relative acceptability can be compared using standard statistical tests (4) and the decision made whether or not a particular model is significantly superior to any other.

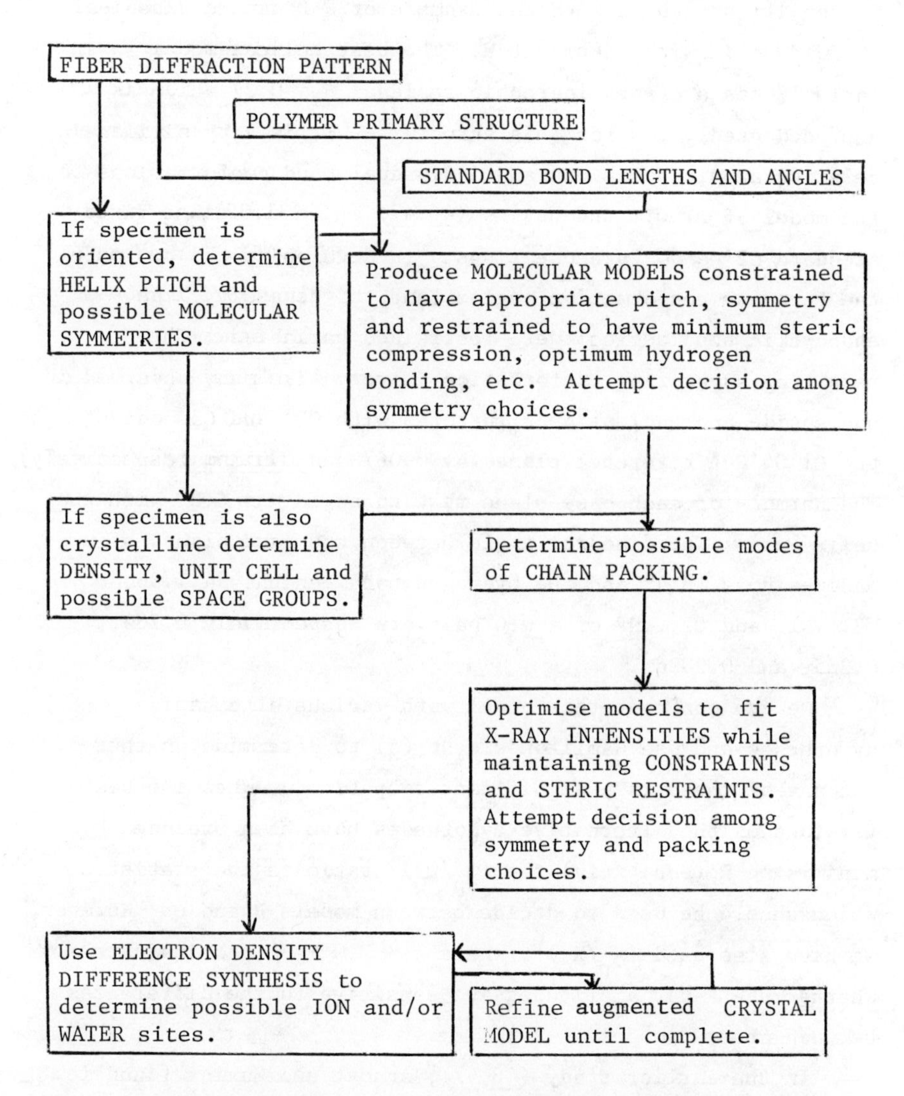

Figure 7. Scheme for defining and refining fibrous structures using x-ray diffraction data and augmented by stereochemical information

Many of our structural analyses of polysaccharides (40-45) have provided examples of applications of this approach. Here we can illustrate it with the example of *B*-DNA, the classical 10-fold helix with 3.4nm pitch. The best helical model (I in Table 1) has a crystallographic residual $R = 0.27$ which is unprecedentedly low for this structure. It is very similar in relative atomic positions and individual bond conformations to the model of Arnott and Hukins (46,47) which contained fixed standard C2'-*endo* furanose rings. In developing the new model the furanose conformation angles (and, of necessity, the endocyclic bond angles) were restrained variables. The resulting sugar ring conformation is very like many observed in nucleoside and nucleotide structures with C2' and C3' out of the C1'O4'C4' reference plane (by 0.067 and 0.011nm respectively). The normals of each base plane make an angle $\gamma = 6.8^{\circ}$ with the helix axis. The dihedral angle between the base planes of a base pair is 13.4° and the hydrogen bond lengths N4...O6, N3...N1, and O2...N2 of a G:C pair are respectively 0.284, 0.285, and 0.283nm.

We can confront this model with various alternative hypotheses and use Hamilton's test (4) to determine whether or not any of these is a significant competitor. When the best versions of the alternative hypotheses have been produced by minimising Ω (eqn. xxiv) the $(\Omega_P/\Omega_Q)^{\frac{1}{2}}$ ratio is the statistic which should be used to decide between models P and Q. However, we have also used $(X_P/X_Q)^{\frac{1}{2}}$, where $X = \Sigma\, w_m\, \Delta F_m^2$, to indicate whether the X-ray amplitudes are sensitive to the differences between the two models.

In the earlier study (46,47), Arnott and Hukins found it difficult to decide between a model with (fixed) C2'-*endo* furanose rings and a related one with (fixed) C3'-*exo* rings. Another difference between these two models was in the rotation about O3'-P which had a *trans* value (-136°) in the former and a

Table 1. Values of Variable Parameters Obtained by Linked-Atom Least-Squares Refinement of Different β-DNA Models

	variable parameter	MODEL [a]					standard value (e.s.d.)[b]
		I	II	III	IV	V	
back-bone conformations (°)	θ(C4'-C3'-O3'-P)	-133.3	-174.5	-166.8	-125.7	-64.7	-123.0(24.0)
	θ(C3'-O3--P-)5')	-156.7	-119.4	-143.5	-159.7	-178.6	180.0(14.0)
	θ(O3'-P-O5'-C5')	-40.7	-29.1	-12.8	-45.9	-124.9	-60.0(14.0)
	θ(P-O5'-C5'-C4')	135.5	171.6	151.6	131.5	132.6	180.9(15.0)
	θ(O5'-C5'-C4'-C3')	37.5	28.1	21.3	41.4	25.5	52.0(8.0)
	θ(C5'-C4'-C3'-O3')	139.4	141.8	146.2	136.9	149.8	148.8(6.3)
glycosidic angle	θ(C2'-C1'-N-C4Pu)	136.5	136.1	141.8	144.8	85.2	114.0(11.0)
sugar ring conformations[c] (°)	θ(C4'-O4'-C1'-C2')	-27.8	-19.2	-23.5	-33.4	-26.2	-20.7(10.2)[c]
	θ(O4'-C1'-C2'-C3')	39.5	28.5	36.9	43.4	45.2	34.1(11.2)
	θ(C1'-C2'-C3'-C4')	-34.6	-27.1	-35.7	-35.7	-44.9	-35.2(3.6)
	θ(C2'-C3'-C4'-O4')	19.9	17.0	22.9	17.9	32.0	23.9(6.3)
	θ(C3'-C4'-O4'-C1')	4.6	1.2	0.1	9.3	-4.1	-1.8(10.1)
sugar ring bond angles[c]	τ(C4'-O4'-C1')	107.3	110.4	109.7	106.3	106.6	109.9(1.0)
	τ(O4'-C1'-C2')	106.3	107.8	105.4	105.7	106.0	106.6(1.2)
	τ(C1'-C2'-C3')	100.7	101.8	101.3	99.2	97.4	102.6(2.4)
	τ(C2'-C3'-C4')	102.4	105.8	102.9	102.8	100.4	102.8(1.9)
	τ(C3'-C4'-O4')	107.8	105.7	106.0	107.4	106.2	105.2(1.6)
base orientation and position	tilt (°)	1.5	-0.3	0.0	6.0	-1.1	
	twist (°)	6.7	6.9	5.5	13.8	2.3	
	position (nm)	-0.034	0.015	0.008	-0.045	-0.040	
2nd mol. posn.[d]	Δz	0.325	0.327	0.327	0.327	0.320	
X-ray	scale	10.3	10.5	10.6	10.3	11.4	
	attenuation (nm^2)	0.025	0.001	-0.004	0.038	-0.115	

Footnotes

a Models I-IV are right-handed helices. Model IV is a left-handed helix. In Models I, IV, V the conformations and bond angles were tied elastically to the standard values shown with elastic constants $1/(e.s.d)^2$. Model II was similarly restrained but the standard value (and e.s.d.) of θ[C3'-O3'-P-O5'] was -60°(14). In Model III the sugar ring conformations and bond angles were not refined.

b From Arnott and Hukins (1972).

c Because C3'-*exo* and C2'-*endo* conformations have been proposed, the standard values and e.s.ds. were calculated for the joint range.

d Fraction of c, the axial translation repeat distance.

gauche minus value (-96°) in the latter. In the new Model I this conformation is *trans*. However, with the sugar ring geometry variable, we need not decide the ring conformations in advance and by tying θ(C3'-O3'-P'-O5') to the standard *gauche minus* value we can examine the properties of the resulting Model II (Tables 1 and 2). The value of $(X_{II}/X_{I})^{\frac{1}{2}}$ is so near unity that it is clear that the X-ray amplitudes cannot discriminate between these two models. On the other hand the value of $(\Omega_{II}/\Omega_{I})^{\frac{1}{2}} = 1.047$ implies that the hypothesis that Model II is as good as Model I by the criteria we have set up can be rejected with better than 99.5% confidence (4).

Model III has furanose rings fixed at the values given by Arnott and Hukins (46,47). As a result the other (varied) conformation angles are somewhat less standard (Table I) although the fit to the X-ray amplitudes is trivially better. Although the X-ray amplitudes detect no difference between Models I and III, the addition of steric considerations permits Model III to be rejected (Table 2) with better than 99.5% confidence.

In Model IV, the refinement was conducted in exactly the same way as for Model I except that the normal to the base planes was required to be 15° to the helix axis. Such a large inclination was a feature developed by Levitt (48). In our final model of this kind this has distributed itself between the tilt and twist parameters so that the former is increased to 6° and the latter to 13.8° implying a dihedral angle of 27.6° between the base pairs. Nevertheless there is a remarkable similarity between the values of all the other parameters and those of Model I: no conformation angle differs by more than 8°; no furanose angle changes by more than 5°. Indeed, it appears that the polynucleotide backbone is perturbed minimally and that the glycosidic conformation is the main variable which has changed to permit the different orientation of the bases. The conservation in Model IV of most of the features of Model I

Table 2. Goodness-of-Fit Indices for the Various β-DNA Models

MODEL	I	II	III	IV	V
$\Omega \times 10^{-2}$	7.47	8.19	8.17	8.19	9.17
$S \times 10^{-2}$	4.75	5.14	5.48	5.45	5.06
$X \times 10^{-2}$	2.72	2.78	2.69	2.73	4.11
$\Omega^{\frac{1}{2}}/\Omega_I^{\frac{1}{2}}$	1.00	1.05	1.05	1.05	1.11
N_Ω^a	299	278	278	316	346
$X^{\frac{1}{2}}/X_I^{\frac{1}{2}}$	1.00	1.01	0.99	1.00	1.23
R''^b	0.30	0.31	0.31	0.31	0.38
R^b	0.27	0.27	0.27	0.28	0.36

a The number of X-ray terms in X is constant (111) but the number of close contacts in S and therefore the number of terms in Ω varies from model to model.

b The significance tests are based on the quantities $\Omega = S + X$ minimised in the least-squres refinements. The traditional X-ray indices, $R'' = (\Sigma\omega\Delta F_m^{\ 2}/\Sigma_o F_m^{\ 2})^{\frac{1}{2}}$ and $R = \Sigma|\Delta F_m|/\Sigma_o F_m$ are given also.

makes the insignificant worsening of the X-ray fit quite understandable. The cost of forcing a large propeller twist on the base pairs is seen in the steric term. However, since $(\Omega_{IV}/\Omega_{I})^{\frac{1}{2}}$ is 1.047 for a model with 1 degree of freedom less than Model I, Model IV can be rejected with greater than 99.5% confidence. The steric compression in Model IV could be reduced by allowing base atoms to be up to 0.02nm out-of-plane and bond angles in the bases to change by up to 7°. These changes are both beyond the limits usually observed in accurate single crystal structural analyses of nucleotides.

A final question raised by some recent studies (49) is whether the *B*-DNA might be left-handed rather than right-handed. Wilkins and co-workers were well aware (50) that conformationally inoffensive models of this kind could be produced. This fact has been rediscovered again (49) and again (51). It enhances the initial plausibility of warped zipper models (52,53) which consist largely of quasi-helical oligonucleotide segments which alternately have right and left-handed chirality. Clearly a quantitative appraisal is needed of just how viable left-handed helical models are. The best left-handed helix we could contrive is Model V (Table 1). The furanose ring is somewhat distorted and other conformation angles are somewhat unconventional but not unduly so. Mainly due to a lack of close *inter*molecular contacts, the lack of steric compression is second only to Model I. All this confirms in a very clear way Wilkins' early contention that left-handed DNA models were not unthinkable. Nevertheless, $(S_{V}/S_{I})^{\frac{1}{2}} = 1.032$, where $S = \Omega - X$, implies that on steric grounds *alone* there is a less than 10% probability that this left-handed model is as good a candidate as Model I. More importantly this is the first of the alternative models considered here against which even the X-ray data by themselves speak decisively. Although the absolute values of the crystallographic residuals ($R = 0.36$, $R'' = 0.38$) are not unimpressive by historical standards the fit with the X-ray amplitudes compares very unfavorably with all the other models. $(X_V/X_I)^{\frac{1}{2}} = 1.229$ implies that this left-handed model can be rejected with rather more than 99.5% confidence. $(\Omega_V/\Omega_I)^{\frac{1}{2}} = 1.108$ supports the same conclusion.

Apparently, the flexible furanose rings of Model I have permitted a better reconcilation of stereochemical expectations and diffraction data than in any *previous* model: only two conformation angles have values more than two standard deviations from the mean observed for monomers, none is more than three standard deviations. By contrast Model III (with rings of fixed shape) has four conformation angles deviating by more than two standard deviations and two of these deviate are more than three. For similar reasons we can now argue that a model with *trans/gauche minus* P-O conformations is superior to one (Model II) with *gauche minus/gauche minus* conformations. On the other hand varying the furanose conformation does not improve the fit to the X-ray amplitudes, nor lead to a strange sugar conformation, nor to noticebly different conclusions about the positions and orientations of the bases. When a high twist between base-pairs is imposed on *B*-DNA (Model IV) the system contrives to localise the perturbation to the interbase interactions causing probably unremediable steric compression. Both steric and diffraction considerations insist on unequivocal rejection of the left-handed helical structural (Model V). This is particularly noteworthy when one considers how tolerant the X-ray data are to the variations among Models I through IV.

Opportunistic Arbitration. Subjecting every postulated model to least-squares analysis may be avoided in many cases. For example, the unit cell of polycrystalline *B*-DNA (54) has a base with $b/a = \tan 2\pi/10$. This is the condition that molecules at (0,0,0) etc. and ($\frac{1}{2}$,$\frac{1}{2}$, Δz) etc. can have additional equivalent interactions if they are 10-fold helices (55,56). This greatly reduces the liklihood that *B*-DNA molecules are warped zippers and not helices. Another example occurs with *A*-DNA duplexes where the molecules are found packed in either of two trigonal forms (57,58), with special arrangements which would maximise equivalent contacts for 11-fold helices. At the time this discovery was used to discount 10-fold helices, then a postulated alternative (59). It can also be used (56) to discount the possibility that the molecules are 1-fold helices, i.e. not helices at all.

Low resolution Fourier syntheses of electron density with phases calculated from an approximate but erroneous model may nevertheless indicate another model and thus provide an unusually compelling indication that the phasing model is imperfect. This strategy has been used to reject Hoogsteen in favor of Watson-Crick base-pairing in DNA (60,61), to eliminate 10-fold in favor of 11-fold helicies in *A*-RNA (62), to favor nested chains rather than double helices in the case of one form of sodium hyaluronate (40).

General Conclusions

Machinery now exists to permit, in many cases, very detailed analyses of fibrous structures using the under-appreciated X-ray diffraction data supplied by the polymers themselves. Some of this machinery can be adapted to tackle the problem of providing unique solutions: statistical tests can be applied to (least-squares) optimised versions of competing models. However, additional or alternative tests of the creditability of different models should not be ignored.

Acknowledgements

I thank my colleague Dr. R. Chandrasekaran for help in preparing this manuscript in which I have taken the opportunity of updating and correcting some of the theory originally presented at the ACA Symposium in Gainsville, FL in 1973. I am indebted to the National Institutes of Health for research support (GM 17371 and GM 20682).

Literature Cited

1. Cochran, W.; Crick, F.H.C.; Vand, V. Acta Cryst., 1952, 5A, 581-586.
2. Stokes, A. R. Progr. Biophys., 1955, 5, 140-167.
3. Klug, A.; Crick, F.H.C.; Wyckoff, H. S. Acta Cryst., 1958, 11, 199-213.
4. Hamilton, W. C. Acta Cryst., 1965, 18, 502-510.
5. Pauling, L.; Corey, R. B. Proc. Natl. Acad. Sci., 1951, 37, 235.
6. Arnott, S.; Wonacott, A. J. J. Mol. Biol., 1966, 21, 371-384.
7. Arnott, S.; Dover, S. D. J. Mol. Biol., 1967, 30, 209-212.
8. Elliot, A.; Malcolm, B. R. Proc. Roy. Soc. A., 1959, 249, 30.
9. Marvin, D. A.; Spencer, M.; Wilkins, M.H.F.; Hamilton, L. D. J. Mol. Biol., 1961, 3, 547-565.
10. Arnott, S.; Selsing, E. J. Mol. Biol., 1975, 98, 265-269.
11. Fuller, W.; Hutcheson, F.; Spencer, M.; Wilkins, M.H.F. J. Mol. Biol., 1967, 27, 519-524.
12. Elliot, A. J. Sci. Instr., 1965, 43, 312-316.
13. Fraser, R.D.B.; McCrae, T. P.; Miller, A.; Rowlands, R. J. J. Appl. Cryst., 1976, 9, 81-94.
14. Fraser, R.D.B.; McCrae, T. P.; Suzuki, E.; Tulloch, P. A. J. Appl. Cryst., 1977, 10, 64-66.

15. Arnott, S.; Scott, W. E. J. Chem. Soc. (Perkin Transactions II), 1972, 324-335.
16. Arnott, S.; Hukins, D.W.L. Biochem. J., 1972, 130, 454-465.
17. Eyring, H. Phys. Rev., 1932, 39, 746.
18. Gauss, C. F. "Theoria Combinationis Observatiorum Erroribus Minimis Obnoxiae"; Gottingen, 1832.
19. Hughes, E. W. J. Am. Chem. Soc., 1941, 63, 1737-1739.
20. Rollett, J. S.; Scheringer, C. Acta Cryst., 1963, 16, A175.
21. Brändén, C.-I.; Holmes, K. C.; Kendrew, J. C. Acta Cryst., 1963, A175.
22. Arnott, S.; Coulter, C. L. Acta Cryst., 1963, 16, A175.
23. Diamond, R. Acta Cryst., 1965, 19, 774-789.
24. Arnott, S.; Wonacott, A. J. Polymer, 1966, 7, 157-166.
25. Smith, P.J.C.; Arnott, S. Acta Cryst., 1978, A34, 3-11.
26. Hall, I. H.; Pass, M. G. Polymer, 1976, 17, 807-816.
27. Desborough, I. J.; Hall, I. H. Polymer, 1977, 18, 825-830.
28. Arnott, S.; Wonacott, A. J. J. Mol. Biol., 1966, 21, 371-383.
29. Arnott, S.; Dover, S. D. J. Mol. Biol., 1967, 30, 209-212.
30. Arnott, S.; Dover, S. D.; Elliott, A. J. Mol. Biol., 1967, 30, 201-208.
31. Arnott, S.; Dover, S. D. Acta Cryst., 1968, B24, 599-601.
32. Fraser, R.D.B.; MacRae, T. P.; Suzuki, E., J. Mol. Biol., 1979, 129, 463-481.
33. Arnott, S. "Secondary Structures of Polynucleotides"; First Cleveland Symposium on Macromolecules, Elsevier Scientific Publishing Company, Amsterdam, 1977, pp. 87-104.
34. Arnott, S.; Guss, J. M.; Hukins, D.W.L.; Mathews, M. B. Science, 1973, 180, 743-745.
35. Arnott, S.; Guss, J. M.; Hukins, D.W.L. Biochem. Biophys. Res. Commun., 1973, 54, 1377-1383.
36. Arnott, S.; Hukins, D.W.L.; Whistler, R. L.; Baker, C. W. Carbohydrate Res., 1974, 35, 259-263.

37. Arnott, S.; Guss, J. M.; Hukins, D.W.L.; Dea, I.C.M.; Rees, D. A. J. Mol. Biol., 1974, 88, 175-184.

38. Arnott, S.; Scott, W. E.; Rees, D. A.; McNab, C.G.A. J. Mol. Biol., 1974, 90, 253-267.

39. Arnott, S.; Fulmer, A.; Scott, W. E.; Dea, I.C.M.; Moorhouse, R.; Rees, D. A. J. Mol. Biol., 1974, 90, 269-284.

40. Guss, J. M.; Hukins, D.W.L.; Smith, P.J.C.; Winter, W. T.; Arnott, S.; Moorhouse, R.; Rees, D. A. J. Mol. Biol., 1975, 95, 359-384.

41. Winter, W. T.; Smith, P.J.C.; Arnott, S. J. Mol. Biol., 1975, 99, 219-235.

42. Winter, W. T.; Arnott, S. J. Mol. Biol., 1977, 117, 761-784.

43. Moorhouse, R.; Winter, W. T.; Arnott, S.; Bayer, M. E. J. Mol. Biol., 1977, 109, 373-391.

44. Winter, W. T.; Arnott, S.; Isaac, D. H.; Atkins, E.D.T. J. Mol. Biol., 1978, 125, 1-19.

45. Cael, J. J.; Winter, W. T.; Arnott, S. J. Mol. Biol., 1978, 125, 21-42.

46. Arnott, S.; Hukins, D.W.L., Biochem. Biophys. Res. Commun., 1972, 47, 1504-1509.

47. Arnott, S.; Hukins, D.W.L. J. Mol. Biol., 1973, 81, 93-105.

48. Levitt, M. Proc. Natl. Acad. Sci., 1978, 75, 640-644.

49. Sasisekaran, V.; Pattabiraman, N. Current Sci., 1976, 45, 779.

50. Fuller, W.; Wilkins, M.H.F.; Wilson, H. R.; Hamilton, L. D.; Arnott, S., J. Mol. Biol., 1965, 12, 60-80.

51. Yathindra, N.; Sundaralingam, M. Nucleic Acids Res., 1976, 3, 729-747.

52. Rodley, G. A.; Scobie, R. S.; Bates, R.H.T.; Levitt, R. M. Proc. Natl. Acad. Sci. USA, 1976, 73, 2959-2963.

53. Sasisekaran, V.; Pattabiraman, N.; Gupta, G. Proc. Natl. Acad. Sci. USA, 1978, 75, 4092-4096.

54. Langridge, R.; Wilson, H. R.; Hooper, C. W.; Wilkins, M.H.F.; Hamilton, L. D., J. Mol. Biol., 1960, 2, 19-37.

55. Dover, S. D. J. Mol. Biol., 1977, 110, 699-700.

56. Arnott, S. Nature, 1979, 278, 780-781.

57. Arnott, S.; Hutchinson, F.; Spencer, M.; Wilkins, M.H.F.; Fuller, W.; Langridge, R. Nature, 1966, 211, 227-232.

58. Arnott, S.; Wilkins, M.H.F.; Fuller, W.; Venable, J. H.; Langridge, R. J. Mol. Biol., 1967, 27, 549-562.

59. Langridge, R.; Gamotos, P. J. Science, 1963, 141, 694-698.

60. Arnott, S.; Wilkins, M.H.F.; Hamilton, L. D.; Langridge, R. J. Mol. Biol., 1965, 11, 391-402.

61. Arnott, S. Science, 1979, 167, 1694-1700.

62. Arnott, S.; Wilkins, M.H.F.; Fuller, W.; Langridge, R. J. Mol. Biol., 1967, 27, 535-548.

RECEIVED May 21, 1980.

2

Problem Areas in Structure Analysis of Fibrous Polymers

E. D. T. ATKINS

H. H. Wills Physics Laboratory, University of Bristol,
Royal Fort, Tyndall Avenue, Bristol BS8 1TL U.K.

Many difficulties are encountered in the elucidation of molecular structure of fibrous macromolecules. Fibrous textures suffer from a variety of faults, distortions and blemishes, the more obvious of which are: limited crystallite size which broadens the diffraction signals, uncorrelated (or worst still partially correlated) azimuthal orientation of the needle-shaped crystallites about their major axis resulting in the overlay of diffraction signals, and misalignment of crystallites with respect to the fibre direction which produces arcing of the signals. Since x-ray diffraction is the principal technique employed, the experimental data, on which the structure is based, is of considerably poorer quality than that obtained in classical single crystal diffraction. The consequences are rather obvious: the available resolution is less than that desired for detailed structure determination and therefore as many as possible reliable (or what are thought to be reliable) stereochemical and rigid-body constraints are injected into the model building procedures in an attempt to reduce the number of conformations for consideration. This number is often further reduced by empirical energy minimization calculations. Assumptions are made at this stage and therefore some degree of uncertainty (which will be a function of the particular fibrous polymer under consideration) will surround the accuracy of the proposed structure. Thus it is always most desirable to also collect additional experimental information using other methods. Professor Tadakoro's recent book(1) illustrates the considerable advantages and benefits to be gained by coupling infra-red spectroscopy with fibre x-ray diffraction. The increasing availability of Fourier transform infra-red spectrometers allows the same thick samples, suitable for x-ray work, to be used in the spectrometer thus ensuring that both sets of information emanate from the same structure. The delightful selected area electron diffraction patterns obtained from polysaccharides by Dr. Chanzy (2), which exhibit such remarkable resolution and definition, indicate the importance and value of the modern application of electron micro-

0-8412-0589-2/80/47-141-031$05.00/0

scopy. Finally solid-state NMR has now developed to a level where we can expect major advances to occur in the near future with respect to its application to fibrous polymers.

I wish to outline a few of the problems that are currently irritating the processes of elucidation in polymer structure determination using fibre x-ray diffraction patterns.

1. Reproducibility and Accuracy of Measured Intensities

The increasing use of computerised model building procedures in the structure determination of fibrous polymers is providing a welcome improvement in the reproducibility and precision of polymer conformation and structure and in their presentation in a convenient form. Coupled with this development is a trend towards more extensive and exhaustive structure refinements which, in some cases, have yielded exceptionally detailed three-dimensional model structures. For example certain polysaccharide structures have been proposed incorporating specific hydrogen-bonded networks. The locations of cations and water molecules within the polymer lattice have apparently been deduced by structure determination using classical single-crystal Fourier difference procedures. However plausible such structures may appear they cannot be convincingly substantiated with respect to the expeimentally available x-ray information, at least not by utilizing the rather crude methods typically used for measuring the intensities of the broad x-ray diffraction signals from fibres. Major improvements are required in the collection, processing and measurement of the diffraction intensities. Even when such improvements are made, and the extent of the improvement tested, the accuracy of each proposed structure will need to be carefully scrutinized. The current vogue of quoting probabilities using statistical testing procedures rather than the reliability index dragged over from single-crystal diffraction perhaps offers a better feel for the confidence placed in a structure, but is no substitute for improvement in accuracy. Proper estimation of the errors involved in the measurement of intensity should now be of prime concern to all serious fibre diffractionists ! The next generation of structure determinations should not rely on a testing system based on a single column of observed intensities, without any error limits and with minimum description concerning their collection and estimation of the background surfaces. Any improvements made will go some way towards balancing the increased precision being wielded in the computerised modelling. There is no need to labour this point since many of the improvements I have outlined are discussed in considerable detail elsewhere (3-6). Dr. Fraser, who should be congratulated for putting his foot in this particular door first (4), highlights the advantages of using a film scanning procedure for processing intensities of fibre diffraction patterns and allowing a more accurate application of correction factors and inclusion of meridional diffraction signals into the experimentally recorded intensity set (3). Two other contributions (5,6) discuss the collection of intensities from photographic films with greater accuracy than current methods. The use of mathematically derived splines to represent background surface profiles is also discussed in some detail and offers scope for better reproducibility (5).

2. Intertwining Helices

Since the discovery in 1953 that DNA consists of a double helix (7) intertwined molecular ropes have attracted considerable attention. The structure of tropocollagen consists of three polypeptide chains intertwining about a common axis stabilized by interchain hydrogen bonds (8) and a number of the α-protein structures and actin have been interpreted in terms of intertwining molecular ropes (see for example 9). Triple strand helices have been established in polysaccharides for the 1,3 linked β-D-xylans and β-D-glucans (10-13) and double strand models for carragennan (14), agar (15) and amylose (16) and particular polymorphs of hyaluronate (17). Structural investigations of the polysaccharide xanthan favoured a 5_1 single helical structure separated from adjacent chains by 1.85nm (18). The density of this model (polysaccharide chains only) is 0.46gm/cm^3 which is over three times less than the measured density of oriented films and the excessive amount of water needed to reduce the difference between calculated and measured densities is not consistent with the behaviour of the x-ray diffraction pattern as a function of relative humidity and drying out of the sample (19). The mass per unit length, measured in solution (20), is twice that calculated for a single chain and results obtained from electron microscopy suggest an intertwining of chains (21). Clearly the single chain 5_1 helix needs to be re-examined giving due consideration to those factors. Further details of the structure for xanthan are given later in the proceedings (22).

Strong support of the triple-strand β-D-xylan and β-D-glucan and the double-strand model for hyaluronate (17) came from observable systematic absences yielding a space group assignment with some confidence. In this latter case systematic absences of the type $h+k+\ell=2n$ in combination with the unit cell dimension and the measured density supported a centred tetragonal lattice with two chains placed at each corner and centre. A similar line of reasoning has been applied by Tadokoro in the case of isotactic poly(methyl methacrylate) the x-ray fibre diffraction pattern of which (i-PMMA) has been reinterpreted in terms of an intertwined double-strand molecule (1,23). The reflections index on an orthorhombic unit cell and density measurements favour four chains running through the cell. Systematic absences of the form $h+k=2n$ support a space group assignment $P2_12_12$. The agreement between calculated and measured structure factors is still rather poor overall but better than alternative models with four single polymer chains (24). Better quality x-ray fibre diffractions have been obtained, as shown in Figure 1, which indicate that a number of reflections split allowing a more accurate determination of the unit cell dimensions, and providing an improved set of observed intensities on which to base a structure refinement.

3. Highly Extended Helix of Isotactic Polystyrene Found in Gels

Natta (25)showed that in crystalline isotactic polystyrene the molecular chains form three-fold helices with an axial advance (h) per styrene monomer of 0.222nm, some 15% below the theoretical maximum extension of h = 0.26nm for the styrene monomer in the all-trans (tt) fully extended chain.

X-ray diffraction patterns of oriented gels of isotactic polystyrene, obtained at high supercoolings in decalin, are quite different (26)as shown in Figure 2a) from the traditional Natta patterns. In particular a pronounced meridional reflection at a spacing of 0.51nm, together with successive orders, required an extended, or nearly extended, chain conformation (26). In addition to the 0.51nm meridional reflection layer lines are observed with six times this spacing at 3.06nm which equate with twelve styrene units (27). The 0.51nm meridional reflection correlates with two styrene monomers rotating about the chain axis to form a six-fold helix, the average advance per styrene monomer being 0.255nm only marginally below the fully extended tt conformation. Extended conformations were ruled out on stereochemical arguments (28)and conformational analysis (29). Thus Atkins (27)considered other configurationally different polystyrene models, in particular syndiotactic and syncephalic (head-to-head) sequences although in conflict with the initial ^{13}C NMR spectra (30)which indicated no departure from isotacticity within 1-2%. Benson (31) has however reported the presence of low concentrations of syncephalic sequences in polystyrene.

The assertion of 98-99% isotacticity by ^{13}C NMR prompted Atkins (32)to examine the stereochemistry and calculate the energy of highly extended isotactic chains. In particular a twelve-fold regular helix with an axial advance (h) of 0.255nm. (Conformationally it is simpler to construct a regular twelve-fold helix and consider slight modifications of environment for alternating monomers, either by variation of torsional backbone angles, or rotation of aromatic appendages, or both). Previous analyses (29,33,34) highlighted only two conformations, viz. the Natta three-fold helix and a near all trans conformation (33,34) where the latter, however, does not lead to an extended chain but to a large circular structure (this is explicit in ref. 34, but not stated as such for the independently derived, basically identical bond rotation sequence in ref. 33. For the distinction of the two kinds of all ttt sequences leading to two basically

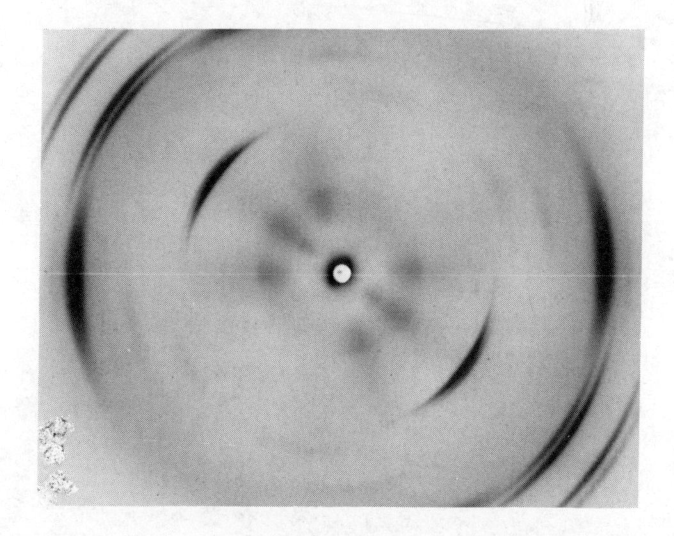

Figure 1. (a) (top) *X-ray fiber diffraction pattern of i-PMMA. Fiber axis vertical. The layer line spacing is 1.036 nm with a meridional reflection on the fifth layer line. The model proposed by Tadokoro (1) has a pitch of twice the value for the observed layer line spacing which becomes halved by the symmetry related intertwining chain. (b)* (bottom) *Higher resolution diffraction pattern of the equatorial region of i-PMMA. The equator is tilted to extend the range on the film. Note that the reflections shown in (a) are now split. (Patterns obtained in collaboration with G. Challa).*

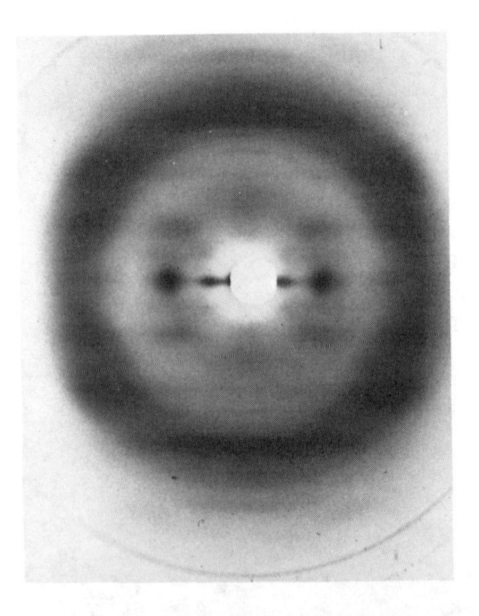

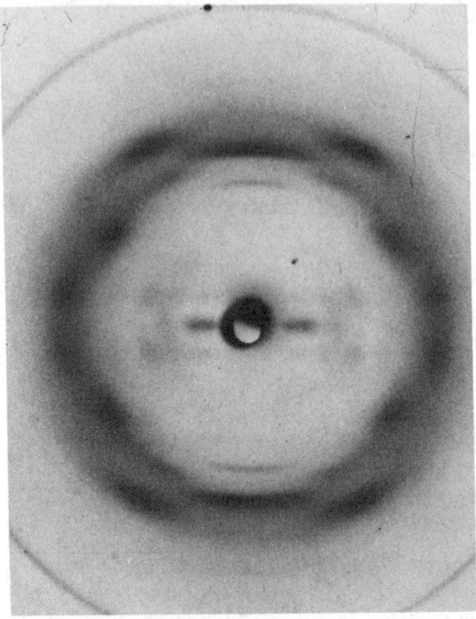

Figure 2. X-ray fiber diffraction patterns of i-PS gels. (a) (top) *i-PS in* trans-*decalin showing rather weak, odd-order layer lines at 3.06-nm spacing. (b)* (bottom) *i-PS in* cis-*decalin showing very strong first order layer line again with 3.06-nm spacing.*

different conformations,i.e. circular and extended chains see Atkins (32)

Conformational analysis by Atkins (32)have shown that a highly extended twelve-fold helix is indeed stereochemically possible and has low energy. Independent calculations by Sundararajan (35) and Corradini (36) have confirmed the essential features of the structure which is shown in Figure 3. In addition Lovell and Windle (37) have proposed a model based on a conformation sequence ttts where s represents a skew of 40°- 60° rotation from the trans position. No full scale energy calculations have been undertaken and the stereochemical feasibility monitored only with CPK space filling models. Thus the proposed model must await more rigorous testing before it can be decided if it is an acceptable alternative to the models discussed above (32,35,36).

As an independent approach Fourier transform infrared studies of isotactic polystyrene gels favours isotactic sequences in an ordered conformation different from the Natta type three-fold helix (38).

Two features of the polystyrene gel obtained from trans-decalin deserve comment. First the meridional reflection at a spacing of 0.51nm (twice the monomer repeat (h) of 0.255nm) indicated that the idealized twelve-fold helix is slightly distorted with differences between the axial projection of contiguous monomers reducing the twelve-fold symmetry to six-fold. Secondly, in the X-ray fibre diffraction patterns (32)it is evident that odd order layers are very weak, in particular the 1st and 3rd. The calculated cylindrically averaged Fourier transform intensities of a twelve-fold helix does not give rise to a pronounced weakness of odd order layer lines, and the 1st layer line is calculated very strong. The weakness of odd order layer lines has prompted speculation regarding the possibility of two polystyrene chains intertwining to form a double helix (35) similar to the well known concept in DNA.

X-ray diffraction patterns obtained from oriented isotactic polystyrene gels prepared from cis-decalin (39) exhibit very strong intensity on the 1st layer line as shown in Figure 2b. Such an observation would not support in any obvious manner a double helical structure for these oriented isotactic polystyrene gels.

X-ray diffraction results from oriented polystyrene gels have shown that highly extended structures of isotactic polystyrene chains can exist. These findings are contrary to the traditional textbook knowledge of polyolefins and also to conclusions of recent conformational analyses (29,33,34). The differences between the latter and our findings arise partly through the particular choice of non-bonded atomic radii and also from freedom of rotation of the aromatic side groups about their linking bond. This highlights the subtle factors which may determine the outcome of conformational calculations in general (even to

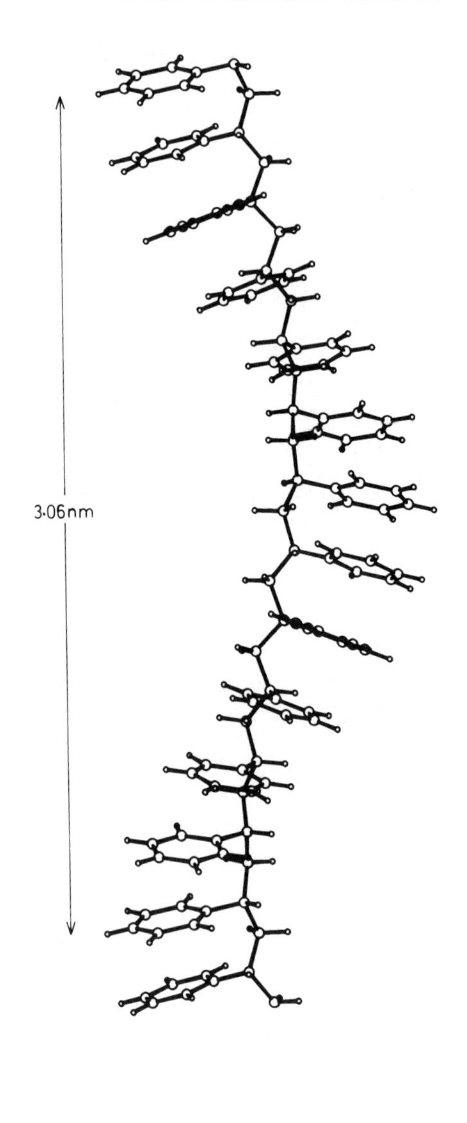

Figure 3. (a) (top) *Projection perpendicular to helix axis of near all-*trans *isotactic polystyrene $\psi 1 = 23.1°$ and $\psi 2 = 11.6°$. The helix has 12 monomers in one turn with an axial advance per dimer of 0.51 nm. (b)* (bottom) *Projection down helix axis.*

the extent of the qualitative nature of the chain shape) and in particular to the open-endedness of conformational possibilities in polyolefins.

4. Summary

Improvements in the collection and processing of observed intensities are now being developed and applied. We should expect in the future to request the two-dimensional densitometer scan from authors if one wishes to re-examine a particular structure determination.

Intertwining chain molecules create considerable problems for the fibre diffractionist. There are two aspects in this area worth particular consideration. It is not difficult to imagine biopolymer chains being biosynthesised together and draping around each other. Once formed it is logical to expect that untwining can occur given the appropriate conditions. For example collagen chains separate on heating to yield gelation. The mounting evidence in the case of the curdlan triple helix (13)is that the chains can untwist slightly (or twist tighter) under different conditions giving rise to quite different x-ray diffraction patterns as the symmetry is destroyed. These changes should not be confused with complete untwining of the chains. This can be accomplished by more drastic chemical treatments where degradation also takes place and untwining occurs of the short segments (40). There is no support at present for these chains to re-intertwine (40). An untwining is envisaged by Sarko (16,41) in the case of A-amylose to V-amylose. Thus the mechanistic problem of untwining and retwining in carrageenan gels is still a major conceptual problem (42). Recent experiments by Smidsrød(43)cast serious doubt concerning the relationship between untwining and retwining of individual chains to form complexes and the actual mechanism of gelation (43). Solution and gelation studies by Smidsrød(44)of carrageenans favour the interaction of double-strand molecules as the basic mechanism of gelation. Certainly such a model eases the topological problems associated with the intertwining mechanism.

If it is convincingly proved that i-PMMA is a double helix this will precipitate much discussion and activity concerning the mechanism of polymerization. Does one chain start forming and then act as a substrate to encourage faster polymerization of another chain on the substrate surface ? Or do two chains come together with the correct polarity and intertwine in order to lower the free energy ? Certainly experiments should be undertaken on i-PMMA to see if there is evidence for melting apart of the duplex, which is only held together by Van der Waals interactions.

Acknowledgements

I wish to thank Professors A. Keller, G. Challa and Dr. P. Lemstra for valuable discussions and the Science Research Council for financial support.

Literature Cited

1. Tadokoro, H. "Structure of Crystalline Polymers", John Wiley & Sons, New York, 1979.
2. Chanzy, H. these proceedings.
3. Fraser,R.D.B; MacRae,T.P; Suzuki,R; Tulloch,P.A. these proceedings.
4. Fraser,R.D.B; MacRae,T.P; Miller,A; Rowlands,R.J. J. Appl. Cryst., 1976, 9, 81.
5. Meader,D; Atkins,E.D.T; Elder,M; Machin,P.A;Pickering,M. these proceedings.
6. Miller,D.P; Brannon,R.C. these proceedings.
7. Watson,J.D; Crick,F.H.C. Nature, 1953, 171, 737.
8. Ramachandran,G.N; Kartha,G. Nature, 1955, 176, 593.
9. Fraser,R.D.B; MacRae,T.P. "Conformation in Fibrous Proteins", Academic Press, New York, 1973.
10. Atkins,E.D.T; Parker,K.D; Preston,R.D. Proc. Roy. Soc.B, 1969, 173,205: Atkins,E.D.T; Parker,K.D. J. Poly. Sci. C, 1969, C28, 69.
11. Bluhm,T.L; Sarko,A. Can. J. Chem., 1977, 55, 293.
12. Marchessault,R.H; Deslandes,Y; Ogawa,K; Sundararajan,P.R; Can. J. Chem., 1977, 55, 300.
13. Fulton,W.S; Atkins,E.D.T. these proceedings.
14. Arnott,S; Scott,W.E; Rees,D.A; McNab, C.G.A. J. Molec. Biol., 1974, 90, 253.
15. Arnott,S; Fulmer,A; Scott,W.E; Dea,J.C.M; Moorhouse,R; Rees,D.A. J. Molec. Biol. 1974, 88, 175.
16. Wu,H.H; Sarko,A. Carbohyd. Res., 1978, 61, 7.
17. Sheehan,J.K; Gardner,K.H; Atkins,E.D.T. J. Molec. Biol., 1977, 117, 113.
18. Moorhouse, R; Walkinshaw,M.D; Winter,W.T; Arnott,S. "Cellulose Chemistry and Technology" edit: Arthur,J.C. ACS Symposium Series , 1977, 48, 133.
19. Isaac,D.H; Atkins, E.D.T. unpublished results .
20. Holzwarth,G. Carbohyd. Res., 1978, 66, 173.
21. Holzwarth, G; Prestridge, F.G. Science , 1977, 197, 757.
22. Arnott, S; Okuyama,K; Walkinshaw,M.D. these proceedings.
23. Kusanagi,H; Tadokoro,H; Chatani,Y. Macromolecules , 1976, 9, 531.
24. Atkins,E.D.T; Meader,D. in preparation.
25. Natta,G; Corradini, P; Bassi,I.W. Nuovo Cimento, Suppl.1., 1960, 15, 68.
26. Girolamo,M; Keller,A; Miyasaka,K; Overbergh,N. J. Polym. Sci., Phys. Ed., 1976, 14, 39.

27. Atkins,E.D.T; Isaac,D.H; Keller,A; Miyasaka,K. J. Polym. Sci., Phys. Ed. 1977, 15, 211.
28. Bunn,C.W; Howells, E.R. J. Polym. Sci., 1955, 18, 307.
29. Liquori,A.M; de Santis,P. J. Polym. Sci. C. 1969, 16, 4583.
30. Cudby,M. unpublished results.
31. Benson,R; Maxfield,J; Axelson,D.E; Mandelkern,L. J. Polym. Sci., Phys. Ed., 1978, 16, 1583.
32. Atkins,E.D.T; Isaac,D.H; Keller,A. J. Polym. Sci., Phys. Ed. 1980, 18, 71.
33. Yoon,D.Y; Sundararajan,P.R; Flory,P. Macromolecules, 1975, 8, 776.
34. Beck.L; Hägele, P.C. Colloid Polym. Sci., 1976, 254, 228.
35. Sundararajan,P.R. Macromolecules, 1979, 12, 575.
36. Corradini,P; Guerra,G; Pirozzi,B. in press (preprint kindly made available by Professor Corradini).
37. Lovell, R; Windle,A.H. J. Poly. Sci. Polym. Letters Ed., 1980, 18, 67.
38. Painter,P.C; Kessler, R.E; Sayder,R.W. J. Polym. Sci. Phys. Ed., in press (preprint kindly made available by Dr. Painter).
39. Atkins,E.D.T; Keller, A; Lemstra,P. in preparation.
40. Norisuye,T; Yanaki,T; Fujita,H. J. Polym. Sci., Phys. Ed., 1980, 18, 547.
41. Sarko,A; Zugenmaier,P. these proceedings.
42. Rees,D.A. "Polysaccharide Shapes", Chapman and Hall, London, 1977.
43. Smidsrød, O; Andersen,I-L; Gresdalen,H; Larsen,B; Painter,T. Carbohyd. Res., 1980 in press.
44. Smidsrød, O. "Structure and Properties of Changed Polysaccharides", in press (preprint kindly made available by Professor Smidsrød).

RECEIVED June 10, 1980.

3

Recent Developments in Structure Analysis of Fibrous Polymers

HIROYUKI TADOKORO

Department of Polymer Science, Faculty of Science,
Osaka University, Toyonaka, Osaka, 560 Japan

As is well known the x-ray diffraction data of polymers are less abundant than in the case of single crystals of low-molecular-weight substances. Therefore the essential process of the x-ray analysis of polymers is still a trial-and-error method (1). To overcome this difficulty, various useful methods have been developed.

The Cochran-Crick-Vand equation for helical polymers (2), the molecular transform of nonhelical polymers (3), the calculation of intramolecular interaction energy (4), molecular conformation parameter equations (5-11), the cylindrical Patterson function (12), and information from infrared and Raman spectroscopy (13) are all important for setting up molecular models. The constrained least-squares method (14,15) the packing energy minimization method (16), and the combination of these two methods (17,18) are very useful for the refinement of atomic parameters. A position sensitive proportional counter (PSPC) has found to be useful in wide angle diffraction as well as in small angle scattering (19). A troidal focusing camera (20,21) and a vacuum cylindrical camera with radius 10 cm (22) have been used to obtain well separated, good intensity data. In this paper useful methods and techniques for structure analysis will be discussed using examples from the author's studies.

Intramolecular Interaction Energy of Typical Isotactic Polymers

The intramolecular interaction energy was calculated for five isotactic polymers, namely, isotactic polypropylene, poly(4-methyl-1-pentene), poly(3-methyl-1-butene), polyacetaldehyde, and poly(methyl methacrylate) (23). The molecular structures of the first four polymers have already been determined by x-ray analyses as (3/1) (24), (7/2) (18,25,26), (4/1) (27), and (4/1) helices (28), respectively. Here (7/2) means seven monomeric units turn twice in the fiber identity period. For isotactic poly(methyl methacrylate) (29), a (5/1) helix was considered reasonable at the time of the energy calculation in 1970, before the discovering of

0-8412-0589-2/80/47-141-043$05.00/0

the double helix structure which will be discussed later (30).

The energy calculations were performed without fixing the fiber identity period, the only assumption being that the chain forms a helical structure, that is, the set of internal rotation angles repeats along the subsequent monomeric units of the chain. For the calculation the internal rotation barriers, van der Waals interactions [mainly after Scheraga (31)], and dipole-dipole interactions were taken into account.

Polypropylene and polyacetaldehyde are the simplest of the above polymers, having only two internal rotation angles, τ_1 and τ_2, in the main chain. Figure 1 shows the potencial energy contour map for polyacetaldehyde. The crosses indicate the potential minima, and the closed circles the x-ray structure determined by Natta et al. (28). The two minima correspond to the right- and left-hand helices.

The other three polymers have additional rotation angles in the side chains, τ_3 and/or τ_4. For poly(3-methyl-1-butene), the minimum was found in the three-dimensional plot. For poly(4-methyl-1-pentene) and poly(methyl methacrylate), the stable conformation of the side chain was first calculated with the fixed main chain conformation corresponding to the (7/2) and (5/1) helices, respectively. The potential energy was calculated against the main chain rotation angles, τ_1 and τ_2, by fixing τ_3 and τ_4 of the side chain at the values thus obtained.

Table I. Stable Conformations (23)

Polymer	Energy calculation			X-ray analysis		
	τ_1(°)	τ_2(°)	N	τ_1(°)	τ_2(°)	N
it-Polypropylene	179	-56	2.91	180	-60	3.0
it-Poly(4-methyl-1-pentene)	163	-71	3.52	162	-71	3.5
it-Poly(3-methyl-1-butene)	132	-83	4.17	149	-81	4.0
it-Polyacetaldehyde	131	-80	3.94	136	-83	4.0

Journal of Polymer Science, Polymer Physics Edition

The results of the first four polymers are listed in Table I; the internal rotation angles of the main chain, τ_1 and τ_2, the number of monomeric units per turn, N, for the calculated stable conformations, and also the values for the structure determined by x-ray analyses. In the case of polypropylene, the number of monomeric units per turn is 2.91, very close to the x-ray value of 3.0. This result for polypropylene is essentially the same as those of Natta et al. (32) and Liquori et al. (33). For the three other polymers, good agreements were also obtained between the predicted models and x-ray structures in spite of the simple assumption of considering only intramolecular interactions. This suggests that the helical structure of these four polymers are determined primarily by the intramolecular interactions, especially the steric hindrance of the side chains.

Figure 2 shows the energy contour map for isotactic poly(methyl methacrylate). The lowest energy minimum was found at the position corresponding to a (12/1) helix contrary to the expectation of the (5/1) helix. The minimum corresponding to the (5/1) helix is higher than the (12/1) helix by 3 kcal/mole of monomer unit. This result led to the postulation of the double stranded helix for this polymer.

These energy calculations can provide suitable and stable molecular models, and have been successfully utilized for the structure analyses of many other polymers, such as poly(tert-butylethylene oxide) (34) and polyisobutylene (35).

Polyisobutylene

The x-ray analysis of polyisobutylene was achieved by the author and his coworkers (35) as shown in Figure 3 by starting with the (8/3) molecular model proposed by Allegra et al. from energy calculations (36). The polyisobutylene molecule has only a twofold screw-symmetry in the crystal lattice, and deviates appreciably from the (8/3) uniform helix as shown in Figure 3(b). The helical molecule has a fiber period consisting of two asymmetric units each containing four monomeric units. The helix axis coincides with the 2_1 screw axis in the lattice ($P2_12_12_1$-D_2^4). The averaged bond angle C-C(CH_3)$_2$-C is about 110°, but the angle C-CH_2-C is much larger, about 128°. The conformation is the alternate sequence of nearly gauche (-54°) and nearly trans (-164°). The overall structure is still similar to the (8/3) helix proposed by Allegra et al. (36).

Poly(ethylene oxybenzoate): α-Form

The x-ray diagram suggested that two helical chains, each fiber period consisting of two monomeric units, pass through an orthorhombic unit cell (4). As shown in Table II, the number of internal rotation angles is large; τ_1 for the virtual bond O-Ph-C, ω for the dihedral angle between the planes of the benzene ring, and the ester group angles, τ_2, τ_3, τ_4, and τ_5. First, the internal rotation angles except for ω were examined by assuming that the angle τ_2 of the ester group is essentially 180° and that the fiber period is 15.60 Å. The possible conformations are limited on the closed surface in the cube defined by the three-dimensional Cartesian coordinates τ_3, τ_4, and τ_5, each covering from 0° to 360° (Figure 4). If τ_3 and τ_4 are given, τ_5 should take two values, upper and lower intersecting points, resulting in a pair of values of τ_1. The intramolecular potential energies for about 5,000 models were calculated, and seven stable molecular models were obtained. Among these, only model 3 was found to give a

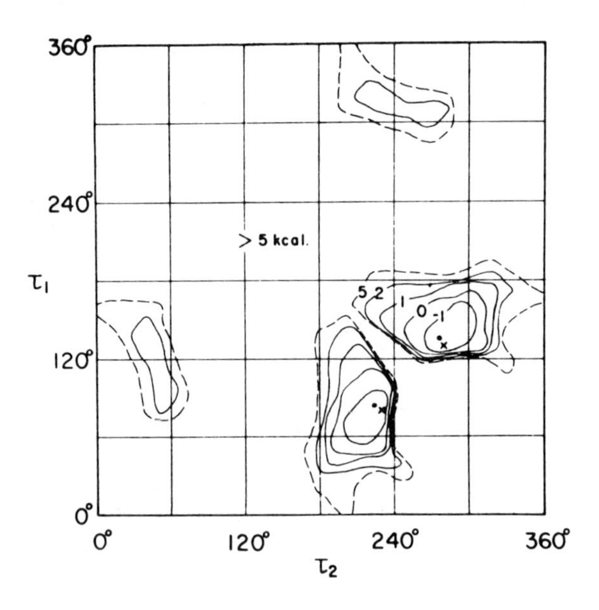

Journal of Polymer Science, Polymer Physics Edition

Figure 1. Potential energy map of isotactic polyacetaldehyde. The energy values are given in units of kilocalories per mole of monomer unit (23).

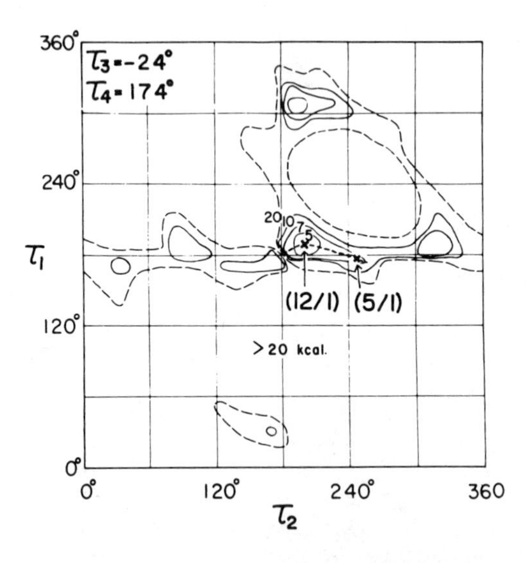

Journal of Polymer Science, Polymer Physics Edition

Figure 2. Potential energy map of isotactic poly(methyl methacrylate) (23)

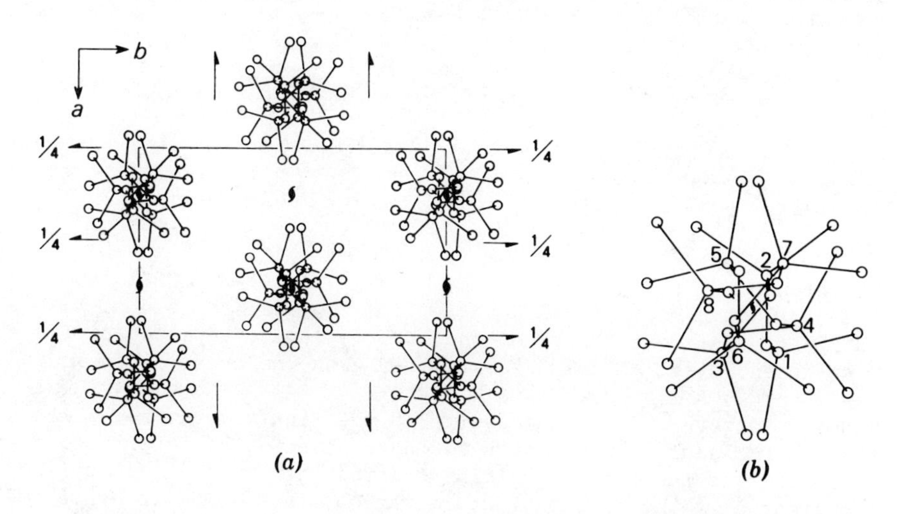

Figure 3. (a) Crystal packing and (b) enlarged view of a single chain in crystal of polyisobutylene (35)

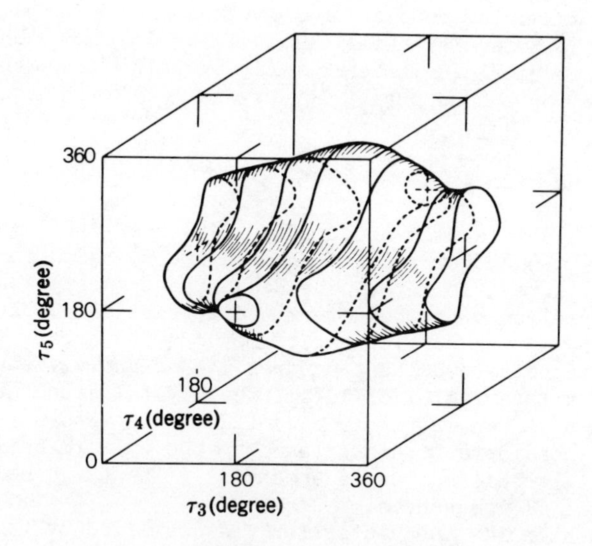

Figure 4. Three-dimensional closed surface for possible conformations of the skeletal chain of poly(ethylene oxybenzoate) with the (2/1) helical symmetry and a fiber period of 15.60 Å (4)

Table II. Internal Rotation Angles of Poly(ethylene Oxybenzoate) α-Form on the Process of Analysis (4)

— O — [—C₆H₄—— C(=O) —— O —— CH₂ —— CH₂ —— O —] —

	τ_1(°)	ω(°)	τ_2(°)	τ_3(°)	τ_4(°)	τ_5(°)
Initial	-41	10	180	-75	-60	-164
Final	-13	-15	-172	-102	-59	-186
Difference	28	25	8	27	1	22

fairly good agreement between the observed and calculated diffraction intensities. Model 3 was refined with the constrained least-squares method (14,15), using a variety of starting points. This model 3, however, did not converge to the final structure. Since the intermolecular interaction was considered to be important, a newly derived packing energy minimization method (16) was applied, and eventually the final structure was obtained. The internal rotation angles of the initial model and the final structure are given in the table together with their differences. The differences in τ_1, ω, and τ_3 are especially large, 25-28°. The rotation angle of the benzene ring ω changed from 10° to -15°, a remarkable change. This example shows both the utility and the limitation of the intramolecular interaction energy calculation as the method to assume the molecular models for structure analysis. As shown in Figure 5, the molecular chain has a zigzag conformation of large scale, one monomeric unit being a zigzag unit. This structure is responsible for the unusually low crystallite modulus of this polymer, 6 GPa (43).

Poly(ethylene Oxide)

In 1961 a molecular model of a (7/2) helix [Figure 6(a)] was proposed by the author and his coworkers (13) based on the information from x-ray, infrared, and Raman spectroscopy, but the crystal structure could not be determined at that time. After ten years, owing to the development of methods and apparatus, especially the constrained least-squares method and a vacuum cylindrical camera with a radius of 10 cm, the crystal structure has been determined as shown in Figure 6(c) (22). The internal rotation angles are considerably distorted from the uniform helix, although the molecular conformation is essentially the (7/2) helix and close to the TTG sequences.

The reason for the distortion was clarified by using a packing energy minimization method (16). Starting from a crystal structure model consisting of the uniform (7/2) helices [Figure 6(a)], the packing energy minimization method without the condition of a uniform helix symmetry resulted in the model shown in

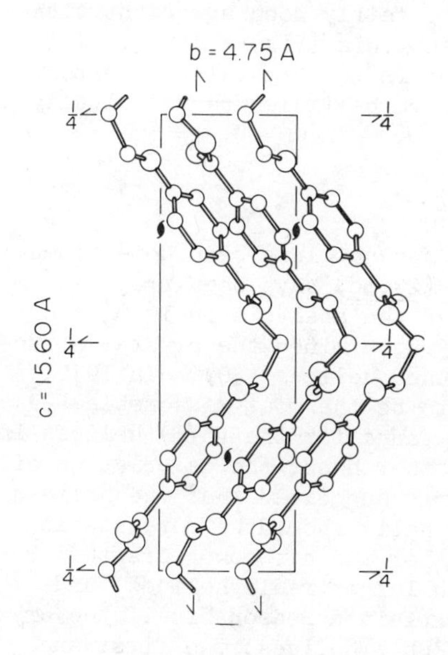

Figure 5. Crystal structure of the α-form of poly(ethylene oxybenzoate) (4)

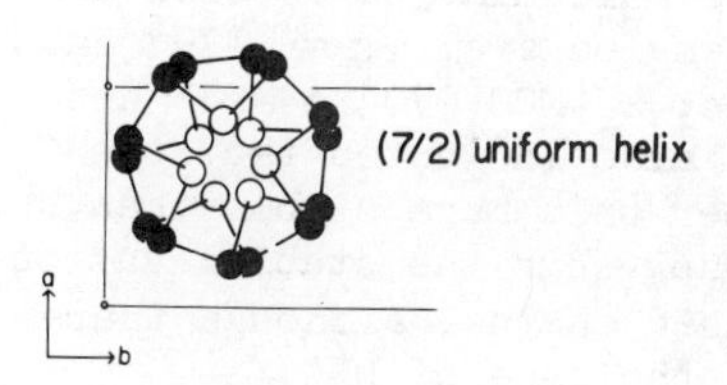

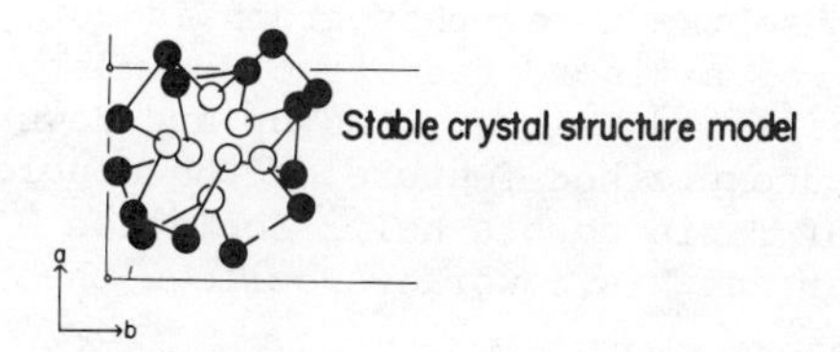

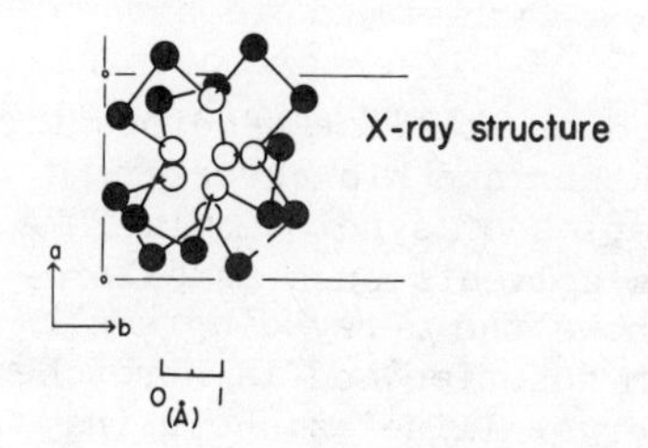

Polymer Journal

Figure 6. Application of packing energy minimization method to poly(ethylene oxide) (16). (a) Starting model of uniform helix; (b) stable crystal structure model obtained by energy minimization calculations; and (c) the structure determined by x-ray analysis.

Figure 6(b). This structure is in a fairly good agreement with the structure determined by x-ray analysis [Figure 6(c)]. This result suggests that the deformation in the crystal of the poly(ethylene oxide) chain from the uniform helix is due principally to intermolecular interaction.

Isotactic Poly(methyl Methacrylate)

The structure of this polymer has not long been determined, and models of (5/2) (37) and (5/1) (29,38) have been proposed. The author and his coworkers started the analysis in 1966, and after a long roundabout quest, they determined the crystal structure which consists of double stranded helices (30). In 1970, they had considered a (5/1) helix to be the most reasonable (29). However, no reasonable packing was found for the (5/1) helices in the orthorhombic lattice. On the other hand, the calculation of the intramolecular interaction energy suggested that the helices with larger radii such as a (12/1) helix should be more stable than the (5/1) helix (23). The author and his coworkers then reexamined the structural models with larger radii helices, and found that the x-ray data can be explained reasonably if the crystal structure consists of (10/1) double helices, the first such case for synthetic polymers (Figure 7). In right-hand helices, the ester group pointing upward and the molecular parameters are as follows. The main chain torsional angles are: τ_1 = -179° and τ_2 = -148°, side chain torsional angles: τ_3[MCC(O)O] = -24° and τ_4 [CC(O)OM] = 174°. Energy calculations also indicated that the double helix is 4.4 kcal/mole of monomer unit more stable than two isolated (10/1) helices. This result suggests the stablilization is due to good fitting of the intertwined chains, although there is no hydrogen-bond between them as in the case of DNA.

Further refinement of the crystal structure consisting of double helices is difficult, because the x-ray photograph is not well-defined, and the possibility of a disordered structure must be considered, e.g., right and left-hand helices, and up and down chains. Although there are some unexplained feature of the double helical model, such as the mode of rapid double helix formation during crystallization, the author and his coworkers believe the result to be essentially correct.

Syndiotactic Poly(methyl Methacrylate)

There has been no report of oriented crystalline syndiotactic PMMA so far as the author knows. The author and his coworkers (39) found that oriented crystalline samples of syndiotactic PMMA can be obtained by adsorption of various solvents such as chloroacetone and diethyl ketone. Figure 8 shows the x-ray diagrams. When the polymer is cast from chloroform solution and is stretched in hot water at 80°C, the sample is noncrystalline as shown in Figure 8. By adsorption of the solvent the fiber diagram can be

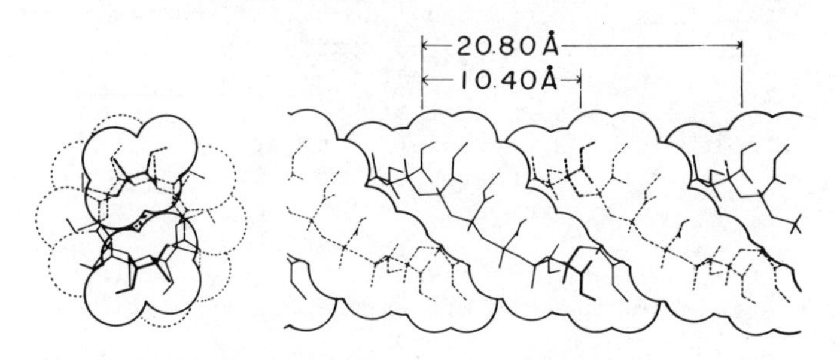

Figure 7. Double-stranded helix of isotactic poly(methyl methacrylate)(30)

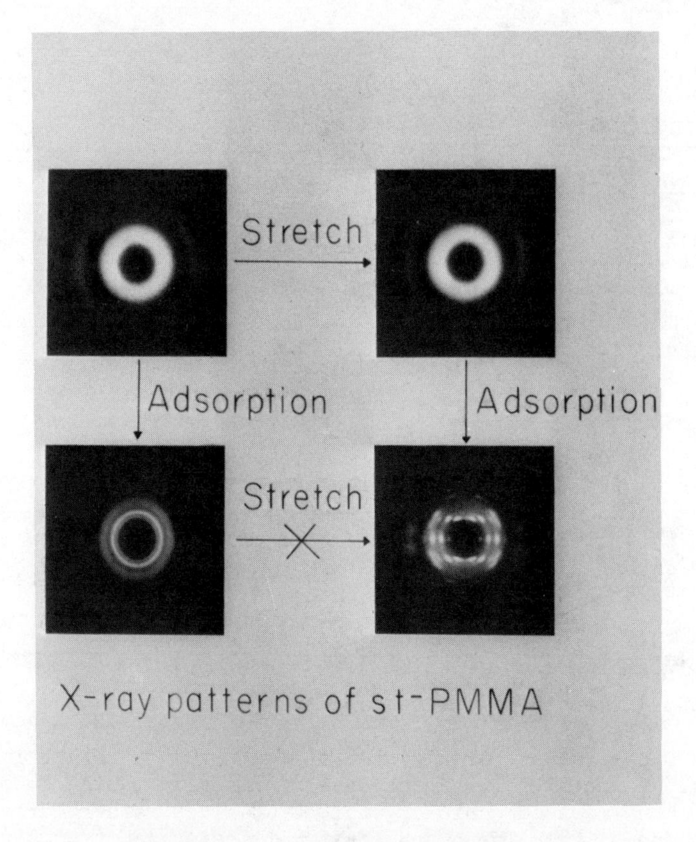

Figure 8. X-ray diagrams of syndiotactic poly(methyl methacrylate)(39)

obtained. By heating at about 90°C, the crystalline sample becomes noncrystalline but oriented, and the process is reversible.

In the fiber diagram, spots corresponding to a long fiber period (35 Å) were observed, in addition to the lines corresponding to 8.8 Å. The structure analysis is in progress but a very complicated molecular and a crystal structure is anticipated.

Figure 9 shows the change of the polarized infrared spectra by adsorption of diethyl ketone. Several bands appear by crystallization especially at 858 cm^{-1}. The bands due to diethyl ketone itself are not observed. The essential features of the x-ray diagrams and infrared spectra are the same irrespective of the kinds of solvents. From these results it may be considered that formation of a complex of syndiotactic PMMA and solvent molecules is not reasonable, and the adsorbed molecules contribute to the stabilization of crystallites of syndiotactic PMMA.

Polyethylene

Recently the author and his coworkers made detailed structure analyses for high density polyethylene samples with various histories using a PSPC (40). The PSPC has been applied to small-angle scattering, and further applications to the wide-angle diffraction of high polymers have also been found. The PSPC has been very useful in experiments with polyethylene. This may be the first case of structure analysis by using a PSPC. Deformation effects on reflection profiles from oblique incidence was considered first. Figure 10 shows reflections from a silicon single crystal taken at various positions on the linear-anode of the PSPC probe. The integrated intensities are constant independent of the positions. Figure 11 shows the similar figures for a polyethylene sample.

The polyethylene samples examined are shown in Table III; slowly cooled or quenched from melt, original monofilament, annealed, over drawn, cold drawn, single crystal, cast film, extended chain crystal, etc. The sample-probe distance can be chosen from 70 to 260 mm. The setting angle ϕ is defined as the angle between the molecular plane and the bc plane according to Bunn (41) as shown in Figure 12.

A summary of the cell constants and setting angles of the samples are given in Table IV. The a axis dimension and the setting angle ϕ are smallest in the sample slowly crystallized from melt; 1A. The a axis dimension is 7.399 Å and ϕ is 44.4°. These two values increase by stretching, from 1A to 3A, to 7.432 Å and 46.2°, and decrease by annealing, from 3A to 4A, to 7.427 Å and 45.8°. These values for a single crystal mat are rather large, 9A, 7.416 Å and 47.1°, and decrease by annealing, 11A, 7.389 Å and 46.3°. The extended chain crystal, ECC, gives rather large values 7.412 Å and 47.2°. The b and c axis dimensions show no significant change.

In Figure 13 cell dimensions are plotted against the lattice

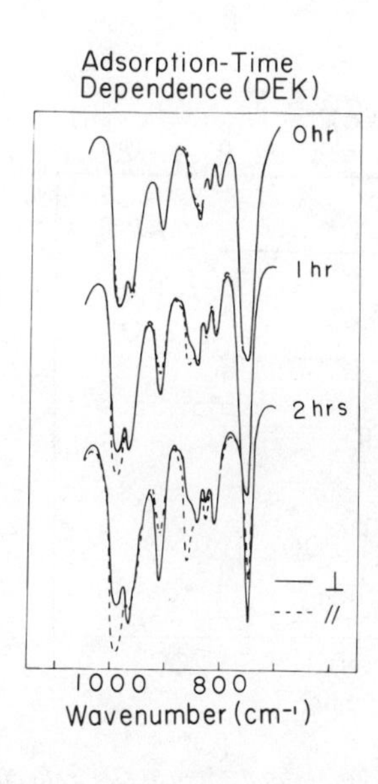

Figure 9. The change of the polarized IR spectra of syndiotactic poly(methyl methacrylate) by adsorption of diethyl ketone (39)

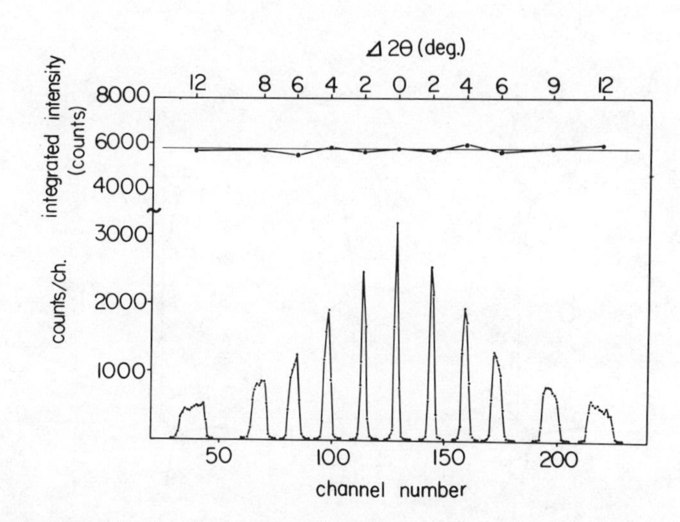

Figure 10. Oblique effect on profile and integrated intensity of reflection of silicon single crystal (40)

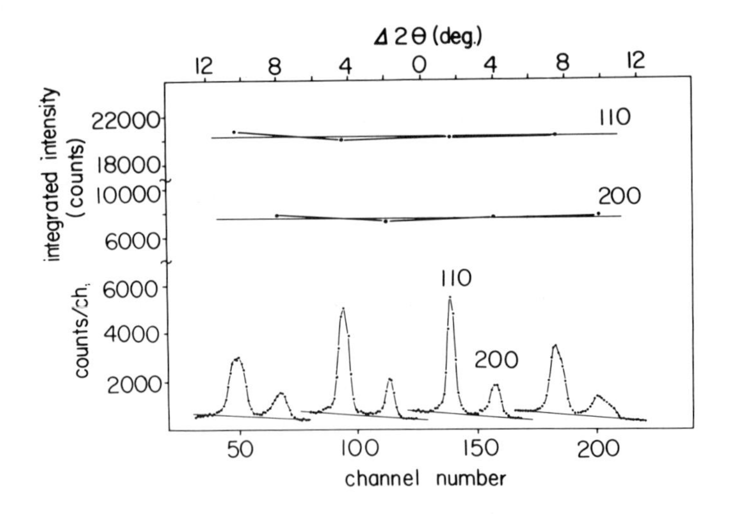

Figure 11. (a) Oblique effect on profile and (b) integrated intensity of reflections of polyethylene (40)

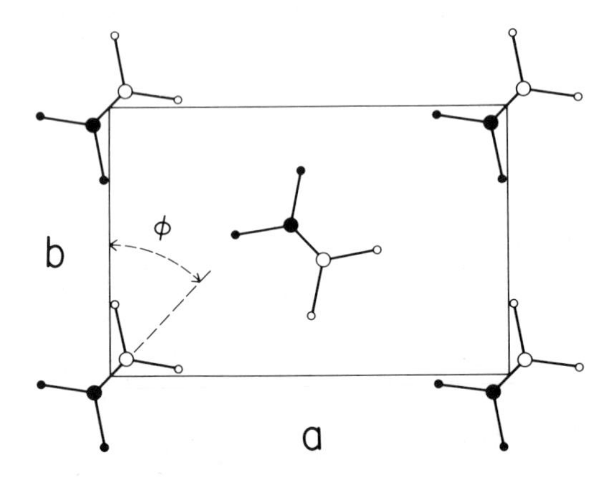

Figure 12. The setting angle ϕ of polyethylene molecule in the unit cell (41)

Table III. Sample Number and Treatment (40)

Sample No.	Crystallization	Drawing	Treatment
1A	From melt	No	Slowly cooled from 160°C and annealed at 110°C
2A		No	Quenched from 160°C in dry-ice/methanol
3A		Yes	Original monofilament (x 11)
4A		Yes	As No. 3A, annealed at 120°C under free tension
5A		Yes	As No. 3A, over-drawn at 70°C (x 23)
6A		Yes	As No. 2A, cold-drawn at room temperature (x 7)
7B		No	Slowly cooled from 160°C and annealed at 110°C
8B		Yes	Cold-drawn and annealed at 110°C from 7B (x 10)
9A	From solution	No	Single crystal grown from 0.01% p-xylene solution at 80°C
10A		No	As No. 9A, annealed at 110°C
11A		No	As No. 9A, annealed at 120°C
12A		No	Cast from 5% tetrachloroethylene solution at 110°C
13A		Yes	As No. 12A, cold-drawn (x 13)
14C	Under high pressure	No	Extended chain crystal (supplied by Professor T. Takemura of Kyushu University)

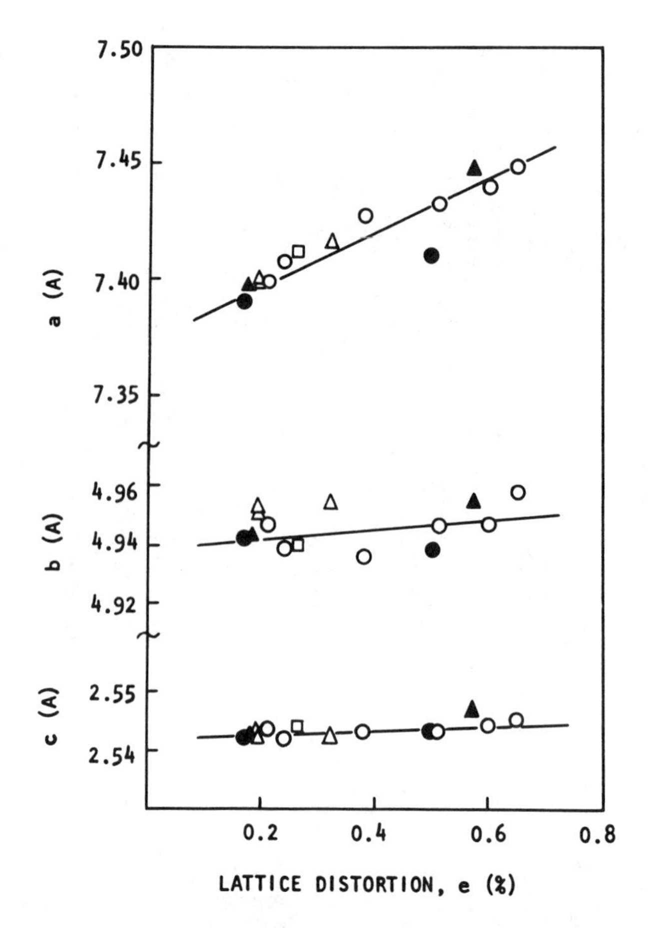

Figure 13. Cell dimensions plotted against the lattice distortion parameter of polyethylene (40) *((○) melt crystallized, A; (●) melt crystallized, B; (△) single crystal; (▲) cast film; (□) high pressure crystallized, C)*

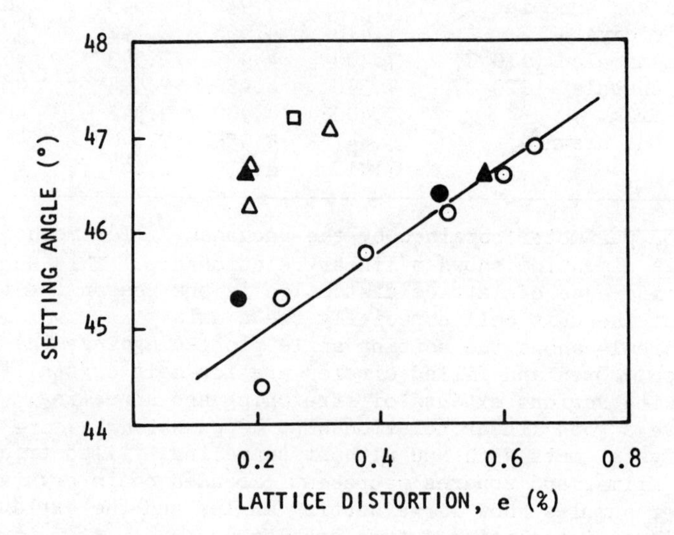

Figure 14. Setting angle plotted against the lattice distortion parameter (40) *((○) melt crystallized; (●) melt crystallized; (△) single crystal from* p-*xylene; (▲) cast film from tetrachloroethylene; (□) pressure crystallized)*

Table IV. Crystallograpic Data of Polyethylene Samples (40)

Sample	Cell constants (Å)			Setting angle (°)
	a	b	c	
1A Cooled from melt	7.399	4.946	2.543	44.4
2A Quenched from melt	7.403	4.936	2.542	45.3
3A Original-drawn	7.432	4.945	2.543	46.2
4A Annealed	7.427	4.932	2.543	45.8
5A Over-drawn	7.440	4.946	2.544	46.6
6A Cold-drawn	7.448	4.959	2.545	46.9
7B Cooled from melt	7.389	4.940	2.542	45.3
8B Drawn and annealed	7.410	4.935	2.543	46.4
9A Single crystal	7.416	4.955	2.542	47.1
10A Annealed (110°C)	7.400	4.935	2.543	46.7
11A Annealed (120°C)	7.398	4.950	2.544	46.3
12A Cast film	7.400	4.940	2.542	46.6
13A Cast and drawn	7.447	4.955	2.546	46.6
14C ECC	7.412	4.937	2.544	47.2

distortion parameter obtained by the Buchanan-Miller method (42). The a axis dimension shows a linear relationship. This suggests that the increase of lattice distortion is one reason for the expansion of the unit cell especially the a axis.

Figure 14 shows the setting angle plotted against the lattice distortion. Open and filled circles are for melt crystallized samples with various extents of stretching and annealing. These plots have a good linear relationship. Open triangles are for single crystal mats with and without annealing, filled triangles are cast films, and squares represent extended chain crystals. The latter samples show large setting angles and the explanation thereof is an interesting future problem.

The detailed analyses of polyethylene samples with various histories, show appreciable variation of the setting angle, the a axis dimension, and lattice distortion. These changes are small but appreciable, and some of them are in a linear relation, but some are not. Similar phenomena will be found in other polymers, and the reason may be interesting.

Further Remarks

In the future, complicated structures which are difficult to be analysed will remain. Further development of this field will require more accurate measurements of diffraction data, consideration of disorders, special sampling techniques, new devices and ideas for analysis of individual samples.

Literature Cited

1. e.g., Tadokoro, H., "Structure of Crystalline Polymers," Wiley-Interscience, New York, (1979).
2. Cochran, W., Crick, F. H. C., and Vand, V., Acta Crystallogr., (1952), 5, 581.
3. Suehiro, K., Chatani, Y., and Tadokoro, H., Polym. J., (1975), 7, 352.
4. Kusanagi, H., Tadokoro, H., Chatani, Y., and Suehiro, K., Macromolecules, (1977), 10, 405.
5. Shimanouchi, T. and Mizushima, S., J. Chem. Phys., (1955), 23, 707.
6. Miyazawa, T., J. Polym. Sci., (1961), 55, 215.
7. Sugeta, H. and Miyazawa, T., Biopolymers, (1967), 5, 673.
8. Yokouchi, M., Tadokoro, H., and Chatani, Y., Macromolecules, (1974), 7, 769.
9. Ganis, P. and Temussi, P. A., Makromol. Chem., (1965), 89, 1.
10. Tai, K. and Tadokoro, H., Macromolecules, (1974), 7, 507.
11. Go, N. and Okuyama, K., Macromolecules, (1976), 9, 867.
12. MacGillavry, C. H. and Bruins, E. M., Acta Crystallogr., (1948), 1, 156.
13. Tadokoro, H., Chatani, Y., Yoshihara, T., Tahara, S., and Murahashi, S., Makromol. Chem., (1964), 73, 109.
14. Arnott, S. and Wonacott, A. J., Polymer, (1966), 7, 157.
15. Takahashi, Y., Sato, T., Tadokoro, H., and Tanaka, Y., J. Polym. Sci., Polym. Phys. Ed., (1973), 11, 233.
16. Kusanagi, H., Tadokoro, H., and Chatani, Y., Polym. J., (1977), 9, 181.
17. Arnott, S., Scott, W. E., Rees, E. A., and McNab, C. G., J. Mol. Biol., (1974), 90, 253.
18. Kusanagi, H., Takase, M., Chatani, Y., and Tadokoro, H., J. Polym. Sci., Polym. Phys. Ed., (1978), 16, 131.
19. Chatani, Y., Ueda, Y., and Tadokoro, H., Rep. Progr. Polym. Phys. Jpn., (1977), 20, 179.
20. Elliott, A., Sci. Instrum., (1965), 42, 312.
21. Hall, I. H. and Rammo, N. N., J. Polym. Sci., Polym. Phys. Ed., (1978), 16, 2189.
22. Takahashi, Y. and Tadokoro, H., Macromolecules, (1973), 6, 672.
23. Tadokoro, H., Tai, K., Yokoyama, M., and Kobayashi, M., J. Polym. Sci., Polym. Phys. Ed., (1973), 11, 825.
24. Natta, G. and Corradini, P., Nuovo Cimento, Suppl., (1960), 15, 40.
25. Frank, F. C., Keller, A., and O'Connor, A., Phils. Mag., (1959), 4, 200.
26. Bassi, I. W., Bonsignori, O., Lorenzi, G. P., Pino, P., Corradini, P. and Temussi, P. A., J. Polym. Sci. A-2, (1971), 9, 193.
27. Corradini, P., Ganis, P., and Petracone, V., Eur. Polym. J., (1970), 6, 281.
28. Natta, G., Corradini, P., and Bassi, I. W., J. Polym. Sci., (1961), 51, 505.

29. Tadokoro, H., Chatani, Y., Kusanagi, H., and Yokoyama, M., Macromolecules, (1970), 3, 441.
30. Kusanagi, H., Tadokoro, H., and Chatani, Y., Macromolecules, (1976), 9, 531.
31. Scott, R. A. and Scheraga, H. A., J. Chem. Phys., (1966), 45, 2091.
32. Natta, G., Corradini, P., and Ganis, P., Makromol. Chem., (1960), 39, 238.
33. De Santis, P., Giglio, E., Liquori, A. M., and Ripamonti, A., J. Polym. Sci. A, (1963), 1, 1383.
34. Sakakihara, H., Takahashi, Y., Tadokoro, H., Oguni, N., and Tani, H., Macromolecules,(1973), 6, 205.
35. Tanaka, T., Chatani, Y., and Tadokoro, H., J. Polym. Sci., Polym. Phys. Ed., (1974), 12, 515.
36. Allegra, G., Benedetti, E., and Pedone, C., Macromolecules, (1970), 3, 727.
37. Stroupe, J. D. and Hughes, R. E., J. Am. Chem. Soc., (1958), 80, 2341.
38. Coiro, V. M., De Santis, P., Liquori, A. M., and Mazzarella, L., J. Polym. Sci. C, (1969), 16, 4591.
39. Takase, M. Higashihata, Y., Tseng, Hsiung-To, Chatani, Y., Tadokoro, H., Miyamoto, T., and Inagaki, H., J. Polym. Sci., Polym. Phys. Ed., submitted.
40. Chatani, Y., Ueda, Y., and Tadokoro, H., Rep. Progr. Polym. Phys. Jpn., (1977), 20, 179. To be published in detail elsewhere.
41. Bunn, C. W., Trans. Faraday Soc., (1939), 35, 482.
42. Buchanan, D. R. and Miller, R. L., J. Appl. Phys., (1966), 37, 4003.
43. Sakurada, I. and Kaji, K., J. Polym. Sci. C, (1970), 31, 57.

RECEIVED February 19, 1980.

4

Simulation of Fiber Diffraction Patterns

E. SUZUKI, R. D. B. FRASER, T. P. MACRAE, and R. J. ROWLANDS

Division of Protein Chemistry, CSIRO,
343 Royal Parade, Parkville, Victoria 3052, Australia

Many polymeric materials have a fibrous texture in which elongated particles with an ordered internal structure are preferentially aligned parallel to a particular direction termed the fiber axis. Diffraction patterns obtained from such materials contain information about both the particles and the matrix in which they are embedded. This matrix may consist of amorphous polymer of the same or different composition to the particle or may be a liquid.

The factors which influence the diffraction pattern and about which it therefore provides information include the following:

1. Intraparticle (a) molecular structure
   (b) random and cumulative disorder
   (c) periodic distortions
   (d) particle dimensions

2. Interparticle (a) distribution function for particle orientation
   (b) interparticle ordering

3. Matrix (a) mean electron density
   (b) radial electron density distribution.

In the past, practical difficulties of data collection have placed severe restrictions on the extraction of this information. However the recent development of a technique for calculating a quasi-continuous map of the specimen intensity transform (1) affords the possibility of realising the full potential of fiber diffraction data.

Comparison of the observed specimen intensity transform with that calculated for a model of the structure of the specimen provides a powerful test of the correctness of the model. In the present contribution we describe some preliminary attempts to simulate fiber diffraction patterns. When the observed and simulated intensity transforms are displayed visually they provide a useful guide to the progress of a structure refinement as well

0-8412-0589-2/80/47-141-061$05.00/0

as providing hints for further improvements. Since the information is stored in digital form it is possible to devise measures of goodness of fit, tailored to the particular problem, which can be used for automated refinement.

The specimen intensity transform I_s is a type of convolution product of the particle intensity transform I_p and the particle orientation density function (1,2). The procedure that we have used to simulate I_p involves firstly the calculation of the intensity transform for an infinite particle, with appropriate allowances for random fluctuations in atomic positions and for matrix scattering. A mapping of I_p is then carried out which includes the effects of finite particle dimensions and of intra-particle lattice disorder, if this is present. A mapping of I_s is then obtained from I_p by incorporating the effects of imperfect particle orientation.

Calculation of Intensity Transform of Particle (I_p)

Two types of diffraction patterns are encountered in practice: in the first, the reflections are discrete indicating that the internal structure of the particle is crystalline, in the second, there is a continuous distribution of intensity along discrete layer lines indicating that the particle is helical. In a particle containing helices arranged on a crystalline lattice the diffraction pattern will of course be of the first type. In the present description we limit consideration to cases where the natural breadth of reflections or layer lines from individual particles does not lead to significant overlap.

Crystalline Particles. Usually, one of the unit cell edges is preferentially oriented parallel to the fiber axis and the particle intensity transform corresponds to a single-crystal rotation pattern and the reflections are confined to layer lines spaced at intervals of l/c where c is the dimension of the unit cell parallel to the fiber axis (3).

It is convenient to use spherical polar coordinates (D,σ,ψ) at some stages of the calculation and cylindrical polar coordinates (R,ψ,Z) at others to define position in reciprocal space (Figure 1). The distance D corresponds to the reciprocal of the Bragg spacing.

Values of $|F_{hkl}|^2$ are calculated and stored for each layer line, where F_{hkl} is the structure factor for the reflection with Miller indices h,k and l. The effects of matrix scattering are approximated by a modification of the atomic scattering factors according to the expression

$$f'(D) = f(D) - v\rho \exp(-\pi v^{2/3} D^2) \quad (1)$$

where $f(D)$ is the normal atomic scattering factor, $f'(D)$ is the modified factor, v is the volume of matrix displaced by the atom and ρ is the mean electron density of the matrix (4). The effects of random displacements of the atoms from their lattice positions are incorporated by further modifying the atomic scattering factors by the inclusion of one or more temperature factors, following standard procedures (5,6).

The spread of the reflections around the reciprocal lattice points due to the finite particle dimensions and to cumulative lattice disorders of the "ideally paracrystalline" type is calculated according to the formula given by Hosemann and Wilke (7). They showed that for a one-dimensional crystal the integral breadth (β) varies with the Miller index (h) of the reflection and can be approximated by the expression

$$\beta = \beta_o + \frac{\pi^2 h^2}{\langle d \rangle} \left[\frac{\langle d^2 \rangle - \langle d \rangle^2}{\langle d \rangle^2}\right] \tag{2}$$

where β_o is the integral breadth in the absence of a paracrystalline disorder, $\langle d \rangle$ is the mean lattice distance and $\langle d^2 \rangle$ is the mean square lattice distance. The values at the reciprocal lattice points are normalized so that the integrated intensity about each reciprocal lattice point remains proportional to $|F_{hkl}|^2$.

The cylindrically averaged particle intensity transform is obtained by evaluating

$$I_p(R,Z) = \frac{1}{2\pi} \int_0^{2\pi} I(R,\psi,Z).\, d\psi \tag{3}$$

where $I(R,\psi,Z)$ is the stationary particle intensity transform.

Helical Particles. The intensity distribution along the layer lines in the intensity transform of a helical particle is continuous but is completely defined by a set of values tabulated at intervals of $1/2d$ where d is the diameter of an exscribed cylinder (8). Values of the cylindrically averaged square of the modulus of the structure factor $\langle FF^* \rangle_\psi$ for one period of the helix, c, are calculated at these intervals according to the expression

$$\langle FF^* \rangle_\psi = \int_0^{2\pi} F(R,\psi,l/c) F^*(R,\psi,l/c).d\psi \tag{4}$$

Procedures used are described in detail elsewhere (5). The effect of matrix scattering is again approximated by modifying the atomic scattering factors according to expression (1). Random fluctuations in atomic positions parallel to the axis of the helix are allowed for by multiplying $f'(D)$ by the term $\exp(-\frac{1}{4}B_z Z^2)$ where B_z is a Debye temperature factor.

The effects of finite particle length are approximated by supposing that the layer lines have a Gaussian profile in the Z direction (9) so that:

$$I_p(R,Z) = I_p(R,l/c) \exp\left[-\pi p^2 (Z-l/c)^2\right] \tag{5}$$

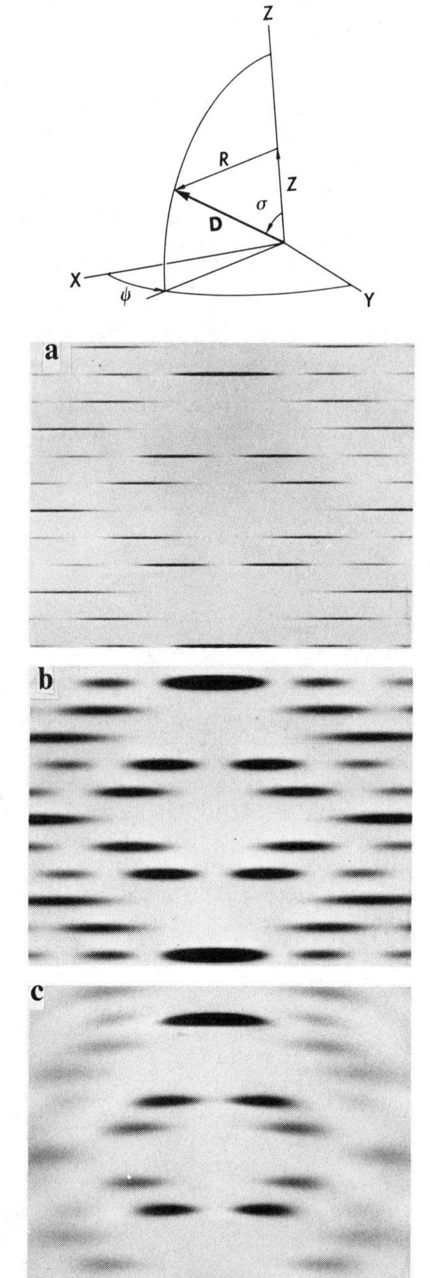

Journal of Applied Crystallography

Figure 1. Relationship between reciprocal space position vector D *and the reciprocal space coordinates (1)*

Figure 2. Intensity transforms for points uniformly distributed on a helix with radius r = *4.5 Å, unit height* h = *3 Å, and unit twist* t = *108°. (a) Particle intensity transform for* p = *500 Å; (b) particle intensity transform for* p = *100 Å; (c) specimen intensity transform for* p = *100 Å,* α_0 = *3°.*

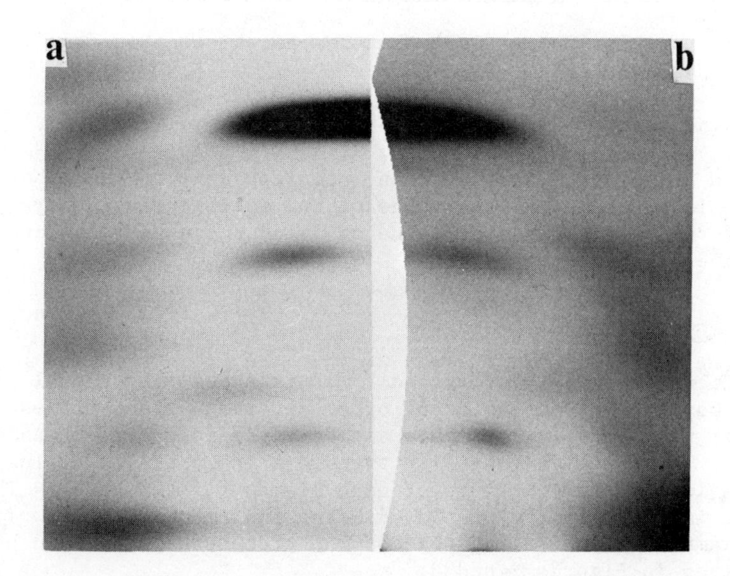

Figure 3. Comparison of (a) simulated specimen intensity transform calculated for a model of the structure of collagen with (b) a mapping of the observed specimen intensity transform derived from a diffraction pattern obtained with the specimen tilted at 15.75° to the normal to the x-ray beam.

The parameters used in the simulation are those derived from the observed pattern (11). No allowance has been made for interparticle interference effects which are responsible for the sampling of the particle intensity transform along the equator in the observed specimen intensity transform.

where p is the particle length and

$$I_p(R, l/c) = (p/c) \langle FF^* \rangle_\psi \tag{6}$$

The effects of paracrystalline type disorder in the z direction can be approximated by decreasing the value of p with increasing Z (7).

Calculation of Intensity Transform of Specimen (I_s)

It has been shown (1,2) that for an orientation density function of the type

$$G(\alpha) = \exp(-\alpha^2/2\alpha_o^2)/2\pi\alpha_o^2 \tag{7}$$

where α is the inclination of the particle axis to the fiber axis and α_o is a parameter, the value of the specimen intensity transform for a particular value of D and σ can be calculated from the expression:

$$I_s(D, \sigma_s) = \frac{1}{\alpha_o^2} \int_o^\pi I_p(D, \sigma_p) \exp\left[-\frac{(\sigma_p - \sigma_s)^2}{2\alpha_o^2}\right] i_o\left(\frac{\sin\sigma_s \sin\sigma_p}{\alpha_o^2}\right) \sin\sigma_p \, d\sigma_p \tag{8}$$

where $i_o(x) = \exp(-x)\, I_o(x)$ and $I_o(x)$ is a modified Bessel function of the second kind of order zero, and the suffixes s and p refer respectively to sample and particle intensity transform space. The specimen intensity transform is mapped by evaluating expression (8) numerically for each element of the array. The result is conveniently displayed using an Optronics Photowrite and examples illustrating some of the procedures described here are given in Figures 2 and 3.

Although not illustrated in these examples the extension of the method to include interparticle interference effects is straightforward. The principles involved and the necessary formulae are given by James (10). Similarly the effects of specimen absorption and finite beam size can readily be incorporated if required.

Literature Cited

1. Fraser, R.D.B., MacRae, T.P., Miller, A., Rowlands, R.J. J. Appl. Cryst. 1976, 9, 81-94.
2. Holmes, K.C., Barrington-Leigh, J. Acta Cryst. 1974, A30, 635-638.
3. Bunn, C.W. "Chemical Crystallography", Clarendon Press: Oxford, 1945.
4. Fraser, R.D.B., MacRae, T.P., Suzuki, E. J. Appl. Cryst. 1978, 11, 693-694.
5. Fraser, R.D.B., MacRae, T.P. "Conformation in Fibrous Proteins", Academic Press: New York, 1973.

6. "International Tables for X-ray Crystallography", Vol. III, Kynoch Press: Birmingham, 1962.
7. Hosemann, R., Wilke, W., Makromol. Chem. 1968, 118, 230-249.
8. Bracewell, R. "The Fourier Transform and Its Applications", McGraw-Hill: New York, 1965.
9. Stubbs, G.J. Acta Cryst. 1974, A30, 639-645.
10. James, R.W. "The Optical Principles of the Diffraction of X-rays", Bell and Sons: London, 1954.
11. Fraser, R.D.B., MacRae, T.P., Suzuki, E. J. Mol. Biol. 1979, 129, 463-481.

RECEIVED May 29, 1980.

Application of the Rietveld Whole-Fitting Method to Linear Polymer Structure Analysis

R. A. YOUNG and J. L. LUNDBERG
Georgia Institute of Technology, Atlanta, GA 30332

A. IMMIRZI
Instituto di Chimica Delle Macromolecole, C.N.R. Via A. Corti 12, 20133 Milano, Italy

During the last five years, a powerful new method of getting crystal structural information from powder diffraction patterns has become widely used. Known variously as the Rietveld method, 'profile refinement', or, more descriptively, whole-pattern-fitting structure refinement, the method was first introduced by Rietveld (1, 2) for use with neutron powder diffraction patterns. It has now been successfully used with neutron data to determine crystal structural details of more than 200 different materials in polycrystalline powder form. Later modified to work with x-ray powder patterns (3, 4), the method has now been used for the refinement of more than 30 crystal structures, in 15 space groups, from x-ray powder data. Neutron applications have been reviewed by Cheetham and Taylor (5) and those for x-ray by Young (6).

One of the reasons the method is so powerful is because it utilizes all of the scattered radiation intensity-vs-angle data in a powder pattern, not just that assigned ab initio to Bragg peaks. (Thus the absences of intensity are also usable data.) A structural model is required, and the parameters in the model are adjusted programmatically, by computer, to give the best least squares fit of the whole calculated powder diffraction pattern to the whole observed pattern. Many things besides crystal structure can be and often need be considered in the calculations, e.g., instrumental profiles, preferred orientation, and contributions from background and from other phases. It is possible to refine simultaneously the structures, or at least amounts present, of two or more phases present (7).

An interesting side benefit is relatively precise determination of lattice parameters, even from forward reflection ($2\theta<90°$) diffraction patterns with broadened reflections. Fig. 1 shows an example. Even though $LaPO_4$ has four lattice parameters (space group P2/n) and the pattern is not well resolved, the precision in the 3 cell-edge parameters was 6 parts in 10^4 and that in the angle was 0.04° (6).

Let us be more specific about the method itself, couching

0-8412-0589-2/80/47-141-069$05.75/0

our discussion first in a context of ionic or molecular crystal structures and later taking up practical applications to linear polymers.

The powder diffraction pattern is first obtained in digitized form as a set of intensities y_i where i identifies a scattering angle increment in the pattern. Typically, the increment size is 0.10 to 0.05°(2θ) for neutrons and 0.05° or less for x-rays. Thus, a pattern running from $20° \leq 2\theta \leq 100°$ will contain more than 2,000 separate intensity data, y_i. The object is to minimize the residual, M,

$$M = \sum_i w_i (y_i(\text{obs}) - \frac{1}{c} y_i(\text{calc}))^2, \qquad (1)$$

i.e., to minimize in a least squares sense the difference between entire calculated and observed patterns. Here c is a scaling factor and w_i is the weight assigned to the ith datum and is usually the reciprocal of the variance, $1/\sigma^2$, based on counting statistics and any other sources of quantifiable errors. The bias of least squares in overweighting data far from the expected fit, "the outliers", is mitigated to some extent by this weighting of each datum.

Many things do, or may, go into making up each y_i. At a minimium there is a background plus contributions from the "tails" of all nearby reflections (Fig. 2). Let $G(\theta_i - \theta_H)$ represent the Hth individual reflection profile centered at the Bragg position, θ_H. Let y_{inH} represent a net intensity, above background, contributed by the Hth Bragg reflection and let y_{ib} be the background level at this i^{th} position. Then

$$\begin{aligned} y_i &= y_{ib} + \sum_H y_{inH} \\ &= y_{ib} + \sum_H G(\theta_i - \theta_H) \; |F_H|^2 \, p_H \, T_H \, (LP)_i \end{aligned} \qquad (2)$$

where F_H is the structure factor for the Hth reflection ($H \equiv h,k,\ell$), p_H is the multiplicity, $T_H \equiv \exp(-P\alpha_H^2)$ is a preferred orientation function wherein α_H is the acute angle between the reciprocal lattice direction identified by H and the preferred orientation direction, P is the preferred orientation parameter and is zero if there is no preferred orientation, and $(LP)_i$ is the usual Lorentz and polarization factor evaluated at θ_i. If two crystalline phases were present, the summation terms in (2) would have to be replaced by similar sums over the two sets of H, say H and H', one for each phase.

The positional and thermal parameters for the J atoms in a unit cell are contained in the structure factor $F_H \equiv F_{hk\ell}$

$$F_{hk\ell} = \sum_j^J M_j f_j \exp[-2\pi i(hx_j + ky_j + \ell z_j)] \exp(-\beta_{jmn} h_m h_n) \qquad (3)$$

where in the last factor m and n are integers, $h_1 = h$, $h_2 = k$, $h_3 = \ell$, and repetition of an index implies summation. Here f_j is the atomic scattering factor (x-ray) or scattering length (neutron) for the jth atom and M_j is a multiplier related to the site occupancy of the jth site. For a stoichiometric material, this would be unity divided by the site symmetry factor. Thus, the crystal structural parameters to be adjusted, to give the best fit between observed and calculated patterns, are the positional parameters x_j, y_j, z_j, the site occupancies, and the thermal parameters β_{jmn}. (It often suffices to replace the six component tensor, β_{jmn}, with a single isotropic temperature factor B_j.)

Adjustments in other factors must also be made. These may include the lattice parameters, a 2θ-zero correction parameter, and a preferred orientation parameter. Principal among the non-structural parameters to be adjusted are (1) any adjustable parameters in the expression used for y_{ib} and (2) the parameters in $G(\theta_i-\theta_H)$. $G(\theta_i-\theta_H)$ is itself a convolution of the instrumental profile function and the intrinsic diffraction profile, both of which are angle dependent. The latter may be directionally dependent also. Rather than dealing with the actual details, almost universal practice to date has been to represent $G(\theta_i-\theta_H)$ by a single function for all cases (e.g., Gaussian or some form of Lorentzian) and to adjust its width at half height, K_H, by optimizing the adjustable parameters U,V,W, in the relation

$$K_H^2 = U \tan^2 \theta + V \tan \theta + W \tag{4}$$

as a part of the overall least-squares refinement process. Although the Gaussian function has served rather well in the neutron cases so far reported, better profile functions for the x-ray case are much needed (6). It is reasonable to hope that, as better profile functions come into use, crystallite size information can be deduced directly from the refined values of the profile breadth parameters.

The success of the pattern-fitting process is normally adjudged from visual comparison of calculated and observed patterns, from the magnitudes of the standard deviations in the adjusted parameters, and from the values of factors such as R-pattern (R_p) and weighted R-pattern (R_{wp}) defined as

$$R_p = \frac{\sum_i |y_i(\text{obs}) - \frac{1}{c} y_i(\text{calc})|}{\sum_i |y_i(\text{obs})|} \tag{5}$$

and

$$R_{wp} = \left[\frac{\sum_i w_i (y_i(\text{obs}) - \frac{1}{c} y_i(\text{calc}))^2}{\sum_i w_i (y_i(\text{obs}))^2} \right]^{\frac{1}{2}} . \tag{6}$$

R_{wp} is the statistically preferred R-factor since the numerator is the quantity being minimized in the least-squares refinement process. R_p, however, is more consistently reported. In the published reports, R_p values have generally been in the range 5-17% with an average near 10% for the neutron cases and 12-28% with an average near 20% for the angle-dispersive x-ray cases.

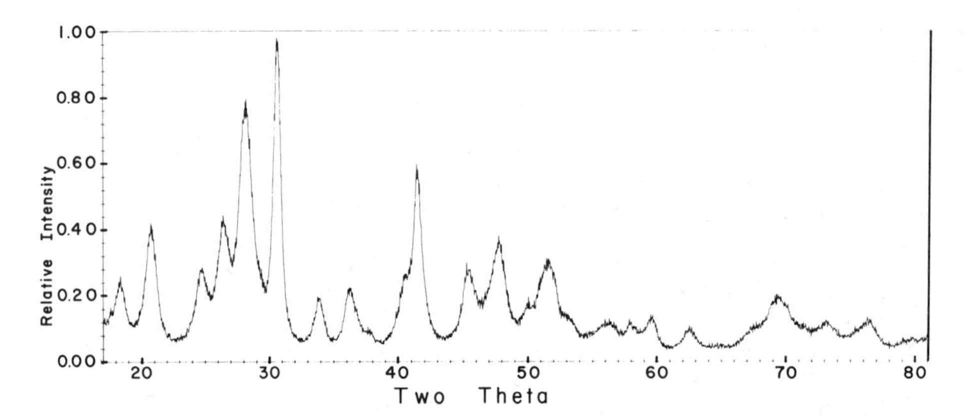

Figure 1. Observed x-ray (CuKα) powder diffraction pattern for $LaPO_4$ showing crystallite-size broadening

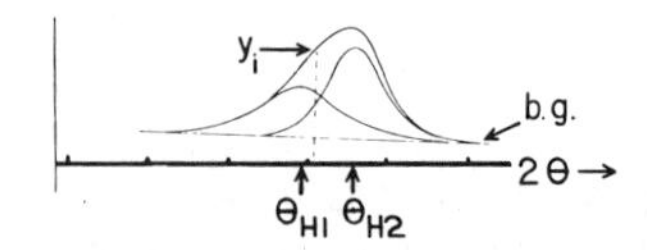

Figure 2. Contributions to y_i from background and two reflections with origins at θ_{H1} and θ_{H2}

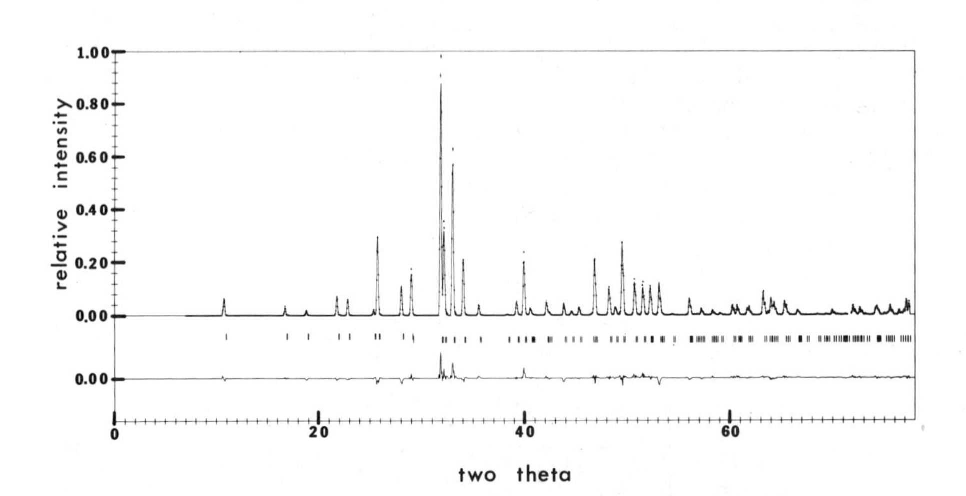

Journal of Applied Crystallography

Figure 3. Rietveld result for fluorapatite (4)

In the upper portion, the observed data are shown by the dots with error bars; the calculated pattern is shown as the solid line pattern. In the central portion, the vertical markers show positions calculated for Bragg reflections. The lower portion is a plot of the y_i difference, observed minus calculated. Gaussian profiles were used, 28 parameters were varied, and $R_p = 12\%$, $R_{wp} = 14\%$.

The number of parameters adjusted was generally in the range 10-30 (7 to 51 in the x-ray cases reviewed by Young (6)). The standard deviations in atom position parameters are typically 1-4 parts in 1,000 in the x-ray cases reviewed. Fig. 3 shows the x-ray patterns for a fluorapatite, $Ca_5(PO_4)_3F$, case in which 28 parameters were varied (20 of them structural), R_p= 12.1%, and R_{wp}= 13.9% (4). In this Fig., the observed pattern is indicated by the data points with vertical error bars, the calculated pattern is the continuous curve overlying them, the difference between observed and calculated patterns is shown by the lower curve, and the positions of Bragg peaks are shown by the vertical bars between the upper and lower curves. In this case the fit both looks to be very good and is so indicated by the R_{wp} value.

It is not always necessary that the fit be as good as that in Fig. 3 for useful results to be obtained. Fig. 4 shows the patterns for Rietveld refinement of a modified hydroxyapatite model for human tooth enamel (17) with both x-ray and neutron data. Here the difference curves show poorer fits than are shown in Fig. 3 and R_{wp} is larger, being 26% for both the x-ray and the neutron cases. Nonetheless, the positions of the principal atoms were verified to ∿0.01 Å, some of their site occupancies suggested the presence of expected substitutions, and information about the location of structurally incorporated water was revealed.

Considerations In Applications To Polymers

Because linear polymer structures can be complex with many atoms per unit cell, and because their powder diffraction patterns generally show even poorer resolution and less character than does Figure 1, it may be hopeless to do Rietveld refinement of the parameters in a detailed atomic model such as is implied by equation 3. Instead, it is appropriate to put into the model all that is already adequately known, such as the identities, sizes and shapes of the molecular units, in order to use the available power of the refinement to determine things that are not known *a priori*. This type of approach has long been used to advantage in single-crystal structure refinements, where it is variously called constrained refinement and rigid body refinement. In this method, a unit such as a PO_4 group, for example, is put into the model with all internal angles and distances fixed; only the position of the origin of the group and its orientation are allowed to be varied in the refinement. An approach similar in spirit is called for in the case of linear polymers. Structures made up of infinite chains, however, give additional complications, as is illustrated later. First, we discuss some other special aspects of Rietveld analysis of polymers.

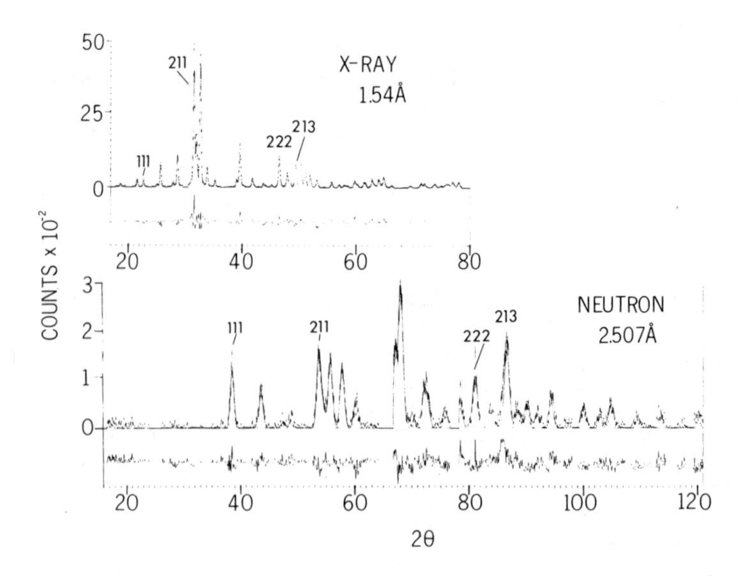

Materials Research Bulletin

Figure 4. Rietveld refinement patterns for human tooth enamel (18). X-ray data were taken at 2θ intervals of 0.375° and neutron data at 0.10° intervals. R_{wp} = *25.5% and 26.7% respectively.*

Amorphous Material. A prerequisite for crystallizability of polymers is regularity both in chemical constitution and in steric configuration (tacticity). In the case of synthetic products made by polymerization of monomers like alkenes or dienes, a number of defects both in chemical constitution (branching of side groups, insertion of head-to-head units, etc.) and in steric configuration (defects in tacticity, i.e., irregular sequencing of 'R' and 'S' monomer configurations) usually are present. Thus, a certain fraction of amorphous material is associated with the crystalline material and the resulting diffraction pattern is the sum of an ordinary pattern with Bragg peaks and a continuous pattern with one or more broad maxima. To illustrate this, x-ray diffraction patterns of some amorphous polymers are given in Fig. 5. The size of the amorphous fraction relative to the crystalline fraction (amorphous to crystalline ratio) does influence the overall spectrum considerably (Fig. 6).

If the aim of the study is just the structure of the crystalline fraction, then the diffraction contribution of the amorphous material ought to be treated as "background intensity", but the precise definition of the background line is troublesome. It is more attractive, therefore, to express the amorphous diffraction pattern in analytical form with a number of parameters to be refined simultaneously with the structural and other parameters. Not only will such a procedure produce an improved estimation of the background, but it also produces, via the refined parameters, an objective measure of the amorphous-to-crystalline ratio. A practical procedure for handling the background this way has now been incorporated in Immirzi's (8) Rietveld-analysis computer program. The calculated intensity y_i(calc) is the sum of the crystal structural contribution plus a "background" (eq'n 2) which is the sum of a number of broad peaks superimposed on a base line made of straight-line segments (Fig. 7). The broad peaks are represented by a Pearson VII peak function (Fig. 8) and all the background parameters (including the m in the Pearson formula) are adjusted in the least-squares refinement. Although this scheme for dealing with the background problem has not yet been applied to polymers, it has been successfully used in the structural analysis of some α-phthalcyanines. These are materials which can be obtained only in the form of very fine powders characterized by large lattice distortions of the paracrystalline type (M. Zocchi & A. Immirzi, unpublished results).

Peak Breadths And Reflection Profile Function. As has been mentioned, the individual reflection profiles tend to be broad for polymers. With both x-ray and neutron radiation the peaks exhibit large peak-widths. In a neutron diffraction pattern of isotactic polypropylene (A. Immirzi, work in progress) the peak width at half maximum, K_H, had values ranging from 0.60° at 2θ= 14° to 1.00° at 2θ= 43° (λ= 1.542 Å), whilst, with the same

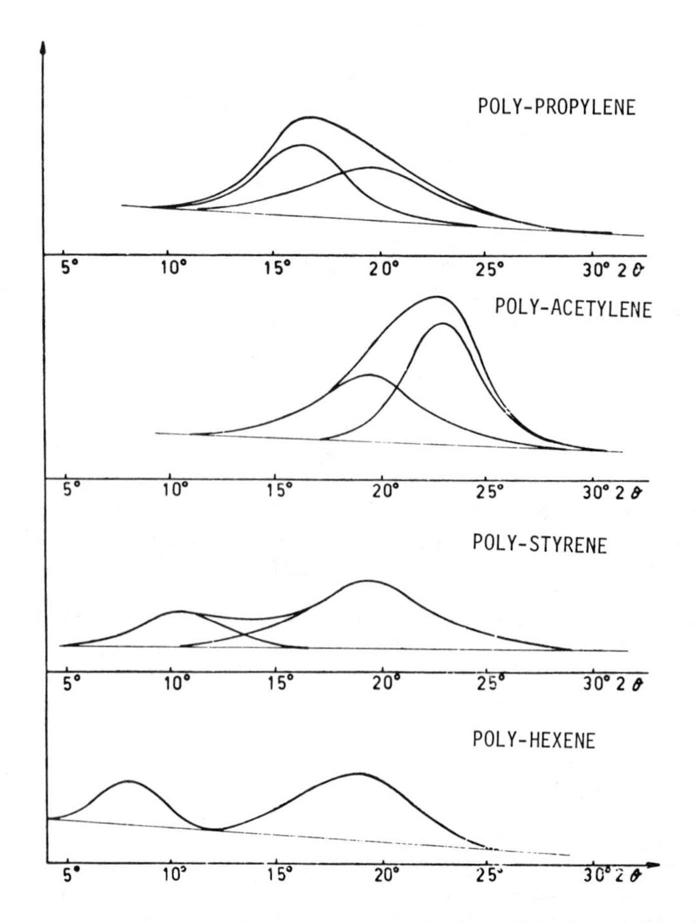

Rendiconti dell Accademia Nazionale dei Lincei

Figure 5. Examples of diffraction patterns of amorphous polymers (19). Observed x-ray intensity is plotted vertically.

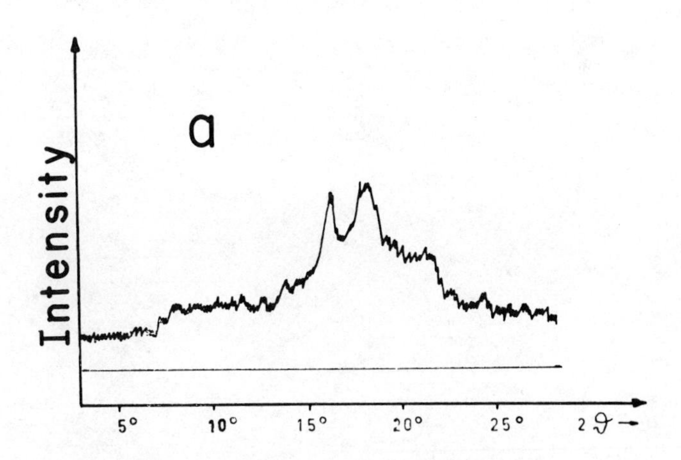

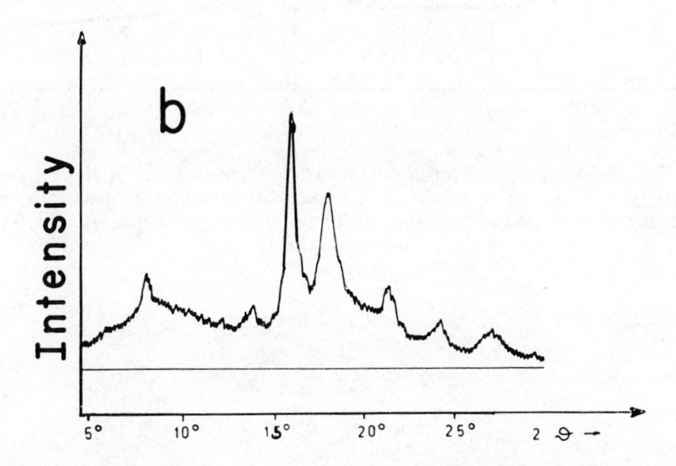

Rendiconti dell Accademia Nazionale dei Lincei

Figure 6. X-ray diffraction patterns of (a) a crude sample of polystyrene and (b) the ether-insoluble fraction of polystyrene after hot-pressing and annealing (20)

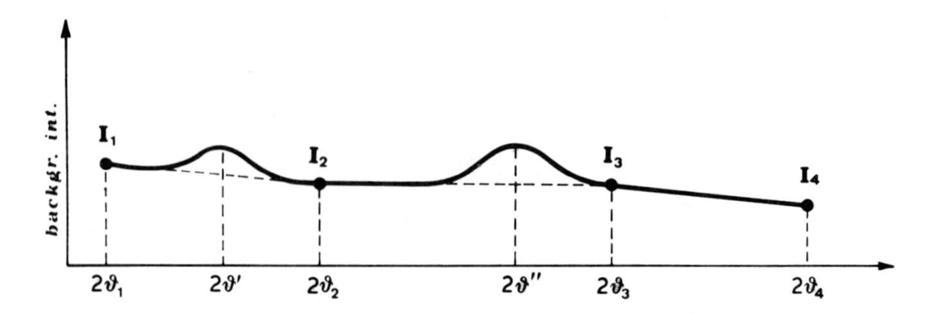

Figure 7. Functional form for the "background" considered in the program by A. Immirzi (8). The curve consists of a number of bell-shaped curves of Gaussian shape super-imposed on a base line made by straight segments. (From Immirzi, in preparation.)

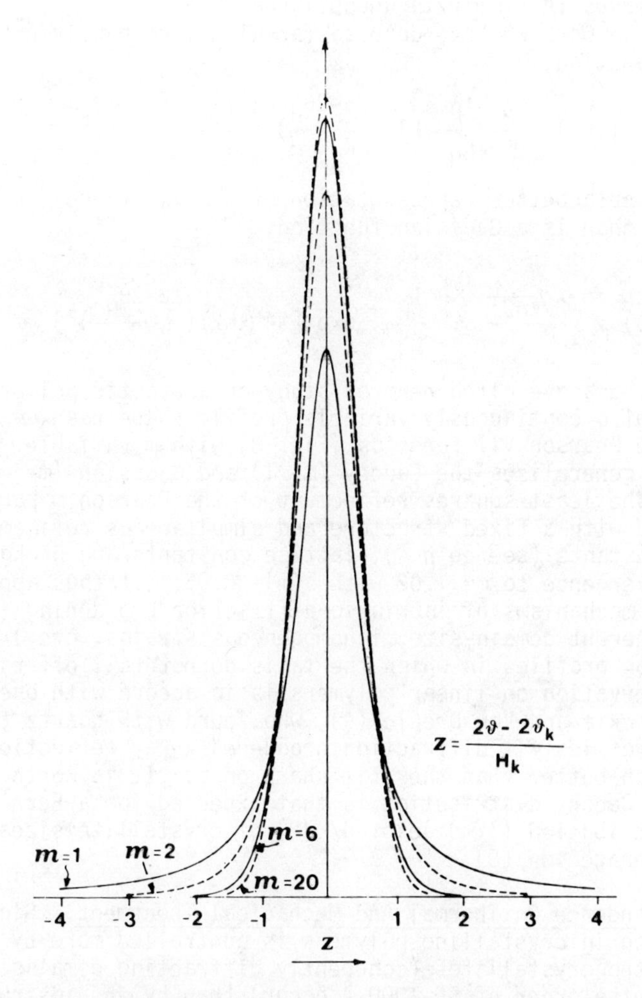

Figure 8. $I_{2\theta} = \frac{2\,\Gamma(m)\sqrt{2^{1/m}-1}}{\sqrt{\pi}\,\Gamma(m-\frac{1}{2})}\frac{I_H}{K_H}\left[1+4\left(\frac{2\theta-2\theta_H}{K_H}\right)^2(2^{1/m}-1)\right]^{-m}$

Normalized Pearson VII Function for various m *values. Note that* m *controls the shape and not the width of the peak.*

instrument and conventional materials (A. Santoro, private communication) typical K_H values were 0.41° and 0.32° respectively. In an x-ray study of the same polymer, but with a different sample and different thermal treatment, values of 0.43° and 0.61° were observed (A. Immirzi, unpublished).

In the work we have done so far with linear polymers, a Cauchy function,

$$I(2\theta) = \frac{2}{\pi} \frac{I_H}{K_H} [1+(\frac{2\theta-2\theta_H}{K_H})^2]^{-1} \tag{7}$$

seems to be a better representation of the individual reflection profiles than is a Gaussian function,

$$I(2\theta) = \frac{2\sqrt{\ell n2}}{\pi} \cdot \frac{I_H}{K_H} \exp[-4(\ell n2)(\frac{2\theta-2\theta_H}{K_H})^2] \tag{8}$$

In the above cited neutron study on isotactic polypropylene, the use of a continuously variable profile shape has been attempted using the Pearson VII function (Fig. 8) with $\underline{m}$ variable. This function generalizes the Cauchy ($\underline{m}$= 1) and Gaussian ($\underline{m}$= ∞) functions. The least squares refinement of the Pearson $\underline{m}$ parameter, performed with a fixed structure and simultaneous refinement of U,V,W constants (see eq'n 4), lattice constants and background gave convergence to $\underline{m}$= 1.02 with $\sigma(m)$= 0.05. It thus appears that the mechanisms of intrinsic diffraction broadening (e.g., small coherent domain size, inhomogeneous strains, etc.) produce reflection profiles in which the tails do not fall off rapidly. This observation on linear polymers is in accord with one by Young, Mackie and Von Dreele (4), who found with quartz that Cauchy profiles fit diffraction broadened x-ray reflection profiles much better than they fit sharp ones. It is worth noting that the Cauchy distribution is that expected for a Bernoullian size distribution $(1/\langle d \rangle)\exp(-d/\langle d \rangle)$ of crystallite sizes around their average $\langle d \rangle$ (9).

Dependence On Thermal And Mechanical Treatment. Since the peak-width in crystalline polymers is controlled more by the size of the crystallites (coherently diffracting domains; dimensions of the order of 50-1000 Å occur) than by the instrumental geometry, the profile depends markedly on thermal and/or mechanical treatment of the material (e.g., hot or cold pressing, stretching, annealing, etc.). Furthermore, the assumption that K_H depends only on the Bragg angle $2\theta_H$ and not on Bragg indices $hk\ell$ is not generally justified. That is to say, crystallite dimensions and microstrains need not be isotropic with respect to crystal directions.

Polymorphism is common in polymers. Thus, thermal and

mechanical treatments may also induce structural changes; perhaps only in a fraction of the material. Programs for performing Rietveld analyses for multi-phase systems are, on this account, very desirable.

Structual Complexity And Generalized Coordinates. We return now to the problem of handling the structural complexity by incorporating in the model all structural information that is available; then the full power of the analysis can be concentrated on the significant unknowns. This general approach has been widely used for crystal structure refinement with Bragg intensity data (10) and is called rigid body refinement or, better, refinement with constraints. In the case of molecular crystals with finite molecules not containing flexible rings, it is easy to describe the structure with only a few independent parameters. Consider, for instance, the simple case of p-sulfanilic acid $^{+}H_3N\text{-}C_6H_4\text{-}SO_3^{-}$. This molecule may be treated with only four variables under the assumption of a rigid and known model for the group $N\text{-}C_6H_4\text{-}S$, known N-H and S-O bond lengths, and local C_{3v} symmetry for NH_3 and SO_3 groups. The four variables are the C-N-H and C-S-O angles and the two torsion angles about C-N and C-S bonds. On the other hand, the bond lengths themselves may be used as variables, also, when their precise values are not known. The number of variables will still be substantially reduced, provided that the symmetry restrictions are maintained. For the description of crystal structures one adds to the preceding molecular parameters an appropriate combination of rotation angles and translations: 6 parameters or less according the space group symmetry.

Diamond (11) has developed a very sophisticated and elegant formalism for dealing with structure refinement of long chain molecules by using torsion angles as refinable variables with restrictions imposed among the angles in order to preserve the α-helix structure of part of the chain. In the case of synthetic linear polymers, as well in some natural homopolymers, however, the chains frequently adopt conformations such that the units are all structurally, though not crystallographically, equivalent. This is typical of the helix structure (see Fig. 9 for some examples), which almost all linear polymers have in order to accommodate substituent groups in crystalline arrays. Moreover, the continuity of the helix, i.e., the repetition by translation after a finite number of chemical units, imposes restrictions among torsion angles which, then, cannot be treated as independent variables. Arnott & Wonacott (12) developed the Linked-Atom Least Squares (LALS) refinement program by introducing Lagrange multipliers to treat the problem of helix continuity, thereby maintaining the independence of the torsion angles. This procedure can lead to large and ill-conditioned normal matrices. A possible alternative is the use of a lesser set of independent

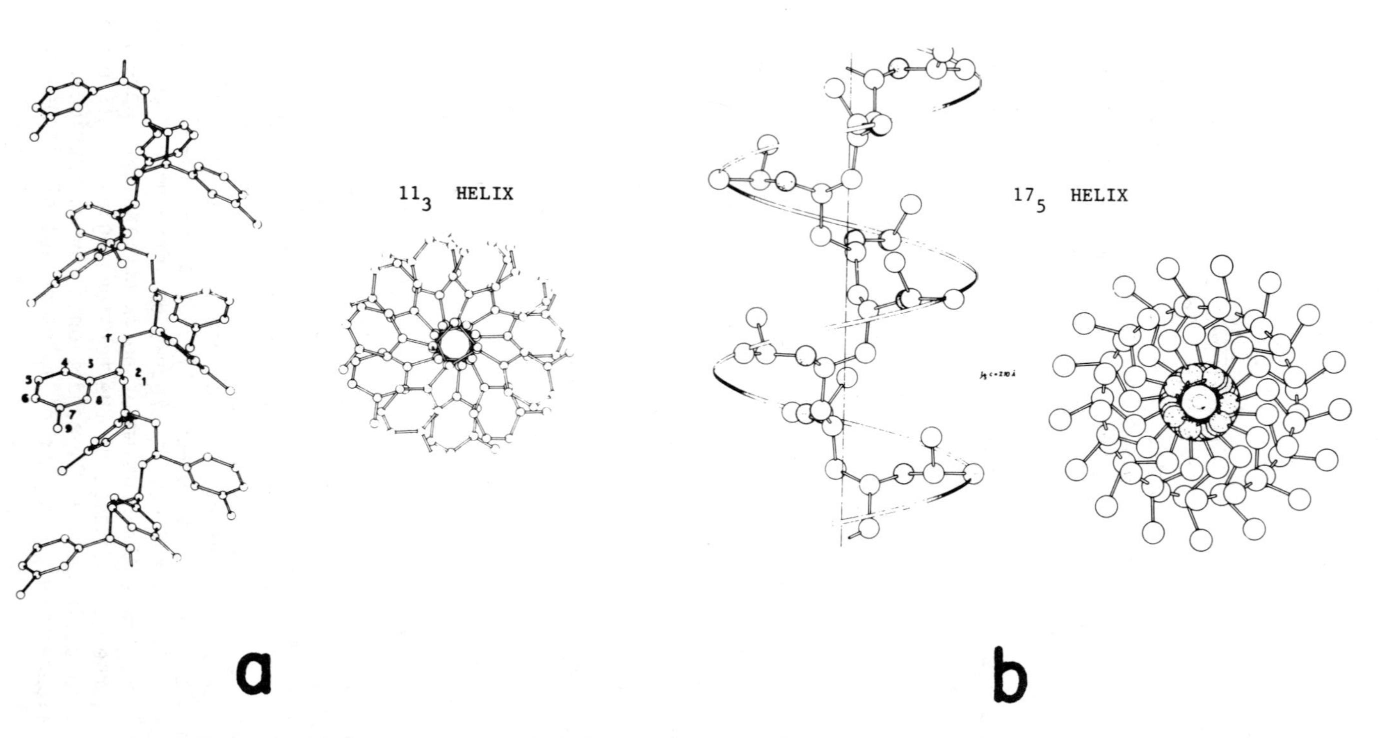

Figure 9. Helix structure: (a) side and end view of the chain of poly-m-methylstyrene (21); (b) side and end views of the macromolecule of isotactic crystalline poly-i-propyl vinyl ether (22)

variables with recourse made, if necessary, to variables of little physical significance. Pawley, et al (14) have successfully used such an approach with two small planar organic molecules, imposing strict constraints which reduced the size of the normal matrix.

This alternative approach has the advantage of allowing the implementation of compact programs of wide applicability. The term "generalized coordinates" ("g.c.") has been proposed (8) for alluding to such a set of variables of any kind, including ones that are inhomogeneous and lacking physical significance.

A possible treatment of helix structures, one which is of restricted applicability but of great simplicity, is illustrated here. It applies to homopolymers with only two kinds of atoms in the chain, e.g., vinyl polymer $-(CH_2-CHR)_n-$, having a regular helix structure and a known ratio m/t (number of monomers/number of turns). By assuming for g.c. the angle ϕ and the ratio $\tau = r_b/r_a$ as defined in Fig. 10 and using the quantities ℓ_1 and ℓ_2 (chain bond lengths, presumably to be kept fixed) and $p = c\, t/m$ (repeat per monomeric unit, c= chain axis period lengths), one can express the cartesian coordinates x,y,z of the chain atoms as analytic functions of the g.c. Let

$$\mu = \cos(2\pi\frac{t}{m} - \phi) - \cos\phi \qquad \nu = 1 + \frac{\ell_1^2 - \ell_2^2}{p^2}$$

then define:

$$q_1 = [\ell_1^2 - \frac{1}{4} p^2 \nu^2] \frac{p^2}{\tau^2 \mu^2}$$

$$q_2 = [1 + \tau^2 - 2\tau \cos(2\pi \frac{t}{m} - \phi) + \tau \mu \nu] \frac{p^2}{2 \tau^2 \mu^2}$$

and find:

$$r_a = \sqrt{\sqrt{q_1 + q_2^2} - q_2} \qquad r_b = r_a \tau$$

Hence

$$x_a = r_a \quad y_a = 0 \quad z_a = 0$$

$$x_b = r_b \cos\phi \quad y_b = r_b \sin\phi \quad z_b = p[1 - \tfrac{1}{2}\nu - \frac{r_a r_b \mu}{p^2}].$$

The other chain atom coordinates are obtained through repeated application of a screw operator. The side groups can be defined as for finite molecules with a proper combination of bond lengths, valénce and torsion angles. Finally, the crystal structure description is completed by an overall rotation Φ and three overall translations x_o y_o z_o (or fewer, depending on the space group).

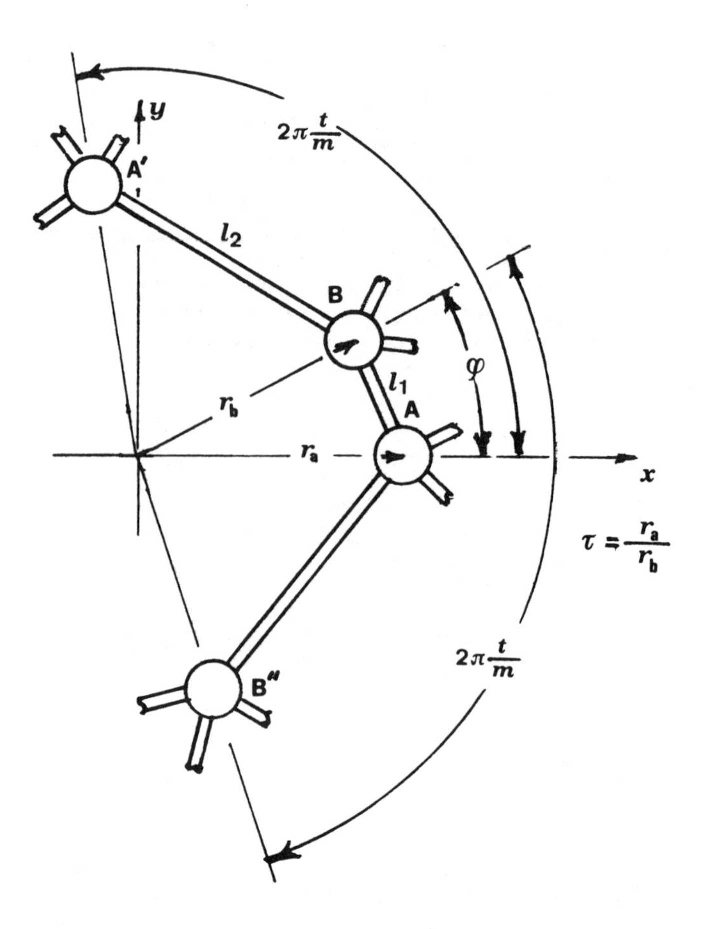

Figure 10. Schematic of a portion of a helix-shaped chain of type—$(A-B)_n$— having m/t ratio of 18/5. The chain structure is described by only 2 g.c., i.e., ϕ and τ if the chain bond lengths are kept fixed.

The use of g.c. has the disadvantage that the actual system must be defined ad hoc for each new type of problem. This drawback is not very troublesome, however, as the computer programs can be so structured that all information relating the g.c. to the fractional coordinates is concentrated in a single small subroutine (50-200 cards depending on the g.c. system). With this contrivance, a very practial computer program for Rietveld analysis has been recently written (8) which lends itself both to analyses of polymer structures and any other type of structural analyses, by the Rietveld method, in which constraints can be used to advantage. The program has the following particular features.

1) use of generalized coordinates
2) use of a "background" consisting of a number of bell-shaped curves superimposed on a segmented base line
3) use of the Pearson VII profile function with the Pearson m parameter adjustable in the least-squares refinement program
4) applicability to both x-ray and neutron diffraction powder patterns
5) other features, i.e., preferred orientation and peak-asymmetry correction, refinement of lattice constants, isotropic thermal parameters, site occupancies, etc. according to Rietveld's formalism
6) Fortran language
7) reflection list generated by a built-in routine
8) option of block-diagonal or full-matrix least squares refinements
9) dynamic allocation of variables to optimize memory utilization and time consumption
10) calculation of serveral different R factors for various 2θ intervals.

Further details of the program are available on request from the author (AI).

Example Of Application: Polypropylene

The program just described, for Rietveld analyses using generalized coordinates, has been used in the structural analysis of isotactic polypropylene recently undertaken both with x-ray and with neutron powder diffraction data. We believe this analysis (Immirzi, in preparation) to be the first Rietveld analysis of a polymer done from x-ray data. Rietveld analyses of polymers from neutron data have been done but, at least in the polyethylene case reported by Willis and co-workers (15), there was no use of generalized coordinates.

For this isotactic polypropylene, different structures based on a 3_1 helix chain structure have been proposed. Natta and Corradini (16) reported that the space group is Cc or C2/c, with preference for C2/c, while Mencick (17) reported that the space group is $P2_1/c$ and ∿25% of helices are disordered (see Fig. 11). In the recent x-ray Rietveld analyses using g.c. (Immirzi, in preparation), these three different models were tried along with a fourth model differing from Mencick's only in the helix disordering.

The x-ray Rietveld analyses (based on counter measurements with a standard powder diffracometer) showed that the models are distinguisable even though the diffraction patterns exhibit poor resolution (Fig. 12). The $P2_1/c$ models give better fits than do the Cc and C2/c models, as judged both visually (Fig. 12, patterns A,B,C,D) and on the basis of R-factors (Table 1). The two $P2_1/c$ models, (Mencick and Immirzi, patterns C and D, respectively) are distinguisable and the Immirzi one seems preferable. It has been demonstrated that appreciable angular aberrations are present in the lower angle part of the pattern, e.g., 2θ below ∿23°. If the data below 2θ = 23.5° are omitted from the refinements, the two $P2_1/c$ models are again distinguishable, again the Immirzi model gives the better fit, but the quality of the fit in the angular region now covered is much improved.

Rietveld (g.c.) analysis of the neutron diffraction data on isotactic polypropylene is still in progress. It has afforded the interesting result, already discussed, that the profiles are better approximated by Cauchy than by Gaussian functions. The structural analysis is now restricted to the fourth model ($P2_1/c$, Immirzi), which gives an excellent agreement between observation and calculation, but with the fraction of reversed helices close to 50% instead of 25% and with less chain symmetry. The other models will be tested for a more complete comparison with x-ray results. We cannot exclude, however, the possibility that the two samples used, which have different chemical, thermal and mechanical history, can really have different structures.

Concluding Remarks

The whole-pattern-fitting structure-refinement method, which was first introduced by Rietveld and used for neutron diffraction powder patterns, does yield from x-ray diffraction patterns correct, refined structural information for linear polymers. Remarkably precise lattice parameters are obtained incidentally in the use of the method. The method lends itself to improved estimations of the fraction of amorphous and crystalline materials, or of two polymorphic forms, present. As improved profile functions come in to use, the method promises to provide crystallite size information, almost as a spin-off benefit.

Use of generalized coordinates, in which accurately known structural information such as certain bond angles and bond lengths and conformation data can be introduced as known parameters, decreases the number of unknown parameters so that the least squares fitting problem is tractable. Some skill is

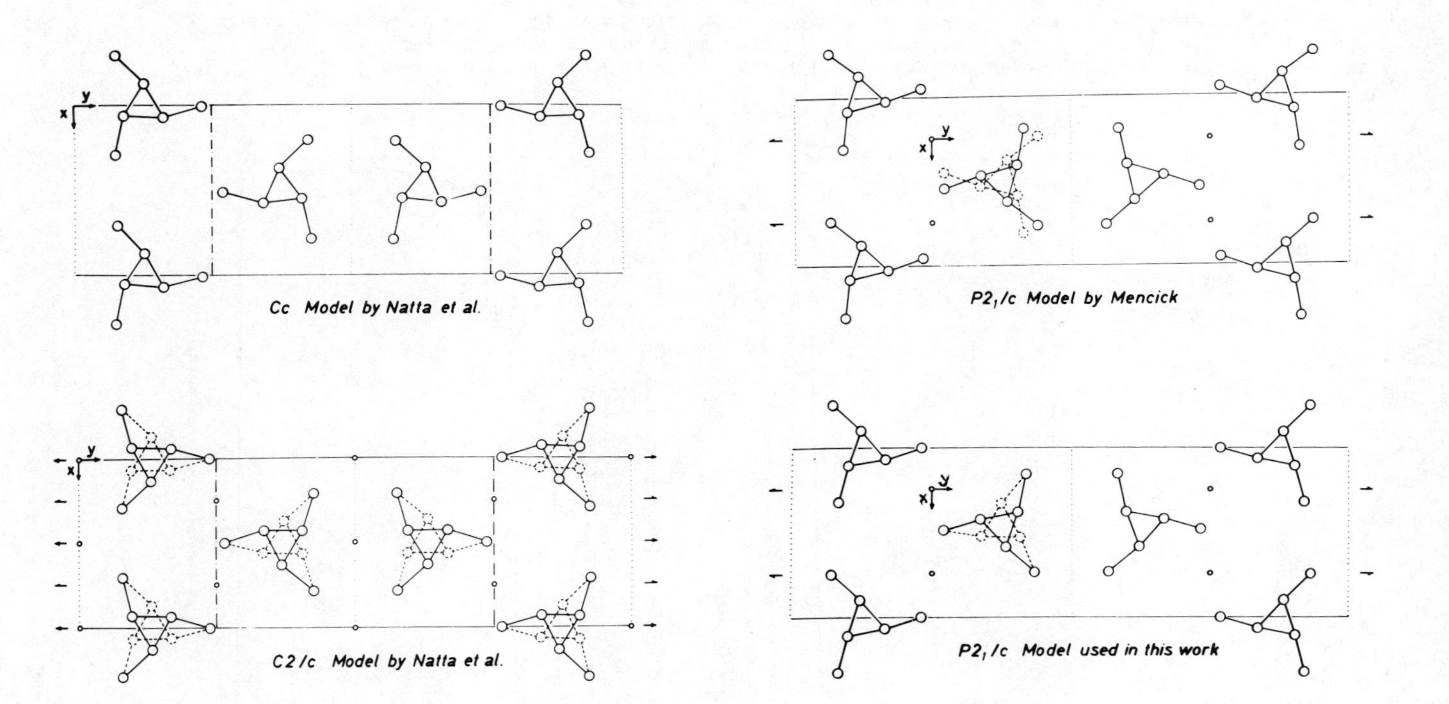

Figure 11. Proposed models for isotactic polypropylene. In the $P2_1/c$ models, a fraction f of helices is disordered through a 180° rotation about the axis (1/4, y, 1/4) (17) or about the axis orthogonal to the chain axis and crossing a side CH_3 group (from Immirzi, in preparation)

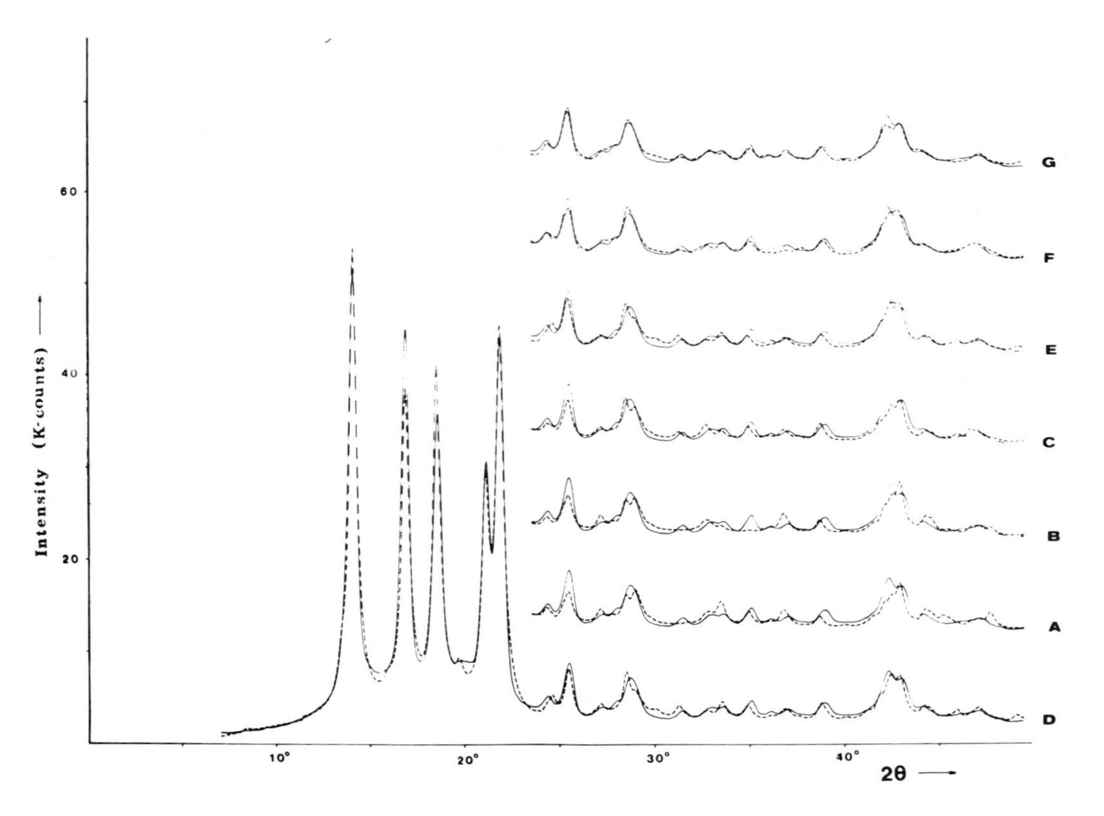

Figure 12. X-ray powder diffraction patterns observed (——) and calculated from Rietveld refinement results (- - -) corresponding to the various models as described in Table I. Refinement E was the same as D except that the reflection profile widths were allowed to vary. The results were not changed. (from Immirzi, in preparation.)

Table I

Results of X-ray Rietveld analyses of isotactic polypropylene (excerpted from Immirzi, in preparation)

Refinement	R_{wp} %	Model		
		space group	Disordered helices (%)	Authors
A	14.3	Cc	-	Natta & Corradini (16)
B	13.9	C2/c	-	Natta & Corradini (16)
C	13.2	$P2_1/c$	21(3)	Mencick (17)
D	11.8	$P2_1/c$	22(2)	Immirzi (in preparation)
F	8.0	same as C but with $2\theta > 23.5°$		
G	7.1	same as D but with $2\theta > 23.5°$		

required in selecting the generalized coordinates to be analytic functions of the atomic co-ordinates, to be few in number, and to involve as a priori, not-to-be-refined, structural parameters only those which in fact are correctly known.

The whole-pattern-fitting structure-refinement method is not limited to specimens containing randomly oriented crystallites. The method does permit an orientation parameter to be refined, though it was not done in this work. Crystallites in the molded polypropylene specimen probably were oriented to some small extent; nevertheless, the refinement method worked well. The sensitivity of the method to crystallite orientation and the applicability of the method to polymers with oriented crystallites such as in drawn films or fibers are yet to be determined.

Summary

The whole-pattern-fitting structure-refinement method can give correct, refined structural information for linear polymers. Use of generalized coordinates, in which accurately known structural information can be introduced as known parameters, makes the least-squares fitting problem tractable. Excellent fits of x-ray diffraction data for isotactic polypropylene permitted the selection of the correct space group and a preferred model.

Literature Cited

1. Rietveld, H.M. Acta Cryst., 1967, 22, 151-152.

2. Rietveld, H.M. J. Appl. Cryst., 1969, 2, 65-71.

3. Malmros, G. & Thomas, J.O. J. Appl. Cryst., 1977, 10, 7-11.

4. Young, R.A.; Mackie, P.E.; Von Dreele, R.B. J. Appl. Cryst., 1977, 10, 262-269.

5. Cheetham, A.K. & Taylor, J.C. J. Solid State Chem., 1977, 21, 253-275.

6. Young, R.A., Proceedings of Symposium on Accuracy in Powder Diffraction held at NBS, June, 1979, Gov. Printing Office.

7. Werner, P.E.; Salomé, S.; Malmros, G.; Thomas, J.O. J. Appl. Cryst., 1979, 12, 107-109.

8. Immirzi, A. Acta Cryst., 1978, A34, S348-S349.

9. Allegra, G.; Bassi, I.W.; Meille, V.S. Acta Cryst., 1978, A34, 652-655.

10. Scheringer, C. Acta Cryst., 1963, 16, 546-550.

11. Diamond, R. Acta Cryst., 1965, 19, 774-789.

12. Arnott, S. & Wonacott, A.J. Polymer, 1966, 7, 157-166.

13. Smith, P.J.C. & Arnott, S. Acta Cryst., 1978, A34, 3-11.

14. Pawley, G.S.; Mackenzie, G.A.; Dietrich, O.W. Acta Cryst., 1977, A33, 142-145.

15. Avitabile, G.; Napolitano, R.; Pirozzi, B.; Rouse, K.D.; Thomas, M.W.; Willis, B.T.M. Polymer Letters Edition, 1975, 13, 351-355.

16. Natta, G. & Corradini, P. Nuovo Cimento, 1960, suppl. 15, 40-51.

17. Mencick, Z. J. Macromol. Sci. Phys., 1972, B6, 101-115.

18. Mackie, P.E. & Young, R.A. Mat. Res. Bull., 1980, 15, 17-29.

19. Natta, G.; Corradini, P.; Cesari, M. Rend. Accad. Naz. Lincei, 1957, (8) 22, 12-17.

20. Natta, G. & Corradini, P. Rend Accad. Naz. Lincei, 1955,(8) 18, 19-24.

21. Corradini, P. & Ganis, P. J. Polym. Sci., 1960, 43, 311-317.

22. Natta, G.; Bassi, I.W.; Allegra, G. Macromol. Chemie, 1965, 89, 81-94.

RECEIVED June 19, 1980.

6

Accurate Fiber X-ray Diffraction Data from Films

Data Array Calculations

DONALD P. MILLER and ROBERT C. BRANNON[1]

Clemson University, Clemson, SC 29631

Although diffraction techniques are often chosen for determining the structure of crystalline and paracrystalline materials, a number of serious difficulties have limited the confidence one may place in the results for fibrous polymers. Such fundamental questions as the relative polarity of adjacent chains and rotational positions of side groups have been afforded little resolution in recent studies (1-7). An excellent example of these difficulties is furnished by recent studies of native and regenerated cellulose. Different studies, from the same photographic data disagreed on chain polarity, though all find the rotatable hydroxymethyl side group in the tg position for which only one single crystal analogue, disordered, has been found. For dissimilar models all yielding R factors below 0.20, data sets are found to disagree by as much as 49%.

The same data collection and reduction techniques are commonly used by the same workers for many different polymers. Therefore, data for these other polymers may contain errors on a similar scale, but that the errors have usually, but not always, gone undetected (8). If more than 500 reflections are observed, from single crystals of simple molecules, recognizable electron-density distributions have been derived from visually estimated data classified only a "weak", "medium" or "strong". The calculation of the structure becomes more sensitive to the accuracy of the intensity data as the number of data points approaches the number of variables in the structure. One problem encountered in crystal structure analyses of fibrous polymers is that of a very limited number of reflections (low data to parameter ratio). In addition, fibrous polymers usually scatter x-rays too weakly to be accurately measured by ionization or scintillation counter techniques. Therefore, the need for a critical study of the photographic techniques of obtaining accurate diffraction intensities is paramount.

[1]Current address: Technical Center, Owens-Corning Corp., Granville, OH 43023

0-8412-0589-2/80/47-141-093$05.00/0

Our general objective has been to develop methods of sufficient accuracy in film data collection to allow direct determination of fibrous polymer structures (rather than indirect, with modeling). The objective can be thought of in four parts: collect x-ray diffraction data arrays using an accurate, stable computer controlled scanning microdensitometer; process the data through the use of a high level computer language chosen for its array (matrix) mathematical capabilities; use the processed data to determine directly the structure of a fibrous polymer and when this had been accomplished to sufficient accuracy; use the results to compare structure determination by the techniques of a) Peak Center method, b) Sum Intensity method, and c) Data Array Integration method (these are discussed below).

Visual Estimation of Intensity

Visual comparison of diffraction spots with a graded intensity scale was by far the most used method to quantify intensities. The human eye, which is a very good comparative photometer, can estimate the relative order of blackness of a series of spots with great accuracy, but not on a true numerical scale. The origin of this method is unknown and seems almost intuitive. Many authors (8-13) suggest ways to obtain an intensity scale. For many films exposed in the region of density <2.5, the optical density of a diffraction spot is often proportional to exposure and inversely proportional to the $\log_{10}$ of transmission coefficient. Thus, accuracy of reflection measurements tended to be reasonably uniform over the full range of density. The technique proved usable in estimating diffraction data on patterns on low background and reflections of moderate intensity, generally of photographic density ≤ 1.0 and has been found to yield an error of 15-20% (14).

Problems encountered with this method include differences in spot size and shape, background scatter, and the fact that a very limited range of densities were measurable. It becomes inaccurate when applied to spots on a dense background or to very dense spots on any type of background.

It must be emphasized that visual matching of the densities tends to give a measure of the peak rather than the integrated intensity. Only when the spots are all of the same size and shape can one obtain fairly reliable integrated intensities. Though no longer used for obtaining data for structure determinations, the technique is still used in routine identification.

Photometric Peak Center Method

Once spot photometers became available (15,16,17,18), many workers determined the intensity at the maximum blackening of the x-ray reflection, (the peak center). To accept this for the integrated intensity is to imply that the integrated intensity is proportional to the peak intensity. Many effects act to limit the degree to which this assumption can be made. The width of diffraction maxima increase with diffraction angle due to α_1 α_2 splitting. In some specimens, absorption may vary the widths of the diffraction maxima at different angles. Overlap of two diffraction maxima at nearly equal diffraction angles alters the maximum peak height and width of both. The effects of incoherent scattering, diffracted beam polarization, the difference in the time the inner and outer sides of the diffraction maxima spend in diffraction condition and the decrease of atom scattering factors with scattering angle all cause the background to vary (decrease with angle) nonlinearly under the diffraction maxima. Other effects due to crystallite size, disorder, fiber tilt distributions also affect the sizes of diffraction maxima.

Many of these effects can be computed and the effects of others at least estimated. Users of this technique measure and average the background on both sides of a diffraction maximum. The mean optical density of this background is then subtracted from that of the peak. Peak intensities, measured of single crystals, are currently claimed accurate to a level of 3% for a range of O. D $\leq$ 2.5. For fiber diffraction data, accuracy is not as good.

Layer Line Scans

A somewhat more complicated method is frequently used to obtain "integrated intensity" values. The photograph is placed on a recording photometer, which provides a graphical record of the transmission along the center line of the diffraction maximum (or along a layer line of the diffraction transform). The ordinates of the trace are proportional to the transmission, T, while the abscissas are distances along the film. The transmission values are then converted to x-ray exposures.

Two methods used to find the area under the photometer trace are peak-height-times-half-width approximations and actual measurements with a polar planimeter. Both methods are time consuming and offer little increase in total accuracy over the peak center method. Another method involves computer fitting an assumed scattering function, usually a Gaussian or Lorentzian (though more exotic functions have also been used) to the scan data. The integrated area under the mathematical curve is then calculated.

Two-Dimensional Scan Methods

Scanning transforms the photographic pattern into an optical density map of the region scanned. Transmission is measured at intervals spaced equally. The photometer output voltage, calibrated against a standard transmission filter set yields values proportional to the energy absorbed by the film, and therefore directly proportional to the number of photons absorbed. A typical map of the region of a diffraction maximum might look like Figure 1.

One current method of obtaining diffraction intensities from these data is by direct summing of map values (SUM INTENSITY). A suitable minimum contour value is chosen and a contour line, dashed curve in the figure, is drawn around each reflection. To find the background, the sum of the numbers around the line is averaged. This is treated as the average background and is subtracted from each number inside the contour line. The numbers inside the contour line are then summed and corrected for the Lorentz, polarization and orientation factors at the location of the peak intensity. One source of error is apparent. Corrections calculated at the peak intensity location do not account for the slope of the background function over the angular breath of the diffraction maximum which, for fibers, is several degrees in size. This error is enhanced where absorption is strong enough to warrant correction. But the data is gathered point by point, so correction should be more easily applied point by point, before integration of the diffraction maxima.

The accuracy of most current scanning microdensitometers is thought to be in the range of ±3%, but enough error sources exist to cast serious doubt upon this estimate.

Improved 2-D Scan Method

Film diffraction data can be obtained by several techniques: flat plate camera, cylindrical camera or precession camera. We chose the latter, because it records a relatively undistorted "slice" of diffraction transform (reciprocal space) (19,20). We felt that this characteristic of the precession technique outweighed the disadvantages of longer exposure times (this was lessened by use of a Helium atmosphere in the camera enclosure). For potassium bromide amylose, the fiber studied, this required exposure of a 79μ x 79μ x 300μ fiber specimen for 45 hours (35kv, Ni filtered Cu radiation 30° precession angle). Three layers of film (Kodak No-screen) were exposed simultaneously so that, if necessary, a longer range of intensities might be measured than with one. The films were chemically processed together.

The scanning microdensitometer was a modified Jarrell-Ash model 23-500. Figure 2. is a block diagram.

The densitometer uses a two beam comparator technique of measurement. A single lamp illuminates, through a variable neutral density filter, the emitter of an eleven stage photomultiplier tube. Because the frequency of measurement is five times the AC line frequency and the amplifiers are extremely high gain, the lamp must be operated from a regulated DC source, otherwise, 120 Hz brightness ripple of the lamp is recorded in the data. This lamp also illuminates, through a focusing optical system, a small region of the x-ray diffraction film. The transmitted

134	138	131	128	126	125	125	125	125	123	122	122	123	122	123
136	146	148	142	136	130	127	129	127	125	123	123	120	122	123
130	145	159	162	158	144	137	134	131	130	125	124	122	121	119
126	137	153	170	161	172	158	143	140	137	134	129	124	122	120
120	128	141	160	162	184	165	173	155	148	142	136	130	128	123
120	121	126	136	160	165	201	197	163	166	151	143	139	133	130
118	121	122	127	137	154	175	191	190	183	164	153	143	138	134
117	120	121	121	123	131	143	159	172	174	167	155	144	136	134
117	116	118	117	121	121	125	132	145	147	153	151	143	135	130
115	116	117	116	118	120	118	123	128	128	132	134	134	128	126
118	115	116	116	116	118	117	117	119	120	125	124	124	123	123
116	114	112	114	115	115	116	116	117	119	117	116	116	116	122
115	116	115	116	115	114	113	116	116	115	117	119	115	118	120
115	113	115	114	114	112	115	112	114	115	117	116	119	117	116
116	115	112	113	116	114	115	116	115	115	115	116	117	116	118

Figure 1. Optical density map of a region near a diffraction maximum

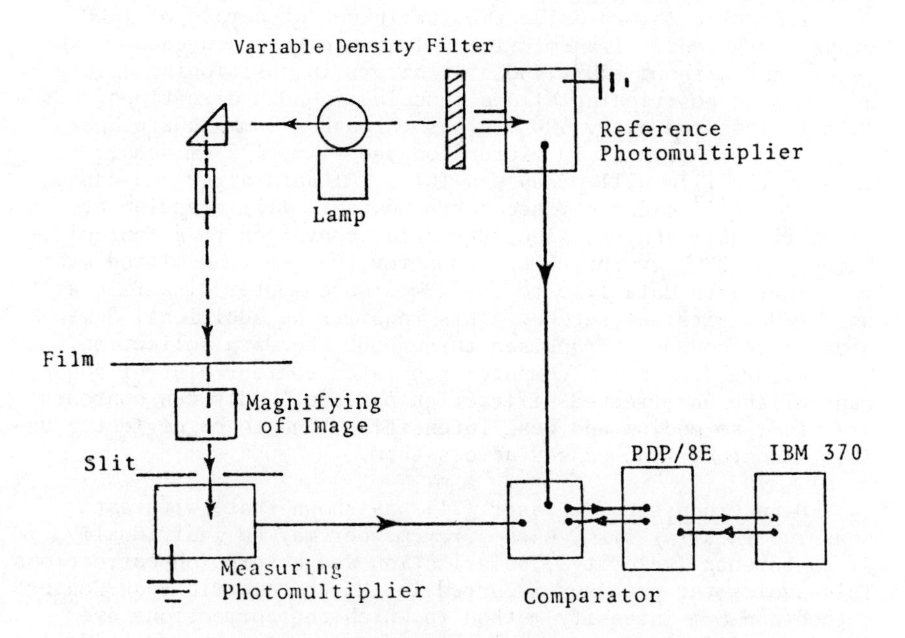

Figure 2. Block diagram of the scanning microphotometer

light then passes through a 100 power magnifying optical system, and perpendicular, adjustable slits onto the emitter of a second matched, eleven stage photomultiplier tube. The outputs of the photomultiplier tubes are then amplified by a pair of solid state, chopper stabilized amplifiers. The pair of voltages from the amplifiers are then compared in a solid state, stabilized comparator. The difference in voltage is amplified again in a stabilized operational amplifier, which is adjustable for zero level and gain. This output is then measured both by a digital panel meter and, through shielded cable, a 12 bit analog to digital converter connected to the Digital Equipment Corporation PDP8/F computer data bus.

The densitometer is adjusted for operation by setting the zero level with the lamp turned off. The least exposed area of the film to be measured is placed in the focused spot of the optical path. The lamp is turned on and the variable neutral density filter is adjusted until the digital meter reads zero. This balances the instrument and eliminates most of the photomultiplier white noise. The slits above the detector are closed and the gain is adjusted for maximum analog to digital input voltage. The slits are then opened and calibrated neutral density filters are placed in the optical path at the sample position to calibrate the instrument. The densitometer has proved stable for periods of up to eight hours at a sensitivity of 5 parts per 4000 over an optical density range of zero to three.

The pulse motors drive the instrument at a rate of 1500 microns per second. Transmission values can be measured 300 times per second without start and stop errors in positioning. This accuracy in positioning allows scanning in both directions. If a film is measured every 500 microns through a 1 mm square aperture, a resolution of 10 microns on the film, a 5 cm square region of the film will produce a 101 x 101 data array and can be measured in 56 minutes 7 seconds. However, film measurement actually takes longer, since the data, converted to a four digit number, 0-4096, by the PDP8/F computer, can be transmitted over the timeshared data line to the IBM 3305 computer disk file at only 120 digits per second. This requires an additional 8 minutes, 31 seconds interspersed throughout the data collection.

Figure 3. shows a computer generated contour plot of a quadrant of the uncorrected diffraction pattern. Only ten contours are used, so medium and weak intensities tend to be neglected unless a contour line passes across them.

Data Processing. Fraser (21) has shown that, with data measured in array form, each array member may be individually given Lorentz (velocity), polarization and absorption corrections. This avoids the problems incurred in the Photometric Peak Center method and Sum Intensity method in which the corrections are applied as if all intensity is recorded at the peak amplitude position. Fraser (22) has also shown that, by choosing data

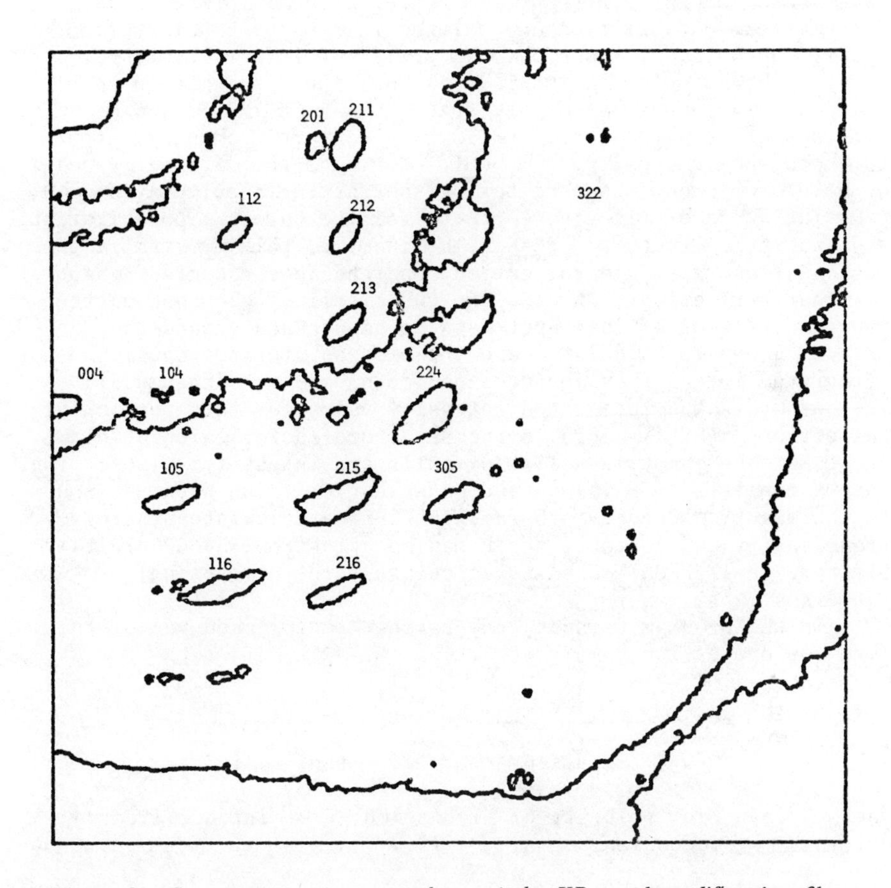

Figure 3. Contour map of one quadrant of the KBr-amylose diffraction film. The highest background intensity contour is at upper left. Note the (1,0,4) maximum, although weak is in medium background, but the (3,2,2) is weak in low background.

array values from background regions between intensity maxima, one may fit a Gaussian background function, which may be evaluated at all array positions. The difference array then exhibits minimal background.

The computer language chosen to perform background corrections and do integrations for programs was SPEAKEASY (23), a language based on Fortran, but with conversational format, array mathematical capabilities and a large resident vocabulary (functions, subroutines, input/output, etc.) It is very fast, for array math, but slow in looping. It has the disadvantage of using a large amount of memory space, 90K-450K IMB 370 words. Programs written (24) are: PHOTODEN - Converts computer output to photographic density. CORSPOT - Corrects the data array point by point for background scatter by subtracting a calculated gaussian surface. It also corrects for the Lorentz, polarization, and absorption factors. PEAK - The input to this program is the region around a single reflection, and the lowest contour value for that reflection. It uses the cubic spline technique on the rows and columns of the input data to smooth and expand the array. Sums are calculated and the process iterates until the integrated intensity value does not change by more than 0.5%. The output is the integrated intensity value for that particular reflection. FHKL - This is the structure factor calculating program. The input is a list of hkl's and intensity values. The output consists of E values and phase angles to be used as input to the electron density program. ELECDEN - Calculates the electron density and contours the E-map on a Tektronix 4662 digital plotter. PATTERSON - Used to calculate three-dimensional Patterson maps.

In the program Corspot, the Lorentz factor used was of the form for precession camera data,

$$L = \frac{1}{\Omega\ \xi \sin \bar{\mu} \sin \eta} * \frac{1}{1 + \tan^2 \bar{\mu} \sin^2 \phi}$$

where Ω = angular velocity of precession, ξ = distance from origin, $\bar{\mu}$ = precession angle,

$$n = \cos^{-1} \left[\frac{\sin^2 \bar{\mu} + \xi - \sin^{-2} \bar{\mu}}{2\ \xi \sin \bar{\mu}}\right] \quad \text{and} \quad \phi = \tan^{-1} (x/y).$$

Because both Br and K are heavy scatterers compared to C, O and H, it was necessary, in the KBr-amylose study, to include absorption corrections. We decided to check the values for cylindrical samples currently available (25). An auxiliary program ABSORCALC was written for this purpose. ABSORCALC sets up a grid of 1,000,000 possible reflection points for each Bragg angle inside the sample cross section. The absorption coefficient for values of 2θ from 0 degrees to 24 degrees in increments of one degree is calculated. A least-squares polynomial fit to the values results in an equation that is dependent only on the Bragg angle and whose coefficients can be used in CORSPOT. The coefficients were also used to calculate values to compare with those currently available. The differences were less than errors from the photometer, etc. A polynomial fitted to the literature values over the region of the fiber diffraction transform should suffice.

The peak intensity, sum of intensities, and the integrated intensity were found for each reflection in the following manner. After general background removal (22), a square array, generally 15x15, was extracted from the corrected data array in the vicinity of each intensity. To further remove background, particularly in the case of maxima that overlie amorphous scattering or the shadowing of the precession camera zone plate, the program PLANES constructed a bent plane by least squares fit to the perimeter as shown in Figure 4. Values of the surface were subtracted from the array values. This reduced the local background to nearly zero. The area was then contoured and the lowest value of a continuous, closed contour that gave a continuous level line was chosen as the bottom of the reflection. Two dimensional cubic splines (26) were fitted to the array values and the array was interactively interpolated and summed until the sum changed by less than 0.5%. The error of the intensity is the numerical integration under the lowest closed contour. If the ratio of error to intensity is less than 0.5% it is set to 0.5%. For each intensity, integrated values from three quadrants of the film were averaged (the beam stop blocks the fourth quadrant). After all correction and integration 55 intensities were obtained for the KBr-amylose study. Many were not unique, of course, since the HKL and KHL intensities, although not equal, are exposed simultaneously on the film.

The effects of applying corrections to and removing background from each point of the data array can be seen in Figure 5. Both intensities show the non-uniform tilt distribution of the fiber bundles. The weak intensity (1,0,4) has been resolved from the background by the procedure. Of the 13 pairs of intensities observed by eye to be overlapped on the unmeasured film, only 2 remained so after correction and smoothing. They were easily separated by assuming both were Gaussian. The overlap was removed extending each function into the overlap region and apportioning the array values accordingly.

Direct Structure Determination: KBr-Amylose: Senti & Witnauer (27,28) determined, from visually estimated intensities, that filaments of hydrated KBr-amylose have the space group $P4_32_12$ (D_4^8) with cell constants as a = b = 10.7A, c = 16.1A. Their Patterson projections revealed the K^+ & Br^- positions and Fourier projections the general helical nature of the polymer chains. They used steric hindrance as determinative in construc-

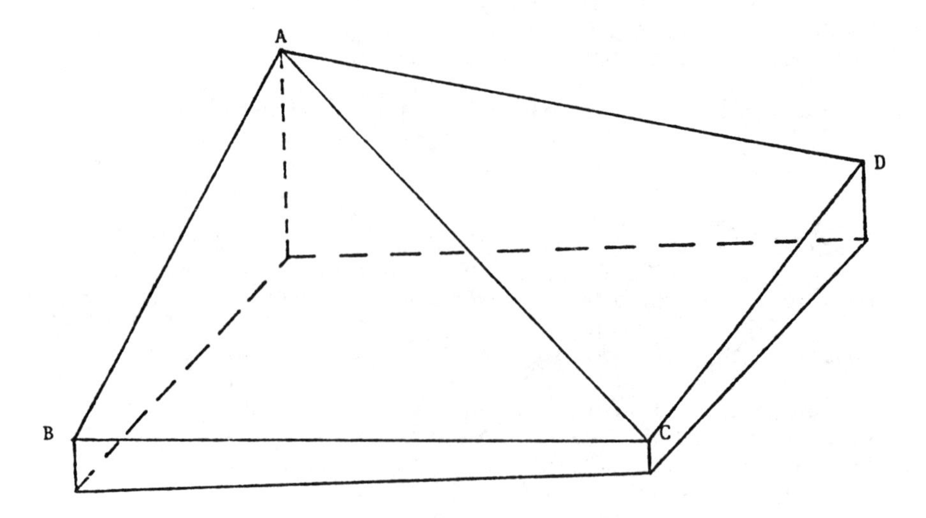

Figure 4. Construction of bent plane for additional background removal. A corresponds to the upper left corner of Figure 1 while C corresponds to the lower right corner.

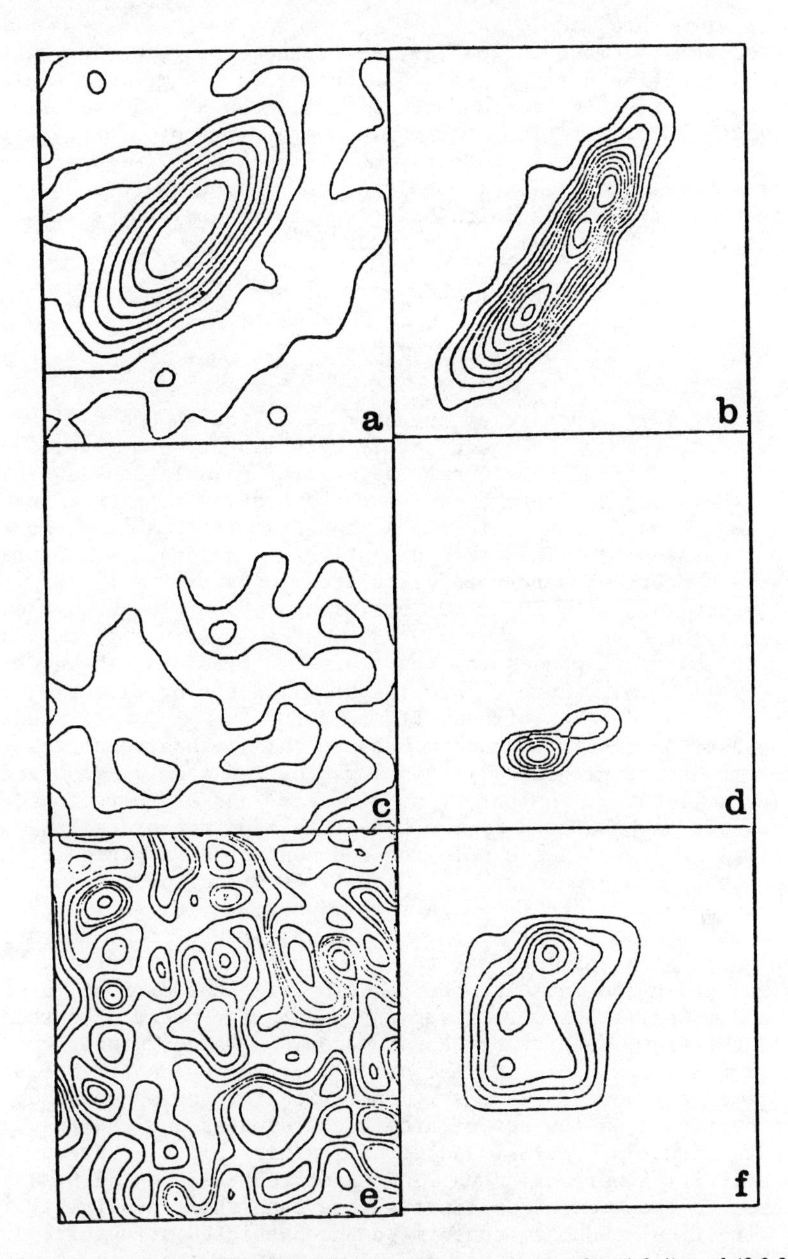

Figure 5. Contours of several diffraction maxima: the (2,1,3), (1,0,4), and (3,2,2) before (a,c,e) and after (b,d,f) corrections and background removal. Smoothing by cubic splines is performed during contouring. Note the nonuniformity of fiber tilt distribution appears in the corrected maxima.

ting a structure model.

Jackobs, Bumb & Zaslow (29) reset the strong, medium, and weak diffraction data of Senti & Witnauer to a numerical scale. They used Patterson sections to determine the K^+ (.54,.54,0) and Br^- (.20,.20,0) positions. The ring parameters of Brown & Levy (30) were used to build computer models packed in various arrangements. CPK (spacefilling) models of the polymer were constructed of the computer models which had given relatively low reliability indices. The best reliability index

$$R = \frac{\sum ||F_{obs}| - |F_{calc}||}{\sum |F_{obs}|}$$

was 0.41. The structure of the hydrate, although generally understood, had not been accurately determined. Using the 55 intensities derived by accurate data array calculations, we determined the space group to be as given by Senti & Witnauer, but with cell constants A_o = 10.55 A° and C_o = 16.12 A° for the film in our possession. This is apparently the anhydrous structure. Patterson sections were then calculated to determine the K^+ (.56,.56,0) and Br^-(.25,.25,0) positions. Figure 6 shows the clarity of the section at z = 0.25. These positions were then used to calculate phases for the scattering factors. E-values $\sum F_{HKL} / \sum f_{i(HKL)}$, where f_i are the atom scattering factors, were then calculated. E-sections, Figure 7 for example, were used instead of an electron density because this method equalizes the effective scattering of all atoms, making low atomic number atoms easier to locate. The positions of the K^+ and Br^- were found to be correct as predicted by the Patterson map interpretation. Sections 2A° by 2A° were selected and contoured in regions away from the K^+ and Br^- ions. Approximate locations of all carbon and oxygen atoms were found in this manner. No oxygens of water of hydration was found, so the sample was judged to be anhydrous. Once a new atom was located, the structure factor program, FHKL, was run again and included the position of the additional atom. All additional atoms were given the scattering power of carbon until the true identity of the atom could be ascertained from the electron density. Many of the atoms located were not members of the same residue. The symmetry operations of the space group were performed on the set of atomic coordinates and the residues were constructed by trial and error graphics plots. Various connectivities were tried until a connected residue chain was apparent. The atomic coordinates were then shifted slightly in the direction needed to conform to the predicted structure. After several cycles of this procedure the R factor was found to be 0.19.

The trial structure was refined by using a modified version of the full-matrix least-squares fortran program SFLS (31). The quantity minimized was

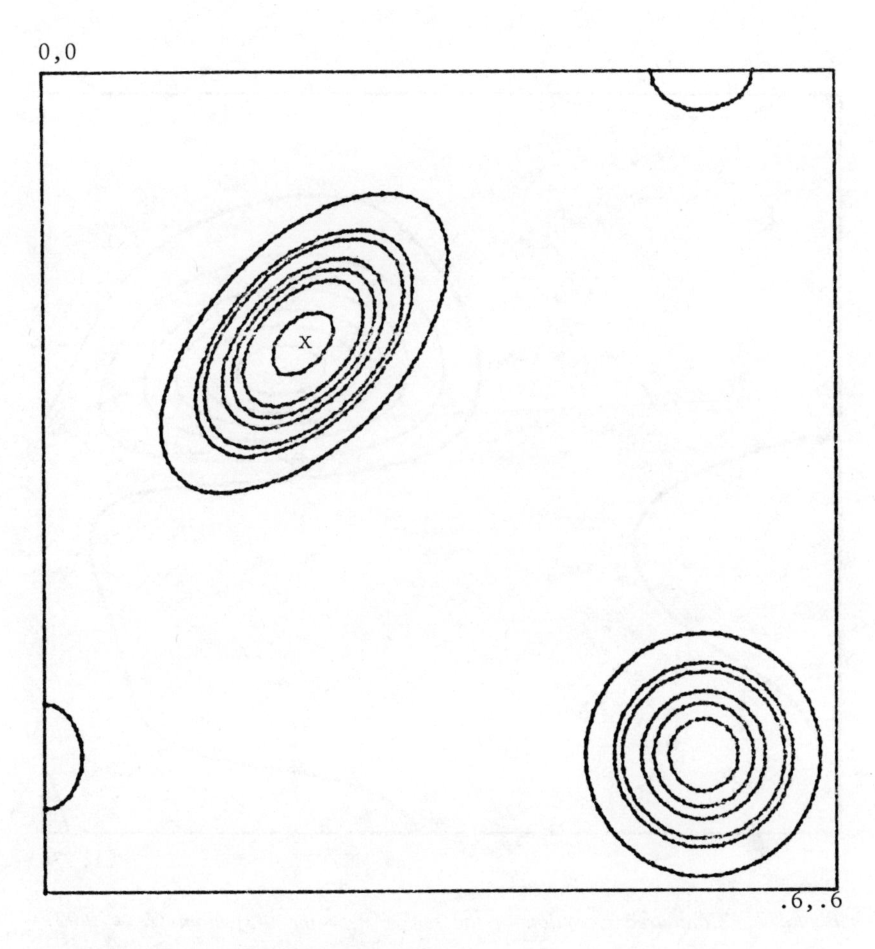

Figure 6. Contoured Patterson section at z = *0.25.* X = $K^+ - Br^-$ *vector peak.*

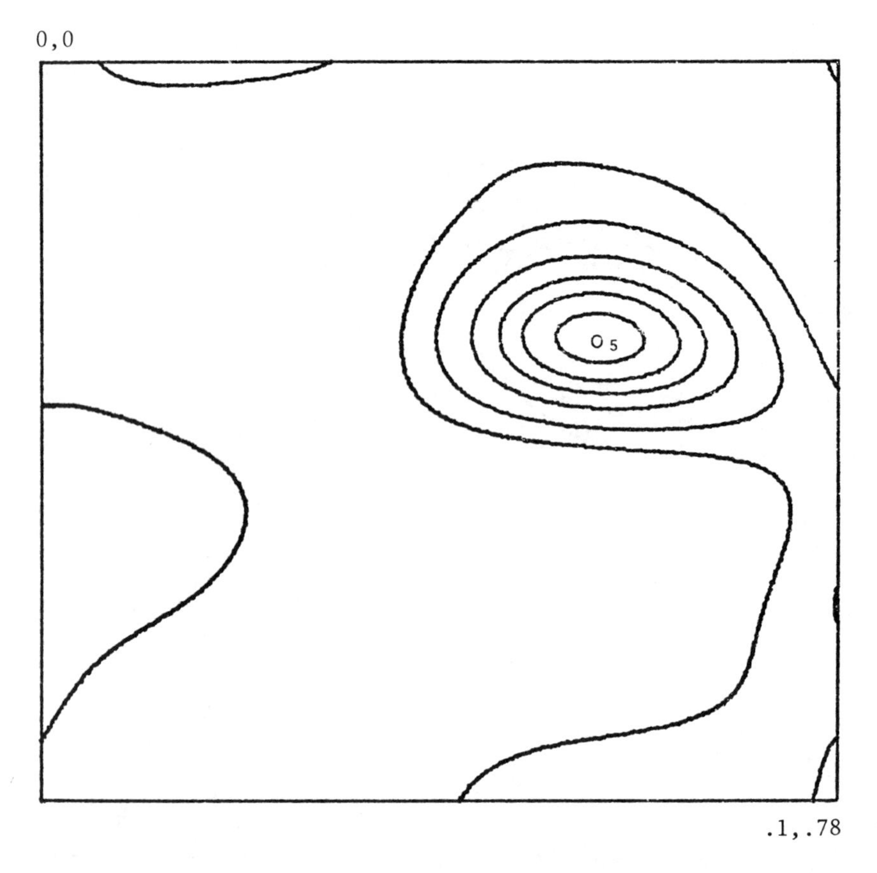

Figure 7. Contoured E-section of the region near the 0(5) atom (Z = 0.02)

$$\sum((F_o^2 - F_c^2)/k),$$

where k is a scale factor. Once the correct scale factor had been calculated, the carbon and oxygen atomic coordinates were allowed to vary until the change in atomic position was less than the calculated error in position. During final cycles atomic positions and isotropic temperature factors allowed to change. The ratio of intensities to variables was 55/37 (the K^+ and Br^- positions were not varied). The final R factor was 0.058. At this R value, not all bond lengths and bond angles were acceptable. When the glucose ring was adjusted to correct this problem, the R value rose to 0.068. The final ring conformation and adjustment will be discussed in future publications.

Table 1. lists the intensities calculated from the final atom locations and temperature factors and the observed intensities scaled to them. An ORTEP (32) plot of one amylose chain in the unit cell and nearby KBr is shown in Figure 8.

Comparison of Data Measurement Techniques. As the integrations of diffraction maxima were performed, the peak (array maximum) value of each was determined and the first sum of the integration process was determined and kept. These two values should adequately represent intensities observed by the Peak Center technique and by the Sum Integration technique (more than adequate, since, at the time they were determined, background had already been removed from each data sub array). These values are shown in Table 1. as $Peak_{obs}$ and Sum_{obs}. The peak intensities are corrected for spot shape and fiber tilt distribution. They may be scaled to the calculated intensities by dividing by 0.0475 and 0.785 respectively. Their R values are R_{PEAK} = 0.40 and R_{SUM} = 0.31. These were not refined, but determined by using the positions determined from the SFLS refinement and varying the scale factors for the two data sets until R minimized. If the atoms were allowed to adjust in refinement with the Sum Intensities, the R value would also reduce somewhat. Figure 9 shows Residuals for the three observed data sets.

It is now clear why Jackobs, Bumb and Zaslow were able to model an essentially correct structure with inaccurate intensities. The heavy scattering of the K^+ and Br^- ions determined their positions in the cell sufficiently to restrict the range on model configurations in the remaining unit cell space. In systems which do not contain heavy atoms or contain more atoms (more parameters) the correct structure was not as easily obtained. Yet many structures have been successfully (apparently) determined by means of modeling techniques.

The Sum Intensity technique yields more accurate atom position values and results in models with R factors, in some cases, as low as 0.15. This success must more be attributed to the input of additional information into the modeling, such as steric

Table 1. Observed (obs) and Calculated (cal) X-Ray Diffraction Intensities of KBr-Amylose (Integrated intensities are in units of electrons; Peak and Sum are Unscaled)

HKL	$PEAK_{obs}$	SUM_{obs}	$INTEGRATED_{obs}$	$INTEGRATED_{cal}$	ERROR(%)
112	27.01	143.35	156.27	115.94	.93
113	20.40	122.56	146.34	141.94	1.10
116	105.68	858.38	1075.68	1074.32	1.75
117	0.56	0.55	0.76	0.54	8.65
102 012	21.20	298.65	335.32	315.96	15.20
103 013	57.74	1578.03	1931.08	1987.72	9.80
104 014	6.02	7.98	10.18	9.06	13.18
105 015	77.23	668.87	966.36	931.18	5.54
106 016	36.76	199.45	237.58	218.64	4.07
107 017	59.98	498.34	559.86	486.32	23.00
201 021	37.62	278.65	310.16	232.50	8.00
202 022	37.50	58.76	69.00	49.62	7.50
203 032	27.04	578.98	871.50	874.78	8.00
205 025	7.16	17.67	35.64	21.48	25.90
206 026	2.00	98.99	102.86	102.62	16.40
313 133	37.68	1014.50	1284.28	1189.11	6.80
315 135	29.04	489.34	527.52	543.51	3.12
316 136	48.69	599.12	677.82	644.07	4.80
317 137	28.99	346.82	460.16	383.32	2.60
321 231	35.00	499.80	571.00	521.35	16.80
322 232	17.05	107.99	178.09	176.08	7.12
323 233	117.05	1012.60	1250.78	1281.18	16.82
324 234	18.12	32.18	320.30	262.42	9.76
325 235	14.76	143.88	169.96	236.74	6.78
401 041	.02	71.82	130.42	136.40	4.71
411 141	170.32	1642.80	1772.68	1684.27	4.90
412 142	6.28	17.78	25.74	22.81	5.20
413 143	244.16	3108.90	3419.16	3421.28	5.50

207 027	0.38	12.67	20.98	22.54	24.90
211 121	313.56	3369.56	4938.14	4686.74	3.00
212 122	17.52	867.56	1085.48	1112.52	3.20
213 123	90.30	2345.46	2611.62	2467.40	8.60
214 124	9.54	858.65	1027.90	1019.68	9.20
215 125	59.53	2895.10	3465.14	3364.60	4.80
216 126	22.69	413.12	619.98	550.69	6.30
217 127	69.25	1409.10	1500.58	1344.28	5.90
221	22.96	16.12	32.62	29.43	10.00
224	124.66	1978.60	2018.80	1950.71	1.87
301 031	28.92	2263.10	2831.80	2841.40	4.60
302 032	30.57	43.18	55.12	51.88	2.78
303 033	12.48	100.10	113.76	109.32	1.15
304 034	.02	2.76	3.88	6.88	12.60
305 035	83.12	967.18	1087.22	1070.10	2.68
306 036	37.12	121.70	134.32	134.82	1.15
307 037	34.44	746.80	818.46	751.76	2.87
312 132	94.40	1312.19	1540.92	1412.58	5.46
414 144	6.02	121.60	128.42	120.36	5.50
415 145	56.13	555.50	630.26	499.25	5.70
416 146	9.68	78.20	101.10	73.39	5.20
120 210	19.88	122.38	176.02	157.26	8.81
220	214.58	3128.00	4053.70	4413.21	1.60
130 310	54.09	191.20	204.42	191.88	7.60
230 320	.05	53.20	69.12	66.58	5.40
140 410	63.13	762.90	824.68	753.18	6.91
004	18.68	2874.60	3275.50	3289.85	.97

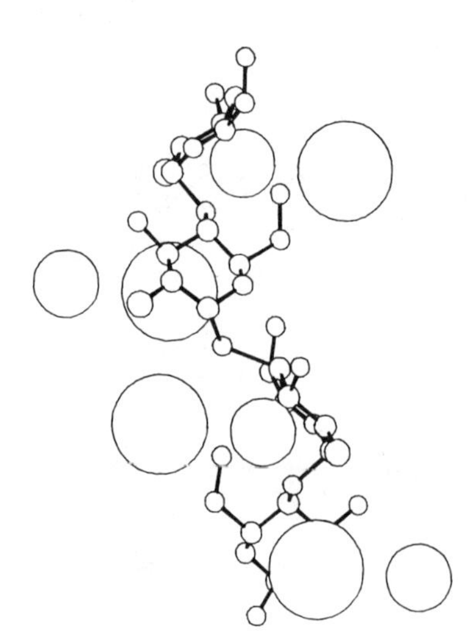

Figure 8. ORTEP drawing of amylose chain and KBr viewed along the b axis. The chain atoms are drawn to 50% probability thermal circles. The K^+ and Br^- ions are drawn to half their ionic radii so as not to obscure the amylose chain.

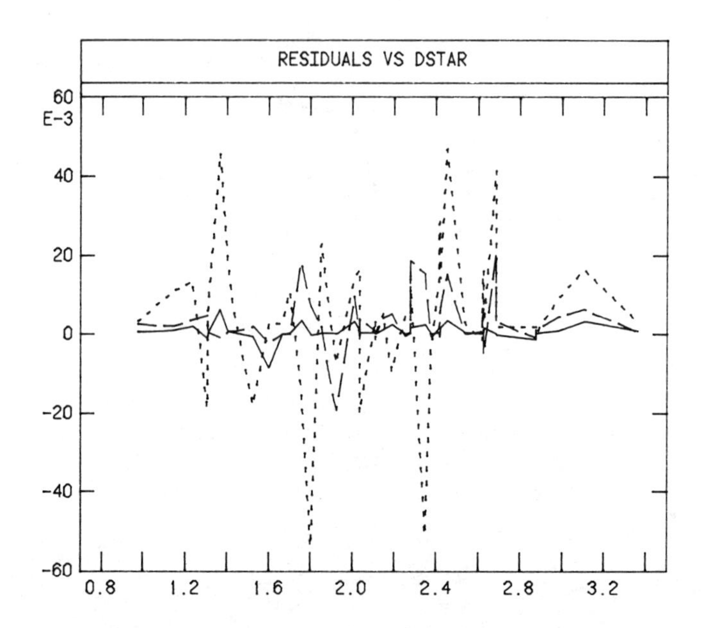

Figure 9. Residuals, $((I - I_{CAL})/\Sigma(I_{CAL}))$, for the three data sets (- - -) PEAK; (— — —) SUM; (———) INTEGRATED). The abscissa, DSTAR = $4\pi s$ in θ/λ.

hindrances, pseudopotentials, N-H mapping and/or linked-atom models with few parametral variables than to the x-ray diffraction data. The increased data to parameter ratios of linked-atom refinements, which should give more least squares accuracy, is offset, however, by the reduced flexibility of the chain conformation in refinement. The effects appear to be compensatory. Comparison of the SFLS refinement of KBr-amylose with LALS and Virtual Bond refinements may prove most informative.

Data array calculations with cubic spline fitting and numerical integration appears to yield data of much greater accuracy. The measurable range is greatly extended (6500:1 for the KBr-amylose data) and errors can be accounted for more accurately in the process of obtaining intensity data. These errors may even be used in weighting of least-squares refinement calculations as is presently done in single crystal structure studies.

Conclusions

Techniques are now available and, through the use of contemporary computer languages such as SPEAKEASY, convenient to extract accurate fiber x-ray diffraction data from films. The accuracy is enough to allow direct determination of structures with data to parameter ratio as low as 1.5 if a phase determining atom (heavy) is present. Problems in simpler polymer systems, for instance the question of chain conformation in cellulose, should now be determinable with greater certainty. In even more complicated systems, the increase in accuracy of the data and the greater data magnitude range should allow modeling to be carried out with much more confidence in the results.

Acknowledgements

The authors wish to thank Dr. Alfred French for the KBr-amylose sample and for numerous helpful suggestions and critical comment, Profs. Edw. Vaughn and J. P. McKelvey for their encouragement, Prof. Marie La Prade for use of her copies of the SFLS and ORTEP (32) computer programs and her excellent suggestions, Mr. James Eubanks for an excellent electronic package design and Mr. James Mann for precision machinecraft.

Literature Cited

1. Sarko, A. and Muggli, R. Macromolecules, 1974, 7, 1447.
2. Gardner, K. H. and Blackwell, J. Biopolymers, 1974, 13, 1975.
3. Kolpak, F. J. and Blackwell, J. Macromolecules, 1976, 9, 273.
4. Stipanovic, A. J. and Sarko, A. Macromolecules, 1976, 9, 857.
5. French, A. D. Carbohydrate Research, 1978, 61, 67.
6. French, A. D. personal communication.
7. French, A. D. and Murphy, V. G. Cellulose Chemistry and Technology, Authur, J. C., Ed. Amer. Chem. Soc. Symp. Series, Washington, DC, 1977, 48, 12.
8. Klug, H. P. and Alexander, L. E. "Xray Diffraction Procedures", John Wiley & Sons, New York, 1954.
9. VanHorn, M. H. Rev. Sci. Instr., 1951, 22, 809.
10. Lukesh, J. S. J. Chem. Phys., 1941, 9, 659.
11. Wood, R. G. J. Sci. Instr., 1948, 25, 202.
12. Bev, K. E. Rev. Sci. Instr., 1953, 24, 103.
13. Iball, J. J. Sci. Instr., 1954, 31, 305.
14. Phillips, D. C. Acta Cryst, 1956, 9, 819.
15. Spiegel-Adolph, M. and Peckham, R. H. Industr. and Engr. Chem.-Analyt. Ed., 1940, 12, 3, 182.
16. Jay, A. H. J. Sci. Instr., 1941, 18, 128.
17. Ronnenbeck, H. R. J. Sci. Instr., 1943, 20, 154.
18. Taylor, A. J. J. Sci. Instr., 1951, 28, 200.
19. King, M. V. Acta Cryst., 1966, 21, 629.
20. Miller, D. P. and Murphy, V. G. J. Appl. Cryst., 1973, 6, 73.
21. Fraser, R. D. B.; Macrae, T. P.; Miller, A.; and Rowlands, R. J. J. Appl. Cryst., 1976, 9, 81.
22. Fraser, R. D. B.; Macrae, T. P.; Suzuki, E.; and Tullock, P. A. J. Appl. Cryst., 1977, 10, 64-66
23. Cohen, S. and Pieper, S. "Speakeasy Manual", Speakeasy Computer Corp., Chicago, IL, 1979.
24. Brannon, R. C. "The Development of An Integrated Intensity Technique and Its Application In Determining the Crystal Structure of Fibrous Polymers", thesis, University Microfilms, Ann Arbor, MI, 1979.
25. "International Tables for X-ray Crystallography, Vol. 2". Kynoch Press, Birmingham, GBI, 1965.
26. Meader, D. this proceedings, pg
27. Senti, F. R. and Witnauer, L. P. J. Am. Chem. Soc., 1948, 70, 1438.
28. Senti, F. R. and Witnauer, L. P. J. Poly. Sci., 1952, 9, 115.
29. Jackobs, J. J.; Bumb, R. R. and Zaslow, B. Biopolymers, 1968, 6, 1959.
30. Brown, G. M. and Levy, H. A. Science, 1965, 147, 1038.
31. Markey, P. Proc. Royal Soc. London, 1954, 226A, 532.
32. Johnson, C. K. "ORTEP: A Fortran Thermal-Ellipsoid Plot Program for Crystal Structure Illustrations", ORNL-3794 Rev., Oak Ridge, TN, 1965.

RECEIVED May 21, 1980.

AXIS

A Semi-Automated X-ray Intensity and d-Spacing Analyser for Fiber Diffraction Patterns

D. MEADER and E. D. T. ATKINS
H. H. Wills Physics Laboratory, University of Bristol, Royal Fort, Tyndall Avenue, Bristol BS8 1TL U.K.

M. ELDER, P. A. MACHIN, and M. PICKERING
Daresbury Laboratory, Science Research Council, Daresbury, Warrington WA4 4AD U.K.

In the last few years rapid advances have been made in the field of computational crystallography, so that it is now possible to produce highly refined computer models of a wide variety of polymeric materials using X-ray diffraction data. Unfortunately, these achievements have been negated to some extent because the techniques used to collect the data for such refinement programs have not advanced at a comparable rate. In this contribution we describe a computer program which facilitates the reduction of intensity and d-spacing data obtained by the multiple film-pack method, and attempts to quantify the errors associated with such measurements.

The program, AXIS, was specifically designed to analyse fibre diffraction patterns similar to that shown in Figure 1, although some of the methods described in this chapter may be extended to other types of pattern if required. The fibre patterns are recorded on flat film using pinhole collimated, nickel-filtered CuKα radiation, and finely powdered calcite is dusted onto the specimen to provide a calibration ring of spacing 0.3035 nm.

Fibrous polymers are, at best, paracrystalline, exhibiting lattice distortions of the second kind which destroy long range order (1, 2). In general, the orientations of the crystallites in such materials are distributed about a particular direction, termed the fibre axis. This disorientation results in azimuthal spreading (or arcing) of the Bragg reflections, which complicates the task of measuring their integrated intensities.

In some instances, intensities are still measured by visual estimation, although in general this practice has now been superseded by the use of one-dimensional microdensitometers. Typically, a radial scan is taken through the centre of each spot, the shape of the background is estimated and sketched in, and overlapping reflections are apportioned. The area under each reflection profile is then measured, and an empirical "arcing factor" is applied. In correcting for the Lorentz and

0-8412-0589-2/80/47-141-113$06.50/0

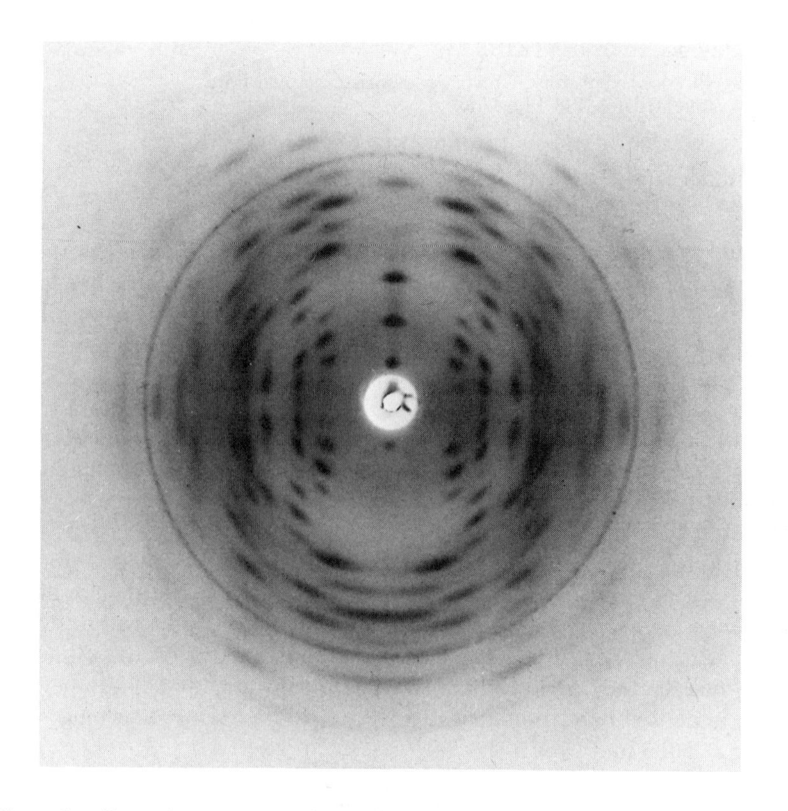

Figure 1. Example of an x-ray fiber diffraction pattern suitable for analysis using AXIS. The spots are discrete, there is no layer line streaking, and the background scatter may be seen clearly. The sample (potassium hyaluronate) was slightly tilted out of the plane perpendicular to the x-ray beam.

polarisation factors it is normally assumed that these remain constant over an entire reflection, an approximation which becomes progressively less accurate as the meridian is approached. Despite the approximations such an analysis is tedious, and the inherent subjectivity of the method can lead to large discrepancies when similar patterns are measured by different experimenters (3). With the introduction of two-dimensional scanning microdensitometers it is now feasible to utilise far more of the information contained in the pattern directly, thus reducing the need for approximations and subjective decisions. Fraser et al. (4) have devised a method by which the digital output from such a microdensitometer may be analysed using a computer. Some of the techniques which they describe have been incorporated in AXIS and, for convenience, we have retained the notation used in that paper, wherever relevant. It is not practicable to automate the analysis completely, because many of the factors which contribute to the measured data are either random or only partially understood. However, by making use of interactive computer graphics it is possible for the user to control the analysis and make any necessary decisions, while the computer takes on the burden of calculation.

1. Equipment

1.1 The Computer. AXIS is written in FORTRAN and runs on a GEC 4000 series multi-user mini computer equipped with a Graphics Option Controller (GOC) model 5250, manufactured by Sigma Electronic Systems Ltd. This device converts an alphanumeric visual display unit into a graphics terminal, allowing independent use of the alphanumeric and graphics screens. In addition, it provides cursor handling facilities and the ability to selectively erase part of the screen. All of these features are utilised by AXIS.

1.2 The Film Scanner. The fibre diffraction patterns are digitized on an Optronics P-1000 Photoscan rotating drum microdensitometer using a scanning raster of 100 x 100 μm. The microdensitometer outputs an integer in the range 0-255 corresponding to the specular density at each raster point. A nominal scale is selected such that a value of 255 corresponds to approximately three optical density units. The precise relationship between the scanner density value and the true optical density of the film must be ascertained by calibration against standard films since there is significant non-linearity in the optical system of the microdensitometer. In practice, different film types will exhibit characteristic non-linear blackening on exposure to X-rays and it is appropriate to combine the non-linearity effects of both film and scanner in a single calibration curve. Suitable calibration objects, consisting of sets of artificial spots produced by exposing the film to CuKα radiation from an Americium-Copper radioactive source through a lead mask, were kindly supplied by Dr. D.J. Gilmore. These spots differ from "real" diffraction spots by having a flatter intensity profile and a lower, more uniform background contribution (5). Figure 2 shows a calibration curve for the Daresbury Optronics microdensitometer using Ilford Industrial G X-ray film. It may be seen that the response of this particular combination of film and scanner is linear up to an optical density of at least 1.4. At an optical density of 1.9, the correction to be applied is approximately 4% but insufficient data are available at present to enable an accurate calibration curve to be plotted. Therefore, until further calibration experiments have been completed, only spots with a maximum optical density of less than 1.7 (including background) may be used for intensity measurements.

The standard deviation of each scanner reading has been estimated to be 0.7 when using a scale of 0-255 units, and the positional accuracy of the instrument is claimed to be $\pm$ 2μm. Data from the microdensitometer are stored in binary format on magnetic tape as an array, typical dimensions being 600 x 600. Prior to analysis the data are reconverted to integer format and copied into a random-access disc file.

2. Facilities for Examining the Data

An X-ray diffraction pattern may be thought of as a three-dimensional "information surface", with two spatial dimensions and the third related to the film blackening. AXIS allows the user to examine this surface by enabling two-dimensional sections to be displayed on the graphics terminal. Such sections fall into three distinct types, each of which assists the user in making particular decisions as the analysis proceeds.

The most versatile method of examining the data is to plot contours of constant optical density. The contouring routines in AXIS are adaptations of those described by Heap (6). Versions suitable for use with the GOC were kindly supplied by Dr. D.C. Sutcliffe. The rectangular region of the pattern to be contoured is defined by specifying ranges, in raster units, along the scanner axes. The pattern may be sampled at regular intervals or blocks of data may be averaged to reduce the effect of film granularity. The contour levels to be plotted are specified by selecting values in the scanner grey scale (0-255) or, if required, the program will automatically contour the region, dividing the observed range of densities into ten equal parts.

The latter option is useful during preliminary analyses but fine control over the contour levels is required when defining a spot boundary or finding the position of the peak. It is clear that a method of selecting the most appropriate contour levels for a particular task is needed; this is provided by two types of section taken normal to the contour map: the radial and azimuthal scans.

As the name suggests, in its simplest form a radial scan is identical to the standard one-dimensional densitometer trace through the centre of the pattern which is commonly used in intensity measurements. However, the user may specify the "origin" of the scan to be at any point on the pattern so that it is a simple matter, for example, to scan through a layer line precisely parallel to the equator. An azimuthal scan is a complete 360° trace taken at constant radius and centred on the centre of the pattern. Such a scan is very useful for determining the extremities of the reflection arcs. Examples of radial and azimuthal scans are shown in Figures 3 and 4, respectively. In practice these scans are displayed on the graphics terminal and interrogated using the cross-hair cursor. The required contour heights may be read directly from the ordinate axis, and the absolute film coordinates of any point on the abscissa may be obtained interactively.

3. The Calibration Ring

Before d-spacings or intensities may be measured, it is necessary to know the position of the centre of the pattern and the specimen-to-film distance. Both these parameters may be determined accurately by making use of the calibration ring (see Figure 1).

3.1 Determining the Position of the Centre. Approximate values for the centre and radius of the calibration ring are estimated from a contour map of the data. Using these values, an annulus is defined which has sufficient width to enclose the entire diffraction ring. Each microdensitometer reading inside the annulus is then compared with the diametrically opposite point relative to the approximate centre. For example, in Figure 5 if the approximate centre is at C5, the scanner reading at raster point X will be compared with that at point P5. A value F(5) is calculated such that

$$F(5) = \sum(\text{Reading}_{X} - \text{Reading}_{P5})^{2} \qquad (1)$$

where the summation is over all the points enclosed by the annulus, using C5 as the centre.

A similar calculation is now performed for the eight raster points surrounding C5. For the point C4, for example, the whole annulus is displaced slightly to the left. If the point X still lies inside the annulus it is compared with P4, forming part of a new sum F(4). When the sums F(1) to F(9) have been computed, they are inspected to find the

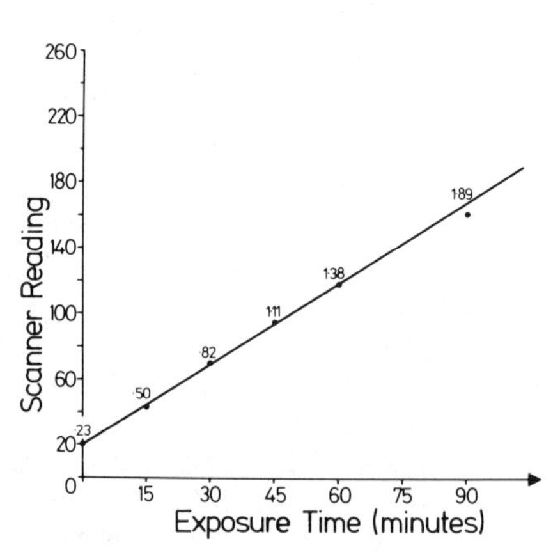

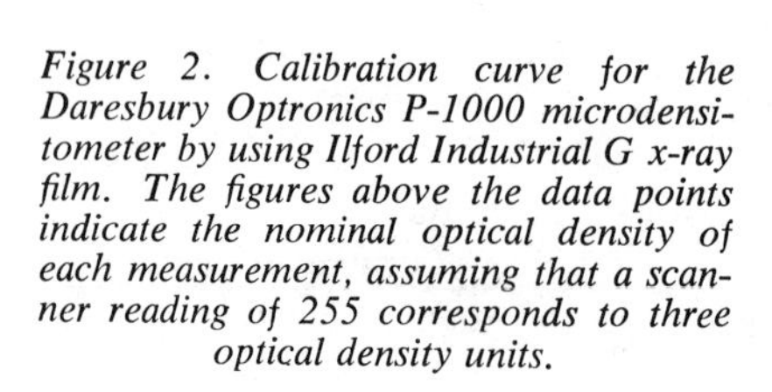

Figure 2. Calibration curve for the Daresbury Optronics P-1000 microdensitometer by using Ilford Industrial G x-ray film. The figures above the data points indicate the nominal optical density of each measurement, assuming that a scanner reading of 255 corresponds to three optical density units.

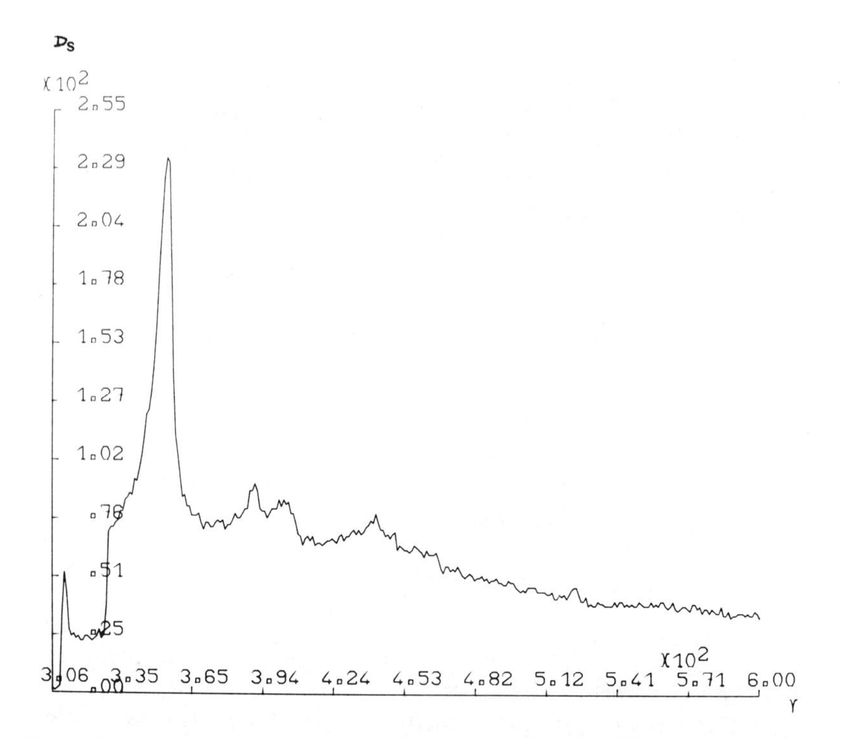

Figure 3. Example of a radial scan. The microdensitometer readings D_S are plotted against a radial coordinate. The signal-to-noise ratio is poor because of the film grain.

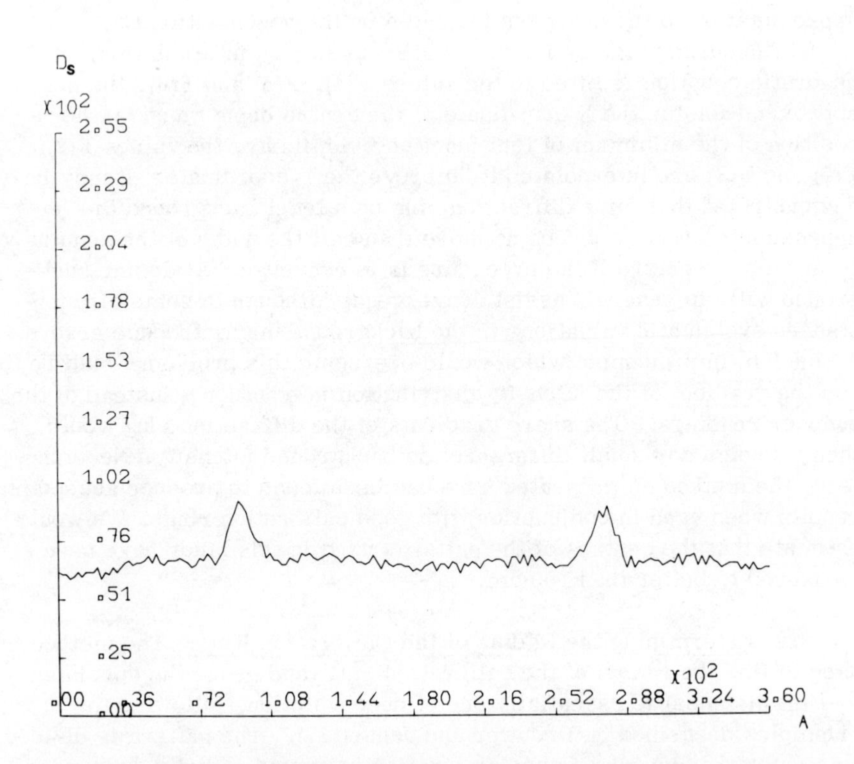

Figure 4. Example of an azimuthal scan. The microdensitometer readings D_S are plotted against an azimuthal coordinate (0-360°). Two symmetry-related equatorial spots are shown.

minimum value. If the centre of the diffraction ring is indeed at C5, the sum F(5) will have the minimum value. Usually, however, the approximate centre will not be correct and the minimum will be found in a different element of F, e.g. F(9). In this situation the position C9 corresponding to the result F(9) becomes the new approximate centre and the entire process is repeated. If the original approximation is sufficiently close to the true centre, the procedure will converge. After the final cycle, F(5) contains the minimum sum, and a more accurate approximation to the centre is indicated by the new position C5.

The accuracy may be improved still further by interpolation. A quadratic equation is fitted to the values F(4), F(5) and F(6), the best approximation for the X coordinate of the centre being given by the position of the minimum of that function. Similarly, the values F(2), F(5) and F(8) are interpolated to improve the Y coordinate. It may be demonstrated that for a diffraction ring on a level background the approximate centre must be no more than half the width of the annulus from the true centre if the procedure is to converge. A sloping background will, in general, assist convergence although in some circumstances systematic variations in the background may introduce errors. A possible improvement which would overcome this problem would be to use the gradient of the intensity distribution in equation 1 instead of the scanner readings. The sharp gradients of the diffraction ring would then outweigh any small differences in background intensity. Nevertheless, the method as presented here has been found to produce acceptable results when used in conjunction with good calibration rings. We would estimate that the centres of the patterns used in this study have been measured to better than 100 μm.

3.2 Determining the Radius of the Calibration Ring. The method used to find the radius of the calibration ring (and hence the specimen-to-film distance) is essentially one of deconvolution, following the principles described by Brouwer and Jansen (7). The pattern is divided into octants and a trial Gaussian function is moved radially outwards within a search range selected to enclose the diffraction ring. The Gaussian function is constrained to fit the observed profile of the calibration ring as closely as possible by selecting an exponent which equates the half-widths at half height. A deconvolution function is calculated at regular radial intervals within each octant using the microdensitometer data and the third derivative of the trial Gaussian function. It may be shown (7) that this is equivalent to seeking the presence of the first derivative of a Gaussian function in the first derivative of the microdensitometer data. This procedure eliminates any distortion produced by a linearly sloping background. It may be shown that the position of the diffraction ring is detected when the deconvolution function changes sign from negative to positive, the precise radius being found by interpolation. In this way, a value for the radius is obtained from each octant, allowing a mean and standard deviation to be calculated. The division of the ring into octants provides a very sensitive check on the position of the centre since a small error will produce unequal radii in each sector. In addition, it is often possible to detect deviations from circularity in the calibration ring (caused by collimator mis-alignment or buckling of the film). For well-defined, undistorted calcite rings the standard deviation in the radius measurement is frequently less than 100 μm.

4. Determining Spot Boundaries and Positions

The position and boundary of each spot are determined interactively from contour maps, using contour heights selected by taking radial and azimuthal scans through the approximate centre as described in section 2.

4.1 Boundaries. It is important to realise that if a spot is superimposed on a sloping background (as is almost always the case), no single contour level will accurately represent its boundary. Figure 6 demonstrates how an inappropriate choice of contour levels can affect the perceived boundary of the reflection. In general, up to four contour levels are necessary to give an accurate indication of a spot boundary, and the radial and azimuthal scans play a crucial role in selecting their values. When a suitable contour map has been displayed on the graphics screen, the cursor is used to construct an irregular polygon around the spot, defining its boundary. All data points enclosed by the polygon are then found, using the method proposed by Reid (8), and "flagged" by adding 256 to their values on the disc. This modification places all of the microdensitometer readings inside the spot boundary on an integer grey scale from 256 to 511, which unambiguously distinguishes them from the rest of the data. Later, if required, the boundary may be "forgotten" and the disc file returned to its original state. In any case, these alterations to the data are ignored by all AXIS routines which do not make specific use of them. The important advantages of this method are firstly that no assumptions are made about the shape of a spot and secondly that, once defined, the boundary is stored and may be used many times. The latter facility is used extensively in the background subtraction procedure which is described in Section 5.

4.2 Positions. The position of the centre of each reflection is located using a contour map in which the levels have been selected to give a clear indication of the diffraction spot. The cursor is used to

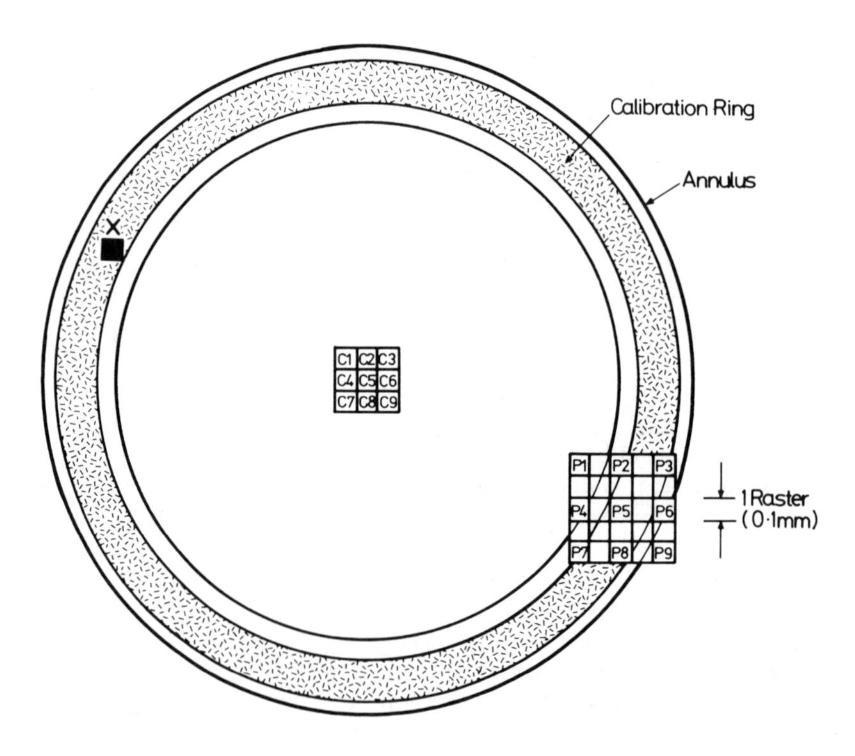

Figure 5. Schematic of the method used to determine the position of the center of the pattern. If the annulus is centered on C5, each data point X is compared with the centrosymmetrically related point P5.

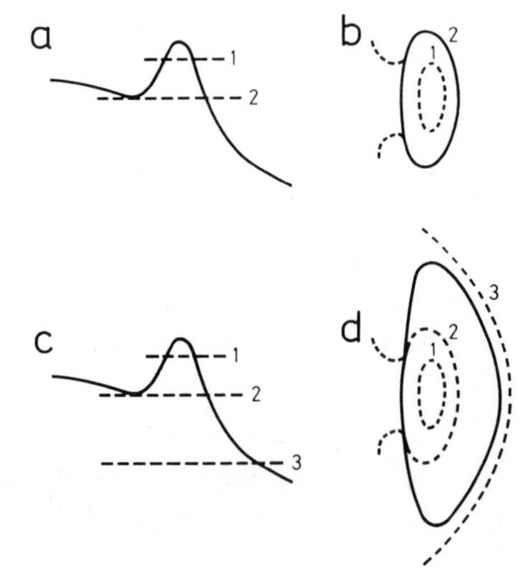

Figure 6. If inappropriate contour levels are selected, the boundary defined will be incorrect. If only contour level 2 is used to define the boundary as in (a), the result (shown as the full line in (b)) is very inaccurate. To construct the correct boundary it is necessary to use contour levels 2 and 3, as in (c). The true boundary is illustrated in (d). Sometimes it is necessary to use up to four contour levels to obtain an accurate indication of the boundary.

indicate the point on the screen corresponding to the centre of a reflection; this is then converted into scanner coordinates and stored in a disc file for future use. The reflection is given a unique number by AXIS; this is used to identify it during the rest of the analysis. Also stored in the file is an indication of how accurately it is possible to measure a reflection position from the current contour map. When a region of the pattern is selected for contouring, AXIS scales the map so that full use is made of the 256 x 256 "pixels" (picture elements) provided by the GOC. It is not possible to measure a position on the screen to better than one half the width of the cursor (1/512 of the screen width), in fact in many cases it is more realistic to double this value. If a large area of the pattern is displayed this error will be more significant than for a small area. This fact must be taken into account later when calculating the errors in the d-spacings. The user may also specify which layer line the diffraction spot lies on. If given, this enables the layer line spacing to be calculated directly, and it is used as a check to ensure that erroneous spot combinations are not entered in response to requests from AXIS (see section 6).

It should be noted that when finding the centre of a reflection it is not sufficient to search for the point of maximum optical density in the peak. The relatively high granularity found in X-ray films such as Ilford Industrial G yields rather a poor "spot-to-noise" ratio (see Figures 3 and 4) and the profile of a reflection is normally very uneven. Hence the maximum optical density value may not occur at the true centre of the reflection. Indeed, the problem of grain noise has proved to be a major limiting factor in measurements of intensities and spot positions. Some weak reflections, although resolvable by the human eye (with its inherent integrating properties) are very poorly defined in terms of optical density and it is frequently impossible to obtain any data from them. In such cases longer exposures must be taken to improve the definition of the peak.

4.3 Overlapping Reflections. When measuring intensities, reflections which overlap by a significant amount (so that the error involved in ignoring the overlapping part of the peak is not acceptable) are treated as composite. We feel that at this stage it is better to measure a smaller number of intensities as accurately as possible than to apply *ad hoc* techniques in an attempt to distinguish the component parts of a spot. Care must also be exercised when measuring the positions of overlapping reflections. A steep background may distort a reflection so that the measured centre does not coincide with the true one. Therefore, background subtraction is usually carried out before a position is measured (see section 5). However, if reflections overlap, the tails of adjacent diffraction spots behave like extremely steep background profiles and may cause the measured position to be erroneous. In such cases it is necessary to treat the adjacent reflections as background and subtract them before any measurements are recorded. When making visual measurements of spot positions, the eye naturally applies a correction to allow for overlap. However, in some cases of severe overlap, or when one diffraction spot is much stronger than the reflections on which it encroaches, the correction applied by the eye may be insufficient and considerable mis-measurement may result.

5. Background Subtraction

Much of the uncertainty associated with the measurement of integrated intensities from a fibre diffraction pattern derives from the method of background correction applied. The background arises from many sources. For example, the film may exhibit fogging due to age, poor storage or the use of contaminated developer. In addition, if the camera is not evacuated there will be a component due to scatter from the atmosphere inside it, e.g. helium and water vapour. The major contribution to the background is provided by the specimen itself and especially by the amorphous regions. The scatter from such non-crystalline portions of the material is characterised by diffuse haloes of intensity which are frequently not azimuthally symmetric. Therefore, a method of background subtraction is required which is both reproducible and sensitive to local variations.

AXIS uses data from the immediate vicinity of a spot (or group of spots) to define a minimal, least-squares bicubic spline surface representing the background. Splines are used because they are less prone to unwanted fluctuations than, for example, high order polynomials. In addition, cubic splines have been studied extensively (9, 10, 11) and algorithms exist which enable them to be evaluated quickly and efficiently. An excellent introduction to surface fitting using splines has been given by Hayes and Halliday (9) and the essential features are summarised below.

5.1 Spline Fitting in One Variable. By definition, a cubic spline function in one variable consists of a set of polynomial arcs of degree three or less joined smoothly end to end. The smoothness consists of continuity in the function itself and in its first and second derivatives. The points at which these arcs are joined are termed "knots". In general, the third derivative of the function will be discontinuous at each knot, and the number and position of the knots may be varied to give an acceptable fit to the observed data.

For most applications the most stable method of calculating a spline is to express it in terms of the set of Basis- or B-splines (12) associated with the knots you have chosen. Figure 7 shows part of a set of B-splines. For simplicity, the knots are shown equispaced and the B-splines have identical shape; in general neither of these conditions is required, but the B-splines are always normalised to enclose the same area. It may be seen that a B-spline is non-zero only over a range of four adjacent knot intervals. This property may be used to reduce substantially the amount of calculation to be performed when evaluating a spline function.

A typical curve-fitting problem is illustrated in Figure 8. A spline function $S(x)$ is required which interpolates a set of observed values (represented by f_1, f_2 etc.) in an interval $a \le x \le b$. If we choose a set of knots $k_1, k_2, \ldots, k_h$ in the interval, then Curry and Schoenberg (12) show that $S(x)$ has a unique representation in $a \le x \le b$ of the form

$$S(x) = \sum_{i=1}^{h+4} \gamma_i B_i(x) \qquad (2)$$

where the γ_i are constants which must be evaluated. The B-splines $B_i(x)$ may be calculated efficiently using a recurrence relation derived by Cox (13). However, in order to define the complete set of B-splines $B_i(x)$, $i = 1, 2, \ldots, h+4$ (see Figure 7) it is necessary to introduce eight additional knots, k_{-3}, k_{-2}, k_{-1}, k_0, k_{h+1}, k_{h+2}, k_{h+3}, and k_{h+4}. In general, the positions of these "exterior" knots may be selected by the user, subject to the conditions

$$\left.\begin{aligned} & k_{-3} \le k_{-2} \le k_{-1} \le k_0 \le a \\ \text{and} \quad & b \le k_{h+1} \le k_{h+2} \le k_{h+3} \le k_{h+4} \end{aligned}\right\} \qquad (3)$$

It has been found experimentally (14) that a stable fit is achieved if we choose

$$\left.\begin{aligned} & k_{-3} = k_{-2} = k_{-1} = k_0 = a \\ \text{and} \quad & b = k_{h+1} = k_{h+2} = k_{h+3} = k_{h+4} \end{aligned}\right\} \qquad (4)$$

despite the fact that this introduces a discontinuity into $S(x)$ at $x = a$ and $x = b$.

From equation 2 it is evident that at least $h+4$ observed values, f_i, are required if the coefficients γ_i are to be evaluated. Figure 8 illustrates the case where this requirement is satisfied exactly but, in general, there will be far more than h+4 observed data and the equations for γ_i will be over-determined. In this situation least-squares methods may be used to obtain the "best" fit.

5.2 Spline Fitting in Two Variables. The methods described in the previous section may be extended to functions of two variables (9). The problem now is to find a surface S(x, y) which interpolates data values in a rectangular region $a \leq x \leq b$, $c \leq y \leq d$. By analogy with the previous section, we specify h "interior" knots in the x direction and ℓ in the y direction. These knots, $k_1, k_2, \ldots, k_h$ and $n_1, n_2, \ldots, n_\ell$ divide the region into rectangular sections or "panels" and may be used to calculate two sets of cubic B-splines, $B_i(x)$, $i = 1,.,h+4$ and $C_j(y)$, $j = 1,.,\ell+4$. As before, four exterior knots are superimposed at the extreme ends of each range so that equations 4 are satisfied. Similarly,

$$\left.\begin{array}{ll} & n_{-3} = n_{-2} = n_{-1} = n_0 = c \\ \text{and} & d = n_{\ell+1} = n_{\ell+2} = n_{\ell+3} = n_{\ell+4} \end{array}\right\} \quad (5)$$

Following de Boor (15) and McKee (10), if we choose to take the "tensor product surface" generalisation, we may write

$$S(x, y) = \sum_{i=1}^{h+4} \sum_{j=1}^{\ell+4} \gamma_{ij} \, B_i(x) C_j(y) \quad (6)$$

The spline surface S(x, y) consists of a set of bicubic polynomials, one in each panel, joined together with continuity up to the second derivative across the panel boundaries. Because each B-spline only extends over four adjacent knot intervals, the functions $B_i(x)C_j(y)$ are each non-zero only over a rectangle composed of 16 adjacent panels in a 4 x 4 arrangement. The amount of calculation required to evaluate the coefficients γ_{ij} may be reduced by making use of this property. As before, least-squares methods may be used if the number of data exceeds $(h+4)(\ell+4)$, which is usually the case.

5.3 Background Subtraction in Practice. Using methods described in section 2 a suitably contoured map of a selected region of the film is displayed on the graphics screen. The boundaries of all spots contained in the region are defined, and the corresponding microdensitometer readings are "flagged" on the disc. The boundary sizes are slightly over-estimated to ensure that the spot tails are completely included.

When this process has been completed, all data in the region which have not been flagged to be background and are used to calculate the coefficients γ_{ij} in equation 6.

To solve the resulting set of observation equations

$$\sum_{i=1}^{h+4} \sum_{j=1}^{\ell+4} \gamma_{ij} B_i(x_r)C_j(y_r) = H_r, \ r=1,2,\ldots,m \qquad (7)$$

where m is the number of unflagged data, H_r, at positions (x_r, y_r) in the region, AXIS makes use of NAG routine E02DAF (14). In some instances the least-squares criterion is not sufficient to define the bicubic spline surface uniquely. In such situations the surface for which the sum of the squares of the coefficients γ_{ij} is smallest (the minimal least-squares solution) is selected. Although this choice is somewhat arbitrary, it yields satisfactory results in practice.

Because, in general, background scatter does not exhibit rapid fluctuations, we find that acceptable results are obtained using a small number of equispaced knots, although frequently the number of knots required in the x and y directions is different. When selecting the knot positions, care must be taken that panels are not inadvertently defined which are completely inside a spot boundary (and hence contain no unflagged data). The fit obtained using such panels is likely to be poor.

When the background surface has been defined, NAG routine E02DBF (14) is used to interpolate under the spots, calculating the background contribution to each flagged microdensitometer reading. AXIS enables the user to inspect the calculated background from several different directions and to decide whether to proceed or not with the subtraction. If the spline surface is satisfactory, each reading in the specified region is corrected by subtracting its background component and the disc record is updated accordingly.

Figure 9 shows a radial scan similar to Figure 3 and includes a section through the bicubic spline background surface calculated by AXIS for this region of the pattern. Careful use of both radial and azimuthal scans when defining the spot boundaries has enabled a reflection which is poorly resolvable in terms of optical density to be distinguished (arrowed). Visual examination of an X-ray diffraction pattern from the same sample but given a longer exposure confirms the existence of this reflection.

6. The Pattern Coordinate System

In the preceding sections, all measurements are made in the

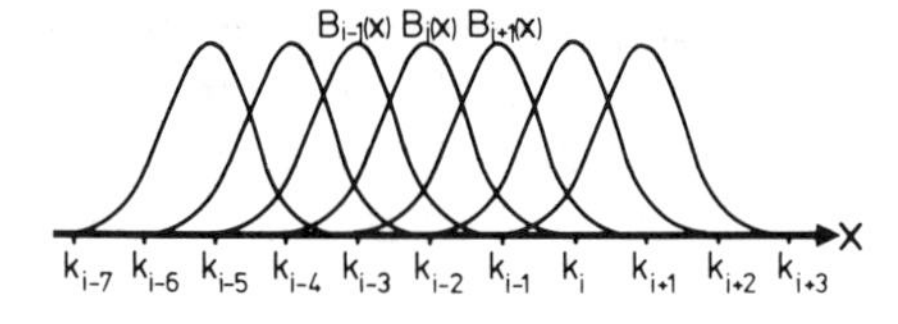

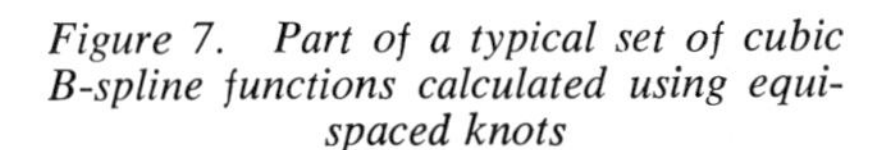

Figure 7. Part of a typical set of cubic B-spline functions calculated using equispaced knots

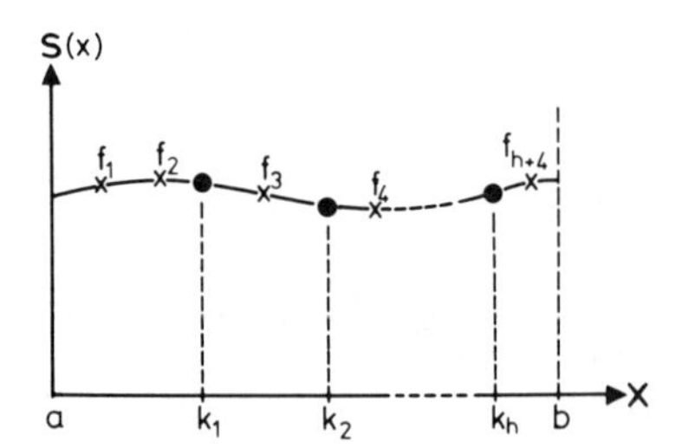

Figure 8. Example of a typical curve-fitting problem in one variable illustrating the use of spline functions ((●) knots; (+) observed data)

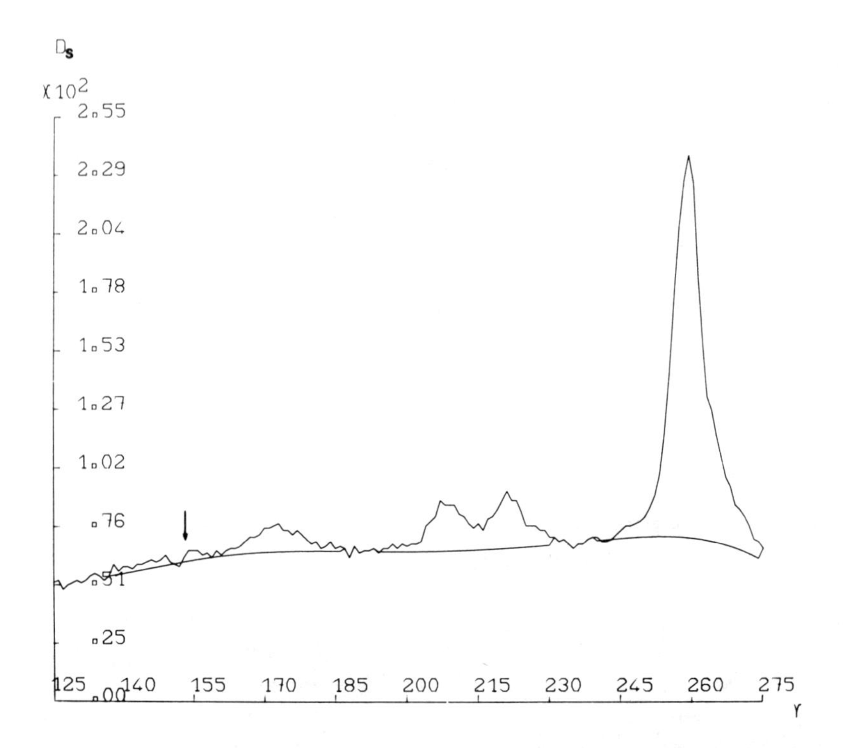

Figure 9. Radial scan showing a section through the minimal, least-squares bicubic spline background surface calculated by AXIS for this region of the pattern.

The background is only shown within the spot boundaries. When an increased exposure time is used, the arrowed reflection becomes clearly visible, but on this pattern it is very weak and poorly defined. However, the background subtraction procedure is sufficiently sensitive to resolve it.

"scanner" coordinate system which is defined solely by the orientation of the film on the microdensitometer. At this point it is convenient to introduce the "pattern" coordinate system (u,v). The origin of this coordinate system is at the centre of the pattern (determined using the methods described in section 3.1), and the axes u and v correspond to the equator and meridian, respectively. In general there is an angle δ between the meridian and the longitudinal axis of the scanner drum. AXIS provides two methods for determining this angle, the first of which assumes that suitable spot positions have already been measured and are stored in a data file (see section 4.2). On request from AXIS, the user supplies a series of pairs of spot positions, specifying each spot by number. Both members of a pair must lie on the same layer line and be related by two-fold symmetry about the meridian. For each pair of spots, AXIS calculates a value for δ by determining the angle between a line joining the specified positions and the scanner "axis" around the drum. This method is poorly conditioned for equatorial and meridional spots but works well for all others, the standard deviation of δ being typically less than half a degree.

The second method is more general and is similar in principle to that used for determining the position of the centre. A strong spot is selected (preferably not lying on the equator or meridian, although such spots may be used if necessary) and, using scanner coordinates, a rectangular box is defined which encloses the entire spot. A value for δ is estimated and an angular search range is defined on either side of this approximation. At regular intervals within the search range a trial "meridian" M_n, $n = 1,2,\ldots,m$ is constructed. For each raster point P inside the box the position of a point Q is calculated which is related to P by a two-fold rotation axis about the trial meridian. A value G(n), $n = 1,2\ldots,m$ is then calculated such that

$$G(n) = \sum (\text{Reading}_P - \text{Reading}_Q)^2 \qquad (8)$$

where the summation is over all the raster points within the bounds of the box. Thus, a trial meridian M_1 is constructed at one extremity of the search range and a value G(1) is calculated. A new trial meridian M_2 is then constructed with a small angular displacement from M_1 and the process is repeated until the search range has been covered. If M_s is the best approximation to the true meridian, G(s) will contain the smallest sum of the set G(n), $n = 1,2,\ldots,m$ and, as the value for δ corresponding to each G(n) is known, the optimum angle may be found by interpolation. If several spots are used to determine δ for each pattern, the standard deviation is, again, usually less than half a degree.

7. The Specimen Intensity Transform

Following Fraser et al. (4), we choose to represent the scattered intensity in terms of a cylindrically symmetric "specimen intensity transform" $I_s(D)$, where D is a position vector in reciprocal space. Figure 10 shows the Ewald sphere construction, the wavelength of the radiation being represented by λ. The angles μ and χ define the direction of the diffracted beam and are related to the reciprocal-space coordinates (R, Z) and the pattern coordinates (u,v) as follows:

If r is the specimen-to-film distance,

$$u = r \tan \mu \tag{9}$$

and

$$v = \frac{r \sin \chi}{\cos \mu \cos \chi} \tag{10}$$

If the fibre axis is inclined at an angle β to the normal to the incident X-ray beam (see section 9), a simple (though tedious) analysis shows that

$$\sin \chi = \frac{1}{\cos \beta}\left[\lambda Z - \frac{\lambda^2 D^2}{2} \sin \beta\right] \tag{11}$$

This corresponds to equation 25 given by Fraser et al. (4) (except that a factor λ has been omitted from that paper). Using Bragg's law,

$$D = \frac{1}{\lambda}\left[2(1 - \cos 2\theta)\right]^{\frac{1}{2}} \tag{12}$$

Hence,

$$\cos 2\theta = \cos \mu \cos \chi = 1 - \frac{\lambda^2 D^2}{2} \tag{13}$$

It is convenient to express D, R and Z in terms of the pattern coordinates (u, v). Noticing that

$$\cos 2\theta = \frac{r}{\sqrt{r^2 + u^2 + v^2}} \tag{14}$$

equation 12 becomes

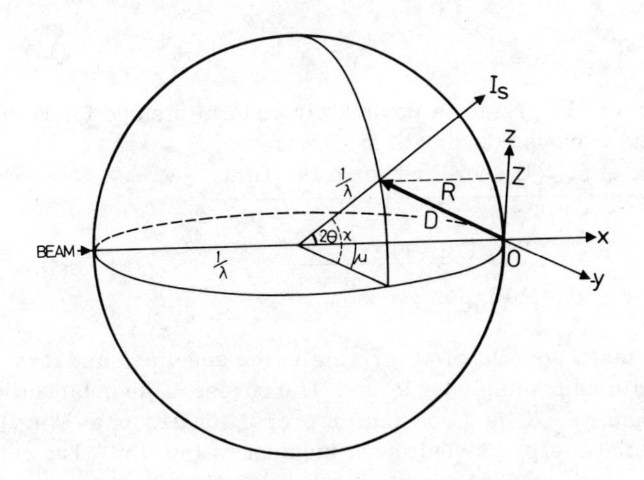

Figure 10. Ewald sphere construction showing the relationship between the reciprocal space position vector D, the reciprocal space coordinates (R,Z), and the angles μ and χ which define the direction of a scattered ray

$$D = \frac{1}{\lambda}\left[2\left(1 - \frac{r}{\sqrt{r^2 + u^2 + v^2}}\right)\right]^{\frac{1}{2}} \tag{15}$$

Substituting 11, 13 and 15 into 10 and simplifying gives

$$Z = \frac{1}{\lambda}\left[\frac{V\cos\beta - r\sin\beta}{\sqrt{r^2 + u^2 + v^2}} + \sin\beta\right] \tag{16}$$

and, by definition,

$$R = (D^2 - Z^2)^{\frac{1}{2}} \tag{17}$$

Equations 9 to 17 provide a completely general method for transforming between the various coordinate systems.

Fraser et al. (4) show that for a flat film

$$I_s = \frac{Kr^2\ E(u,v)}{APL\ \cos^3\mu\ \cos^3\chi} \tag{18}$$

where A takes account of absorption in the specimen and may be neglected for thin samples. P and L represent the polarisation and Lorentz factors, K is a constant of proportionality and E(u, v) is related to the energy per unit area incident on the film. The calculation of E(u, v) is discussed in more detail in section 8. It may be shown by suitable substitution in equation 32 given by Buerger (16) that a satisfactory approximation to the Lorentz factor is given by

$$\frac{1}{L} = \frac{\lambda}{2}\left[R^2 - (R^2 + Z^2)(\sin^2\beta + \frac{\lambda^2}{4}(R^2 + Z^2) - \lambda Z\sin\beta)\right]^{\frac{1}{2}} \tag{19}$$

and the polarisation factor may be written in the form

$$P = \tfrac{1}{2}(1 + \cos^2\mu\ \cos^2\chi) \tag{20}$$

8. Corrections for Oblique Incidence

By simple geometry, in a flat-film diffraction camera the thickness of film traversed by the scattered X-rays increases with the Bragg angle, 2θ. This has two important consequences. Firstly, X-rays scattered through different Bragg angles will be attenuated unequally as they pass through each layer of the film-pack, thus affecting the intensity incident on the next film. Secondly, because the path length in the film emulsion varies with 2θ, the blackening produced will not be proportional to the incident energy per unit area. The latter problem was first considered by Cox and Shaw (17) who derived an approximate correction factor. A more exact treatment is discussed by Whittaker (18) and Wonacott (19). It may be shown (18) that a suitable correction factor is of the form

$$N_1(2\theta) = \frac{(1-e^{-k_e})(1+e^{-(k_e+k_b)})}{(1-e^{-k_e \sec 2\theta})(1+e^{-(k_e+k_b)\sec 2\theta})} \qquad (21)$$

where k_e and k_b are, respectively, the absorption coefficients for a single layer of emulsion and for the film base.

Usually, the front of the film-pack is covered with black paper and additional sheets may be introduced between the films to act as "spacers". Effects due to absorption in these extra layers must also be taken into account. If the absorption coefficient for a spacer is k_{sp}, a fraction $N_2(2\theta)$ of the radiation incident at an angle 2θ to the normal will be transmitted, where

$$N_2(2\theta) = e^{-k_{sp} \sec 2\theta} \qquad (22)$$

Similarly, we may define a factor $N_3(2\theta)$ for the X-ray film such that

$$N_3(2\theta) = e^{-(2k_e + k_b) \sec 2\theta} \qquad (23)$$

In principle, therefore it is possible to define a correction factor for the nth film in the pack of the form

$$T_n(2\theta) = \frac{N_1(2\theta)}{N_2(2\theta)^m \cdot N_3(2\theta)^{n-1}} \qquad (24)$$

where m is the number of spacers preceding film n. This expression enables all effects due to absorption to be calculated directly.

AXIS can calculate this quantity if required, but it is usually better to scale the intensities using selected equivalent spots from successive films. The reason for this is that measured values for the absorption coefficients k_e and k_b show marked variations. Typical values (19) for Ilford Industrial G X-ray film are $k_e \simeq 0.5$, $k_b \simeq 0.1$ and $k_{sp} = 0.05 - 0.1$ for paper.

When inter-film scaling is used, it is convenient to define parameters $N_4(2\theta)$ and $N_5(2\theta)$ such that

$$N_4(2\theta) = \frac{N_2(0)}{N_2(2\theta)} \tag{25}$$

and

$$N_5(2\theta) = \frac{N_3(0)}{N_3(2\theta)} \tag{26}$$

We may then write:

$$T'_n(2\theta) = N_1(2\theta).\ N_4(2\theta)^m.\ N_5(2\theta)^{n-1} \tag{27}$$

Equation 7.33 given by Wonacott (19) is incorrect. At present, spots used for inter-film scaling by AXIS may not have a maximum optical density, including background, greater than 1.7 (see section 1.2).

When the microdensitometer data have been corrected for oblique incidence, they are in a form suitable for substitution into equation 18 for the specimen intensity transform. In practice, the correction is made as the integration is performed (see section 10.2).

9. Determining the Fibre Tilt

It is very difficult to mount a specimen in an X-ray fibre camera such that it is precisely normal to the beam. Indeed, frequently the fibre must be tilted deliberately by a nominal amount to observe specific meridional reflections. Because the tilt angle β features in the Lorentz factor and other expressions given in section 7, it is necessary to obtain a value for it using the microdensitometer data. AXIS provides two methods for calculating β; both are similar to procedures outlined by Fraser et al. (4) and, superficially, resemble those described in section 6 for calculating δ. The first method again requires the user to specify the positions of pairs of equivalent spots from a data file. This time, however, both members of a pair must be on the same side of the meridian. A value for β is calculated for each pair of spots using the relationship given by Franklin and Gosling (20)

$$\tan \beta = \frac{|Z_2| - |Z_1|}{\lambda D^2} . \qquad (28)$$

Z_1 and Z_2 refer to the reflection above and below the equator respectively, and are calculated using equation 16 with β set to zero. The reciprocal spacing D may be calculated using equation 15 and AXIS checks that both members of a pair generate similar values for this parameter. In addition, if a layer line number q has been specified for each pair of spots, the following relationship (20)

$$\frac{q}{\ell} = \left(\frac{|Z_1| + |Z_2|}{2}\right) \cos \beta \qquad (29)$$

may subsequently be used to calculate values for the layer line spacing ℓ.

This method is most useful for patterns in which the spots are sharply defined, similar to that shown in Figure 1. In general, the value of β obtained is extremely sensitive to small inaccuracies in the measured spot positions. Although this is useful for detecting mis-measurement, in our experience it is rarely possible to obtain a standard deviation for β of less than two degrees, even when many spots are used. However, this degree of precision is often adequate for our purposes.

The second method is, in general, rather more accurate. A rectangular box is defined which encloses a suitable spot, a value for β is estimated (possibly using the procedure described above) and a search range is defined. At regular intervals within the search range a value W(n), n = 1,2,...,m is calculated such that

$$W(n) = \sum (I_s(R, Z) - I_s(R, -Z))^2 \qquad (30)$$

where the summation is over all raster points within the box. The optimum value for β occurs at the minimum of the function defined by W(n) and is found by quadratic interpolation. Using this method the tilt angle may usually be determined to better than one degree.

10. Calculating d-Spacings and Relative Integrated Intensities

10.1 d-Spacings. AXIS calculates d-spacings from the stored spot position measurements using the inverse of equation 15. In addition, two error estimates are calculated. The first of these is the "most likely error" which assumes that the position of each spot has been indicated as accurately as possible, given the limited resolution of the GOC. The second allows for a small misjudgement in the positioning of the cursor during measurement and this "maximum likely error" can be useful when indexing the pattern. Both error estimates also take account of the precision with which it has been possible to measure the radius of the calibration ring. At this stage it is possible to detect any systematic trends in the d-spacings due to small errors in determining the position of the centre of the pattern. Using iterative procedures it is frequently possible to refine the position of the centre to a precision of 20 to 30 μm.

Because it is expected that AXIS will only be used to analyse patterns which exhibit a well defined layer line structure, no direct provision is made for measuring meridional spacings. However, if necessary, the relation

$$d_{mer} \cong \lambda \left[\frac{r^2 + u^2 + v^2}{u^2 + v^2} \right]^{\frac{1}{2}} \tag{31}$$

may be used to calculate such spacings from the measured meridional positions.

10.2 Relative Integrated Intensities. After the background has been subtracted it is usually a simple task to define the boundary of a spot (or group of overlapping spots) using a contour map. The specimen intensity transform I_s is then computed for each raster point within the boundary and the integration is performed numerically. Because all correction factors are applied point-by-point, no special procedures are necessary when measuring meridional intensities and there is no need to invoke the undesirable approximations inherent in traditional methods of data reduction.

Werner (21) has made a detailed study of the errors associated with the measurement of intensities from a protein single-crystal X-ray diffraction pattern. He has demonstrated that when a two-dimensional scanning microdensitometer is used, an individual intensity measurement differs by approximately 4% from the mean value of the four symmetry-related reflections. In general, it will not be possible to measure a fibre pattern to this degree of precision because of the various factors which complicate the analysis. If Werner's criteria are applied to a series of intensity measurements made from fibre patterns using AXIS, the precision attained lies between 7% and 10%. While we do not have enough data to draw any firm conclusions at present, this result is sufficiently close to 4% to be encouraging.

11. Program Availability

11.1 GENS. Versions of the subroutines for determining δ, the position of the centre of the pattern and the radius of the calibration ring are also included in a program called GENS (General Scanner Utilities) which runs on an IBM 370/165 computer. The program is written in modular form and provides many additional facilities, including the ability to print selected regions of the data, produce contour maps and plot circularly-averaged radial density. Copies of the FORTRAN source are available on request from M. E.

11.2 AXIS. For reasons of copyright it is not possible to release the complete AXIS source. However, some of the subroutines are available on request from D.M.

Acknowledgements

We thank the Science Research Council for support, and the Interactive Computing Facility, Rutherford Laboratory, Didcot, Oxford, U.K. for provision of the GOC.

Literature Cited

1. Hosemann, R. and Bagchi, S.N. "Direct Analysis of Diffraction by Matter"; Amsterdam: North-Holland, 1962.

2. Vainshtein, B.K. "Diffraction of X-rays by Chain Molecules"; Amsterdam, London and New York: Elsevier, 1966.

3. Jeffrey, G.A. and French, A.D. in "Molecular Structure by Diffraction Methods"; vol. 6, The Chemical Society (specialist periodical reports), London: Bartholomew Press, 1977, Ch. 8.

4. Fraser, R.D.B; Macrae, T.P.; Miller, A. and Rowlands, R.J. J. Appl. Cryst., 1976, 9, 81.

5. Arndt, U.W.; Gilmore, D.J. and Wonacott, A.J. in "The Rotation Method in Crystallography"; Amsterdam, New York and Oxford: North-Holland, 1977, Ch. 14.

6. Heap, B.R. National Physical Laboratory (U.K.) Report NAC 47, 1974.

7. Brouwer, G. and Jansen, J.A.J. Anal. Chem. 1973, 45, 2239.

8. Reid, J.K. United Kingdom Atomic Energy Authority Report AERE-R7298, 1972.

9. Hayes, J.G. and Halliday, J. J. Inst. Maths. Applics., 1974, 14, 89.

10. McKee, J.M. in "Applications of Computer Methods in Engineering"; (L. Carter Wellford, Ed.), vol.2, University of Southern California, 1977, p.747.

11. Greville, T.N.E., Ed. "Theory and Applications of Spline Functions"; New York and London: Academic Press, 1969.

12. Curry, H.B. and Schoenberg, I.J. J. d'Anal. Math., 1966, 17, 71.

13. Cox, M.G. J. Inst. Maths. Applics. 1972, 10, 134.

14. Numerical Algorithms Group (NAG) FORTRAN Library Manual Mk 6, 1977, Section E02.

15. de Boor, C. J. Maths. Phys., 1962, 41, 212.

16. Buerger, M.J. "Crystal-structure analysis"; New York and London: Wiley, 1960, Ch. 7, p.160.

17. Cox, E.G. and Shaw, W.F.B. Proc. Roy. Soc., 1930, A127, 71.

18. Whittaker, E.J.W. Acta Cryst., 1953, 6, 218.

19. Wonacott, A.J. in "The Rotation Method in Crystallography"; Amsterdam, New York and Oxford: North-Holland, 1977, Ch.7.

20. Franklin, R.E. and Gosling, R.G. Acta Cryst., 1953, 6, 678.

21. Werner, P.-E. Acta Cryst., 1970, A26, 489.

RECEIVED February 19, 1980.

8

Resolution of X-ray Intensities by Angular Deconvolution

LEE MAKOWSKI

Rosenstiel Basic Medical Sciences Research Center,
Brandeis University, Waltham, MA 02254

The resolution to which X-ray diffraction intensities can be measured from fiber diffraction patterns is often limited by the disorientation of particles in the specimen. Imperfect orientation of the particles has the effect of spreading out the diffracted intensity along Debye-Scherrer rings. That is, the reflections are arced about the center of the diffraction pattern at a constant diffraction angle. The distribution of intensity as a function of angle about the center of the pattern can be treated as the convolution of the reflections occuring at a given radius with an angular intensity distribution function which is related to the distribution of particle orientations within the specimen (1). At small angles of diffraction, a small degree of disorientation will not significantly affect data collection. However, at higher diffraction angles, any disorientation will ultimately cause diffraction from neighboring reflections (at the same diffraction angle) to overlap. This limits the resolution to which reliable data can be collected.

The intensities in a fiber diffraction pattern are usually measured either by linear densitometer scans along the centers of layer lines and between the layer lines (to determine background) for the case of non-crystalline fibers, or by the numerical integration of the reflections from crystalline fibers (e.g., 2, 3). Where disorientation causes adjacent reflections to overlap, it is extremely difficult to separate the contributions from different reflections or to make quantitative measurements of background using the conventional procedures. This problem has been dealt with to some extent by the use of an interactive display system to determine approximate base lines for each reflection (4), and several investigators have briefly discussed methods for dealing with overlapping reflections (3,5). A numerical angular deconvolution procedure making use of all the data in the diffraction pattern has been developed for separating reflections that are overlapping due to disorientation. This method has been used in the measurement of diffrac-

0-8412-0589-2/80/47-141-139$05.00/0

tion from a wide range of biological macromolecular assemblies. Refinements of this method can be used to extract additional information from X-ray patterns. In this paper, it is shown that the small splitting of layer lines in diffraction from gels of Tobacco Mosaic Virus (TMV) can be used to separate Bessel function terms that are superimposed on the layer lines.

Theory

It will be assumed here that the X-ray diffraction data were collected on flat films with a point focus camera. This simplifies the theoretical presentation. The TMV data analyzed in the results section were collected on cylindrical films with Guinier cameras, but positions on the cylindrical films can be mapped onto positions on a flat film by a simple geometric transformation. In general, the form of the optical density, $D(r,\phi)$, in a fiber diffraction pattern can be expressed in film coordinates as the sum of contributions from all reflections, $I_i(r,\phi_i)$, plus a background term, $B(r,\phi)$:

$$D(r,\phi) = \sum_i I_i(r,\phi_i)\ f(\phi,\phi_i) + B(r,\phi) \tag{1}$$

where the sum is over all reflections contributing at a radius, r, r is the distance from the center of the diffraction pattern, and ϕ is the angle about the center of the diffraction pattern. The angular intensity distribution function, $f(\phi,\phi_i)$, describes the spreading out of the intensity as a function of angle, ϕ. This angular smearing causes a reflection centered at (r,ϕ_i) to contribute to the optical density at (r,ϕ). Except near the meridian, the intensity distribution function is simply a function of the angle $(\phi-\phi_i)$, so that equation (1) can be rewritten as a convolution (1),

$$D(r,\phi) = \sum_i I_i(r,\phi_i)\ f(\phi-\phi_i) + B(r,\phi) \tag{2}$$

of the intensities with the angular intensity distribution function.

The distribution of particle orientations about the fiber axis can be defined in terms of a disorientation function, $N(\alpha)$, where $N(\alpha)d\Omega/4\pi$ is the probability of the axis of a particle being at an angle α to the fiber axis in an element of solid angle $d\Omega$. For a Gaussian distribution of particle orientations,

$$N(\alpha) = (2/\sigma^2)\exp(-\alpha^2/2\sigma^2). \tag{3}$$

where σ is the standard deviation of the angular disorientation. Holmes and Barrington-Leigh (2) showed that except in the immediate vicinity of the meridian, the angular intensity distribution function was approximately a Gaussian,

$$f(\phi-\phi_i) = \exp(-(\phi-\phi_i)^2/2\sigma^2). \tag{4}$$

Many real angular intensity distribution functions have been found to closely approximate a Gaussian and for the remainder of this paper a Gaussian distribution will be used. It should be noted, however, that the angular deconvolution can be used on patterns from specimens with any form of disorientation function. For a Gaussian distribution, equation (2) becomes

$$D(r,\phi) = \sum_i I_i(r,\phi_i) \exp(-(\phi-\phi_i)^2/2\sigma^2) + B(r,\phi). \tag{5}$$

The background presents additional problems, for without knowledge of the form of the background the problem is indeterminant. Two possible strategies are available for dealing with the background. It can be measured on a diffraction pattern taken with a blank replacing the specimen, in which case it need not appear in equation (5), or it can be expanded as a set of orthogonal functions, increasing the number of unknowns in (5). Accurate experimental measurement of the background is extremely difficult. Except for regions close to the center of the diffraction pattern, the experimental background can often be made almost entirely circularly symmetric, in which case it is possible to assume that the background is a constant at a given distance from the center of the pattern; that is, $B(r,\phi) = B(r)$. This is a good approximation for the backgrounds in most of the diffraction patterns taken in this laboratory and analyzed by angular deconvolution. In all cases studied to date, one constant and one slowly varying (as a function of angle) background term have been sufficient to characterize the background. Use of more than two background terms severely limits the accuracy and resolution of the method (1).

Assuming that the background is a constant at any given distance from the center of the diffraction pattern, all the optical density on the film is describable using the form

$$D(r,\phi) = \sum_{i=1}^{n} I_i(r,\phi_i) \exp(-(\phi-\phi_i)^2/2\sigma^2) + B(r) \tag{6}$$

There are (n+1) unknowns in this equation where n is the number of reflections contributing at radius r on the film. The intensities, $I_i(r,\phi_i)$, in this equation are equal to the background subtracted optical densities at the centers of the reflections. These intensities, once calculated, must be corrected for geometric factors in the same way as intensities measured by conventional procedures.

Methods

Optical densities on the films are measured on a square raster over the entire region of interest using an Optronics rotating drum microdensitometer. The optical densities must be placed on a polar coordinate system in order to be used in the solution of equation (6). To do this, the center of the diffraction pattern is located by determining the positions of symmetrically equivalent reflections on the film. Each optical density is assigned a position (x,y) relative to the center of the diffraction pattern with the equator and meridian chosen as the principal axes. The optical densities in small regions $(\Delta r, \Delta\phi)$ are then averaged,

$$D(r,\phi) = \frac{1}{\ell} \sum_{j,k} D(x_j, y_k)$$

for all (x_j, y_k) within $(r \pm \Delta r/2, \phi \pm \Delta\phi/2)$, where ℓ is the number of data points in the sum. For the TMV films analyzed here, Δr was chosen to be 0.1 mm (0.00058 $Å^{-1}$ for an 11.2 cm specimen-to-film distance) and the $\Delta\phi$ used was 1°. Note that at higher diffraction angles, the region of the film averaged to obtain $D(r,\phi)$ will be larger. This averaging improves the signal-to-noise ratio at high diffraction angles.

Assuming that the positions (r, ϕ_i) of all the reflections are known, the intensities can be determined by a simple inversion of equation (6). For m optical density measurements at a radius, r, equation (6) represents a set of m linear equations with (n+1) unknowns. The intensities, $I_i(r,\phi_i)$ are determined by a matrix inversion (1). The eigenvalues of the matrix provide a measure of the reliability of the separation at each radius. For specimens with equally spaced layer lines corresponding to an axial repeat distance, c (in Å), the highest spacing (in Å) to which the data can be measured is approximately $c(\sin(1.5\sigma))$. Data collected at higher spacings will have significantly lower signal-to-noise ratios (1). This limitation is not specific to the method used here, but is, rather, a general limitation on the amount of information that can be extracted from any diffraction pattern containing noise.

Variations of this method can, in some cases, provide other kinds of information about the diffracting particles. For instance, TMV is a helical virus with 49.02±0.01 subunits in the three turns of its helix in its 69 Å axial repeat. This deviation from an integral number of units in three turns causes the layer lines to split slightly (6). The first two Bessel function terms that contribute to the first layer line of TMV do not fall at a spacing of 69 Å above the equator, but rather, at 67.6 and 71.5 Å, which corresponds to a splitting of about 0.15 mm on the film at typical specimen-to-film distances (11 cm). Figure 1 shows the calculated positions of the Bessel function terms on the TMV layer lines and the slight splitting to be expected. Depending on the positions of the reflections relative to the meridian and the center of the diffraction pattern, this can, at a given radius, cause a shift in the observed angular position of a layer line of as little as one or two-tenths of a degree up to as much as several degrees near the meridian. Under these conditions, the positions of the reflections, ϕ_i, in equation (6) are not known, but a determination of these positions will provide information about the relative intensities of the Bessel function terms superimposed on the layer line.

The positions, ϕ_i, were determined as follows. Since the ϕ_i are known approximately, the linear equations (6) were solved assuming no layer line splitting. Then, for each reflection, a limited, one-dimensional search was made for a position which fit the measured intensities better. The angular widths, σ, were also allowed to vary, since splitting leads to an apparent increase in the measured angular widths of the layer lines. Once a new set of positions and widths were determined, the set of linear equations was solved a second time. The search was repeated and the linear equations solved again. No substantial changes were found in either the intensities or positions of reflections after the second round of fitting the non-linear parameters.

The refined angular positions of the layer lines at each radius provide a method for estimating the relative intensities of the contributing Bessel function terms. When only two Bessel function terms contribute, a quantitative separation can be made. The sum of two Gaussians with standard deviation σ that are centered at positions ϕ_0 and ϕ_1 (where $|\phi_1-\phi_0| << \sigma$) will be to a first approximation, equal to a single Gaussian of height (A+B) centered at the position $(A\phi_0 + B\phi_1)/(A+B)$, where A and B are the heights of the Gaussians at positions ϕ_0 and ϕ_1, respectively. The heights A and B are determined from the apparent position and height of the layer line. Where more than two Bessel function terms contribute, the observed layer line position may provide an indication of the relative strengths of the terms, but a quantitative separation cannot be made using only this measured position.

Results

Figure 2 shows the results of the non-linear fitting procedure on a part of the equator, first and second layer lines of TMV. The diffraction pattern used was taken by Dr. S. Warren and Dr. G. Stubbs at the Max-Plank-Institut in Heidelberg, Germany. This pattern was taken on a Guinier camera and the arcsdue to disorientation are not circular. The natural coordinate system (3)of the camera (rather than polar coordinates)was used in the deconvolution procedure. The standard

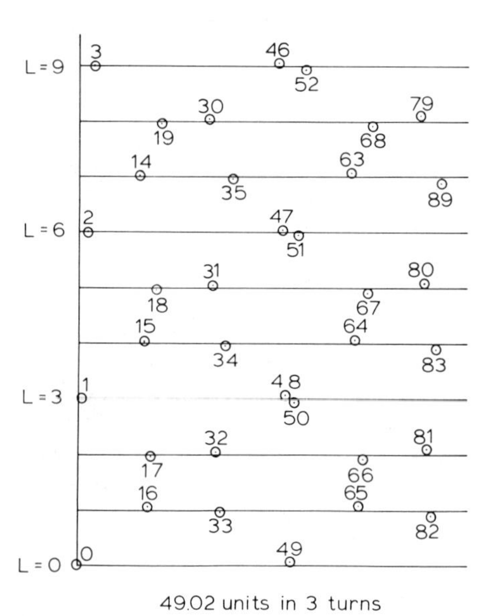

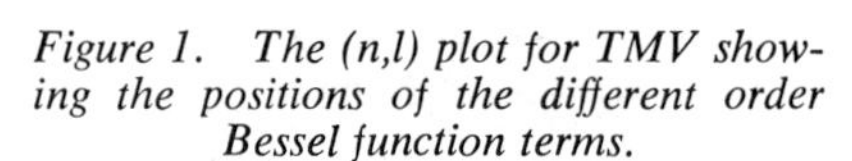

Figure 1. The (n,l) plot for TMV showing the positions of the different order Bessel function terms.

The horizontal lines indicate the position on which the layer lines would fall if there were exactly 49 subunits in three turns in the TMV axial repeat. The circles show the positions the Bessel function terms take assuming 49.02 subunits in three turns. The vertical deviations from the solid lines give an indication of the magnitude of the layer line splitting. The horizontal positions indicate the relative positions of the Bessel function terms along the layer lines; each term can only contribute to diffraction further from the meridian than the positions marked.

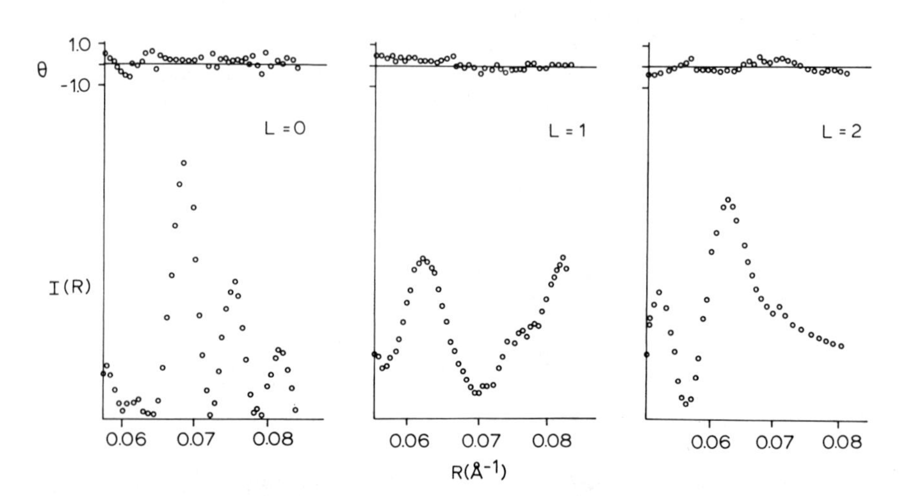

*Figure 2. Intensities (*bottom*) and positions (*top*) on the equator and layer lines 1 and 2 as measured by the fitting of nonlinear terms during angular deconvolution.*

Each data point was calculated from data falling at a different radius about the center of the diffraction pattern. The intensities, I(R), have not been corrected for geometric factors. The positions, Θ, are plotted relative to the position expected for the layer line assuming 49 units in three turns.

deviation of the orientation of the particles in the specimen was about 1.4°. The intensities determined by angular deconvolution are shown in the bottom part of Figure 2. The layer line positions relative to their expected positions, as determined by a nonlinear fitting of the data are in the top part. Each point along a given layer line has been calculated from optical density data at a different radius in the diffraction pattern. The position of the equator should be zero degrees and should not vary even at radii where higher order Bessel function terms contribute. These higher order terms contribute symmetrically; falling both slightly above and slightly below the equator; broadening it but not shifting it. Variations in the calculated position of the equator provide a measure of the accuracy of this method. Where the measured intensities are large the calculated positions do not vary significantly. Where the intensity is small the calculated position varies widely; the fitting procedure cannot accurately determine peak positions in regions where the noise is of magnitude comparable to that of the data. The weighted average of positions calculated for the equator is 0.14°, indicating that the measured orientation of the film in the densitometer was in error by that amount. The positions of the other layer lines must be corrected for this error before they can be used to calculate the relative contributions of the Bessel function terms.

The intensities and angular positions of layer lines 1 and 2 are also shown in Figure 2. Again, the noise in the positions calculated is lowest at points corresponding to the peaks of intensity. For these layer lines there are also systematic variations in the position with subsequent peaks falling at different positions relative to that which would be expected if there were exactly 49 subunits in the 69 Å axial repeat. This layer line splitting was used to calculate the relative contributions of the Bessel function terms on layer lines 1 and 2.

Figure 3 shows the contributions from the first two Bessel function terms on layer lines 1 and 2 as calculated from the layer line splitting. This separation is compared to a previous one made using heavy atom derivatives (7). The data points calculated from the splitting are quite noisy. This is to be expected since the layer line splitting amounts to not more than 1.0° in all the data analyzed. The values determined using heavy atom derivatives have been smoothed by several rounds of box-function refinement (Stubbs, unpublished). On the first layer line, the contribution from the 33rd order term starts with a weak peak between R = 0.06 and 0.07 $Å^{-1}$, where R is the distance from the meridian, but does not become significant until R> 0.075. The general features of the heavy atom separation on this layer line are confirmed by the analysis of layer line splitting.

On the second layer line, the agreement between the two methods is not as close as on the first layer line. The analysis of layer line splitting shows that the 32nd order term has one peak in the region analyzed which extends from 0.06 to 0.075 $Å^{-1}$. The heavy atom derivative separation includes a second peak at about 0.075 $Å^{-1}$. This discrepancy is due, at least in part, to the fact that the heavy atom derivative separation was made on a different data set than the one used here. That data set was obtained by conventional methods of data collection, and comparison with the results of angular deconvolution indicates that the intensity on the second layer line in the region of 0.07 $Å^{-1}$ was underestimated in that data set. A reanalysis of the heavy atom derivative data using angular deconvolution is in progress.

Discussion

The method of angular deconvolution provides a means of measuring the intensities of X-ray reflections in diffraction patterns from partially oriented specimens to the highest possible resolution. Refinement of the non-linear parameters specifying the position and angular profile of reflections has been used to separate contributions from Bessel functions superimposed on layer lines in diffraction patterns from oriented gels of TMV. This has been done by taking advantage of the slight splitting of layer lines in the patterns. The results of this separation confirm the general features of the separation made using heavy atom derivatives and suggest that a reanalysis of the heavy atom derivative data using angular deconvolution will improve the estimates of the contributions of the Bessel function terms.

The separation of Bessel function terms using layer line splitting is confined to regions where only two terms contribute. At higher diffraction angles, where more terms contribute, the separation of Bessel function terms should, ideally, utilize both heavy atom derivative data and the apparent positions and widths of the layer lines. The combined use of both types of information may be possible using a linear relationship between the layer line position and the relative intensities of the Bessel function terms.

Other strains of TMV (for instance, U2) exist that show considerably greater layer line splitting. This larger splittingshould improve the signal-to-noise ràtio of the derived Bessel function terms. The data shown in Figure 3 is from the region of layer line splitting closest to the center of the diffraction pattern. The splitting as a function of angle about the center of the diffraction pattern is greatest for these layer lines. Since the splitting as a function of angle becomes smaller at distances further from the center of the pattern, the signal-to-noise ratio will be smaller at higher diffraction angles. Thus, analysis of diffraction from strains of virus with larger layer line splitting may provide additional

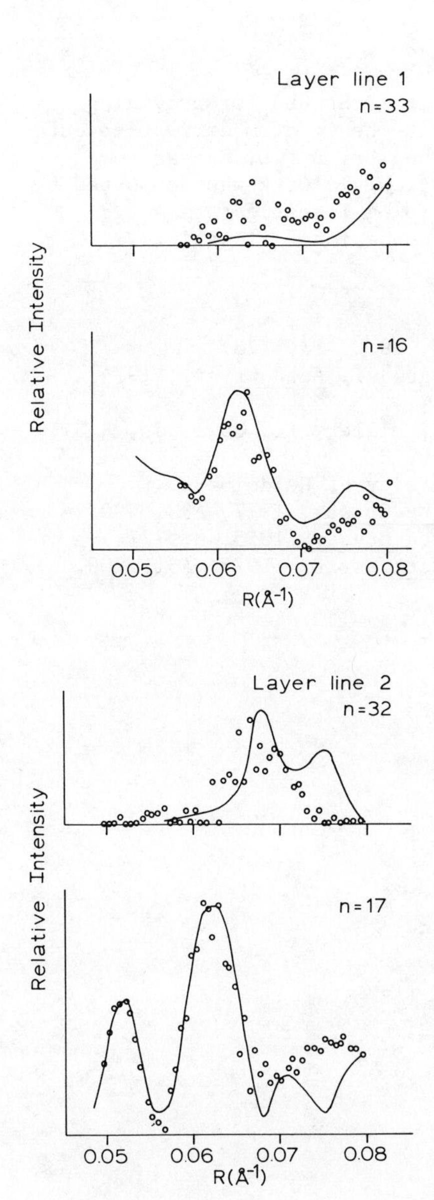

Figure 3. Intensities for separated Bessel function terms on a portion of layer lines 1 and 2.

The circles represent the relative intensities calculated from the observed intensities using the layer line splitting. The solid lines are the results of the analysis of heavy-atom derivative data (7).

information about the relative contributions of superimposed Bessel function terms at higher resolution.

Acknowledgements

I would like to thank Dr. Gerald Stubbs for providing the diffraction patterns of TMV and the heavy atom derivative data and for many useful discussions and Dr. D.L.D. Caspar for support, advice and encouragement. This work was supported by a National Institutes of Health Young Investigators Award CA24407 and an Alfred P. Sloan Foundation fellowship.

Literature Cited

1. Makowski, L. J. Appl. Cryst., 1978, 11, 273.
2. Holmes, K.C., Barrington-Leigh, J., Acta Cryst., 1974, A30, 635.
3. Fraser, R.D.B., MacRae, T.P., Miller, A., Rowlands, R.J., J. Appl. Cryst., 1976, 9, 81.
4. Mandelkow, E., Thesis, University of Heidelberg.
5. Lovell, R., Windle, A.H., Acta Cryst., 1977, A33, 390.
6. Franklin, R.E., Klug, A., Acta Cryst., 1955, 8, 777.
7. Stubbs, G.J., Diamond, R., Acta Cryst., 1975, A31, 709.

RECEIVED February 19, 1980.

9

Computational Methods for Profile Resolution and Crystallite Size Evaluation in Fibrous Polymers

A. M. HINDELEH[1], D. J. JOHNSON, and P. E. MONTAGUE

Textile Physics Laboratory, Department of Textile Industries,
University of Leeds, Leeds, LS2 9JT U.K.

X-ray diffraction patterns from fibres generally contain a few closely overlapping peaks, each broadened by the contributions of crystallite size, crystallite-size distribution, and lattice distortion. In order to achieve complete characterisation of a fibre by X-ray methods, it is first necessary to separate the individual peaks, and then to separate the various profile-broadening contributions. Subsequently, we can obtain measures of crystallite size, lattice distortion and peak area crystallinity, to add to estimates of other characteristics obtained in complementary experiments.

Four major computational steps are necessary to separate the individual peaks and the different profile-broadening components: (i) correction and normalisation of the diffraction data, (ii) resolution of the total peak scattering from the so-called background scatter, and resolution of crystallographic, paracrystalline, and amorphous peaks from each other, (iii) correction of the resolved profiles for instrumental broadening, (iv) separation of the corrected profiles into size and distortion components. In this paper we will discuss these steps in turn, but most attention will be paid to the hitherto largely neglected step of profile resolution.

Experimental

A modified Hilger and Watts Y115 diffractometer mounted on a Hilger and Watts Y90 constant output X-ray generator, utilizing either $CuK\alpha$ or $MoK\alpha$ radiation was used throughout. All specimens were mounted in special holders at the centre of the specimen table; bundles of parallel fibres were examined in the symmetrical transmission mode, powder specimens were examined in the symmetrical reflection mode. Counting was usually carried out at 15 steps per degree, occasionally at 30 steps per degree

[1] Current address: Department of Physics, University of Jordan, P.O. Box 13093, Amman, Jordan.

0-8412-0589-2/80/47-141-149$08.50/0

(two theta). The diffractometer traces were recorded on punched paper tape via a scintillation counter, pulse-height analyser, and timer-scaler system. For setting up purposes, output can be directed to a chart recorder via a ratemeter.

All computation was carried out on the University of Leeds Computing Service's ICL 1906A computer. The programs were written in ALGOL 60.

Computational Methods

(i) Correction and Normalisation

The standard method for normalisation of diffracted intensity data into electron units, is to compute both the mean square atomic scattering factor and the mean incoherent scatter for the particular molecular repeat over a range of high two theta values (say 40°-60°) where their total value can be considered to be equivalent to the actual diffraction from the molecular system concerned. An appropriate normalisation factor is then applied to the experimental intensity data after geometrical correction and, finally, incoherent scatter is subtracted (1).

This method, although well established, is not entirely satisfactory, since the experimental scatter is difficult to measure accurately over a small range of the two theta scale. We prefer to use a normalisation procedure based on Vainshtein's law of conservation of intensity. This law states that total scatter over identical regions of reciprocal space will be equal despite different degrees of lattice order (2).

In terms of a randomly oriented specimen, for example a powder or finely divided fibre, we use the relation

$$4\pi\int_0^\infty I(s)s^2ds = 4\pi\int_0^\infty \overline{f^2}s^2ds$$

and for equatorial traces from a fibre bundle with cylindrical symmetry we use the relation

$$2\pi\int_0^\infty I(s)s\,ds = 2\pi\int_0^\infty \overline{f^2}s\,ds$$

where

$$\overline{f^2} = \frac{\Sigma N_i f_i^2}{\Sigma N_i}$$

is the mean square atomic scattering factor for X-rays of the molecule concerned. N_i is the number of molecules of type i in the molecular repeat, and s is the reciprocal space vector, $s = \sin\theta/\lambda$, where λ is the wavelength of the radiation, and I is the corrected intensity (3).

The atomic scattering factors f_i are evaluated from the expression

$$f_i(x) = A_i \exp(-a_i x^2) + B_i \exp(-b_i x^2) + C_i$$

where $x = \sin\theta/\lambda$ and the constants A_i, B_i, C_i, a_i, b_i, are given by Lee and Pakes (4). It is worth noting that $A_i + B_i + C_i = Z_i$, the atomic number of the atom of type i.

The mean incoherent scatter $\overline{C}$ is given by

$$\overline{C} = \frac{\Sigma N_i C_i}{\Sigma N_i}$$

where C_i is the incoherent scatter for atoms of type i and N_i is again the number of atoms of type i in the molecular repeat. Values of $\overline{C}$ were obtained from tables and best-fit third degree polynomials computed for subsequent use in the normalisation program.

In practice we evaluate the total scatter in terms of either the area AR1(1) or the area AR1(2), where

$$AR1(1) = \int_{2\theta_1}^{2\theta_2} (\overline{f^2} + \overline{C})\ s\ ds = \int_{2\theta_1}^{2\theta_2} (\overline{f^2} + \overline{C})\ \frac{\sin\theta\ \cos\theta d(2\theta)}{\lambda^2}$$

for a parallel bundle of fibres, specimen type (1), and

$$AR1(2) = \int_{2\theta_1}^{2\theta_2} (\overline{f^2} + \overline{C})\ s^2 ds = \int_{2\theta_1}^{2\theta_2} (\overline{f^2} + \overline{C})\ \frac{\sin^2\theta\ \cos\theta}{\lambda^3}\ d(2\theta)$$

for powders or finely divided fibres, specimen type (2).

The array of experimentally observed intensities $I_1(2\theta)$ must first be corrected for polarization to give

$$I_2\ (2\theta) = I_1(2\theta)\ \ 2/\ (1 + \cos^2 2\theta)$$

and for the Lorentz geometric factor to give

$$I_3\ (2\theta) = I_2(2\theta) \sin 2\theta \qquad \text{specimen type (1)}$$

$$\text{or} \quad I_3\ (2\theta) = I_2(2\theta)\ 2 \sin^2\theta\cos\theta \qquad \text{specimen type (2)}$$

Cella, Lee and Hughes (5) have studied the Lorentz correction for fibres in some detail; they consider that the correction $2 \sin^2\theta\cos\theta$ is most appropriate for the equatorial scatter from a parallel bundle of fibres.

The area AR2(1) or AR2(2) is then evaluated, where

$$AR2(1) = \int_{2\theta_1}^{2\theta_2} I_3 (2\theta) \frac{\sin\theta\cos\theta}{\lambda^2} d(2\theta) \qquad \text{specimen type (1)}$$

$$AR2(2) = \int_{2\theta_1}^{2\theta_2} I_3 (2\theta) \frac{\sin^2\theta\cos\theta}{\lambda^3} d(2\theta) \qquad \text{specimen type (2)}$$

so that the normalisation factor is then

$$\frac{AR2(1)}{AR1(1)} \quad \text{or} \quad \frac{AR2(2)}{AR1(2)}$$

and is applied to the array of intensities $I_3(2\theta)$ to give the intensities $I_4(2\theta)$ in electron units. The incoherent scatter $\overline{C}$ computed earlier is then subtracted to give the array $I_5(2\theta)$.

The standard version of this program is identified as STEPSCAN; variations are available for use with data obtained from X-ray or electron diffraction photographs via a Joyce-Loebl autodensidater, and contain appropriate corrections for optical density. The input data are usually averaged in groups of 3 or 5 steps, single step data being used less frequently. A correction for air scatter may be carried out, and the true background radiation, as measured separately, is subtracted. Because the correction for absorption within the specimen is negligible over the range of two theta normally covered, this has been omitted. Expected peak positions are also input (for peak-resolution purposes) together with graph plotting parameters. The output is in the form of arrays $\overline{f^2}$, $\overline{C}$, $I_1(2\theta)$ to $I_5(2\theta)$, the area calculations, estimates of peak height and width, and a graph of the type shown in Figure 1. Appropriate data are held in a file for subsequent profile resolution.

Results. Typical normalised equatorial X-ray diffraction traces from the same poly(ethylene terephthalate)(PET) specimen are shown in Figures 1 and 2. The mean square atomic scattering factor $\overline{f^2}$, the incoherent scatter $\overline{C}$, and the total scatter $\overline{t}$, are also included. The corrected intensity data are evidently dependent on the method of normalisation employed. Figure 1 was obtained with two theta limits of 5° and 40°, Figure 2 has limits 30° and 40°; it is evident in this case that fitting to the coherent scatter over a high two theta range gives almost double the intensity compared with fitting to the total scatter within the range covered.

The effect of the Lorentz correction is illustrated in Figure 3. Here the equatorial trace of a viscose rayon specimen is shown uncorrected (LOR 0), corrected in the normal way for fibres (LOR 1), and for powders (LOR 2). Table I shows the results of profile analysis on the viscose rayon specimen using the different

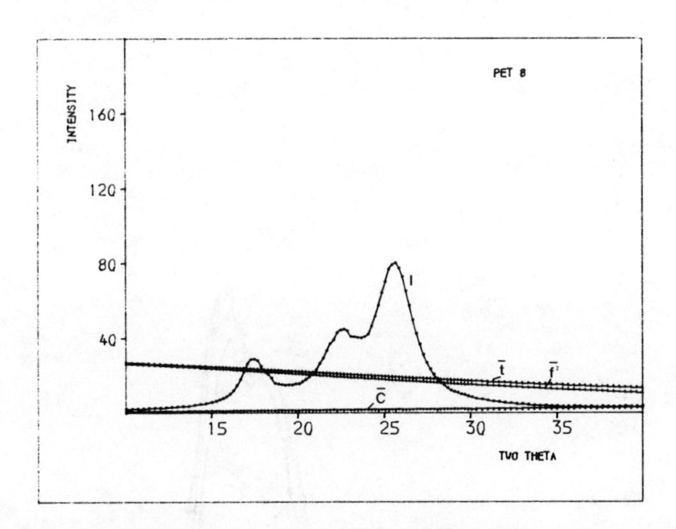

Figure 1. Corrected equatorial trace in electron units for a PET specimen normalized over the full two theta range ($\overline{f}^2$ mean-square atomic scattering factor; $\overline{C}$ incoherent scatter; $\overline{t}$ total scatter)

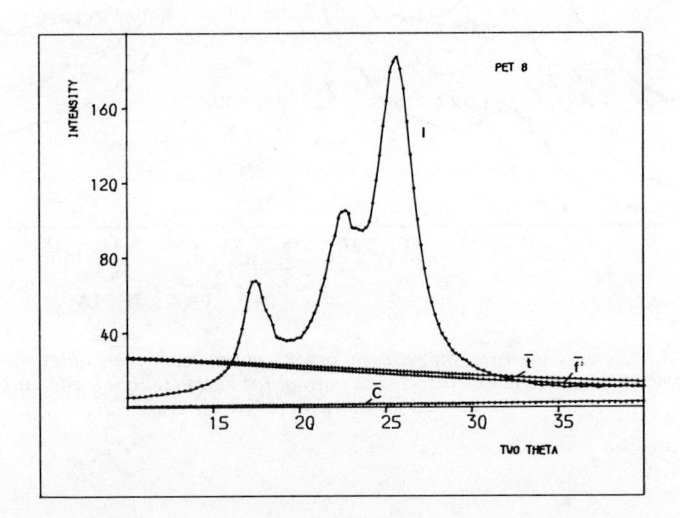

Figure 2. Corrected equatorial trace in electron units for the same PET specimen as Figure 1, normalized over the two theta range 30°–40° ($\overline{f}^2$ mean-square atomic scattering factor; $\overline{C}$ incoherent scatter; $\overline{t}$ total scatter)

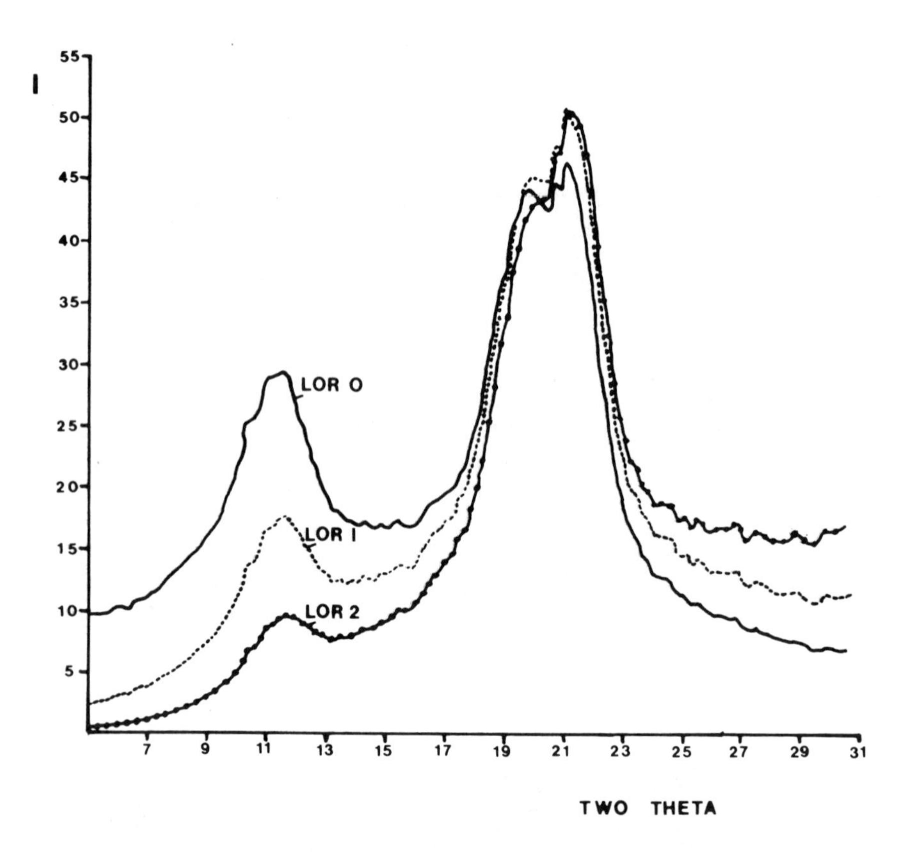

Figure 3. Equatorial trace for viscose rayon, normalized over full two theta range (uncorrected (LOR 0); corrected as normal for fibers (LOR 1); and corrected as normal for powders (LOR 2))

TABLE I

Resolved parameters for an untreated viscose rayon with different Lorentz corrections

Parameters	LOR 0 No correction	LOR 1 $(\sin 2\theta)^{-1}$	LOR 2 $(\sin 2\theta \sin\theta)^{-1}$
f_{101}	0.2	0.6	0.7
A_{101}	14.5	8.0	4.4
W_{101}	2.3	2.1	2.1
P_{101}	12.0	12.1	12.1
$f_{10\bar{1}}$	-0.2	0.0	0.2
$A_{10\bar{1}}$	32.1	30.0	26.1
$W_{10\bar{1}}$	2.4	2.3	2.2
$P_{10\bar{1}}$	20.0	20.0	20.1
f_{002}	0.7	0.5	0.3
A_{002}	25.6	30.6	33.8
W_{002}	1.8	1.9	1.9
P_{002}	21.7	21.8	21.8
%A	48.6	47.1	46.1
%B	51.4	52.9	53.9

f - profile function parameter, A - peak height (eu), W - peak width (two theta), P - peak position (two theta) %A - percentage area under the peaks, %B - percentage area under the background.

Lorentz corrections. As can be seen, there are significant changes in the resolved parameters, particularly the peak positions and peak heights; the effect on peak width is less marked. The standard fibre correction (LOR 1) gives the most reasonable results, the correction (LOR 2), suggested for the equatorial trace of a fibre specimen, is not considered to be realistic.

(ii) Resolution

Our standard resolution program RESOLVE finds the optimum fit of t separate peaks and a polynomial background to the corrected and normalised intensity trace (6). Each peak profile is considered to have the form

$$f_t G_t + (1 - f_t C_t)$$

where G_t is the Gaussian function

$$A_t \exp\left\{ - \ln 2 \left[\frac{2(X - P_t)}{W_t} \right]^2 \right\}$$

and C_t is the Cauchy function

$$A_t \Big/ \left\{ 1 + \left[2(X - P_t)/W_t \right]^2 \right\}$$

The peaks are defined by the parameters A_t the peak height, W_t the width of the peak at half height, and P_t the peak position; f_t is the profile function parameter,which can vary between -0.5 and 1.0 for sensible profiles and effectively describes the tail region of the profile (7). The scatter from disordered molecules in the fibre is considered to have the polynomial form

$$a + bX + cX^2 + dX^3$$

and corresponds to the so-called background scatter. X may be either the two theta or the s scale. The function to be minimised in terms of these parameters is

$$S = \sum_{i=1}^{n} (I_{(calc)i} - I_{(norm)i})^2$$

Input is in the form of arrays of two theta and I_5, expected peak positions, peak heights and widths estimated by STEPSCAN, and appropriate background parameters. The latter may all start at zero, alternatively they may be found to a first approximation by putting a small number of points (from an assumed background) in a subsidiary program BASELINE. Output is in the form of parameter lists, the areas and coordinates of the resolved peaks, the total area under the peaks, and the area under the background.

Optimization. We have recently tried several methods of optimization, but for many years made satisfactory use of the minimization procedure due to Powell (8), which employs the method of conjugate directions and ensures efficient convergence in most cases. Although no bounds can be put on the parameters, it is possible to constrain any of the parameter sets during a single optimization cycle by appropriate ordering of the parameter list. This procedure (POWELL 64) has recently been withdrawn from the NAG library of algorithms and we have searched for a suitable alternative although retaining the procedure in our own program. The nearest equivalent (E04CEA) employs a quasi-Newton approach with difference approximations to the gradient. Again no bounds are possible, and constraint depends upon the parameter list.

Another useful program (E04HAA) provides constrained optimization with bounds for each parameter using a sequential penalty function technique, which effectively operates around unconstrained minimization cycles.

Versions of RESOLVE which use either POWELL 64 (NUSOLVE G), E04CEA (NUSOLVE E) or E04HAA (NUSOLVE F), and graph plotting versions are available. It is also possible to deal with asymmetric profiles (9). In this case, two separate parameters for each of f_t and W_t are evaluated, one for the left-hand side of the profile, the other for the right-hand side. We have also tried the Pearson VII function suggested by Hueval, Huisman and Lind (10) in place of our combined Gaussian-Cauchy function. Unfortunately, because the tail-fitting parameter m tends to infinity for a Gaussian function, we have experienced computational difficulties where profiles approximated to a Gaussian form. However, in all cases where resolution was successful, the final peak parameters were identical in terms of both Gaussian-Cauchy and Pearson VII functions.

Results. Where X-ray diffraction peaks are sharp and relatively well defined, and where there is no ambiguity in the number of peaks, there are few problems in achieving satisfactory resolution with any of the three programs discussed above. Typical results achieved with the crystalline cellulose I fibre, Ramie, and with the crystalline cellulose II fibre Fortisan, are illustrated in Figures 4 and 5. The 101, $10\bar{1}$ and 002 profiles resolved by NUSOLVE G are given, together with the best-fit polynomial background. The parameters are listed in Table II, together with the total peak area expressed as a percentage. This latter parameter can be considered as the peak-area crystallinity within the two theta limits employed.

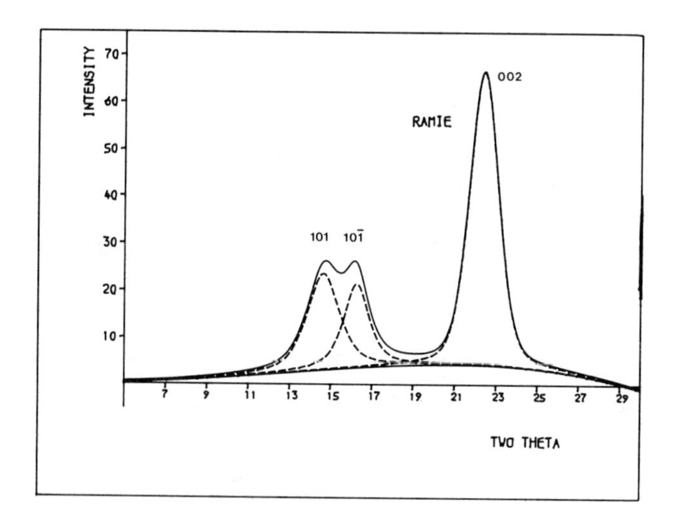

Figure 4. Smoothed equatorial trace for Ramie (Cellulose I) fibers with resolved 101, 10$\bar{1}$, and 002 profiles, and best-fit background

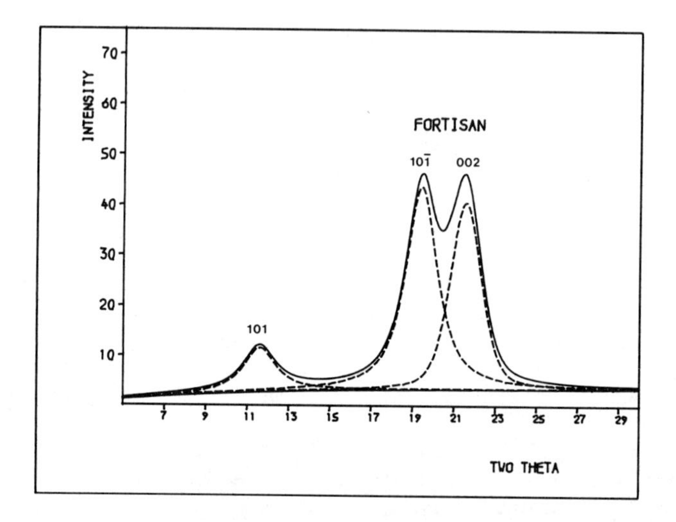

Figure 5. Smoothed equatorial trace for Fortisan (Cellulose II) fibers with resolved 101, 10$\bar{1}$, and 002 profiles, and best-fit background

TABLE II

Resolved parameters for Ramie (Cellulose I) and Fortisan (Cellulose II)

Parameters	Ramie	Fortisan
f_{101}	0.4	0
A_{101}	20	8.9
W_{101}	1.6	1.9
P_{101}	14.6	11.6
$f_{10\bar{1}}$	0.3	0.2
$A_{10\bar{1}}$	18	40
$W_{10\bar{1}}$	1.5	1.9
$P_{10\bar{1}}$	16.3	19.4
f_{002}	0.7	0.6
A_{002}	62	37
W_{002}	1.7	1.8
P_{002}	22.4	21.5
a	1.45	-0.84
b	-0.5	0.5
c	0.068	-0.021
d	-0.0018	0.0003
%A	77	75
%B	23	25

f - profile function parameter, A - peak height (eu) W - peak width (two theta), P - peak position (two theta), %A - percentage area under the peaks, %B - percentage area under the background.

We will now consider three cases where satisfactory peak resolution was much more difficult to achieve. Case (a), a specimen of PET fibre used for texturising, having poorly defined crystalline peaks and possibly an additional peak due to an intermediate phase (11); case (b), a specimen from a range of PET fibres with different shrinkages (12), again an additional intermediate phase peak was a possibility; case (c), cold drawn polypropylene fibres (13), an additional paracrystalline peak was most likely here.

Case (a). When the input parameters were ill chosen, see Table III, input (1), particularly with background parameters zero, the versions of NUSOLVE based on POWELL 64 and E04CEA (NUSOLVE G and NUSOLVE E), which have no parameter constraints, gave identical results. However, the profile function parameter of the 101 peak is -1.3 which gave a distorted profile. The constrained parameter version NUSOLVE F, being unable to distort the 010 peak, does not find a very good solution (S=67.7) within the limits imposed. A reasonable fit was only found when the background parameters were initially well chosen, i.e. they were found from the subsidiary program BASELINE. Then, input(2), Table III, all three versions of NUSOLVE gave similar output parameters (S=7.2 to 7.7) and realistic profiles, Figure 6. Although there is some saving in cpu time with E04CEA over POWELL 64 when ill-chosen parameters were input, the output parameters were identical. The cpu times required with well-chosen input parameters were almost identical for the three programs.

Because of the possibility that an intermediate phase exists in PET, see Case (b), the resolution was carried out with four peaks and zero background parameters. Eventually, and only after several attempts, a good fit was found with three crystalline peaks, one very broad noncrystalline peak, and a polynomial background. On adding the broad peak to the new background, an identical result to that of Figure 6 (Table III) was obtained, casting doubt on the existence of a true intermediate phase.

Case (b). The polyester specimen PET06 was resolved into three crystalline peaks with almost identical results using all three NUSOLVE programs, both when input parameters for the background were zero, or when background parameters from BASELINE were used, see Figure 7. However, because of the suspected presence of an additional peak, further attempts were made to resolve the trace into four peaks. When the additional peak parameters were ill-chosen (all zero), it was impossible to achieve sensible resolution without constrained optimization; however, when the input parameters for the additional peak were well-chosen, all programs eventually gave similar results, see Figure 8 and Table IV. The use of constrained optimization decreased the time required to find a satisfactory solution.

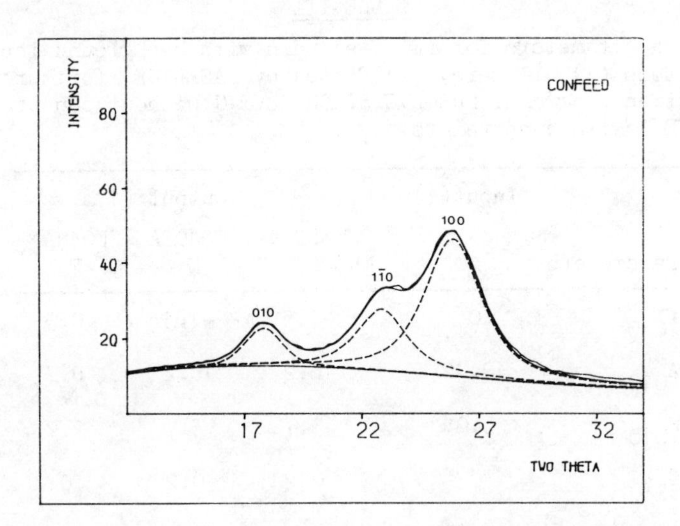

Figure 6. Good resolution with well-chosen background parameters. Corrected and smoothed equatorial trace for PET feed yarn resolved into realistic 010 profile, together with 1$\bar{1}$0, 100 profiles, and new best-fit background.

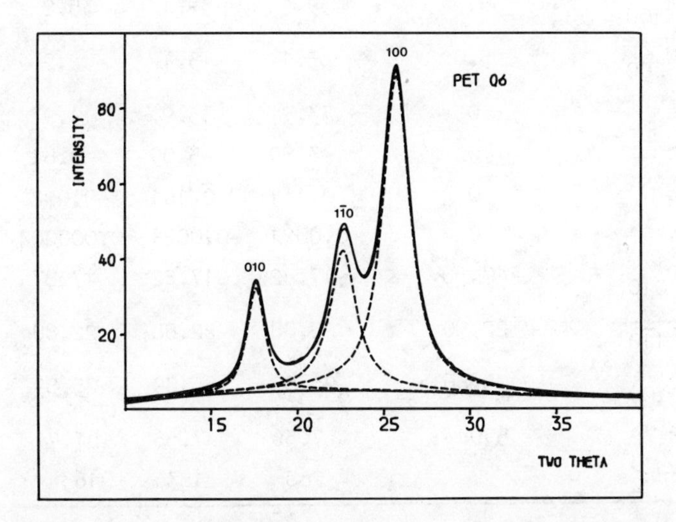

Figure 7. Crystalline peak resolution. Corrected and smoothed equatorial trace of a PET fiber specimen resolved into three crystalline peaks, 010, 1$\bar{1}$0, 100, and the best-fit background.

TABLE III

Resolved parameters for PET feed yarn with background input parameters (1) all zero, (2) fitted by BASELINE, for three resolution programs, NUSOLVE E, NUSOLVE G without constraints, NUSOLVE F with constraints

Parameters	Input(1)	Output (1) POWELL64 NUSOLVE G	E04CEA E	E04HAA F
f_{010}	0	-1.3	-1.3	0.3
A_{010}	19.42	18.2	18.2	8.7
W_{010}	5.0	3.7	3.7	2.2
$f_{1\bar{1}0}$	0	0.2	0.2	0
$A_{1\bar{1}0}$	25.75	17.1	17.1	19.0
$W_{1\bar{1}0}$	5.0	2.6	2.6	2.2
f_{100}	0	0.2	0.2	0.6
A_{100}	38.94	39.1	39.1	35.2
W_{100}	5.0	3.1	3.1	2.9
a	0	33.9	33.9	4.97
b	0	-3.89	-3.89	0.81
c	0	0.161	0.161	-0.022
d	0	-0.0021	-0.0021	0.000004
P_{010}	17.50	17.82	17.82	17.87
$P_{1\bar{1}0}$	22.50	22.88	22.88	22.88
P_{100}	25.86	25.89	25.89	25.89
S	5904	7.56	7.56	67.7
cpu s		268	183	163

f - profile function parameter, A - peak height (eu), W - peak width (two theta), a,b,c,d - background parameters, P - peak position (two theta), S - least sum of squares, cpu s - computing time.

TABLE III (continued)

Parameters	Input(2)	Output (2) POWELL64 NUSOLVE G	E04CEA E	E04HAA F
f_{010}	0.2	0.5	0.7	0.6
A_{010}	10.0	10.3	9.6	10.0
W_{010}	1.8	1.9	1.9	1.9
$f_{1\bar{1}0}$	0.6	0.2	0.2	0.2
$A_{1\bar{1}0}$	15.0	16.7	16.1	16.4
$W_{1\bar{1}0}$	2.4	2.6	2.5	2.6
f_{100}	0.3	0.3	0.3	0.3
A_{100}	37.0	36.2	36.5	36.5
W_{100}	2.9	2.9	2.9	2.9
a	-14.95	-10.20	-21.09	-16.42
b	3.41	3.08	4.69	4.02
c	-0.125	-0.130	-0.202	-0.173
d	0.00124	0.0016	0.0026	0.0022
P_{010}	17.80	17.82	17.82	17.82
$P_{1\bar{1}0}$	22.85	22.86	22.86	22.86
P_{100}	25.89	25.90	25.90	25.90
S	50.26	7.71	7.19	7.28
cpu s		148	134	151

f - profile function parameter, A - peak height (eu),
W - peak width (two theta), a,b,c,d - background parameters
P - peak position (two theta), S - least sum of squares,
cpu s - computing time.

TABLE IV

Resolution of polyester specimen PET 06 into three crystalline peaks, and three crystalline peaks and a paracrystalline peak

Peak	f	A	P	W	L_{hko}	%Area	Total area	Background area
010	0.2	28	17.5	1.3	7.3	9.8		
$1\bar{1}0$	0.0	37	22.6	1.9	4.9	20.6		
100	0.1	84	25.7	1.9	4.7	46.4		
							76.8	23.2
010	0.3	27	17.5	1.3	7.2	9.2		
$1\bar{1}0$	0.3	31	22.6	1.7	5.5	14.5		
100	0.2	83	25.7	1.9	4.7	44.8		
Para	-1.0	9.2	26.7	5.3	-	14.7		
							83.2	16.8

f - profile function parameter, A - peak height (eu),
P - peak position (two theta), W - peak width (two theta)
L_{hk0} - crystallite size normal to (hk0)(nm).

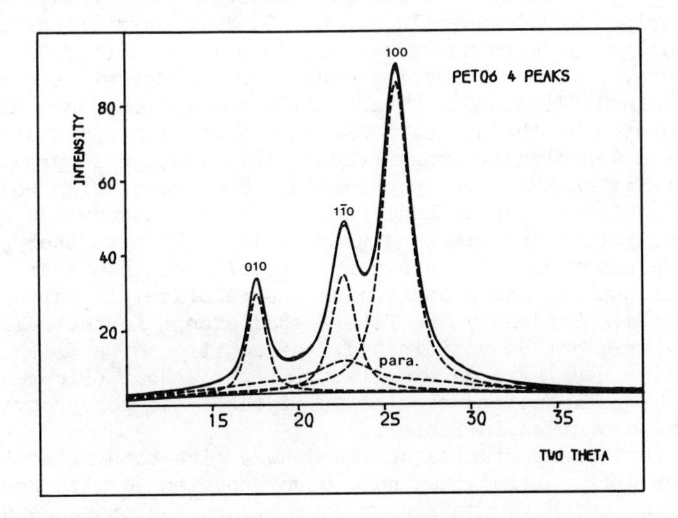

Figure 8. Paracrystalline peak resolution. Corrected and smoothed equatorial trace of a PET fiber specimen resolved into three crystalline peaks, 010, 1$\bar{1}$0, 100, a paracrystalline peak and the new best-fit background.

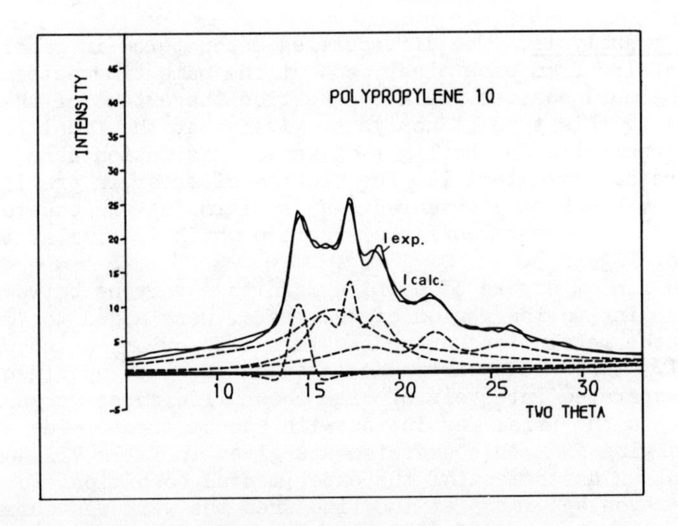

Figure 9. Unrealistic resolution with unconstrained optimization. Corrected equatorial trace of a polypropylene fiber specimen cold drawn ×5.5. Total theoretical trace comprised of unrealistic profiles.

Case (c). Attempts to resolve the equatorial trace of a polypropylene specimen (cold drawn x10) into crystalline peaks and a background were unsuccessful, despite a variety of input parameters. Without parameter constraint, nonsensical profiles were obtained (Figure 9). The profile resolution could not be improved even by the use of BASELINE. Even with constraints, resolution was clearly wrong (Figure 10), although it produced the best-fit mathematically possible. Good resolution was finally achieved when an additional peak was introduced, but only with constrained optimization (Figure 11). With another polypropylene specimen (cold drawn x5), resolution into four crystalline peaks and a background was relatively straightforward (Figure 12). Evidently the fit is good except in the region 15-16°; as might be anticipated, the addition of a fifth paracrystalline peak gave an excellent fit throughout (Figure 13). Details of satisfactory resolution of these two polypropylene specimens are given in Table V.

The three case studies outlined here give some indication of the worst difficulties that have been experienced with peak resolution, and have one factor in common; the presence of an additional amorphous or paracrystalline peak was suspected in each case. All three optimization programs worked well when the input parameters were well chosen, but constraints were necessary in the more difficult cases and when broad peaks were present. Judicious choice of input parameters always speeds the ultimate solution and the use of BASELINE is most helpful.

Error Analysis. The difficulties encountered in resolving peak profiles from each other, and at the same time determining the background scatter, call to question the extent of the error involved in this type of analysis, given that the final resolution appears realistic in the light of known information about the structure of the material. The problem of error in profile resolution has been considered (14) in terms of the equatorial trace of a viscose rayon fibre specimen which is similar to Fortisan, Figure 5.

When random errors of average magnitude varying between 1 and 8%, according to the region of two theta, were added to 106 data points, the resolution gave parameters with errors as shown in Table VI. Tests made on traces taken on the same specimen at widely separated intervals of time show differences very similar to those in the trial resolution with random error. The average error margins from this exercise are given in Table VII and represent an assessment of the experimental conditions for viscose rayon specimens at the time when the work was carried out. For our apparatus, and for similar parameters in other specimens, the error margins will be comparable.

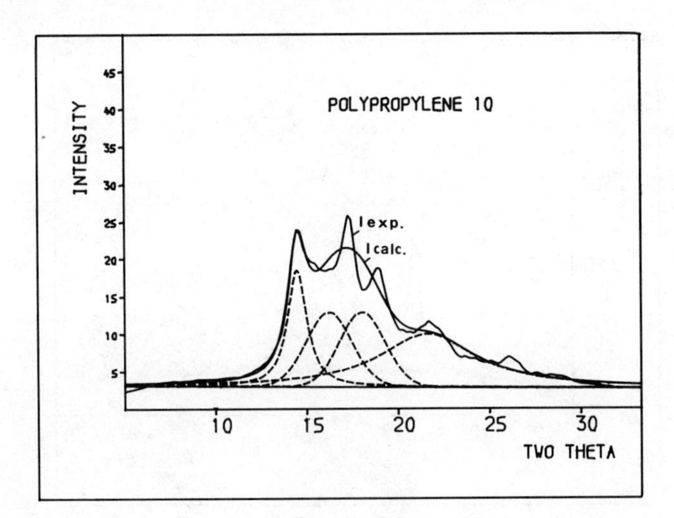

Figure 10. Improved but still unrealistic resolution with constrained optimization. Corrected equatorial trace of a polypropylene fiber specimen cold drawn ×5.5. Total theoretical trace still comprised of unrealistic profiles. Compare with Figure 9.

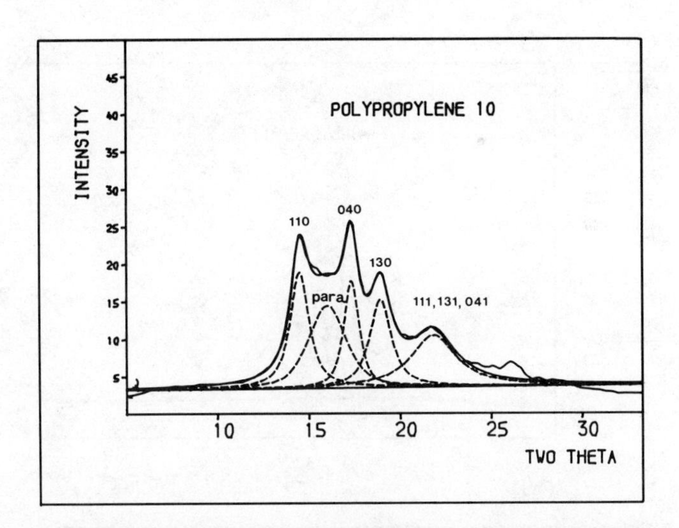

Figure 11. Good resolution with constrained optimization and addition of a paracrystalline peak. Corrected equatorial trace of a polypropylene fiber specimen cold drawn ×5.5. Total theoretical trace now comprised of realistic crystalline peaks 110, 040, 130, 111, 131, 041, and paracrystalline peak at 15.5°. Compare with Figures 9 and 10.

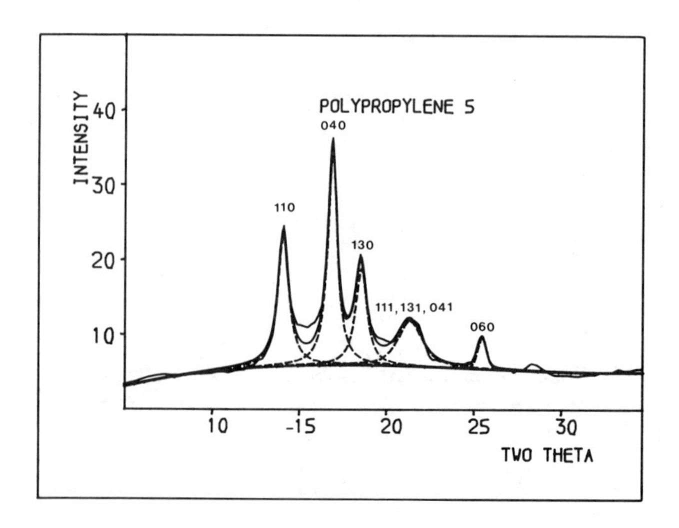

Figure 12. Good resolution with constrained optimization except in 15° region. Corrected equatorial trace of a polypropylene fiber specimen cold drawn ×2.75. Total theoretical trace comprised of realistic peaks 110, 040, 130, 111, 131, 041, 060, but poor fit in 15° region.

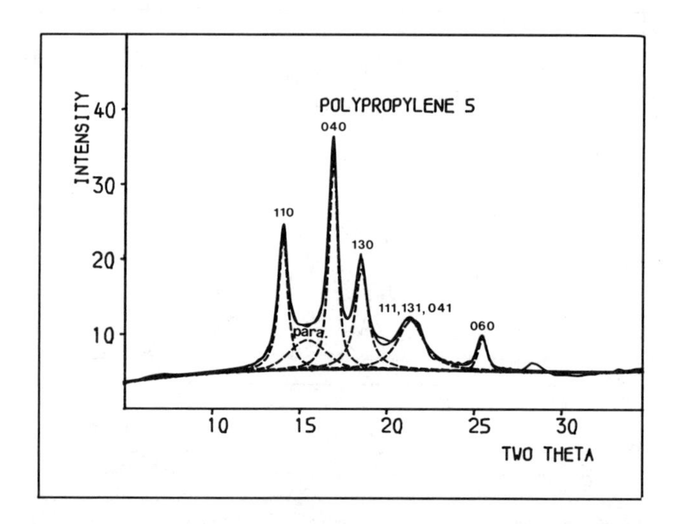

Figure 13. Good resolution with constrained optimization and addition of a paracrystalline peak. Corrected equatorial trace of a polypropylene fiber specimen cold drawn ×2.75. Total theoretical trace comprised of realistic crystalline peaks and a paracrystalline peak at 15.5°. Compare with Figure 12.

TABLE V

Resolution of cold drawn polypropylene specimens into crystalline and paracrystalline peaks

Peak	f	A	P	W	L_{hkl}	%Area	Total area	Background area
X10								
110	0.1	15	14.0	1.2	7.5	11.9		
040	0.3	11	16.9	1.0	9.3	8.6		
130	0.1	14	18.4	1.4	6.6	10.2		
111,131, 041	0.0	12	21.3	2.0	3.0	12.8		
Para	0.5	7.0	15.5	2.5	-	14.6		
							58.1	41.9
X5								
110	0.1	18	14.1	0.6	17	7.2		
040	0.3	28	16.9	0.6	18	10.4		
130	0.0	13	18.6	0.9	11	7.9		
111,131, 041	0.2	7	21.4	1.8	5.0	7.5		
060	0.4	4.6	25.5	0.7	16	1.7		
Para	0.7	4.0	15.6	2.5	-	5.0		
							39.7	60.3

f - profile function parameter, A - peak height (eu), P - peak position (two theta), W - peak width (two theta), L_{hkl} - crystallite size normal to (hkl)(nm).

TABLE VI

Known parameters after resolution

Parameters	Starting values	Known values	Final values	Error
f_{101}	0.0	0.0	0.05	+0.05
A_{101}	6.5	6.0	5.81	-3.3%
W_{101}	4.5	4.0	3.90	-2.5%
$f_{10\bar{1}}$	0.0	0.1	0.04	-0.06
$A_{10\bar{1}}$	14.0	12.0	11.00	-8%
$W_{10\bar{1}}$	4.0	3.25	3.06	-6%
f_{002}	0.0	0.5	0.51	+0.01
A_{002}	17.0	15.0	15.83	+5.5%
W_{002}	4.0	3.25	3.30	+1.5%
%A	100%	64.9%	62.9%	-5.7%
%B	0%	35.1%	37.1%	+5.7%
P_{101}	12.2	12.15	12.145	-0.005
$P_{10\bar{1}}$	20.1	20.0	19.89	-0.11
P_{002}	21.9	22.0	21.92	-0.08
S	442.96	3.720	3.304	

f - profile function parameter, A - peak height (eu),
W - peak width (two theta), P - peak position (two theta),
S - sum of least squares,
%A - percentage area under the peaks,
%B - percentage area under the background.

TABLE VII

Error Margins

Parameter	Margin
f	± 0.15
$A_{101,10\bar{1}}$	$\pm 10\%$
A_{002}	$\pm 5\%$
W_{101}	$\pm 8\%$
$W_{10\bar{1},002}$	$\pm 6\%$
%A, %B	$\pm 6\%$
$P_{10\bar{1}} - P_{101}$	± 0.1
$P_{002} - P_{10\bar{1}}$	± 0.05

f - profile function parameter, A - peak height (eu)
W - peak width (two theta), P - peak position (two theta),
%A - percentage area under the peaks,
%B - percentage area under the background.

(iii) Instrumental Broadening Correction

The two most common methods used to correct resolved peak profiles for the broadening imposed by the finite width of the X-ray beam in the diffractometer, are due to Jones (15) and Stokes (16). Both are essentially unfolding or deconvolution methods, but the Jones method defines specific functions for both the uncorrected and the instrumental broadening profile. If the uncorrected profile is Gaussian, then

$$\beta^2 = B^2 - b^2$$

and if the uncorrected profile is Cauchy, then

$$\beta = B - b$$

where β is the corrected breadth, B the observed breadth, and b the instrumental breadth. The approximations due to Jones are based on integral breadths although many investigators use half-widths instead. Since we are dealing with Gaussian/Cauchy functions, we have made use of the correction

$$\beta = f(B^2 - b^2)^{\frac{1}{2}} + (1 - f)(B - b)$$

The Stokes method is essentially a Fourier transform method making use of the entire profile, and is a reasonably straightforward computation, although limits have to be applied to the profile in transform space to achieve correct results. The main peak from hexamethylene tetramine crystals compacted at 85°C (17) has been used as our standard for the instrumental broadening peak.

Results. We have made a detailed study of the effect of using both the Jones and the Stokes corrections on crystallite size measurements obtained from the most crystalline samples of cellulose I and II, Ramie and Fortisan (18). Our conclusion was that the Jones corrections were always within 3% of the Stokes corrections for both half-width and integral breadth. Current practise is always to use a Stokes deconvolution procedure for the correction of all resolved peak profiles, evaluating an apparent crystallite size $L_{(hkl)}$ in terms of the relationship

$$L_{(hkl)} = \frac{K}{\beta} = \frac{K}{ds} = \frac{K\,\lambda}{\cos\theta\; d(2\theta)}$$

where ds and d(2θ) represent the integral breadths (or half widths) in units of s or two theta. The Scherrer parameter, K, is usually taken as 1 for integral breadths and 0.89 for half widths. Figure 14 illustrates the deconvolution operation for a PET 100 peak; Table VIII lists the appropriate parameters of the observed, instrumental, and corrected peaks, with the crystallite size evaluated as above and by the appropriate Jones factor. Again, the values are within 3%. Apparent crystallite size values for various specimens, evaluated from the above equation with K = 1, are given in Tables IV and V.

(iv) Separation of the resolved and corrected profiles into size and distortion components

The methods in general use for separating the size and distortion broadening components of the resolved and corrected peak profiles can be separated into two groups, non-transform and transform methods. The non-transform methods are essentially similar to the Jones method, being approximations to a convolution. The transform method discussed here makes use of the Fourier coefficients found after the Stokes correction.

Non-transform methods. The non-transform methods make use of Jones type relations; thus, if β_S and β_D are the integral breadths due to size and distortion respectively, then, the observed profile β is given by

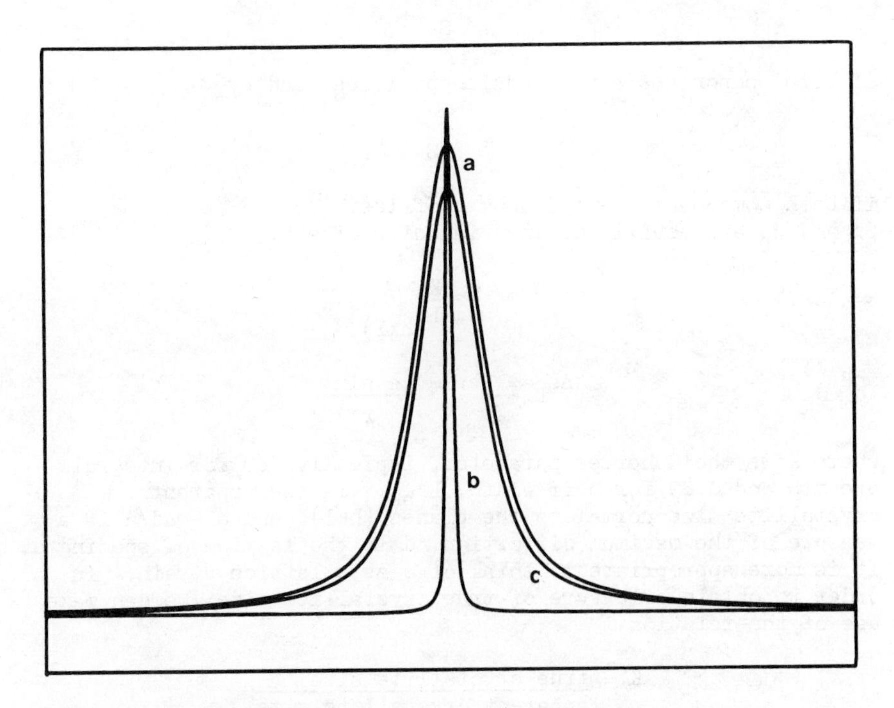

Figure 14. Deconvolution correction for instrumental broadening—simulation. Profile b unfolded from Profile a to give Profile c.

TABLE VIII

Correction of 100 peak of PET 06 specimen for instrumental broadening by Stokes method and by Jones method

Profile	f	A	W	P	IB	β	Δs	L_S	L_J
Uncorrected	0.2	83	1.9	25.7	2.74	0.0478	0.303	3.3	
Instrumental	0.5	89	0.2	25.7	0.30	0.0052	0.033	30	
Corrected	0.2	75	1.8	25.7	2.56	0.0447	0.282	3.5	3.6

f - profile function parameter, A - peak height (eu),
W - peak width (two theta), P - peak position (two theta),
IB - integral breadth (two theta), β - integral breadth (rad),
Δs - integral breadth (nm^{-1}), L_S - crystallite size (nm) Stokes,
L_J - crystallite size (nm) Jones.

$$\beta^2 = \beta_S^{\,2} + \beta_D^{\,2}$$

if both components have Gaussian profiles, and by

$$\beta = \beta_S + \beta_D$$

if both components have Cauchy profiles.
β_S and β_D are usually defined in units of s by

$$\Delta s_S = \frac{K}{L_{(hkl)}}$$

and

$$\Delta s_D = 2es = \frac{4e \sin\theta}{\lambda}$$

where K is the Scherrer parameter, typically 1.0 for integral breadth and 0.89 for half width, $L_{(hkl)}$ is the apparent crystallite size normal to the planes (hkl), and $e = \Delta d/d$ is a measure of the maximum distortion Δd in the lattice of spacing d. It is more appropriate to think of e as a lattice strain. In order to obtain a measure of true crystallite size, we can make use of the relation

$$K = \frac{\text{true crystallite size}}{\text{apparent crystallite size}}$$

provided that we have a useful value of K.

The above equations yield the following relationships for a series of reflections (00l for example):

$$(\Delta s)^2 = (K/L_{(00l)})^2 + 4\,e^2 s_l^{\,2} \qquad \text{..... (1)}$$

$$\Delta s = K/L_{(00l)} + 2\,e s_l \qquad \text{..... (2)}$$

where $s_l = s_0 l$ and $s_0 = 1/d_{(00l)}$.

As an alternative, Thrower and Nagle (19) have adapted a suggestion by Ergun (20) and used the relationship

$$\Delta s = K/L_{(00l)} + 2\pi^2\sigma^2\, s_0^{\,2} l^2 \qquad \text{..... (3)}$$

Here the root mean square lattice distortion is $\langle\epsilon^2\rangle^{\frac{1}{2}}$, and it can be assumed that $e = 1.25\langle\epsilon^2\rangle^{\frac{1}{2}} = 1.25\,\sigma$.

Another relationship, based on Hosemann's theory of paracrystallinity, is (17)

$$(\Delta s)^2 = (K/L_{(001)})^2 + 1/2d_{(001)} \left[1 - \exp\left(-2\pi^2\sigma^2 l^2\right) \right]^2$$

which expands to

$$(\Delta s)^2 = (K/L_{(001)})^2 + \pi^4 \sigma^4 l^4 s_0^2 \quad \ldots\ldots (4)$$

In all cases we require several orders, or pseudo-orders, of a reflection in order to find an appropriate relationship between Δs or $(\Delta s)^2$ and s_1, s_1^2, or l^4, and, if straight-line, the slope and intercept for distortion and size. Many of the methods used for size and distortion separation have been discussed at length by Buchanan and Miller (17); however, they were unable to state which method gave the most realistic evaluations. Here we shall test the methods by comparison with electron-microscope measurements, and by analysis following the computation of simulated profiles.

Fourier transform method. The method used most widely for the separation of size and distortion in peak profiles from metals and inorganic materials is the Fourier analysis method introduced by Warren and Averbach (21). The peak profile is considered as a convolution of the size-broadening profile f_S and the distortion broadening profile f_D, so that the resolved and corrected profile f(x) is given by

$$f(x) = \widehat{f_S(x) f_D(x)}$$

where x is a measure of either two theta or s.
The corresponding transforms are related by

$$F(t) = F_S(t)\ F_D(t)$$

t being a measure of length normal to the diffraction planes. An assumption must be made about the distortion transform; this is taken to be Gaussian so that

$$F_D(t) = \exp(-2\pi^2 s^2 t^2 \sigma^2)$$

where σ is again $\langle \epsilon^2 \rangle^{\frac{1}{2}}$. Hence

$$\ln F(t) = \ln F_S(t) - 2\pi^2 s^2 t^2 \sigma^2 \quad \ldots\ldots (5)$$

and by plotting ln F(t) against s^2 the values of $F_S(t)$ and σ^2 can be found for each t, provided we have enough orders of reflection (usually three) to extrapolate to s = 0. Warren and Averbach showed that the first derivative of a plot of Fourier coefficients gives a measure of crystallite size, thus

$$\left.\frac{dF_S(t)}{dt}\right|_{t=0} = \frac{1}{\bar{N}}$$

where $\bar{N}$ is the number average crystallite size. In practise, it is difficult to follow this procedure with polymers, even if enough orders of a reflection are available. Buchanan and Miller showed that for isotactic polystyrene the plots of ln F(t) against s^2 were not constant and revealed a marked upward concavity as t approaches zero, the implication being that the distortion distribution is not Gaussian. We have followed the Fourier transform method for a few suitable materials, using the Fourier coefficients available from the Stokes deconvolution procedure, and have also found that the ln F(t) plots are non-linear. Because of this, the extrapolations are not well suited to computational methods and the process of finding $\bar{N}$ is somewhat tedious.

Buchanan, McCullough, and Miller (22) found that the non-transform integral breadth methods effectively give a 'weight' average $\langle L_i \rangle$ rather than a number average L_i. The weight average is defined by

$$\langle L_i \rangle = \frac{\sum_i L_i^2 n(i)}{\sum_i L_i n(i)}$$

and the number average by

$$L_i = \frac{\sum_i L_i n(i)}{\sum_i n(i)}$$

Experimental results. Some carbon fibre specimens reveal several orders of 001 particularly in electron diffraction patterns; Figure 15 shows a plot of β against l^2, equation (3), for an electron diffraction pattern from the skin region of a high-modulus material. $L_{(001)}$, usually referred to as L_c, is 3.5 nm and σ = 2%. A full description of electron-diffraction analysis in several similarly heterogeneous carbon fibres has been published (23). Figure 15 also includes a plot from the 001 electron diffraction profiles of a carbon whisker, an exceptionally perfect graphite material. This specimen, with an L_c of 10 nm, has zero distortion, and represents the only case where we have found no distortion in a fibrous specimen.

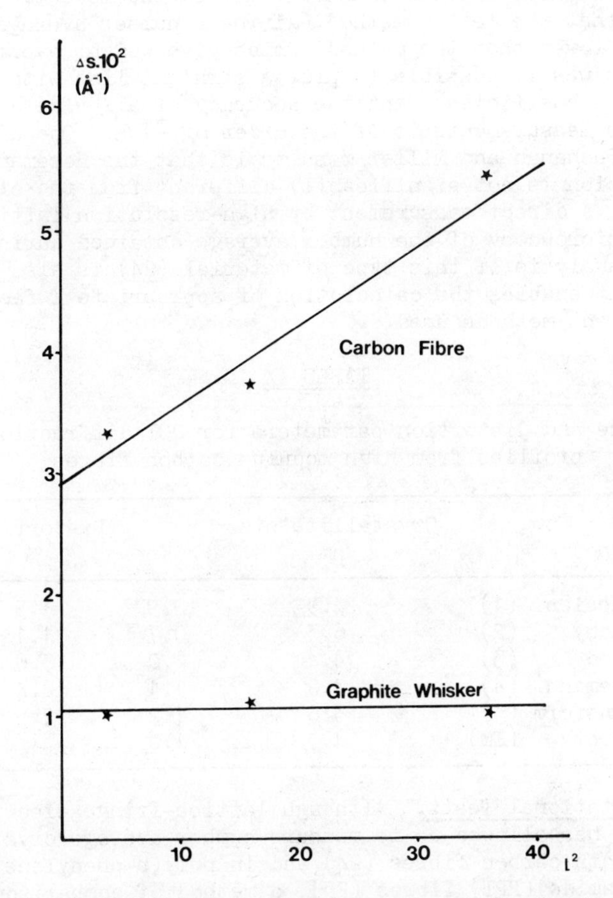

Figure 15. Size and distortion separation. Integral breadth Δs against l^2 for the 001 reflections of carbon fiber and graphite whisker. Electron diffraction traces with 001 reflections resolved from hkO reflections.

A comparison of size and distortion found by X-ray diffraction methods is most valid in the context of this paper, and Table IX shows results obtained recently for a high-modulus PAN-based carbon fibre heat treated at 2500°C. The differences in the crystallite size values are in line with those found for polystyrene by Buchanan and Miller, thus the Cauchy method gives the highest value and the transform method the lowest. It is expected that the latter method, giving a number average, will indeed be lower than the methods which give weight averages. In no case was it possible to plot a straight line with a high correlation coefficient, and the accuracy of all size and distortion measurements is of the order of ±10%. One difference from the Buchanan and Miller results is that the Hosemann measure of distortion is not significantly different from the other measures. A direct measurement by high-resolution lattice-fringe electron microscopy of the number average obtained during a thorough analysis of this type of material (24) is also included; this result enables the calculation of appropriate K factors for the different methods used.

TABLE IX

Size and distortion parameters for 001 diffraction profiles from high modulus carbon fibres

Method	Crystallite size nm	K	Distortion %
Gaussian (1)	4.9	0.9	1.5
Cauchy (2)	6.3	0.7	1.1
Mixed (3)	4.4	1.0	3.6
Hosemann (4)	4.2	1.1	2.4
Transform (5)	3.8	1.2	2.5
Direct (EM)	4.5	1.0	-

Computational Tests. Although lattice-fringe electron microscopy has allowed us to measure number average crystallite sizes in both carbon fibres (24) and in poly(p-phenylene terephthalamide)(PPT) fibres (25), a method of comparison entirely free of all measurement bias can only be obtained by computing intensity profiles and carrying out the various analyses on them. Full details of a computational investigation to carry out this exercise will be reported elsewhere; here it is sufficient to report that a slightly skewed crystallite size distribution based on a number average size of 4.0 nm, with 501 crystallites and 5007 layer planes, and with an r.m.s. distortion of 5%, gave the three orders of reflection shown in Figure 16, which is best considered as a simulated profile of a carbon fibre diffraction pattern with 002,004, and 006

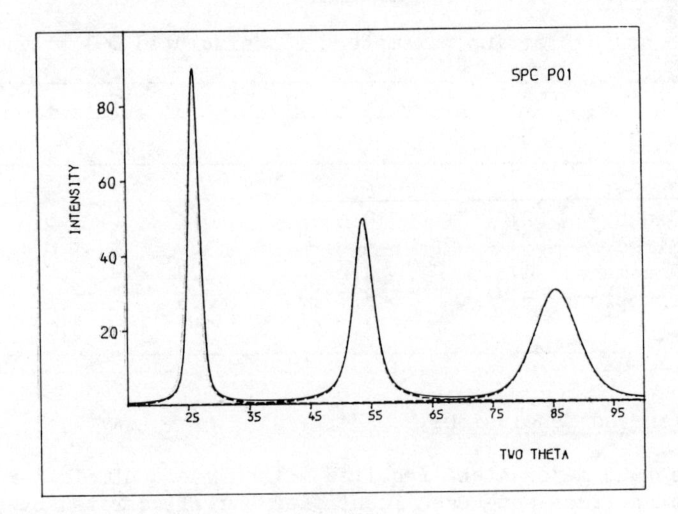

Figure 16. Computed intensity profiles for the first three orders of a 001 reflection for 501 crystallites with 5007 layer planes of mean spacing 0.4 nm, the number average crystallite size being 4.0 nm and the r.m.s. distortion 5%

reflections only.

The results of applying all of the non-transform and the transform test to the three profiles after application of a profile resolution step, are given in Table X. Again correlation in terms of straightline plots was poor and no method gave an adequate solution for both crystallite size and distortion. The mixed method was best for size, and the Hosemann method best for distortion. For a similar crystallite size distribution without distortion, the Fourier transform method gave a crystallite size of 9.2 nm, more than double the true value. The Scherrer K values are again given as the ratio of true to apparent crystallite size.

TABLE X

Size and distortion parameters for simulated 001 profiles

Method	Crystallite size nm	K	Distortion %
Gaussian (1)	6.2	0.6	1.6
Cauchy (2)	10.5	0.4	3.6
Mixed (3)	3.8	1.0	7.0
Hosemann (4)	4.4	0.9	5.5
Transform (5)	4.7	0.9	2.2
True	4.0	1.0	5.0

Discussion and Conclusions

The four major steps required to analyse a simple X-ray diffraction trace have been identified and illustrated by appropriate examples. Correction and normalisation of an X-ray diffraction trace are best carried out over as wide a range of two theta as possible with the normal geometrical corrections for fibres.

Peak resolution is usually easier if well chosen background parameters are input and if constrained optimization methods are utilised. Misleading results can be obtained if the constraints are too limited and tests with unconstrained optimization are desirable if at all possible. In particular, the possible presence of paracrystalline or intermediate phase peaks must be tested with extreme care in order to avoid ambiguity. It is not sufficient to have a good mathematical resolution alone, all peaks must be significant in crystallographic or structural terms. The incidental measurement of peak-area crystallinity is considered to be of secondary importance to the resolution of overlapping peaks.

If the Jones deconvolution approximation is used to correct

for instrumental broadening, it should include the profile function parameter otherwise the error involved can be substantial (18); in general, the Stokes method is preferred.

Both transform and non-transform methods for separating size and distortion broadening effects have been found to be somewhat unsatisfactory, leading to errors of 15-20% in estimates of apparent crystallite size and lattice distortion. We have attempted to find meaningful Scherrer K parameters to correct the estimates of apparent crystallite size obtained by various methods in order to provide true crystallite size. To this end, we have utilised both lattice-fringe images of crystallites and computer simulations, see Tables IX and X. Although the two sets of parameters do not match as well as was hoped, they are reasonably close when considering the limitations of the methods which probably evaluate K to ± 0.2 (20%). If true crystallite sizes are found in this way they will in fact be number averages, even if the apparent crystallite size was a weight average.

Unfortunately another factor complicates the measurement of true crystallite size. The K parameter is also a function of both crystallite size and of lattice distortion. We recently studied the effect of crystallite size and distortion on the K parameter using optical transform methods with simulated lattice images drawn by a computational method (26). The Scherrer K parameters given in Table X may be used to obtain a true number average crystallite size by any of the methods quoted, but will only be valid for crystallites with a number average size in the range 8 to 15 layer planes and a lattice distortion of 4-6%. The mixed function method (3) appears to give the best estimate of true crystallite size ($K = 1.0$) and the Hosemann method (4) the best estimate of lattice distortion.

Methods for estimating lattice distortion generally require two or more orders of a particular reflection to be present, and most polymers have only one order available. A method for estimating both crystallinity and lattice disorder, which does not need higher orders of a reflection, and indeed takes into account the whole of the diffraction trace, is that due to Ruland (27). This method has been applied to many different fibres by Sotton and his colleagues, who have discussed their results both here (28) and elsewhere (12). The major problem with Ruland's method is that an arbitrary separation of the crystalline scatter from the non-crystalline scatter must be made; other restrictions are that the method cannot be used to measure crystallite size and cannot give any indication of the presence of paracrystalline or intermediate-phase material.

New methods of computational analysis will be sought in order to provide simultaneous measures of crystallinity, crystallite-size and lattice distortion in fibrous polymers from X-ray diffraction patterns, but these must pass the test of application to fibres which can be partly characterized by means of electron-microscopy.

Acknowledgements

We would like to acknowledge the experimental and computational work done by A.G. Agaoglu, S.C. Bennett, C. Oates, and A. Wilson, together with the financial assistance provided by the Science Research Council, U.K., the University of Jordan, and the Jordan Science Research Council.

Literature Cited

1. Alexander, L.E., "X-ray Diffraction Methods in Polymer Science"; Wiley Interscience, London, New York, 1969; p.32.
2. Vainshtein, B.K., "Diffraction of X-rays by chain molecules"; Elsevier, London, 1966; p.178.
3. Hindeleh, A.M. and Johnson, D.J. Polymer, 1972, 13, 27.
4. Lee, J.D. and Pakes, H.W. Acta Cryst, 1969, A25, 712.
5. Cella, R.J., Lee, B. and Hughes, R.E. Acta Cryst., 1970, A26, 118.
6. Hindeleh, A.M. and Johnson, D.J. J.Phys.D: Appl.Phys., 1971, 4, 259.
7. Hindeleh, A.M. and Johnson, D.J. Polymer, 1974, 15, 697.
8. Powell, M.J.D. Comput.J., 1964, 7, 155.
9. Bennett, S.C., Johnson, D.J. and Montague, P.E. "Proc. Fourth London Int. Carbon and Graphite Conf. 1974"; Soc.Chem.Ind., London, 1976, p.503.
10. Huevel, H.M., Huisman, R. and Lind, K.C.J.B. J.Polym.Sci., (Polym.Phys.Edn.), 1976, 14, 921.
11. Hindeleh, A.M. and Johnson, D.J. Polymer, 1978, 19, 27.
12. Sotton, M. Arniaud, A.M. and Rabourdin, G. J.Appl.Polym.Sci., 1978, 22, 2585.
13. Agaoglu, A.Y. M.Sc. Thesis, University of Leeds, 1978.
14. Montague, P.E. M.Phil. Thesis, University of Leeds, 1975.
15. Jones, F.W. Proc.Roy.Soc., 1938, A116, 16.
16. Stokes, A.R. Proc.Roy.Soc., 1948, A61, 382.
17. Buchanan, D.R. and Miller, R.L. J.Appl.Phys., 1966, 37, 4003.
18. Hindeleh, A.M. and Johnson, D.J. Polymer, 1972, 13, 423.
19. Thrower, P.A. and Nagle, D.C. Carbon, 1973, 11, 663.
20. Ergun, S. Phys.Rev., 1970, B1, 3371.
21. Warren, B.E. and Averbach, B.L. J.Appl.Phys., 1950, 21, 595.
22. Buchanan, D.R., McCullough, R.L. and Miller, R.L. Acta Cryst., 1966, 20, 922.
23. Bennett, S.C. and Johnson, D.J. Carbon, 1979, 17, 25.
24. Bennett, S.C., Johnson, D.J. and Murray, R. Carbon, 1976, 14, 117.
25. Dobb, M.G., Johnson, D.J. and Saville, B.P. J.Polym.Sci., 1977, 58, 237.
26. Hindeleh, A.M. and Johnson, D.J. Polymer, accepted for publication.
27. Ruland, W. Acta Cryst., 1961, 14, 1180.
28. Sotton, M. This symposium.

RECEIVED June 30, 1980.

10

Diffraction from Nonperiodic Structures

The Molecular Conformation of Polytetrafluoroethylene (Phase II)

EDWARD S. CLARK[1], J. J. WEEKS, and R. K. EBY

Polymer Science and Standards Division, National Bureau of Standards, Washington, D.C. 20234

The determination of crystal structure in synthetic polymers is often made difficult by the lack of resolution in the diffraction data. The diffuseness of the reflections observed in most x-ray fiber patterns results from the small size and imperfect lattice nature of the polymer crystallites. Resolution of individual reflections is also made difficult from misorientation of the crystallites about the fiber axis. This lack of resolution leads to poor accuracy in measurement of peak positions. In particular, this lack of accuracy makes determination of layer line heights difficult with a corresponding loss of significant figures in evaluation of the repeat distance for the molecular conformation. In the case of helical conformations, the repeat distance may be of considerable length or, as we shall show, indeterminate and, in effect, nonperiodic. This evaluation requires high accuracy in measurements of layer line heights.

In this paper we examine electron diffraction fiber patterns of the homopolymer polytetrafluoroethylene $(-CF_2-CF_2-)_n$, PTFE, in which the resolution is sufficient to yield much more accurate values of layer line heights than were available from the previous x-ray diffraction experiments (1) on the crystal structure of Phase II, the phase below the 19°C transition (2). On the basis of x-ray data, the molecule was assigned the conformation 13/6 or thirteen CF_2 motifs regularly spaced along six turns of the helix. This is equivalent to a 13_2 screw axis. The relationship between the molecular conformation and the helical symmetry has been studied by Clark and Muus (3) and is illustrated in Figure 1. The electron diffraction data of high resolution enabled us to determine if this unusual 13-fold symmetry was exact or an approximation of the true symmetry. We have also

[1]Polymer Engineering, The University of Tennessee, Knoxville, TN 37916.

0-8412-0589-2/80/47-141-183$05.00/0

developed mathematical expressions to simplify interpretation of these data.

Theoretical

A convenient method for defining helical symmetry and calculating the distribution of intensity in a fiber pattern was devised by Cochran, Crick and Vand (CCV) (4). As indicated in Figure 1, the molecular conformation is treated as a regular series of diffraction units uniformly spaced along a helix of pitch, P, with axial separation, s. In PTFE, one helix defines the carbon positions; two helices define fluorine positions. If there is a meaningful translational identity, c, it follows that P/s will be the ratio of small whole numbers:

$$r^* = u^*/t^* = P^*/s^* \qquad [1]$$

for a commensurable helix having t turns in u motif units. The asterisk indicates a commensurable helix. The identity period is:

$$c = u^*s^* = t^*P^*. \qquad [2]$$

As shown by CCV, the fiber pattern can be interpreted in terms of Bessel functions governing the intensity distribution in each layer line according to the selection rule:

$$\ell = t^*n + u^*m \qquad [3]$$

where ℓ is the layer line number, n is the order of Bessel function and m is any integer (+, -, or 0). With translational identity, the layer lines will be confined to heights:

$$\zeta_\ell = \frac{\ell}{c} = \frac{\ell}{u^*s^*} \qquad [4]$$

with uniform spacings between the layer lines in reciprocal space:

$$\Delta\zeta = \frac{1}{t^*P^*} = \frac{1}{u^*s^*} \qquad [5]$$

The structure factor governing the distribution of intensity in a layer line for a commensurable helix may be approximated closely by the cylindrically symmetrical transform of a helical molecule with atoms at cylindrical coordinates, (r_j, ϕ_j, z_j) for the asymmetric unit repeating along the helix axis in accordance with equation [1]. The cylindrically averaged intensity function has been given in convenient from by Davies and Rich (5):

$$\overline{F}^2_{\Psi}(R,\ell/c) = \sum_n (A_n^2 + B_n^2)$$

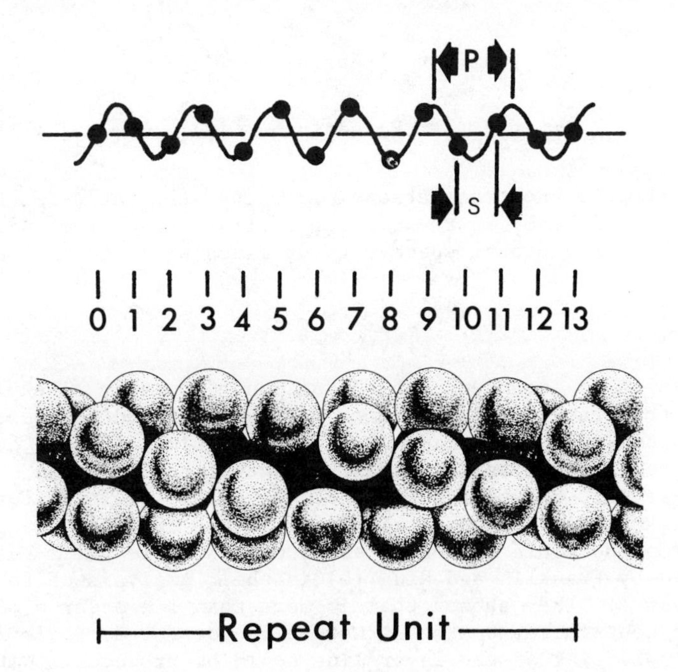

Figure 1. Conformation of PTFE molecule showing 13/6 helical symmetry

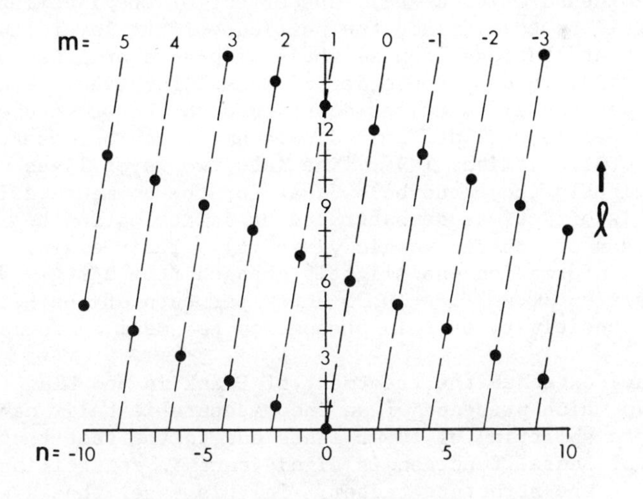

Figure 2. Selection rule for 13/6 commensurable helix

where

$$A_n = \sum_j f_j J_n(2\pi r_j R)\cos[n(\frac{\pi}{2} - \phi_j) - \frac{2\pi\ell}{c} z_j]$$
$$B_n = \sum_j f_j J_n(2\pi r_j R)\sin[n(\frac{\pi}{2} - \phi_j) - \frac{2\pi\ell}{c} z_j] \ . \qquad [6]$$

If there is no translational identity, the ratio P/s will be incommensurable and, in theory, the diffraction pattern will fill the whole of reciprocal space. Layer lines will occur at heights of:

$$\zeta_\ell = \frac{n}{P} + \frac{m}{s} \qquad [7]$$

where P/s is an incommensurable fraction. However, as explained by CCV, the transform, in effect, will be confined to those layer lines of non-uniform spacing close to the values for the commensurable helix which approximates the actual conformation. This is because only Bessel functions of low order, say $n<6$ (for small $2\pi r_j R$) make significant contributions to the intensity calculations. One possible effect of an incommensurable helix has been discussed by Franklin and Klug (6) in their analysis of tobacco mosaic virus. They showed that if more than one order of Bessel function contributes to the structure factor for a single layer line, a splitting of the layer line could be produced. Within a nominal layer line, the contributions from each Bessel function can be considered to arise from a slightly different origin on the meridian in reciprocal space. If the helix is incommensurable but close in conformation to the ratio of small whole numbers (commensurable, u*/t*), the effect of the incommensurability will be to displace the portions of the layer line receiving contributions from slightly displaced origins on the meridian as governed by the order of Bessel function. Thus, if two low order Bessel functions contribute to the structure factor of the nominal layer line, this displacement can be observed as a splitting of the nominal layer line into two layer lines with heights slightly above and below that for the commensurable helix. This effect was demonstrated by splitting in the x-ray fiber pattern of tobacco mosaic virus (6). Their example, in which the conformation was slightly changed from u*/t* = 31/3 = 10.33 to u/t = (31.05)/3 = 10.35 units per turn of the helix, shows the sensitivity of this phenomenon to small conformational changes.

We have extended the treatment of Franklin and Klug (6) to the case in which presence of an incommensurable helix cannot give rise to splitting of layer lines due to the fact that only one order of Bessel function is significant in controlling the amplitude of the structure factor. In this case, the layer line will be displaced slightly from the height calculated for the commensurable helix. Our example is the crystal structure of the low temperature form of polytetrafluoroethylene. The conformation of the helical molecule was determined by Bunn and Howells (1) to be $u^*/t^* = 13/6 = 2.1667$. The graphical solution of the selection rule, equation [3] is given in Figure 2. As shown by Franklin and Klug, a modification of this plot may be used to indicate the general nature of the intensity distribution in a fiber pattern as illustrated in the right hand portion of Figure 3 for the commensurable 13/6 helix. In this ℓ vs n plot, the relative positions of the spots approximate the relative positions of the Bessel function maxima controlling the intensity distribution in the layer line. Thus, the ℓ vs n plot is a rough presentation of the general appearance of the diffraction fiber pattern. For our purposes, it is important to note that a commensurable helix, $r^* = u^*/t^*$, will produce layer lines equally spaced along the reciprocal lattice meridional coordinate, ζ. If the helical conformation is slightly changed from the commensurable 13/6 = 2.167, in this case untwisted by 8° per 13 units to r = 2.159, the layer lines undergo small displacements in the ζ direction as indicated by the left hand portion of Figure 3; the layer lines are no longer equally spaced.

We have developed an expression for the displacement of the layer lines, $\Delta\zeta_\ell$, in terms of the difference in conformation between the actual incommensurable helical conformation and the conformation of a commensurable helix closely approximating this conformation using the definitions described above:

$$\begin{aligned} r^* &= u^*/t^* \quad \text{(commensurable helix)} \\ r &= u/t \quad \text{(incommensurable helix)} \end{aligned} \qquad [8]$$

As described by CCV, this displacement may be expressed:

$$\Delta\zeta_\ell = n\Delta(1/P^*) + m\Delta(1/s^*) \qquad [9]$$

where m and n satisfy the selection rule for the commensurable helix. This expression may be modified in terms of the heights of the layer line, ζ_ℓ, for the commensurable and incommensurable conformation:

$$\zeta^*_\ell = \frac{\ell}{t^*P^*} \;; \quad \zeta_\ell = \frac{n}{P} + \frac{m}{s} \qquad [10]$$

leading to another useful expression for layer line displacement:

$$\Delta\zeta_\ell = \frac{\ell}{u^*s^*} - \frac{n}{rs} - \frac{m}{s} \qquad [11]$$

It should be noted that the solution of this equation requires a separate determination of the value of the spacing, s, between consecutive motifs along the molecular axis.

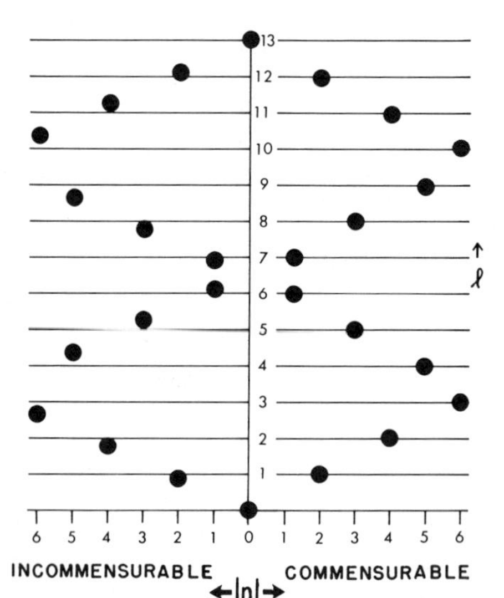

Figure 3. Plot of |n| vs. l for PTFE helix

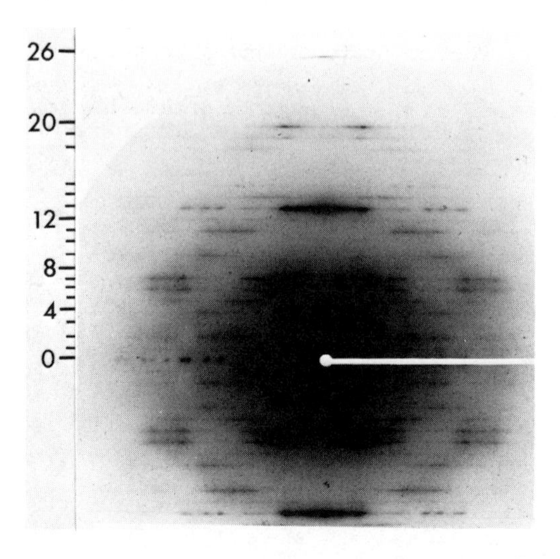

Figure 4. Electron diffraction pattern of polytetrafluoroethylene

Experimental

Fibrous crystals of polytetrafluoroethylene of essentially 100% crystallinity and nearly perfect c-axis orientation were prepared for electron diffraction by the following technique: Several grams of virgin "Teflon" tetrafluoroethylene molding powder (DuPont) (7), were compressed in a pill press at about 30 MPa to form a disk about 3 mm thick. The exact pressure is not critical and trial and error was involved in securing a suitably compressed disk. The disk was cut with a razor blade to form a square pillar about 20 mm long. The pillar was notched with the razor blade and fractured by application of gentle elongational force. Observation of the space between the broken ends of the pillar under an intense light beam revealed what appeared to be bluish "smoke" which was evidently light scattering from an enormous number of fibrils smaller than the wavelength of light, say about 20 nm. A special sample holder of copper (to conduct electric charge) was constructed to hold the pillar with its fibrils within the large sample chamber of an RCA electron diffraction (only) unit (manufactured in late 1940's). The sample to film distance was 500 mm. Since PTFE exhibits hysteresis effects in the transformation of Phase IV (25°C) to Phase II (below 19°C), the copper sample holder was placed for several hours in a container of dry ice. The cold sample and the copper holder were then placed into the diffraction unit which was pumped down rapidly. The diffraction patterns were photographed before the transition to Phase IV could occur. No provision was available for measurement of sample temperature. Many experiments were necessary to obtain patterns suitable for analysis. In particular, it was difficult to get the fiber axes exactly perpendicular to the diffraction beam.

A typical electron diffraction pattern is shown in Figure 4. The plate selected for analysis showed minimal distortion of the pattern from fiber tilt. The pattern was corrected for this distortion as well as for the minor distortions from the slight curvature of the Ewald sphere. The pattern was calibrated for d-values using a comparison of the equatorial reflections in the electron diffraction pattern with the d-values determined by Clark and Muus (3) from careful x-ray diffraction measurements at 0°C. The d-values determined by this method were in excellent agreement with the wavelength of 5.7 pm (picometers) calculated for the 45 kV wavelength. The layer line heights, ζ_ℓ, measured from the electron diffracion plate are given in Table 1. The value of the CF_2 motif spacing along the c-axis, s, needed for solution of equation [11] was determined from x-ray measurements of the 0,0,13 reflection position at 0°C. This value is s = 0.1299 nm ± 0.0001. For each layer line, (nominal ℓ), the value of $\Delta\zeta_\ell$ was calculated from the difference between ζ^* calculated for the commensurable helix:

TABLE 1

EXPERIMENTAL DATA AND CALCULATIONS

ℓ	n	m	ζ(meas) (nm^{-1})	ζ^*(calc) (nm^{-1})	$\Delta\zeta_\ell$ (nm^{-1})	r	$\Delta\zeta_\ell$(calc) (nm^{-1})
0	0	0	0	0	0	-	0
1	-2	+1	0.570	0.592	+0.022	2.1597	+0.024
2	-4	+2	1.138	1.184	+0.046	2.1596	+0.047
3	-6,+7	+3,-3	-	-	-	-	-
4	+5	-2	2.424	2.369	-0.055	2.1599	-0.059
5	+3	-1	2.995	2.961	-0.034	2.1596	-0.035
6	+1	0	3.565	3.553	-0.012	2.1588	-0.012
7	-1	+1	4.131	4.145	+0.014	2.1581	+0.012
8	-3	+2	4.705	4.737	+0.032	2.1600	+0.035
9	-5	+3	-	-	-	-	-
10	-7,+6	+4,-2	-	-	-	-	-
11	+4	-1	6.560	6.514	-0.046	2.1596	-0.047
12	+2	0	7.128	7.106	-0.022	2.1597	-0.024
13	0	+1	7.698	7.698	0.000	2.1591	0.000
14	-2	+2	8.269	8.290	+0.021	2.1600	+0.024
15	-4	+3	8.835	8.883	+0.048	2.1593	+0.047
					avg. =	2.1594 ±	0.0006

TABLE 2

MOLECULAR CONFORMATION OF POLYTETRAFLUOROETHYLENE

Example	r	u*/t*	c(nm)
Phase II (Table I)	2.1594	948/439[a]	123.2
Phase II (+1σ)	2.160	54/25	7.015
Phase II (Ref. 1)	2.1667	13/6	1.69
Phase IV (Ref. 3)	2.1429	15/7	1.95

a. Best fit for t*<500

$$\zeta^*_\ell = \frac{\ell}{13s^*} = \frac{\ell}{13 \times 0.1299} \qquad [12]$$

and the value of ζ measured on the film, ζ_{meas}:

$$\Delta\zeta_\ell = \frac{\ell}{1.689} - \zeta_{meas} \qquad [13]$$

These results are given in Table 1. Equation [11] was solved for each layer line, assuming s = s*, to determine the conformation value, r. The averaged value was r = 2.1594, significant to 2.159. In addition, the value of $\Delta\zeta_\ell$ was calculated using this r value and is given in Table 1.

Results and Discussion

The conformation of the polytetrafluoroethylene molecule in the low temperature form (Phase II) has been determined to be 2.159 CF_2 units per turn of the helix within the limits of experimental error. This conformation is slightly untwisted from the previously assigned 13/6 = 2.167 value but is substantially different from that for the 25°C form (Phase IV) in which the conformation is 15/7 = 2.143. By comparison, the planar zig-zag is 2/1 = 2.000.

In conclusion, we wish to comment on the practice of expressing a helical conformation in terms of some commensurable ratio, u*/t*, and its corresponding repeat distance, c. If the helical conformation is not expressed accurately in terms of simple <u>small</u> numbers, we feel it is preferable to define the conformation in terms of the ratio r = P/s, limited to the number of significant figures. Our point is made tellingly in Table 2 where a change of one standard deviation in r changes u*/t* from 948/439 to 54/25. Furthermore, there are 82 other possibilities for t*<500.

Acknowledgements

We wish to thank E. I. DuPont de Nemours & Co., Inc. for their generous release of the electron diffraction data recorded by one of us (ESC) in 1957 with the assistance of Dr. O. E. Schupp, Jr. The helpful assistance of Dr. Howard Starkweather and Dr. Kenn Gardner of the Central Research and Development Department of DuPont is gratefully acknowledged.

Abstract

Remarkable electron diffraction patterns of the low temperature form (Phase II) of polytetrafluoroethylene have been obtained which exhibit layer lines sharply resolved to the 26th order. These patterns permit accurate measurements of the layer line heights. Equations are developed to relate the layer line heights to a molecular conformation defined in terms of an incommensurable helix of ratio, r = u/t, defining a conformation of u motifs regularly spaced along t turns of the helix. The helix parameter, r, is not the ratio of small whole numbers for an incommensurable helix and the repeat period, c, becomes a crystallographically useless large number. Solution of the equations for the electron diffraction data yields a molecular conformation of r = u/t = 2.159 for Phase II, with the molecule slightly untwisted from the previously assigned conformation of r = 13/6 = 2.167.

Literature Cited

1. Bunn, C. W. and Howells, E. R. Nature (London), 1954, 174, 549.
2. Hirakawa, S. and Takemura, T. Japan J. Appl. Phys., 1969, 8, 635.
3. Clark, E. S. and Muus, L. T. Z. Krist., 1962, 117, 108.
4. Cochran, W., Crick, F. H. C., and Vand, V. Acta Cryst., 1952, 5, 581.
5. Davies, D. R. and Rich, A. Acta Cryst., 1959, 12, 97.
6. Franklin, R. and Klug, A. Acta Cryst., 1955, 8, 777.
7. Certain commercial materials and equipment are identified in this paper in order to specify adequately the experimental procedure. In no case does such identification imply recommendations or endorsement by the National Bureau of Standards, nor does it imply necessarily the best available for this purpose.

RECEIVED May 29, 1980.

11

Crystallinity and Disorder in Textile Fibers

MICHEL SOTTON

Laboratory of the French Textile Institute,
35 rue des Abondances, 92100 Boulogne sur Seine, France

Crystallinity and disorder are important structural parameters for understanding relationships between structure and physical properties. Flaws and distortions are the main features that limit the ultimate properties of textile fibers. Some of these crazes, cracks and voids are revealed under the electron microscope, either on the surface or in cross sections stained with heavy metals (1,2). However, these staining techniques (that reveal the main morphological features) make it much more difficult to determine the degree of distortion of the crystalline fraction. Theoretically, line profile studies permit separation of effects due to crystalline size from those due to structural distortions. However, the lack of peaks in semicrystalline fiber x-ray patterns hinders that approach.

Nevertheless, when we carry out x-ray crystallinity measurements on textile fibers, we must consider distortions that always affect crystalline material. Even in a completely crystalline material, the scattered x-ray intensity is not located exclusively in the diffraction peaks. That is because the atoms move away from their ideal positions, owing to thermal motion and distortions. Therefore, some of scattered x-rays are distributed over reciprocal space. Because of this distribution, determinations of crystallinity that separate crystalline peaks and background lead to an underestimation of the crystalline fraction of the polymer. In this paper, we attempt to calculate the real crystallinity for textile fibers from apparent values measured on the x-ray pattern. This is done by taking into account the factor of disorder following Ruland's method (3).

Theoretical Review-Ruland's Method

The basic equations proposed by Ruland for the calculation of the crystallinity of polymers are :

0-8412-0589-2/80/47-141-193$05.25/0

$$x_c = \frac{\int_0^\infty s^2 I_{cr(s)} ds}{\int_0^\infty s^2 I_{(s)} ds} \cdot \frac{\int_0^\infty s^2 \bar{f}^2 ds}{\int_0^\infty s^2 \bar{f}^2 D\, ds}$$

where x_c = weight fraction of the crystalline material in the polymer ; $s = 2 \sin \theta / \lambda$, magnitude of the radial vector s in the reciprocal space (2θ = diffraction angle, λ = wavelengh expressed in Å) ; $I_{(s)}$ = coherent scattered intensity ; $I_{cr(s)}$ = part of the coherent scattering which is concentrated into the crystalline peaks ; $\bar{f}^2$ = mean square of the scattering factors of the atoms in the polymer ; and D = disorder function.

This method uses the factor K for the "apparent" crystallinity, which is, itself, a function of the disorder parameter, D,

$$K = \frac{\int_0^\infty s^2 \bar{f}^2 ds}{\int_0^\infty s^2 \bar{f}^2 D\, ds}$$

D includes the disorder resulting from both thermal motion and lattice imperfections. Ruland showed that these two kinds of disorder could be represented approximately as one and the same function

$$D = \exp^{-ks^2}$$

under the assumption of isotropic disorder ; that is, the average atom moves away from its ideal position in all directions.

For a given polymer that has variable amounts of crystallinity, the scattered intensity over a large range of reciprocal space may be integrated over a number of intervals $s_o - s_p$. Such intervals can be defined experimentally in such a way that the following equation be verified independently from the crystallinity of the material

$$\int_{s_o}^{s_p} s^2 I_{(s)} ds \simeq \int_{s_o}^{s_p} s^2 \bar{f}^2 ds$$

s_o and s_p being the lower and upper limits of integration. On these angular intervals, equations can be used for any samples, under the form

$$x_c = \frac{\int_{s_o}^{s_p} s^2 I_{cr(s)} ds}{\int_{s_o}^{s_p} s^2 I(s) ds} K(s_o, s_p, D, \bar{f}^2) = C^{st}$$

$$K = \frac{\int_{s_o}^{s_p} s^2 \bar{f}^2 ds}{\int_{s_o}^{s_p} s^2 \bar{f}^2 \exp^{-ks^2} ds}$$

The integration intervals having been defined, this system of equations can be solved by calculating the nomogram of K values that maintain as a constant the crystallinity, x_c for a given function of disorder D.

We have used this approach for several textile fibers, but before showing results, the experimental conditions will be described.

Experimental

Sample Preparation. Fiber crystallinity studies require the preliminary making of a "global sample" in which all preferential orientation has been removed. The general way is to get a "powder" made of fiber cross sections. Our cross sections have been cut with an automatic microtome (4) made especially for this use, allowing us to make regular cuttings. The lenght of the cross sections can be adjusted from 20 μm to 200 μm, and we generally choose a length of 80 μm for the different textile fibers. Cross sections are then sifted and pelleted inside sample holders that are of different sizes, depending on whether the transmission or reflection mode is used. In the symmetrical transmission mode, where scans are performed from 7° to 75° 2θ,the sample thickness is less than 1 mm. In the symmetrical reflection mode, with scans from 70° to 130° 2θ, the sample thickness is 3 mm.

Careful work is necessary to remove all preferred orientation from powder samples. Figure 1 shows results obtained with polyethylene terephthalate (PET) fibers. Curve a is a typical azimuthal scan of the 010 peak (2θ = 17,5°) for a bundle of parallel fibers placed perpendicularly to the x-ray beam. Curve b is the same scan carried out on a "powder" sample, showing that all preferred orientation is removed in our conditions of moulding (350 kg/cm2). For each kind of fiber, it is necessary to do preliminary trials to find the best experimental conditions. For PET fibers, we show on Figure 2 the relative crystallinity index and the residual orientation plotted against the cut-lengh. (5).

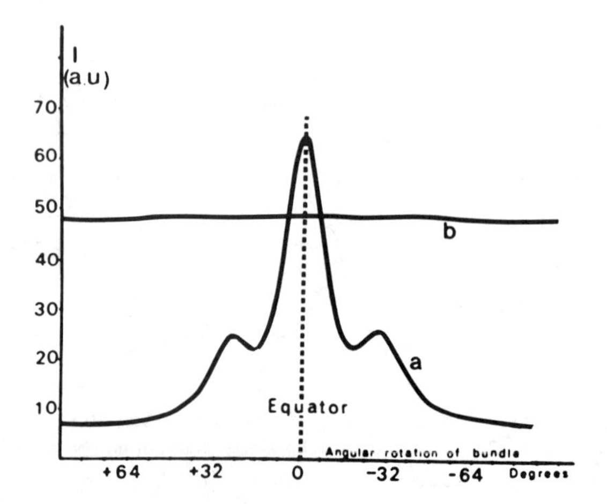

Figure 1. X-ray azimuthal scans of (010) peak of PET fibers.

Curve a: bundle of parallel fibers put perpendicularly to the x-ray beam; Curve b: "powder" made of cut fibers. It can be ascertained that all preferential orientation is practically removed when a powder sample made of regular small cross section is used.

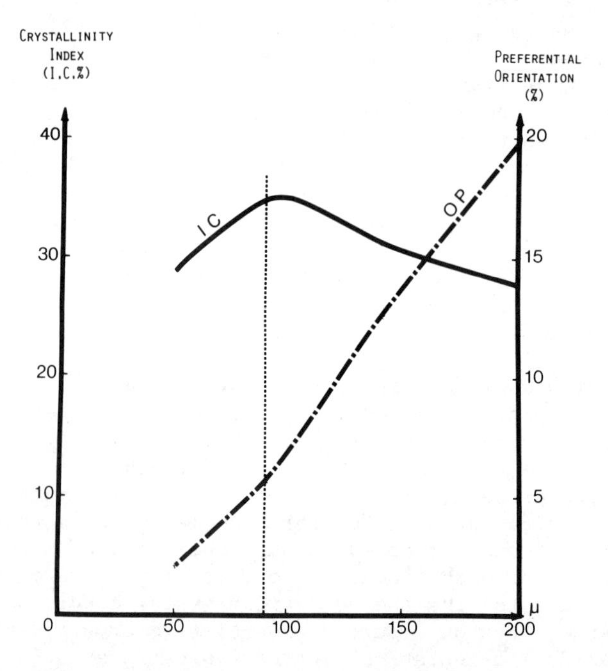

Figure 2. Effect of cross-section length on crystallinity index (——) and on preferential orientation (– – –)

Preferential orientation increases rapidly when the cut-length is greater than 80 μm, and the crystallinity index goes through a maximum at the same cut-length. To explain such results we suggest :

- the weight of the decrystallized material which could appear at the both ends of cuts, becomes greater and greater as the cut-length decreases.
- preferential orientations, noticed for the longest cuts have for effects to reduce the overall crystallinity of the sample (peak extinctions...)

Therefore, 80-μm cut-lengths were used for PET studies. For other more crystalline fibers, experimental conditions seem less critical. Table I shows the crystallinity indices for several cellulosic fibers cut at 30μm and at 80 μm. Those indices do not seem sensibly different in this range of cut-lengths.

Table I

Polynosic Fibers : Effect of cut length		
Samples	Crystallinity Index %	
	Cut length 30 μm	Cut length 80 μm
Polynosic ref. A	68 %	72 %
Polynosic ref. B	62 %	62 %
Polynosic ref. C	58 %	60 %

Treatment of Experimental Values. The experimental values are corrected for air scattering, polarization, but absorbtion - geometric (Lorentz) corrections are not made. After the variable 2θ is transformed into $s = \frac{2 \sin \theta}{\lambda}$, the experimental curves are normalized, in electronic units, by adjustment to a theoretical curve. Theoretical curves (total scattering power, summing up coherent and incoherent scatterings) are calculated from the stoichiometric composition of polymers.

In order to obtain good adjustement between experimental and theoretical curves, we correct for the absorbtion discrepancy between the coherent and incoherent scattering, which becomes larger at wide scattering angles. Because of this discrepancy, corrections are successively made to the ratio $\frac{I\ (incoh)}{I\ (coh)}$ (3) for the absorbtion effects in the sample in reflection mode, the air path, and the Ross filters.

After all corrections are completed, the diagrams of $s^2 I_{(s)}$ vs. s are drawn. Convenient integration intervals are determined for calculating the nomogram of K values. For instance,

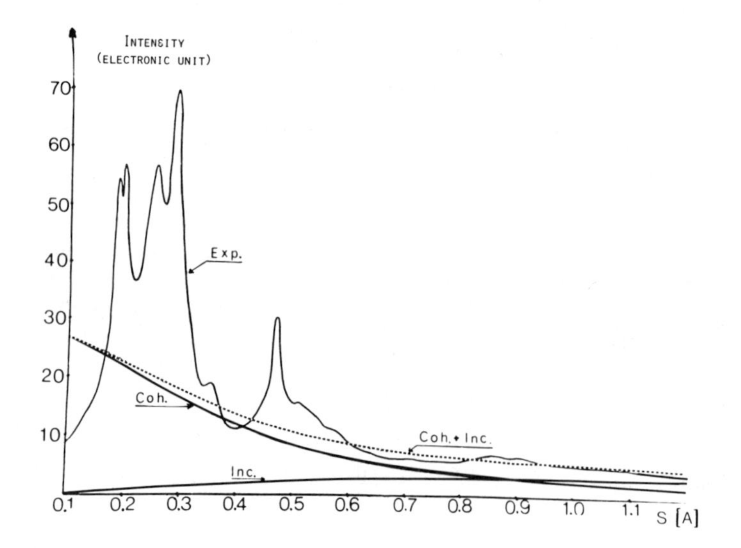

Figure 3. Theoretical curves of scattered intensity by PET and experimental normalized intensity by PET fibers

Figure 4 corresponds to a polypropylene sample and Figure 5 to a well-crystallized PET sample. Four integration intervals chosen for the diagrams of the main textile fibers. For PET fibers, they are :

s_o to s_p = 0.1 to 0.4 0.1 to 1.0
0.1 to 0.67 0.1 to 1.2

It is therefore possible to calculate the K values over these intervals with different k values. The results for PET are shown on Figure 6. It can be ascertained, as Ruland has already shown for polyethylene, that starting from a disorder function with spherical symmetry, the plots of K vs. s_p (the upper limits of integration) can be reduced to a set of straight lines for the different k values.

Finally, before carrying out the calculation, it is necessary to sketch the boundary between the crystalline peaks and the amorphous background. This line can be calculated if an amorphous sample has been used as a reference, such as for PET and cellulose fibers. If no amorphous standards are available, the background is drawn manually, following a line parallel to the theoretical curve (4,5) (total scattering power summing up coherent and incoherent scattering).

Table II shows effects of the disorder parameter on the calculated crystallinity of cotton, nylon 66 (PA 66) and PET fiber samples. When k = 0, no correction for distortion is made during calculation of crystallinity. Accordingly, values of X_c become smaller as the integration intervals increase. On the nomogram of K values, it is possible to determine the disorder parameter value that maintains as approximately constant this crystallinity when using the different intervals : the deviation from the constancy is used, in the computing program to determine the best value of k and to estimate the errors of these analyses. The disorder parameter k is higher in cotton and PA 66 than in all PET fibers. Besides, one can see that apparent crystallinity values (disorder parameter not considered) are lower than the true ones.

Experimental Results

Polyester Fibers. The first example involves two industrial PET yarns. The first yarn has a residual shrinkage of 8 % (Figure 7) and the second yarn has undergone industrial stabilizing heat treatment and has almost no residual shrinkage (0,6 %). We see that the second sample is, of course, more crystalline than the former but that it has a higher disorder parameter (given in brackets). Both samples were annealed for 1 hour at 220°C in a slack state (shrinkage during annealing was allowed).

The sample that shrinks 8 % crystallizes at a level close to that of the industrially stabilized yarn. But simultaneously with

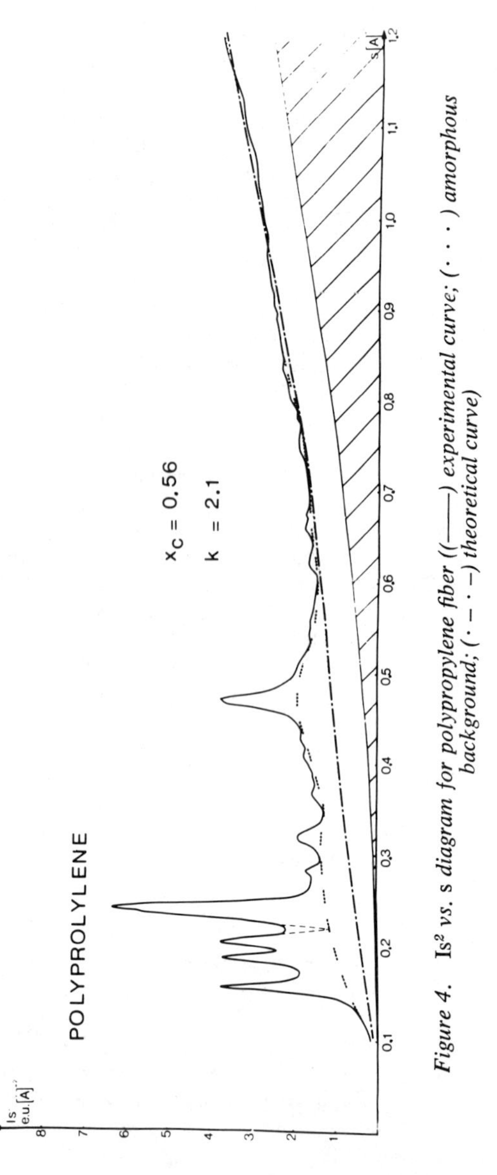

Figure 4. Is^2 *vs.* s *diagram for polypropylene fiber ((——) experimental curve; (· · ·) amorphous background; (· – · –) theoretical curve)*

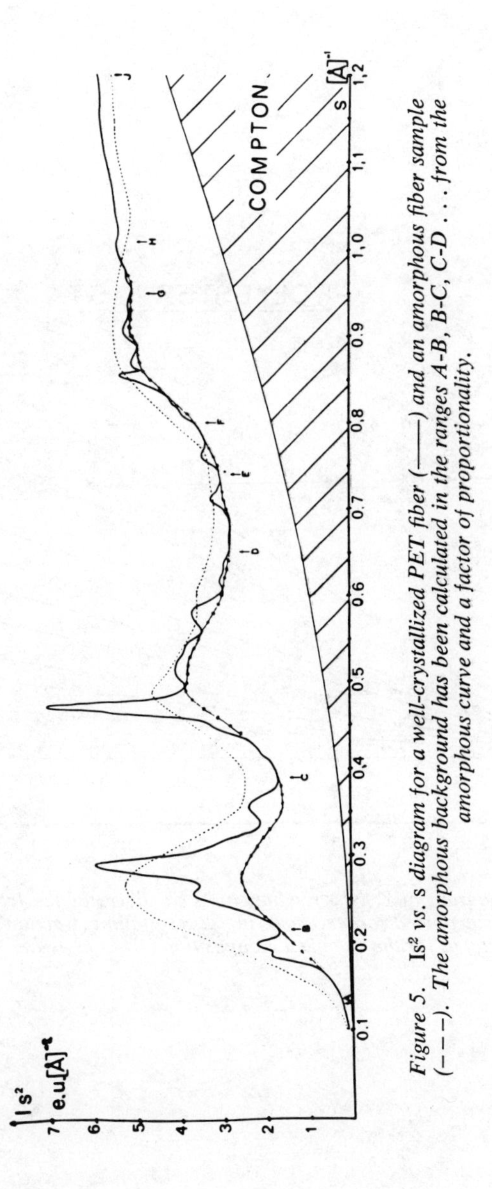

Figure 5. Is² vs. s diagram for a well-crystallized PET fiber (——) and an amorphous fiber sample (– – –). The amorphous background has been calculated in the ranges A-B, B-C, C-D . . . from the amorphous curve and a factor of proportionality.

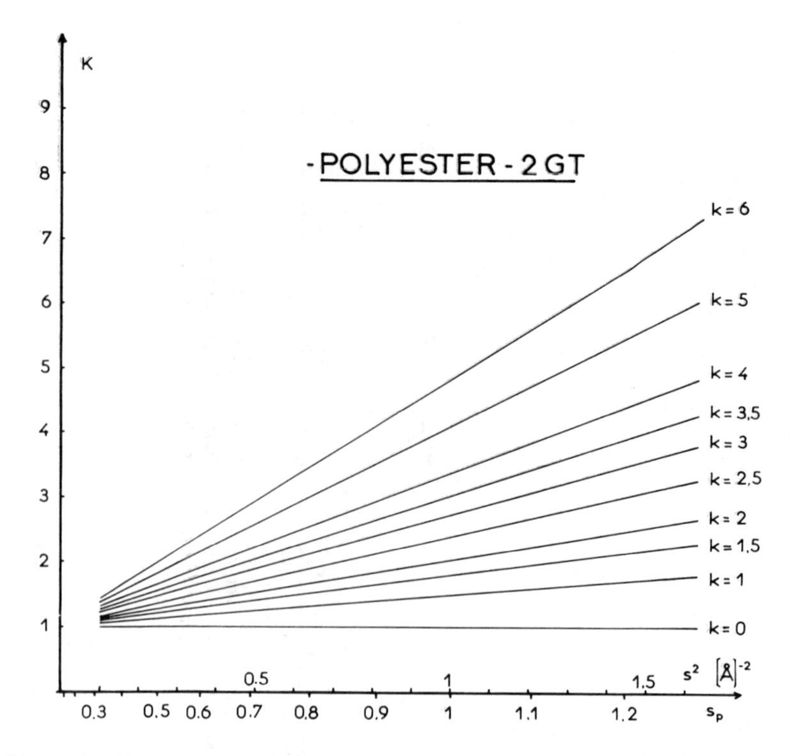

Figure 6. Nomogram of K (s^2_pk) *values for PET fibers in the hypothetical ideal case where an isotropic disorder exists in the crystalline fraction of those fibers and calculated for the chemical composition* $(C_{10}O_4H_8)_n$ *and* $s_o = 0.1$

Table II

Crystalline Fraction X_c as Function of k and Integration Intervals

$s_o - s_p$ Intervals	Cotton Fibers		Polyester Yarn (PET)		Polyamide Yarn (66)	
	A.C. k = 0	T.C. k = 4	A.C. k = 0	T.C. k = 2,5	A.C. k = 0	T.C. k = 3,3
1st : $s_o - s_p$	0,317	0,422	0,204	0,242	0,474	0,586
2nd : $s_o - s_p$	0,184	0,381	0,146	0,232	0,260	0,508
3rd : $s_o - s_p$	0,122	0,401	0,100	0,231	0,191	0,555
4th : $s_o - s_p$	0,101	0,434	0,08	0,239	0,161	0,601
X_c Average		0,41		0,236		0,563
CV %		5,8		2,3		7,3

A.C. : Apparent Crystallinity

T.C. : True Crystallinity

C.V. : Variation Coefficient

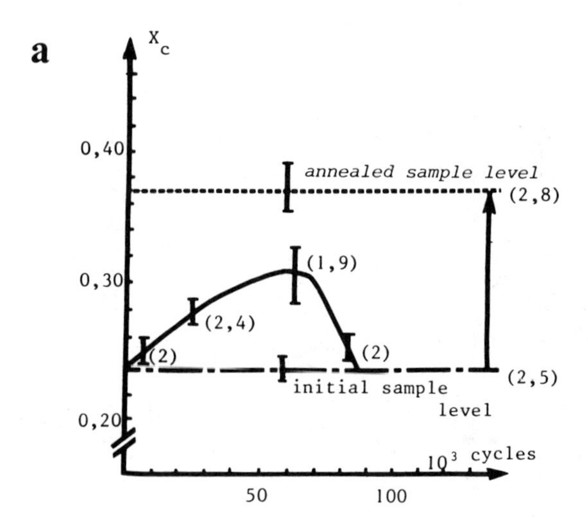

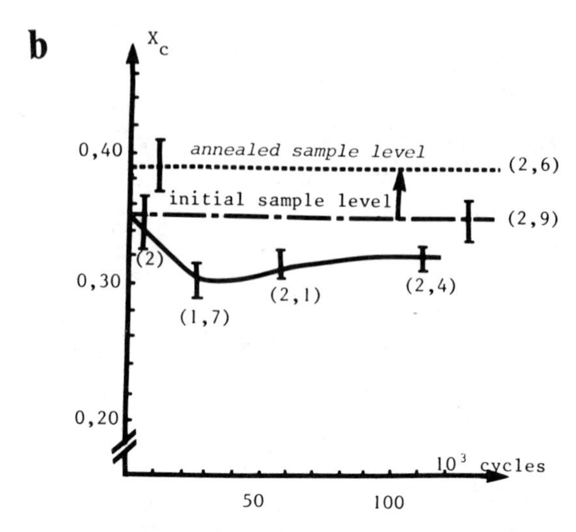

Figure 7. Effect of annealing treatment and mechanical fatigue on crystallinity and disorder for two samples of PET fiber.

(a) PET single filament yarn (industrial sample): 110 tex f 220 S 28 tpm (Tergal 5 dtex), shrinkage capacity 8%; (b) PET single filament yarn (industrial sample): 116 tex f 220 S 32 tpm (Tergal 5,25 dtex), stabilized sample (conditions of stabilization are unknown), residual shrinkage 0.6%. The values of k *appear between brackets. Annealing treatment (1 hr at 220°C) produces an increasing of crystallinity: the range of this change is represented by arrows on the figures for the two samples of PET fibers. Mechanical fatigue produces an increase in crystallinity for the less crystallized sample—8% shrinkage (Figure 7a), and a decrease in crystallinity for the stabilized, well crystallized sample—0.6% shrinkage (Figure 7b). In both cases fatigue produces a decrease in the disorder parameter.*

this neocrystallization, the disorder parameter increases from k = 2.5 to k = 2.8. Also, the annealing treatment does not greatly improve the crystallinity of the stabilized sample but does allow distortion to be removed (k decreases from 2.9 to 2.6).

These two samples have also undergone mechanical fatigue (thousands of cycles of extension in the Hooke's Law zone between two levels of stress). The curves (Figure 7) show changes of crystallinity and changes of the disorder parameter when plotted against the number of cycles. The sample that was initially less crystalline registered an increase in crystallinity and a reduction in distortions after the fatigue treatment. This result is unusual. For most crystalline materials, fatigue reduces the crystalline fraction and the amount of distortion.

Paradoxically, such a mechanical fatigue apparently acts as a treatment for relaxation of stress and allows flaws, located at the stress areas to be partly dissipated. Also, mechanical fatigue treatments can eliminate effects of a previous thermal treatment. Two samples that were initially different became more similar with regard to their crystallinity after 50×10^3 cycles. They had a medium level of crystallinity characterized by disorder parameter values that are particularly low.

Figure 7 also shows that the two samples could become identical after the annealing treatments, this time at a high level of crystallinity and characterized by strong distortions. It seems difficult to remove, by annealing, these distortions (chain foldings) that appear during a thermal treatment with the sample allowed to shrink. In the case of an amorphous sample annealed several hours to get a crystalline standard, real crystallinity was as high as 0.66 but k remained as high as 2.6 (5).

These results are not the only ones that could be registered, and extensive changes occur in the amorphous fraction. For instance, the orientation function of the molecules in the amorphous zone changes strongly after each treatment. The results in Table III and IV will not be discussed here in detail, but simultaneous determinations of the amorphous morphology and crystalline perfection could lead to a better understanding of fiber properties. They could give a better understanding of SAXS patterns, the intensity differences of which are difficult to explain on the basis of changes in crystallinity.

Table V shows results obtained on tetramethylene terephthalate fibers, heat treated at 220°C for different times in a relaxed state. There again, we observe a great improvement in the amount of the crystalline material but also increasing distortions.

Table III

Changes in crystalline and amorphous fractions of PET fibers after annealing treatment and mechanical fatigue.

POLYESTER Yarn - 2 GT - Residual Shrinkage : 0,6 %								
Sample	Crystallinity Parameters						Amorphous	
	X_c	k	V_c 10^3 Å^3	S A X S	L_p Å	f_x	f_{am}	L_a Å
- initial	0,35	2,9	189	R x 3.6	150	0,984	0,790	82
- annealed 220°C	0,39	2,6	217	R x 6,9	143	0,984	0,745	75
- fatigue 4.500 cycles	0,35	2,0	203	R x 2,3	150	0,985	0,831	78
26.000 "	0,30	1,7	216	—	—	0,985	0,840	—
57.000 "	0,32	2,1	203	R x 2,9	150	0,984	0,839	81
110.000 "	0,33	2,4	193	R x 3,6	152	0,982	0,843	87

V_c : Crystal Volume calculated from the 010 - 100 and 105 line-breadth (5)

S A X S : Intensity of the S A X S diagrams, in arbitrary units, and expressed compared to the intensity R of the initial (8% shrinkage) fiber diagram.

L_p : Long Period by S A X S.

f_x : Orientation function of the crystalline fraction (12).

f_{am} : Orientation function of the amorphous fraction calculated following Dumbleton's method (J. Polym. Sci. A_2 Vol. 6 795), 1968.

L_a : Amorphous length.

Table IV

Changes in crystalline and amorphous fractions of PET fibers after annealing treatment and mechanical fatigue.

POLYESTER Yarn - 2 GT - Residual Shrinkage : 8 %								
Sample	Crystallinity Parameters						Amorphous	
	X_c	k	V_c 10^3 Å^3	S A X S	L_p Å	f_x	f_{am}	L_a Å
- initial	0,24	2,5	111	R au	138	0,987	0,832	77
- annealed 220°C	0,37	2,8	181	R x 5,4	131	0,976	0,690	70
- fatigue								
4.500 cycles	0,25	2	109	R x 1,9	137	0,987	0,854	76
25.000 "	0,28	2,4	99	R x 1,5	138	0,985	0,832	81
63.800 "	0,31	1,9	96	—	137	0,985	—	77
82.000 "	0,25	2	119	R x 2	137	0,991	0,854	76

V_c : Crystal volume calculated from the 010 - 100 and 105 line-breadth (5)

S A X S : Intensity of the S A X S diagrams, in arbitrary units, and expressed compared to the intensity R of the initial (8 % shrinkage) fiber diagram.

L_p : Long Period by S A X S.

f_x : Orientation function of the crystalline fraction (12)

f_{am} : Orientation function of the amorphous fraction calculated following Dumbleton's method (J. Polym. Sci. A_2 Vol. 6 795) 1968.

L_a : Amorphous length.

Table V

POLYESTER 4 GT Yarn		
Sample	X_c	k
- initial	0.22	1.9
- annealed 220°C		
shrink : 30 %		
time : 2'	0.43	2.6
15'	0.41	2.8
30'	0.43	2.6

Nylon Fibers. Table VI shows that PA 66 fibers are different, that heat treatments of 2 seconds at 220°C, with shrinkage allowed, do not modify the crystallinity. Such modification requires annealing for 3 minutes, with shrinkage allowed. The real crystallinity of PA 66 fibers is higher than that of PET fibers, but in contrast with results obtained with PET fibers, heat treatments decrease the k parameter.

Variations of k must be compared with variations in a Perfection Crystalline Index (IP_c) obtained according to Dumbleton's method, from the 010 and 100 peaks (7). IP_c increases rapidly after heat treatments. An increase of this index is proof of a progressive transformation from a pseudohexagonal phase in the initial sample (disturbed as seen from the k value) decreasing to give a less disturbed triclinic phase after thermal treatment(8)

We also notice a considerable increase of the SAXS intensity that could not be explained by the change of crystallinity (which remains practically constant) but could partly be explained by a decreased k.

Acrylic Fibers. Table VII shows that crystallinity of poly acrylonitrile is only slightly modified by heat treatments. The fraction of crystalline material seems rather distorted, as judged by the k values. Wet treatments alone allow substantial amounts of distortions to be removed. Water molecules could enter the ordered regions and relax dipole-dipole interactions, allowing some molecular motion.

Table VI

PA 66 Yarn - 78 dtex				
Sample	X_c	k	PI_c %	SAXS Integrated intensity
- initial	0.56	3.3	47	164 a.u.
- heat tr. 220°C-2"				
+ 2 %	0.56	2.5	69	235
0 %	0.59	2.9	67	350
- 2 %	0.55	2.9	66	545
- 4 %	0.56	2.8	67	473
- 6 %	0.56	2.6	72	493
- annealed - 3' 220°C - 10 %	0.61	2.7	80	-

Table VII

CRYLOR (homopolymer)		
Sample	X_c	k
- initial	0.34	3.8
- dry heat treated		
140°C	0.37	3.6
170°C	0.39	3.6
- wet heat treated		
130°C	0.33	3.3
180°C	0.38	3.3

Cotton Fibers. Figure 8 shows the way the background was drawn on the x-ray diagram, in reference to a standard amorphous pattern. As with PAN fibers, the crystalline fraction in cotton seems rather disturbed by distortions, judged by the high k values (Table VIII). A mercerizing treatment with caustic soda solution, which transforms cellulose I to Cellulose II, produces a reduction of the overall crystallinity, with a relaxation of the initial crystalline distortions. Percentages of cellulose I and cellulose II, as well as the overall crystallinity determined according to a relative method (5), are shown in the right part of Table VIII. We could assume that the decrease of k through mercerization could be accounted for by the following simultaneous effects. The first effect is the removal of the most disturbed fraction of the initially crystalline, fibrillar surface of the cellulose. The second effect is a regeneration of a less disturbed phase. If we recall that chain-folding caused disorder in PET, the regeneration of a less disordered phase suggests that the formation of cellulose II occurs without chain-folding.

Table VIII

COTTON 53/2 - Menoufi					
Sample	X_c	k	Cell.I %	Cell.II %	Cr. %
- initial	0.41	4	80	-	80
- mercerized	0.27	3.1	29	48	77

Kevlar Fibers. Table IX shows results obtained with Kevlar 950 fibers. The crystallinity of this aromatic polyamide is only slightly higher than that for the aliphatic polyamide samples that we studied (Table VI). But in Kevlar, as for PET fibers, the disorder parameter k is smaller than in PA 66 fibers. Thermal treatment at 220°C for 1 hour in slack conditions does not substantially increase the average crystallinity of the Kevlar 950 fibers and produces more distortions. Consequently, we conclude that the behavior of this Aramid fiber during our annealing treatment is similar to PA 66 as far as crystallinity is concerned. Kevlar 950 is also similar to PET fiber with regard to the behavior of the disorder parameter.

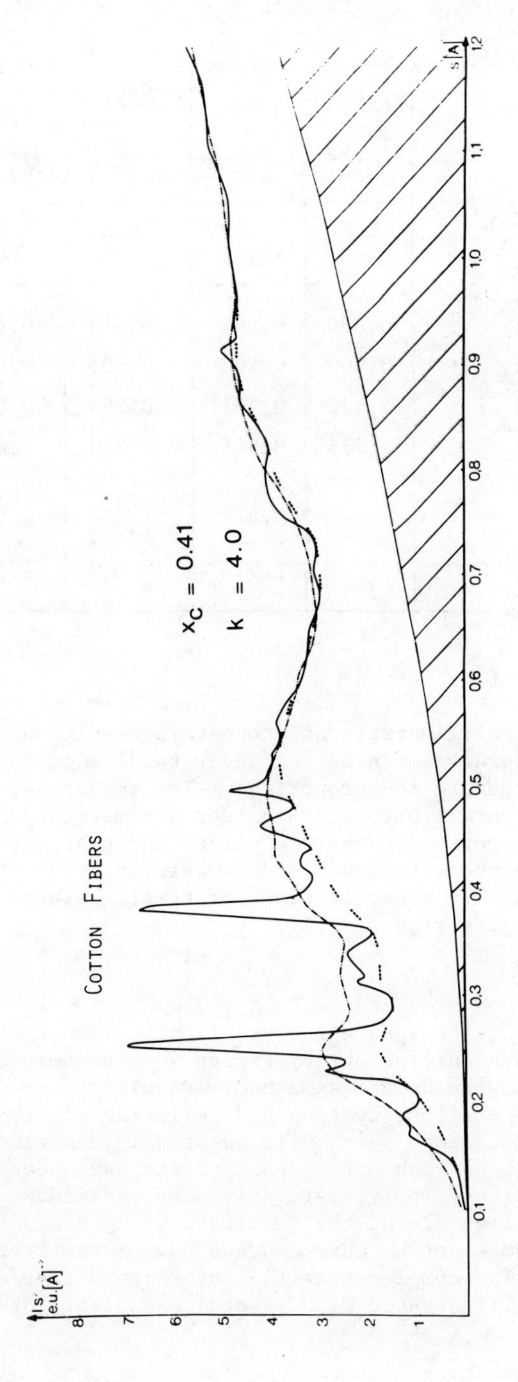

Figure 8. Is^2 *vs.* s *diagram for cotton fibers (——) and amorphous cellulose (– – –). The amorphous background has been calculated from the amorphous curve and a factor of proportionality. (——) Experimental curve; (· – · –) amorphous cellulose sample; (· · ·) amorphous background.*

Table IX

Intervals $s_o - s_p$	Kevlar type 950		Kevlar type 950 annealed 220°C - 1 h (slack)	
	C.A. k=0	C.V. k=1,8	C.A. k=0	C.V. k=2
1st int s_o-s_p	0,486	0,564	0,514	0,606
2nd int s_o-s_p	0,409	0,593	0,401	0,604
3rd int s_o-s_p	0,328	0,591	0,294	0,560
4th int s_o-s_p	0,235	0,561	0,240	0,615
Moy. X_c		0,577		0,596
CV %		2,9 %		4,1 %

Conclusion

These examples illustrate how to obtain results about crystallinity and disorder for a better understanding of the relationships between structure and properties. The reader is, however, cautioned that crystallinity and disorder parameters determined by x-ray diffractometry are average values and that they should be carefully compared with local order measured by electron diffraction on ultra-thin cross sections of textile fibers (9) with differing crystallite sizes (10,11).

Abstract

Corrections of the apparent crystallinity values of fibers materials have been carried out by taking into account a disorder parameter k, following Ruland's method. Peculiar care was taken about samples preparation (cutting and pelleting of fibers), data collection and reduction, which will be briefly described. Crystallinity and disorder parameter measurements have been performed on main textile fibers (polyester, polyamide, aramid, polypropylene, cellulosic fibers) and the results will be discussed comparatively, with those got by more conventional x-ray crystallinity determinations. The complementarities of these different approaches will be illustrated with several examples. For instance, polyester fibers exhibit, after heat treatments in the slack state, higher crystallinity but also higher disorder parameter, and after mechanical fatigue, an increasing of the crystalline fraction with this time a reduction of the k value. For example, after 65,000 cycles of extension between two levels of stress - choosen in the Hooke's zone - the crystallinity in a Poly(ethylene terephthalate) fibers increases from 23 % to 30 % and k decreases from 2.5 to 1.9. On the contrary in 66 polyamide fibers, after heat treatments, the crystalline fraction does not increase a lot, but a significant decrease of the k parameter is registered simultaneously with greater percentage of the gamma to alpha phase transformation. As far as cotton fibers are concerned,mercerizing treatments in caustic soda, produce a significant decrease of the overall crystallinity and a removing of distortions, in proportion as transformation cell.I to cell.II increases.

Literature Cited

1. SOTTON, M. C.R. Acad. Sc. Paris, 1970, 270, série B, 1261.

2. SOTTON, M. Textile Research Journal, 1971, 41, 834.

3. RULAND, W. Acta Crystallogr., 1961, 14, 1180.

4. SOTTON, M. ; ARNIAUD, A.M. ; RABOURDIN, C. J. Appl. Polym. Sci., 1978, 22, 2585.

5. SOTTON, M. ; ARNIAUD, A.M. ; RABOURDIN, C. Bulletin Scientifique ITF, 1978, 7, 265.

6. ALEXANDER, L.E. "X-Ray Diffraction Methods In Polymer Science"; Wiley, 1969.

7. DUMBLETON, J.H. J. Appl. Polym. Sci., 1968, A 2, 2067.

8. RHÔNE POULENC TEXTILE - Private Communication.

9. HAGEGE, R. "Diffraction Methods for Structural Determination of Fibrous Polymers" ; American Chemical Society : Washington, 1979, in Press.

10. HINDELEH, A.M. ; JOHNSON, D.J. Polym. 1972, 13, 27.

11. JOHNSON, D.J. ; "Diffraction Methods for Structural Determination of Fibrous Polymers" ; American Chemical Society : Washington, 1979, in Press.

12. URBANCZYK, G.W. ; Kolloid Z., 1960, 2, 128.

RECEIVED May 21, 1980.

12

The Structure of the Amorphous Phase in Synthetic Polymers

An X-ray Approach

GEOFFREY R. MITCHELL, RICHARD LOVELL, and ALAN H. WINDLE

Department of Metallurgy and Materials Science, University of Cambridge, Pembroke Street, Cambridge CB2 3QZ U.K.

Previous studies of the local conformation and packing of polymer chains based on measurements of wide-angle scattering (X-ray or electron) and radial distribution function analysis have apparently justified widely different models, albeit from similar data (for example 1,2). We present a method which enables detailed and consistent conclusions to be made through systematic comparison of the experimental scattering with the scattering calculated for a wide range of models. This method may be applied to polymeric glasses, melts and rubber and we also show in principle how it may be applied to the non-crystalline component of semi-crystalline polymers.

The determination of the structure of the crystalline component of semi-crystalline polymers is simplified if fibre diffraction patterns may be obtained, enabling the chain repeat distance to be calculated from the layer line spacing. For a polymer glass, however, the degree of orientation that may be introduced is limited : Figure 1 shows the scattering from extruded polycarbonate (extension ratio, λ = 3), and although there is an anisotropic distribution of intensity there are no fibre type features. Attempts to improve the apparent alignment by azimuthal deconvolution techniques have not been wholly successful (3,4). For a semi-crystalline polymer, in general, the degree of orientation that may be induced into the amorphous component in a fibre is very low (Figure 2), and for a polymer melt the introduction of any orientation at all for the duration of an X-ray experiment would require particular ingenuity! These problems indicate that any general method for evaluating the structure must be based on the isotropic scattering from un-oriented samples.

Method

The information from the oriented glasses is not redundant since it suggests in broad terms the different origins of the diffuse peaks in the scattering curve. Those which are weighted

0-8412-0589-2/80/47-141-215$05.00/0

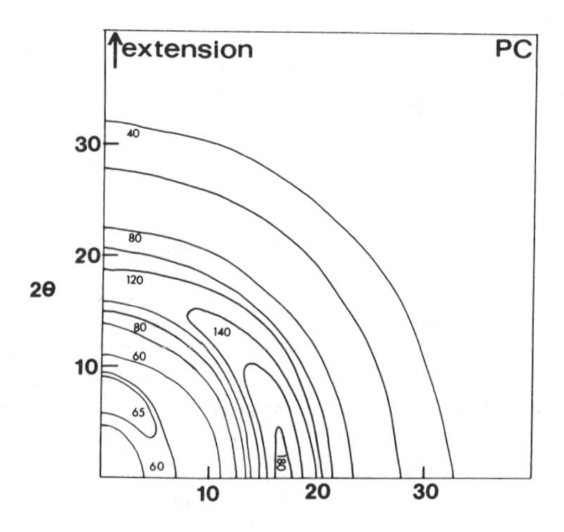

Figure 1. Contour map of the uncorrected scattering (CuKa radiation) from extruded polycarbonate (λ = 3)

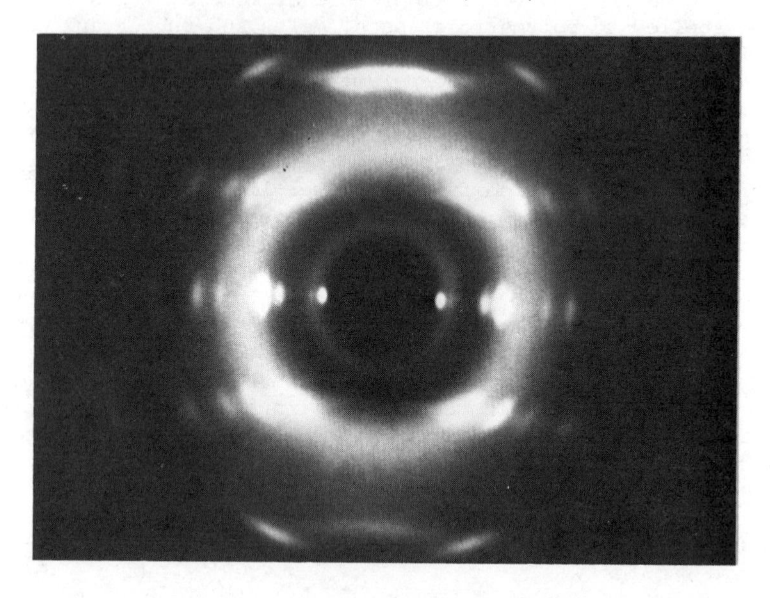

Figure 2. Fiber pattern for isotactic polystyrene

towards the equator must arise from the interference between the aligning segments and those weighted towards the meridian result from correlations along such segments. In general these segments will be portions of the chains rather than sidegroups and so we may distinguish between interchain distances and intrachain distances. Of course if the side groups form a significant proportion of the chain (for example polystyrene) then the interpretation may not be so straightforward. Similar information may be derived from evaluating peak shifts with temperature (3,5) and by comparing, for crystallisable polymers, the positions of the crystalline and amorphous peaks in the scattering curve.

The separation of the scattering curve is facilitated by the distribution of the peaks. Typically the interchain peak is at low scattering vector ($s = 1.0\text{-}1.5\text{Å}^{-1}$, where $s = 4\pi\sin\theta/\lambda$) while the intrachain components are at higher scattering vectors ($>2.0\text{Å}^{-1}$). The division of the scattering in this way enables the structure determination to proceed in two stages: first, using the intrachain scattering to resolve the persistent chain conformation, and then the interchain scattering to explore the possible packing arrangements of this conformation. We have shown (6,7) that when the width of the first interchain peak is correctly modelled using either parallel or randomly oriented (i.e. non-aligned) chains, the predicted intensity of the second and subsequent orders is so low as to be insignificant. Hence the separation into interchain and intrachain effects as outlined appears justified; It is however difficult to achieve in real space since the Fourier transformation of spherically averaged data causes some superposition of interchain and intrachain information, and for oriented data cylindrical distribution functions are necessary which have limited applicability to the scattering from partially aligned materials. For these reasons modelling in reciprocal space is much preferred.

The 'isolated' intrachain scattering is used to determine the persistent local conformation by comparison between the scattering from single chains and the experimental data. The experimental data needs to be corrected for absorption, polarisation, multiple and incoherent scattering and in addition the intensities normalised to electron units, before quantitative intensity comparisons are possible. The s-weighted reduced intensity function $s\,i(s)$ is a better choice for such comparisons where:

$$i(s) = k\,I(s) - \Sigma f_j^2\,(s)$$

and $I(s)$ is the fully corrected data, k is the normalisation factor, and $\Sigma f_j^2\,(s)$ is the independent scattering from a composition unit. The scattering may be calculated from atomic coordinates using the Debye equation (8):

$$i(s) = \sum_i \sum_j f_i f_j \sin sr_{ij}/sr_{ij}$$

where f_i and f_j are the scattering factors for the i and j atoms, and r_{ij} is the distance between them.

The atomic coordinates are generated for a chain of particular chemical configuration by specifying the bond lengths, bond angles and bond rotation angles. For a particular polymer we may assume that the bond lengths are effectively constant as are bond angles, which except for some polymers such as PMMA (9,10) have near tetrahedral values. Thus the local conformation of a chain will be defined by a set of rotation angles about the skeletal bonds (plus sidegroup rotations if applicable). The length of chain used depends upon the regularity of the sequence of rotation angles and is of course one of the parameters to be determined. To minimise the end-effects a chain of more than 6 skeletal bonds is required. If there is considerable disorder in the chain then 500-1000 bonds may be required to fully represent the distribution of persistent conformations. To explore all of the conformational possibilities for such chains is not only time consuming but unnecessary since many of the bond rotations may be excluded as they will lead to 'overlapping' atoms.

Steric information is most easily summarised in a conformational energy map. Such calculations have been made by a number of workers but they require semi-empirical potentials or the effective van der Waals radii which are subject to error and dispute. We have carried out calculations of this sort for polyethylene units using different potentials and find that although the relative energies of the minima vary considerably, their locations are not particularly sensitive to the different potentials. Thus the conformations for which the scattering will be calculated may be restricted to those of low energy without loss of generality.

By comparing all arrangements of these rotation angles in varying lengths of chain the most likely persistent conformation may be determined. We have successfully applied this technique to several polymers (6,7,9,10), but we will illustrate it here with the specific example of molten polyethylene.

Polyethylene melt at 140°C

The published conformational energy calculations for n-alkanes (for example see 11,12) have minima at approximately the following pairs of rotation angles:

0°,0°	trans,trans	(tt)
0°,120°	trans,gauche+	(tg+)
120°, 120°C	gauche+,gauche+	(g+g+)
120°,-120°	gauche+,gauche-	(g+g-)

Other permutations may be generated by changing signs. The pairs are given in order of increasing energy, for although the calculations for n-alkanes do not agree on the energy differences between states they do agree on the order. The (gauche+, gauche-) minimum is split into two minima but this effect has usually been ignored (11) as the energy of this conformation above (trans, trans) is considerable, and so the occurrence of (gauche+, gauche-) pairs is likely to be slight. Figure 3 shows the si(s) curves for different lengths of regular sequences of the lower energy pairs of conformations, compared to the experimental si(s) curve. The agreement between the calculated and experimental curves is poor for any of the conformations, however, the peaks for the (trans,trans) conformation are at the correct scattering vectors although too sharp. These curves indicate that the melt does not contain any significant amount of long runs in simple regular conformation.

The next stage in the proposed method is to take the most promising regular conformation and introduce defects in the form of other conformations. This is achieved by 'building' the chain according to a set of unconditional probabilities. The probability of the next rotation being trans is p_t, the probability of it being gauche+(or gauche-) is $(1-p_t)/2$. Figure 4 shows the si(s) curves for chains containing 500 skeletal bonds built using different values of p_t. Reasonable agreement is obtained for p_t values of 0.5 - 0.6.

The set of probabilities may be varied in several ways, for example by the exclusion of (gauche+,gauche-) pairs or by adjusting the probabilities dependent on the state of the preceding bond. This introduces conditionality to the probabilities, and these and other improvements such as fluctuations in angles are considered in more detail elsewhere (7).

The energy levels in the semi-empirical conformational energy calculations may be used to derive conditional probabilities using the statistical weights method described by Flory (13). This approach takes into account the neighbouring bonds and their effect on the populations of the rotation states. The scattering from PE chains built according to conditional probabilities obtained from the statistical weights proposed by Abe, Jernigan and Flory (11) is in good agreement with the experimental data (6). The energy calculations also indicate that the exact values of the rotation angles are dependent on neighbouring rotation states, and this refinement has been incorporated in the model chain. Such sophistication in the chain building does indeed improve the agreement between the calculated si(s) and the experimental curve (Figure 5a,b). Thus our approach is able to give a good independent indication of the basic statistical parameters for a polyethylene chain, and in addition it can provide some direct experimental justification for the detailed statistical proposals (such as those in 11) obtained from modelling the molecular trajectory.

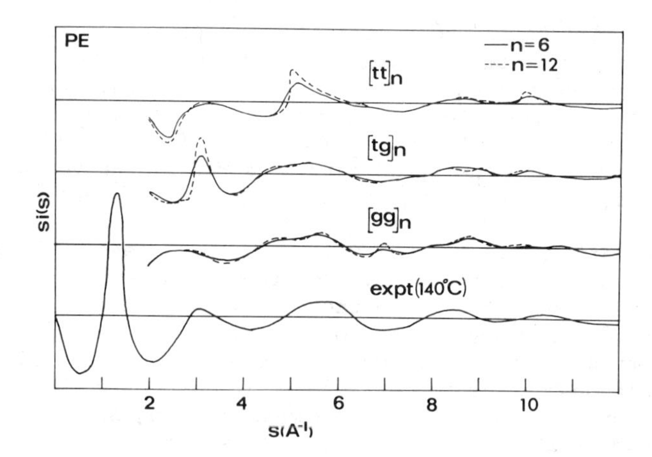

Figure 3. The experimental s-weighted intensity function si(s) for molten polyethylene at 140°C compared with the calculated si(s) curves for single chains of simple regular conformations

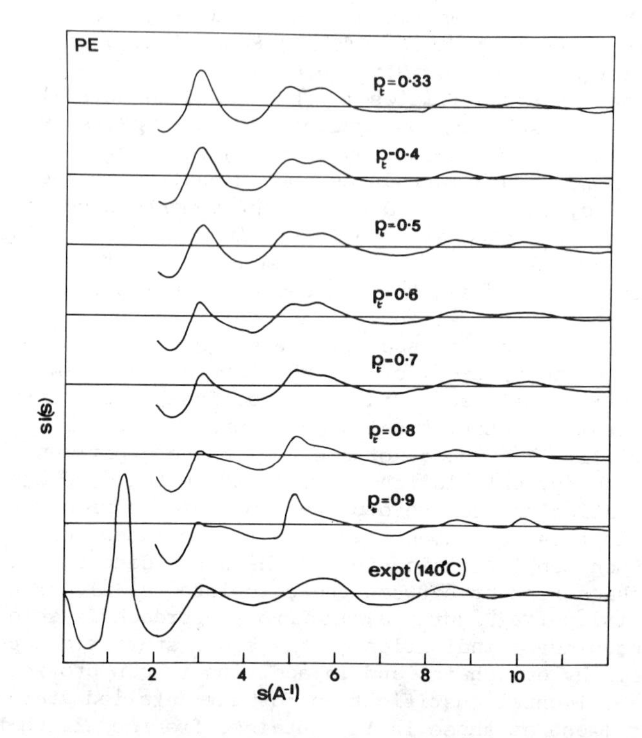

Figure 4. Calculated s-weighted intensity function si(s) for chains with rotation states randomly distributed, the probability of trans being p_t

Having determined the type of persistent conformation we must consider likely packing arrangements that are consistent with bulk density, and the position, height and width of the interchain peak. Since the weighted-average sequence length of trans, for the chains providing the best agreement with experimental data, is about 3, the typical 'segment' in a chain is globular in nature. A possible approach that we have presented in detail elsewhere (6,7) is to consider the centroids of these globular segments to be distributed as the centres of randomly close-packed spheres. By assuming no orientational correlation between neighbouring segments the scattering intensity is simply the product of the scattering of an average segment and the structure factor for an assemblage of random spheres. The si(s) curve for such a representation is shown in Figure 5c.

This model quantitatively accounts for all the features in the si(s) curve for molten polyethylene, and supports the random coil concept.

Semi-crystalline polymers

Investigations of the complete structure of semi-crystalline polymers using X-rays or neutrons have in general been restricted to small scattering vectors ($< 0.5 \mathring{A}^{-1}$). These studies have been generally successful in explaining the essential features of the two phase structure. This has been achieved by assuming the structure of the amorphous component to be similar to the melt. Since in the wide-angle scattering curve we can distinguish which of the peaks arise from the crystalline component by their width, there is the opportunity to extract these peaks utilising the advances made in curve-fitting routines (14,15) and leave the scattering from the amorphous component to be studied using the method described above. Curves illustrating this procedure are shown in Figure 6 and investigations are in hand to explore fully the potential of this approach.

Recent advances in the methods used for evaluating fibre diffraction patterns place importance on matching the whole pattern (16). The methods used for calculating the scattering for the models described above could be extended to model the amorphous part of the pattern prior to subtraction or matching the calculated structure.

Conclusions

This work has shown that wide-angle X-ray scattering is a sensitive tool for evaluating the structure of non-crystalline polymers. The structure determination is only successfully approached by considering first the persistent conformation and then the packing which must to some extent be a consequence of such conformations. The method does not need to place any reliance on the semi-empirical conformational energy calculations,

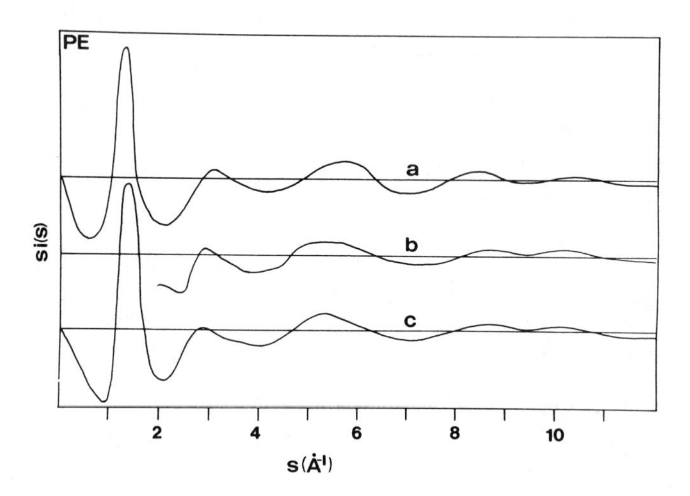

*Figure 5. Experimental s-weighted intensity function (a) compared with calculated si(s) curves for (b) "random chain" according to conditional probabilities derived from the statistical weights of Abe, Jernigan, and Flory (11) with near-neighbor dependence of rotation angles, and (c) an assemblage of randomly packed spheres with "average segments" placed at their centroids (*see *text)*

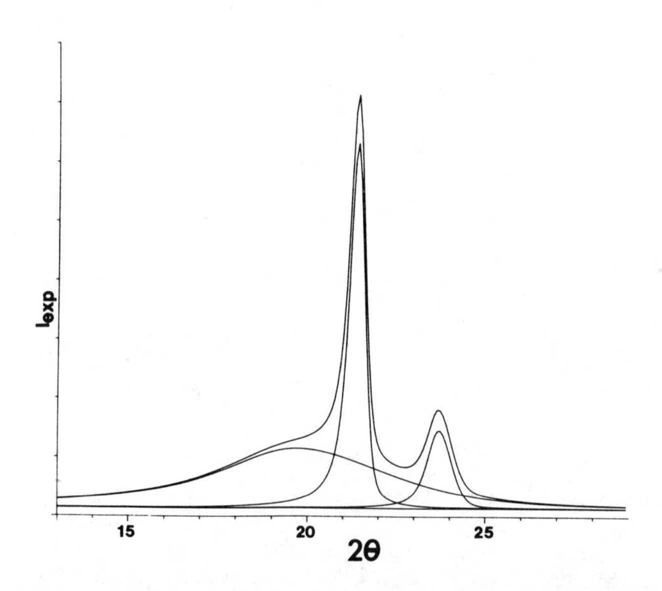

Figure 6. An illustration of the separation of the experimental scattering (outer curve) from low density PE into curves representing the scattering from the non-crystalline (broad peak) and crystalline (sharp peaks) components

although for the case of polyethylene the predictions of such calculations were found to be in accord with the results of this much simpler direct approach.

The method described also offers an approach to the study of molecular conformation and packing of the amorphous component of semi-crystalline polymers.

We are indebted to P.B McAllister and T.J. Carter (BICC) for providing the data in Figure 6, to John Saffell for help in preparing Figure 1, and to the Science Research Council for funds.

Abstract

A method is described which enables structural parameters of non-crystalline polymers to be determined. It is based on the analysis of wide-angle X-ray scattering and where possible incorporates the additional information obtainable from oriented specimens. Particular emphasis is placed on the analysis of the structure of molten polyethylene. The potential application of the approach to the structure of the 'amorphous' phase in semi-crystalline polymers is discussed.

Literature Cited

1. Ovchinnikov, Yu. K., Markova, G. S. and Kargin, V. A. Polym. Sci. USSR, 1969, 11, 369
2. Voigt-Martin, I. and Mijlhoff, F. C. J. Appl. Phys., 1976, 47, 3942
3. Lovell, R. and Windle, A. H. Polymer, 1976, 17, 488
4. Lovell, R. and Windle, A. H. Acta. Cryst., 1977, A33, 598
5. Ovchinnikov, Yu. K., Antipov, E. M., Markova, G. S. and Bakeev, N. F., Makromol. Chem. 1976, 177, 1567
6. Lovell, R., Mitchell, G. R. and Windle, A. H. Faraday Disc., 1979, 68
7. Mitchell, G. R., Lovell, R. and Windle, A. H. to be submitted to Polymer
8. Warren, B. E. "X-ray Diffraction" (Addison-Wesley, 1969)

9. Lovell, R. and Windle, A. H.
"Diffraction Studies on Non-Crystalline Substances",
ed. I. Hargittai and W. J. Orville-Thomas
(Akademiai Kiado, Budapest)

10. Lovell, R. and Windle, A. H.
Polymer, in press

11. Abe, A., Jernigan, R. L. and Flory, P. J.
J. Amer. Chem. Soc., 1966, 88, 631

12. Scott, R. A. and Scheraga, H. A.
J. Chem. Phys., 1966, 44, 3054

13. Flory, P. J.
"Statistical Mechanics of Chain Molecules", (Interscience, 1969)

14. Hindeleh, A. M. and Johnson, D. J.
J. Phys D. (Appl. Phys.), 1971, 4, 259

15. McAllister, P. B. and Carter, T. J., private communication

16. Fraser, R. D. B., MacRae, T. P., Suzuki, E. and Tulloch, P.A.
this symposium

RECEIVED March 21, 1980.

13

The Variable Virtual Bond

Modeling Technique for Solving Polymer Crystal Structures

PETER ZUGENMAIER

Institute of Macromolecular Chemistry, University of Freiburg,
D-7800 Freiburg i.Br., West Germany

ANATOLE SARKO

Department of Chemistry, State University of New York,
College of Environmental Science and Forestry, Syracuse, NY 13210

Although various procedures are available for the model analysis of fibrous polymers, methods based on the *virtual bond* representation of the asymmetric residue may be of advantage in many cases. In the following, we describe one such method that began with simple procedures applied to polysaccharides, but has now been refined into a flexible and powerful model analysis tool that is simple to use with any class of polymer. Its use in the present case, however, is illustrated with examples drawn from the structure analysis of polysaccharides.

The Virtual Bond Method

The earliest attempts at model analysis of polysaccharides - typified by the x-ray crystal structure analysis of amylose triacetate - were usually conducted in three steps (1). In the first step, a model of the chain was established which was in agreement with the fiber repeat and the lattice geometry, as obtained from diffraction data. In the second step, the invariant chain model was packed into the unit cell, subject to constraints imposed by nonbonded contacts. This was followed, in the third step, by efforts to reconcile calculated and observed structure factor amplitudes. It was quickly realized that helical models of polysaccharide chains could be easily generated and varied using the *virtual bond* method. Figure 1 illustrates the generation of a two-fold helical model of a (1→4)-linked polysaccharide chain. The *virtual bond, VB,* is the vector linking successive glycosidic (bridge) oxygens. The starting point of this vector has coordinates $x_1 = 0$, $y_1 = -\dfrac{(VB^2 - h^2)^{1/2}}{2 \sin |\Delta/2|}$, $z_1 = 0$; whereas for the end-point they are $x_2 = -y_1 \sin \Delta$, $y_2 = y_1 \cos \Delta$, $z_2 = h$. (VB = length of the *virtual bond;* h = axial rise per residue, here $c/2$; n = number of residues per helix repeat, here 2; $\Delta = 2\pi t/n$; $\pm t$ = number of turns in repeat (+ right- and - left-handed). Provided a suitable model exists for the monomer residue, a reasonably correct conformation of the chain is obtained simply by rotating the

0-8412-0589-2/80/47-141-225$05.00/0

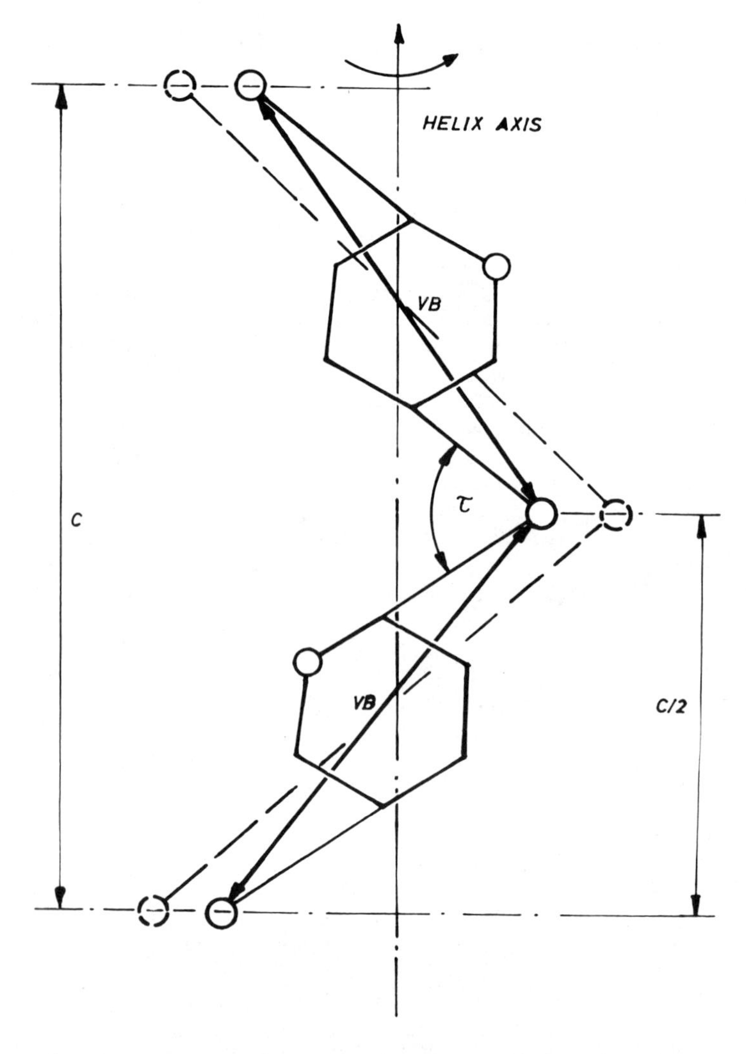

Figure 1. Construction of a two-fold helical model of a polysaccharide with the virtual bond method. Increasing the length VB *of the virtual bond is shown by the dashed line.*

entire residue about the *virtual bond* until the bridge angle τ is within the expected range. For most polysaccharides, such rotation yields two conformations with the proper angle τ, as shown in Figure 2. One of the two choices is generally ruled out because of excessive, short nonbonded contacts within the chain.

Although other methods are available to construct models of polysaccharide helices, such as by rotations about the two bonds leading to the bridge oxygen (the ϕ and ψ rotations), the *virtual bond* method possesses several advantages. With it, a helix with given n and h can be constructed easily, and only one variable - rotation about the *virtual bond* - is needed for gross changes of conformation. The consequences of changing the length of *VB*, such as changes in the helix diameter and the bridge angle τ, are easily predictable, as shown in Figure 1. Most importantly, model refinement with this method is simple, as described in the following sections.

Model Building and Refinement with the Virtual Bond Method

As shown in Figure 1, only the positions of the repeat atoms of the monomer residue are specified by the *virtual bond*. Within this constraint, considerable latitude is available for the positions of all other atoms of the residue. These atoms can be described in two alternate ways, as shown in Figure 3. In the first method, a string of connected atoms extends from the lower to the upper atom of the *virtual bond* (*i.e.*, from O4 to O1 in Figure 3A). All other atoms of the residue not in this main string are placed in separate strings, which are attached as branches to the main string. (Fixed hydrogens, *e.g.*, those attached to ring carbons, could be treated identically. However, it is simpler to calculate their positions when needed, in accordance with preselected C-H bond lengths and associated bond angles). The position of each atom in a string is expressed by polar coordinates r, θ, ϕ, where r is the bond length, θ is the bond angle, and ϕ is the conformation angle, all relative to previous atoms. (These coordinates are illustrated for atom O5 in Figure 3A). Conversion between the polar and cartesian coordinates is easily accomplished, whenever needed. When a bond and its angles cannot be expressed by polar coordinates associated with an atom, such as the "open" bond shown by a dashed line in Figure 3A, its length and all desired angles can still be explicitly defined. All bond lengths and angles, including the length of the *virtual bond* and the angles associated with it, can now be treated as variables during refinement.

The model of the residue can also be described by a second procedure, shown in Figure 3B. Two strings of atoms are used, beginning at separate ends of the *virtual bond*. The strings are not connected to one another, leaving two "open" bonds. This method is useful when the length of the *virtual bond* is to remain fixed during refinement.

The goal of model building is to produce a polymer chain that

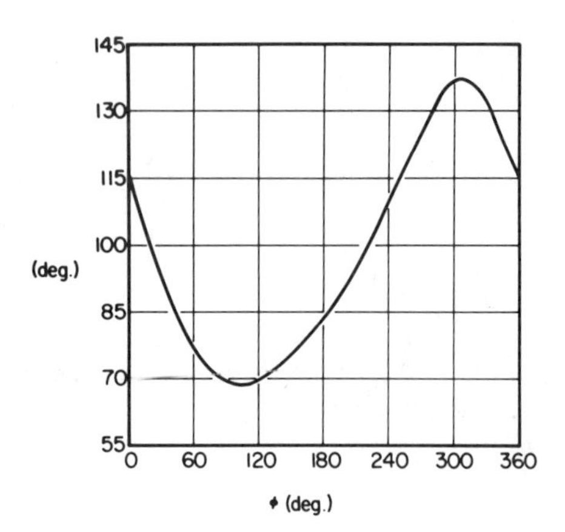

Figure 2. Effect of rotation of the residue about the virtual bond ϕ on the bridge angle τ (for trimethyl amylose (5))

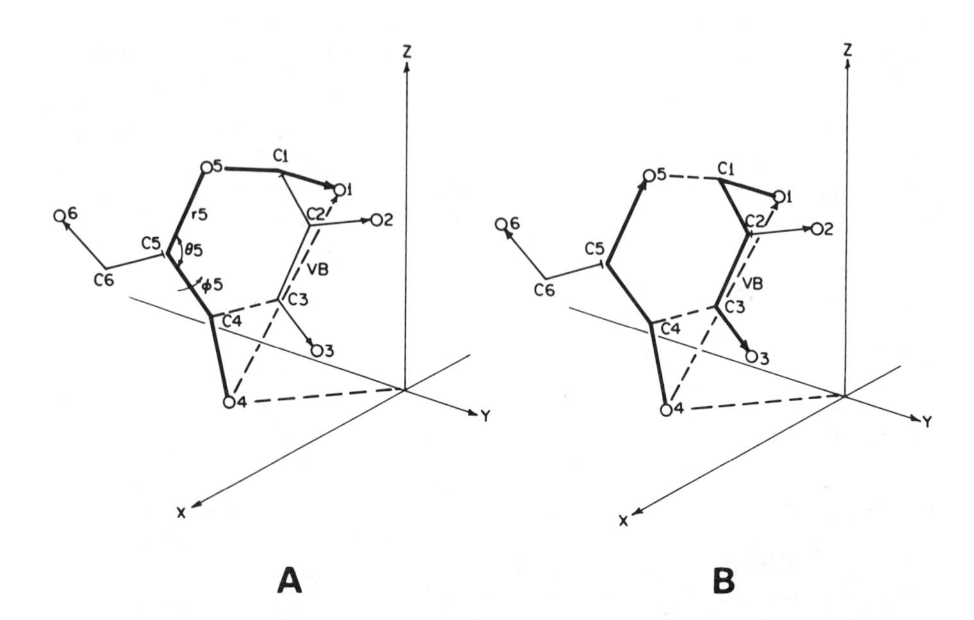

Figure 3. Two alternate methods of describing an α-D-glucose residue using the variable virtual bond method. The bond length, bond angle, and conformation angle for atom 05 are shown as r_5, θ_5, ϕ_5.

possesses sound stereochemical features and, at the same time, is in good agreement with diffraction data. These requirements can be met, as shown in Figure 4, by proper use of information available from experiment, and model refinement based on theoretical principles (2,3). In the refinement of the model, all bond lengths, bond angles and conformation angles are optimized relative to a set of standards, simultaneously with the position of the chain in the unit cell. The refinement is carried out by minimizing the function (4,5):

$$Y = \sum_{i=1}^{N} STD_{oi}^{-2} (A_i - A_{oi})^2 + W^{-2} \sum_{\substack{i=1 \\ j=1}}^{n} w_{ij}(d_{ij} - d_{oij})^2 \qquad (1)$$

where the first term represents the bonded and the second term the nonbonded interactions. In this equation, A_i is any calculated bond length, bond or torsion angle; A_{oi} is an average or standard value of A_i; STD_{oi} is a weight or standard deviation of A_{oi}; N is the number of bonded interactions in the refinement; d_{oij} is the nonbonded equilibrium distance between atoms i and j; d_{ij} is the actual nonbonded distance between atoms i, j; n is the number of nonbonded contacts and W^{-2} is the overall weight factor which balances the bonded and nonbonded interactions.

The standard values A_{oi} for the bond lengths, bond angles and conformation angles can be obtained by averaging from single-crystal structures of carbohydrates (6). This also yields the corresponding standard deviations STD_{oi}. The equilibrium nonbonded distances d_{oij} and their associated weights w_{ij} have likewise been determined from known crystal structures of carbohydrates (7). The actual values of the constants used for polysaccharides are given in Tables I and II. For a good balance between the two terms of eq. (1), a value of 0.5 is appropriate for the overall weight W.

Table I

Constants for the nonbonded repulsion term of Eq. (1). (When $d_{ij}>d_{oij}$, w = 0, except for the hydrogen bond).

Interaction type	d_o, Å	w
C....C	3.70	3.00
C....O	3.60	3.00
C....H	3.30	1.35
O....O	3.60	3.00
O....H	3.25	1.40
H....H	3.20	0.50
O....O (H-bond)	2.80	20.00

It should be clear that a refinement based on the minimization of the function Y results in a structure of minimum steric

Table II

Average bond lengths, bond angles and torsion angles for an α-D-glucose residue shown in Figure 3. Included are lower and upper limits and average standard deviations (6).

Bond Lengths (Å)	
C(1)-O(1)	1.415 (1.405-1.435)
C(1)-O(5)	1.414 (1.392-1.428)
C(5)-O(5)	1.436 (1.425-1.464)
C(4)-O(4)	1.426 (1.409-1.446)
C(3)-C(4)	1.523 (1.509-1.537)
C(2)-C(3)	1.521 (1.508-1.536)
C(2)-O(2)	1.423 (1.411-1.440)
C(3)-O(3)	1.429 (1.410-1.446)
C(5)-C(6)	1.514 (1.495-1.534)
C(6)-O(6)	1.427 (1.415-1.442)
O(4)..O(1)	variable (4.10-4.60)
Ave. STD	0.01 Å

Bond Angles (degrees)	
O(4)-O(1)-C(1)[b]	74.0 (71.0-77.0)
O(1)-C(1)-O(5)	111.6 (109.8-112.7)
C(1)-O(5)-C(5)	114.0 (113.2-114.7)
O(1)-O(4)-C(4)[b]	45.5 (42.5-48.5)
O(4)-C(4)-C(3)	105.5 (103.6-112.4)
C(4)-C(3)-C(2)	110.5 (106.0-113.6)
C(3)-C(2)-O(2)	110.8 (106.4-113.2)
C(4)-C(3)-O(3)	109.7 (106.5-112.5)
O(5)-C(5)-C(6)	106.9 (106.8-107.9)
C(5)-C(6)-O(6)	111.8 (109.4-113.8)
Gycosidic bond angle	variable
Ave. STD	1.5°

Torsion Angles[a] (degrees)	
C(4)-O(4)-O(1)-C(1)[b]	-2.0 (±5)
O(4)-O(1)-C(1)-O(5)[b]	-57.6 (±5)
O(1)-C(1)-O(5)-C(5)	57.7 (±5)
O(1)-O(4)-C(4)-C(3)[b]	-60.1 (±5)
O(4)-C(4)-C(3)-C(2)	168.0 (±5)
C(4)-C(3)-C(2)-O(2)	-177.9 (±5)
O(4)-C(4)-C(3)-O(3)	-69.0 (±5)
C(1)-O(5)-C(5)-C(6)	-174.4 (±5)
O(5)-C(5)-C(6)-O(6)	variable[c]

Table II (continued)

Torsion Angles[a]	
Ave. STD	3.0°

[a] Convention for torsion angles: 0° when bonds A-B and C-D are *cis*; clockwise rotation of bond C-D relative to A-B is positive.

[b] Involves the *virtual bond*.

[c] Any desired value can be used; however, the three staggered conformations, denoted *gg*, *tg* and *gt*, have torsion angles -60°, 180° and 60°, respectively.

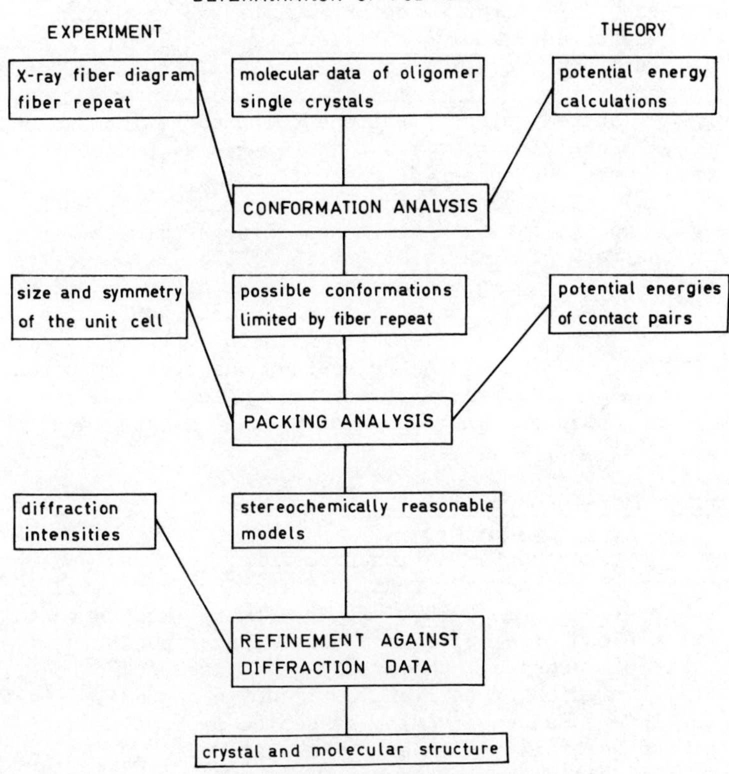

Figure 4. The strategy of determining the crystal and molecular structure of polymers based on model refinement

energy. This is true even for the function as written in eq. (1) which yields an empirical, unit-less value of the "energy" Y. A true energy can be obtained by substituting proper force constants for STD_{oi}, using torsional potentials for conformation angles, and substituting Lennard-Jones or Buckingham potentials for the quadratic nonbonded term. Even though such functions add rigor to the procedure, experience has shown that they add little to the refinement, while increasing demands for computer time.

The assumption underlying the prediction of a minimum-energy structure, of given n and h, is that it is identical with the crystal structure. This is generally true, although agreement may not be present in all details. At times, more than one minimum-energy structure may exist. The refinement should, therefore, continue with bringing the calculated and observed structure factor amplitudes into agreement, as shown in Figure 4. This is done by refining the same parameters optimized in the stereochemical refinement, except that the criterion of refinement is the minimization of the residuals $R = \sum||F_c|-|F_o||/\sum|F_o|$ or $R'' = [\sum W(|F_c|-|F_o|)^2/\sum W|F_o|^2]^{1/2}$. In these equations, F_c and F_o are the calculated and observed structure factor amplitudes, respectively, and W are the weights assigned to individual reflections. The residual R'' is preferred over R, because it allows the use of reflection weights.

Finally, there are cases where a combined refinement based on simultaneous minimization of Y and R'' (or R) may be necessary. For example, when the number of reflections is small, pure x-ray refinement may introduce unacceptable stereochemical features. This is guarded against by minimizing a linear combination of Y and R'' in the form of a function $\Phi = fR'' + (1-f)Y$, where the fractional weight f is chosen to balance the two terms. Good results have been obtained with f ranging from 0.9 to 0.985, which usually weights the R'' term at least equally with the Y term.

The model description and refinement based on the *virtual bond* method need not be restricted to a single monomer residue. Any number may be used, with only one *virtual bond* needed to span all of the residues.

Method of Constrained Optimization

The constrained optimization procedure, originally developed from the simplex method and first described by Box, is ideally suited to model refinement (8). It is a search method that searches for the minimum of a multidimensional function within given intervals. It possesses all the advantages of search methods, among them that calculation of derivatives is not necessary, a test to assure the independence of variables can be omitted, and diverse variables can be easily included. These are exactly the requirements of model refinement where bond lengths, bond angles, torsion angles, and other parameters are used within experimentally defined limits.

The principles of the method can be understood with the help of Figure 5. The minimum of a function $F(x)$ is determined within the limits x_{min} and x_{max}. The variable x represents a set of n variables (x_1, x_2, ..., x_n) to which additional constraints other than the limits may apply. For instance, x_1 may be one point fulfilling all the conditions, thus $F(x_1)$ is the value of this function at x_1. Additional k points (x_2, x_3, ..., x_k) are generated in a random manner within the given limits and the values of the function $F(x_2)$, ..., $F(x_k)$ are calculated. The largest function value, $F(x_1)$ in Figure 5, is replaced by a new $F(x_1)$ for a trial point which is at $x_s + \alpha(x_s - x_1)$, where x_s is the centroid of the remaining points (a good value for α is 1.3). If this trial point represents no improvement, it is moved halfway towards the centroid to give a new trial point x_1'. The procedure is then repeated. If the trial point is reflected outside an interval limit, it is reset to just the inside of the limit. As long as the points have not collapsed into the minimum, the procedure is repeated with all points in turn.

The use of $k > n+1$ points ensures that the complex does not collapse into a subspace. False minima are normally eliminated through the procedure of reflecting a point about the centroid, and because the set of points is distributed over the whole interval. If a false minimum presents particular difficulties, it can usually be eliminated by repeating the optimization with different sets of trial points.

In terms of the refinement illustrated here, the function Y is the function $F(x)$. The first point, x_1, is represented by all variable bond lengths, bond angles, conformation angles, chain position parameters, coordinates of the solvent of crystallization, etc., of the initial model. All other points x_2, ..., x_k, represent trial values for the same n variables within the desired interval limits and subject to any other constraints, such as coupling of variables or hydrogen bond formation. Clearly, the number and type of variables, and their limits and constraints are easily changed in this procedure, as is the form of the function. The search procedure is also relatively rapid and does not suffer from a slowdown in the vicinity of the minimum, as may occur in steepest-descent methods.

The "PS79" Computer Program

The model refinement procedures described in the previous sections have been assembled, over a period of years (4, 9, 10), into a computer program, known as "PS79" in its current version. Although the program was principally developed for use with polysaccharide crystal structures, it is equally applicable to other polymers. It can be used to refine a model with respect to stereochemistry only, using eq. (1), or with respect to diffraction data only, using the residuals R or R'', or as a combination of the two, using the function Φ. In addition, the refinement strategy can be

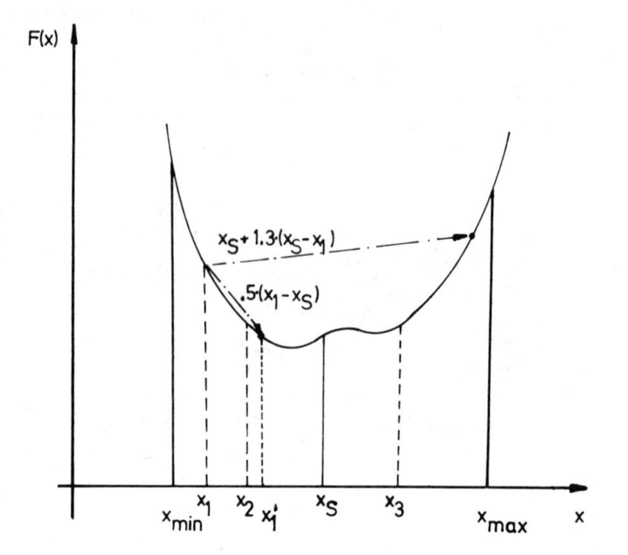

Figure 5. Schematic of the constrained optimization procedure

very flexible. For example, only the chain conformation can be refined by omitting all intermolecular contacts from the nonbonded term of eq. (1). In such cases, the range of variables could extend from a single one of rotation of the residue about the *virtual bond* to all bond lengths, valence-bond angles, and conformation angles of the monomer residue. Conversely, only the chain packing in the unit cell could be refined by completely eliminating the first term of eq. (1), and using only intermolecular nonbonded contacts in the second term. Additionally, any desired combination of the previous extremes could be used, extending to the case where all conformational and packing variables are simultaneously refined. The same applies regardless of whether the refinement criterion is Y, R'' (or R), or Φ.

The initial description of the model is simple, as shown in Figure 3. The atomic coordinates of any suitable structure can serve as the input trial structure, even including a wrong monomer residue. The polar coordinates are calculated from the trial structure, adjusted and modified as necessary, and then subjected to refinement in accordance with the selected list of variables, limits and constraints. Any set of standard values and nonbonded potential function parameters can be used. Hydrogen bonds can be defined as desired, variables can be coupled, and the positions of solvent molecules can be individually refined. Single and multiple helices are equally easily handled, as are a variety of space groups.

The calculation can be terminated when: (a) the minimum of the function does not improve within a given accuracy, (b) a certain time has elapsed, (c) the number of desired iterations is exceeded, or (d) when no improvement is obtained after 20 calculational steps. The program provides varied output, including coordinates written on file that can be used as input to succeeding runs.

The "PS79" program is currently in operation on several major computers - IBM 370 series, UNIVAC 1100 series, CDC 6000 series, and DEC-10 - and in the majority of cases refinement runs can be completed within the fast turnaround job limitations of individual shops. Usually, only the final x-ray runs will demand more time.

Conclusions

Since its introduction several years ago, the *virtual bond*, constrained optimization method has proved very useful in studies of polysaccharide crystal structure. Notable among the successes that can be ascribed to it are the structural determinations of the double-helical amylose (11), the cellulose polymorphs of different chain polarities (12, 13), and of a number of other polysaccharides and their derivatives. As described in a review of amylose structures elsewhere in this volume, the use of this refinement method has produced structural detail that has previously been unavailable (11). These results have provided much-needed insight into how polysaccharides crystallize, and into such aspects of structure as symmetry, relationships between helix conformation and packing, hydrogen-bonding, water and other solvents of crystallization, and effects of chemical substitution on the structure. An added benefit has been the realization that the crystal structure can in many instances be predicted from the stereochemistry alone.

As is to be expected, this optimization method possesses both some advantages and disadvantages. Among the advantages are: (1) Molecular models are generated and refined within desired limits of bond lengths, bond angles and torsion angles. (2) The generation of models is simple and flexible, and is applicable to different polymers. (3) *Virtual bonds* can be used to describe one or more monomer residues, or even atoms in a branch string. For example, the planarity of an acetyl group $\text{-O-}\overset{\text{O(A)}}{\overset{\|}{\text{C}}}\text{(A)-C(M)-H}_3$ can be kept intact by placing the atoms in the following string: O → C(A) → O(A) → C(M) → H and by keeping all distances between the atoms and all conformation angles responsible for planarity constant. (4) The constrained optimization procedure is fast, as long as the calculation of the function is fast. This is true for both conformation and packing refinement using eq. (1), even with a large number of variables. (5) Variables and constraints can be chosen at will, without regard to their number or type.

Among the chief disadvantages is the fact that when the calculation of the function is slow, the refinement proceeds slowly. For instance, in x-ray refinement the computation of R'' (or R) is lengthy, particularly when the number of reflections and the number of variables are both large. This disadvantage is, however, not serious as the increased demand on computer time is still within reasonable limits established by most computer shops.

A more serious limitation placed on this method may occur when the diffraction data are poor. For example, a correct unit cell is a necessary prerequisite for any refinement, yet in many cases its determination from fiber x-ray data may be questionable. This limitation may be avoided by obtaining good electron diffraction diagrams from polymer single crystals. Similar limitations arise from an inability to record diffraction intensities correctly, resulting in poor agreement of x-ray and stereochemical refinements. However, as described by other authors in this volume, two-dimensional recording techniques hold out a great deal of hope for improving the quality of the intensity data. With this improvement, the structure analysis of crystalline polymers may yet approach the reliability of single-crystal structure determinations.

Acknowledgments

This work has been supported by National Science Foundation grant CHE7727749 (to A.S.) and a grant from Deutche Forschungsgemeinschaft (to P.Z.). Cooperative efforts of this work have also been supported by a NATO Research Grant No. 1386, to both authors.

Literature Cited

1. Sarko, A.; Marchessault, R.H., J. Amer. Chem. Soc., 1970, 89, 6454-6462.
2. Kitaigorodskii, A.I., Acta Crystallogr., 1965, 18, 585-590.
3. Williams, D.E., Science, 1965, 147, 605.
4. Zugenmaier, P.; Sarko, A., Biopolymers, 1976, 15, 2121-2136.
5. Zugenmaier, P.; Kuppel, A.; Husemann, E., in "Cellulose Chemistry and Technology", J.C. Arthur, Jr., Ed. ACS Symposium Series No. 48, American Chemical Society: Washington, D.C., 1977, pp. 115-132.
6. Arnott, S.; Scott, W.E., J. Chem. Soc. Perkin Trans. 2, 1972, 324-335.
7. Zugenmaier, P.; Sarko, A., Acta Crystallogr., 1972, B28, 3158-3166.
8. Box, M.J., Comput. J., 1965, 8, 42-52.
9. Zugenmaier, P.; Sarko, A., Biopolymers, 1973, 12, 435-444.
10. Zugenmaier, P., Biopolymers, 1974, 13, 1127-1139.
11. Sarko, A.; Zugenmaier, P., This symposium.
12. Sarko, A., Tappi, 1978, 61, 59-61.
13. Woodcock, C.; Sarko, A., Macromolecules, to be published.

RECEIVED February 19, 1980.

N–H Mapping for Polymers

ALFRED D. FRENCH
Southern Regional Research Center, P.O. Box 19687, New Orleans, LA 70179
WALTER A. FRENCH
Interactive Data Corporation, 350 California St. Suite 1450, San Francisco, CA 94104

Helical shapes of polymers are conveniently, if only approximately, described by *n*, the number of monomeric residues per helix pitch *p*, and *h*, the rise per residue (Figure 1). The present paper addresses the problem of determining which shapes, or values of *n* and *h* can be attained by different polymers. Determination of allowed values of *n* and *h* is important in the early stages of diffraction analysis, because all appropriate trial models should be investigated. Such knowledge is also important for understanding the nuclear magnetic resonance spectra of polymers, the interactions of polymers with other materials, and the behavior of polymers in solution. We propose that the study of regular helices also applies to irregular helices by extension of Natta's monomer equivalence postulate (1). From a review of the geometric features of monomeric residues of cyclic oligomers of glucose, it seems that the average geometric features of the monomeric units of an irregular helix are approximately equal to the monomeric geometry of the most equivalent regular helix.

Previously, the Ramachandran technique (2) was used to learn the range of allowed molecular shapes. We now propose a new representation of the allowed shapes of a polymer. The new representation solves serious problems inherent in the Ramachandran representation. In addition, our method expands the utility of conformational analysis.

The idea for *n-h* mapping occurred in Japan, and the first printed *n-h* map was discussed in Japanese (3). The *n-h* approach was also discussed in a review paper, in English (4). The present paper is more complete treatment of the *n-h* approach and an announcement of the impending availability of a new computer program, NHMAP.

0-8412-0589-2/80/47-141-239$05.00/0

Problems with the Ramachandran Method

In the Ramachandran method, allowed conformations are studied by rotating the monomers through the Φ and Ψ angles about the bonds to the linking atom (Figure 2). At each increment of monomer rotation, the model is tested for conflicts between nonbonded atoms. Models with no serious conflicts or with low potential energy are considered to be stereochemically feasible. The stereochemical energy, or feasibility, is then reported on a grid of Φ and Ψ values (Figure 3). After *n* and *h* are determined for each Φ-Ψ gridpoint, contours of *iso-n* and *iso-h* are laid over the allowed zones and the Φ-Ψ grid to produce the Ramachandran map (Figure 4).

However, representing allowed shapes on a Φ-Ψ grid has fundamental limits. These limits arise when one converts from the linking conformation, expressed in terms of Φ and Ψ, to the polymer conformation, described by the values of *n* and *h*. The limits result because calculation of *n* and *h* values depends on the exact monomeric geometry and linking bond angle used for the stereochemical calculations.

In the original Φ-Ψ work, monomers were assumed to be rigid. Now, however, a large number of single-crystal studies have shown that monomers are more flexible in crystalline environments than was originally thought. Therefore, the Φ and Ψ values of a Ramachandran map do not give unique values of *n* and *h* if the necessary variety of monomeric geometries is considered. The only remedy possible, if Φ-Ψ grids are retained, is to prepare enough Φ-Ψ maps to allow each variant of monomeric geometry and linking bond angle to be combined and tested. This collection of Φ-Ψ maps would be less useful than an overall summary of the allowed conformations.

α-Glucose, a flexible monomer of considerable interest, has a chair form that was thought to be rigid (Figure 5). However, single-crystal studies show a range of 0.6 Å for the distances between the O(1) and O(4) atoms in various crystals containing the glucose molecule. This range results from the different angles at which the linking C(1)-O1) and C(4)-O(4) bonds depart from the bulk of the ring (*5*). As we show later, this monomeric variability, when coupled with a smaller variability in linking bond angle, permits feasible helices with similar values of Φ and Ψ but values of *n* ranging from 5 to 10.

The n-h Map

A simple summary map is a grid of *h* vs. *n*. As shown in Figure 6, we indicate helix chairality by positive and negative values of *n*. Currently, we are relying on hard-sphere analysis for determining feasibility, giving allowed and disallowed zones. The atomic radii are given in Table I.

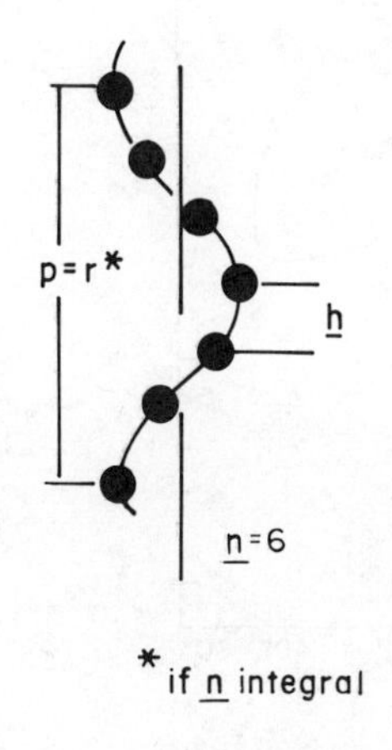

Figure 1. The helical parameters n, h, p, *and* r *(crystallographic repeat) for a right-handed helix. If the helix is integral,* p = r, *and it is always true that* p = n × h.

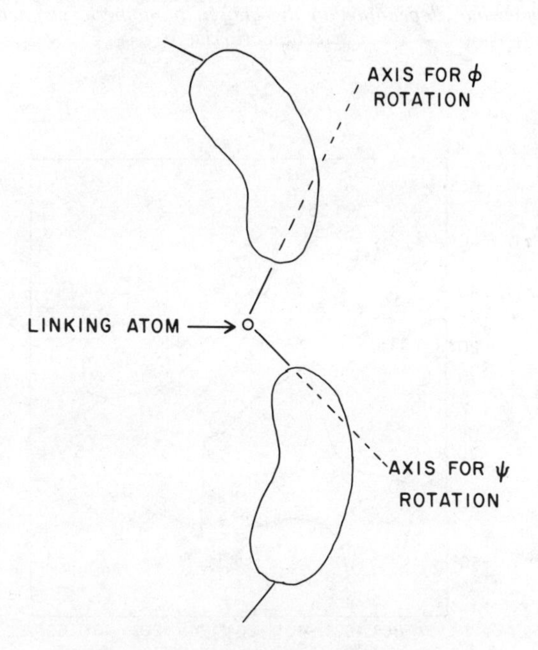

Figure 2. The axes for Φ *and* Ψ *rotation*

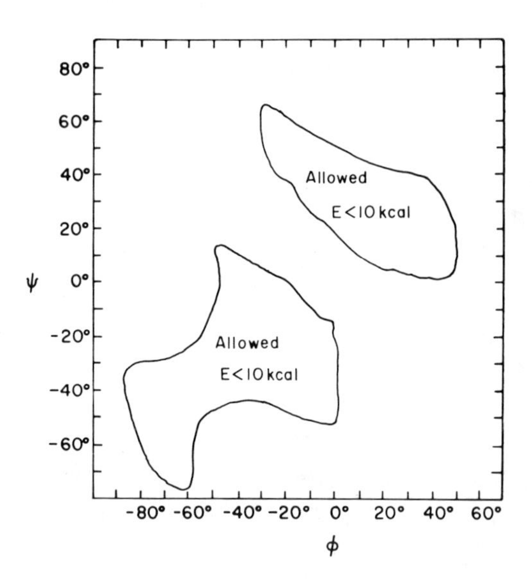

Figure 3. Stereochemically allowed zones, drawing adapted from Ref. 10. *These zones vary somewhat, depending on the chosen monomeric geometry as well as selection criteria.*

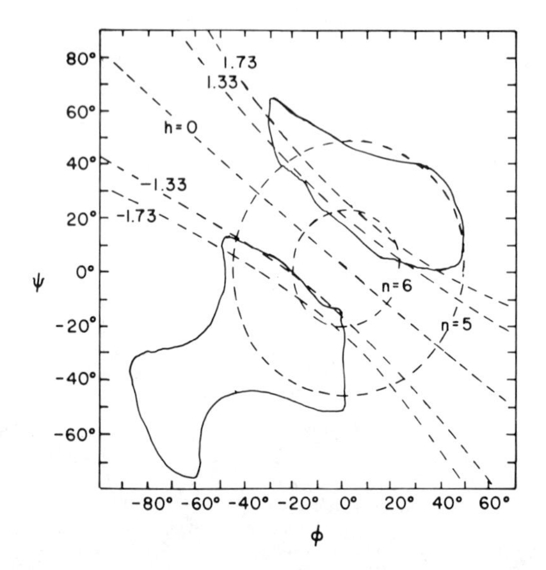

Figure 4. Combination of stereochemically allowed zones (Figure 3) and contours of iso-n *and* iso-h *values. Adapted from Ref.* 11. *The positions of the contour lines vary substantially with different monomeric geometries.*

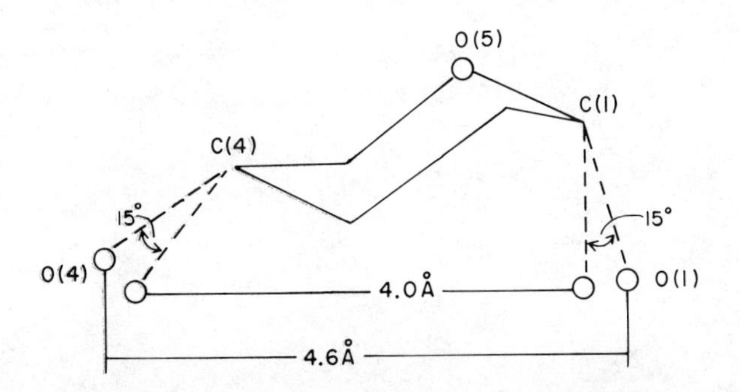

*Figure 5. Illustration of monomeric flexibility. In this α-pyranosic ring, substantial variation occurs but the C(1)–C(*D*) distance stays nearly constant. The virtual angles O(1)-C(1)–C(4) and C(1)–C(4)-O(4) each have 15° ranges and change equally for a given change in O(4)–O(1) distance.*

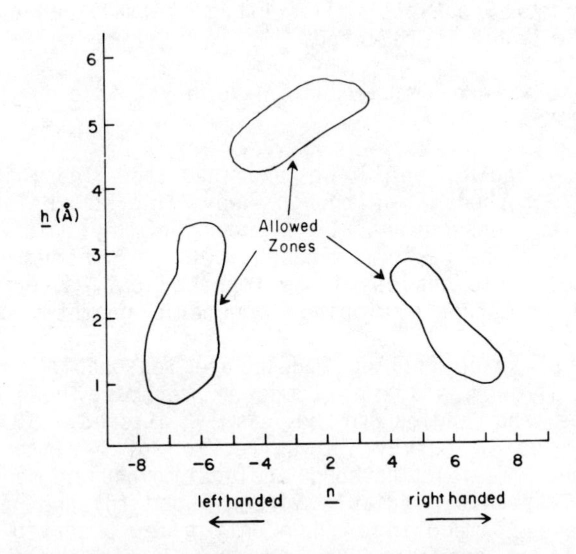

Figure 6. The n–h *map for cellulose, determined by hard-sphere criteria (*see *text and Table I). Note the zones of helices of small* h *and large* n *as well as the more usual conformations with small* n *and large* h. *The short contacts found in models for intermediates between the allowed zones would not be so severe as to prevent interconversion of the two types. Increments for* n *and* h *were 1 and 0.2 Å, respectively.*

TABLE I

Atom Pair	Fully Allowed* Distance (Å)	Minimally Allowed** Distance (Å)
C-C	3.2	2.96
C-O	2.9	2.66
C-H	2.4	2.16
O-O	2.8	2.56
O-H	2.4	2.16
H-H	2.1	1.86

* If six or more contacts between fully allowed and minimally allowed are found, the model is rejected.

** If one or more contacts smaller than minimally allowed are found, the model is rejected.

Feasible models were required to have glycosidic angles between 113° and 122°.

Each *n-h* point could be examined for feasibility by a variety of methods. Although we discuss below several techniques for these examinations, our primary point is not an evaluation of these various methods. Our main point is that the ranges of polymeric shape must be reported only after a thorough test of the important variables: monomeric geometry and linking bond angle.

Physical, space-filling models are reasonable tools in the search for allowed shapes. Computer methods, however, permit bond lengths and angles to be easily altered. One computer method would be to survey the collection of Φ-Ψ maps mentioned above. Other feasible methods include molecular mechanics and its related variants, variable virtual bond (6) and linked-atom, least squares (7) modeling. When only a few observed structures are available to represent the monomer, a technique similar to these can be used to provide a reasonable range of monomeric shapes. This range of monomeric geometries could then be used in a rigid-monomer, computer-modeling method. Modeling of a range of different monomeric shapes is discussed in another contribution (8).

If a flexible, but constrained, modeling technique is used for the entire polymer, the shapes depicted as feasible should represent all the conformations that are actually allowed for regular helices. The difficulty with these methods is, however, that we do not yet know precisely the limits of monomeric flexibility. That point is the subject of some other current research (9).

Program NHMAP

We are currently developing a computer program for *n-h* mapping. Previous experience led us to the rigid-monomer, virtual-bond method (10). Because virtual-bond modeling operates directly with the values of *n* and *h*, it is the simplest method, and it is rapid. Also, modeling with a variety of observed monomers has a number of advantages. For one, it answers the question: What are the allowed shapes of a given polymer, based on the known geometries?

Our program is currently written to run with an IBM level G compiler. Deviations from ANSI Fortran IV are limited to character representation specific to the IBM 360 computer. The system is command driven; users input appropriate monomeric coordinates, atomic radii, desired limits on angles, and ranges of *n* and *h* to be tested. After a run, some of the input values may be changed while other input remains the same.

Because the program is designed to be used interactively by the researcher, the user interface is as flexible as possible. Input values may be entered either by the minimum, maximum, and increment size, or by the actual desired values.

All commands are in free format. Incorrect input syntax is detected and its location marked without affecting other input or the success of the run. The purpose of interaction is to enable the user to test in closer detail around critical regions.

Capabilities of *n-h* Maps

We have structured the computer program for *n-h* mapping to test for effects on the allowed zones caused by different monomeric geometries, different ranges of linking bond angle, and different ranges of Φ and Ψ. Figure 7 shows the domain of allowed left-handed shapes of the α-1,4 glucan, amylose. The smaller allowed zones resulting from two different monomers are superimposed on the total allowed zone. Inadequate coverage of the total allowed zone results from using only one or two monomeric geometries. In particular, Figure 7 shows that only a small fraction of conformational space is available to the monomer with an O(4)--O(1) distance of 4.25 Å. Previously, Goebel, Dimpfl, and Brant reported the two Φ-Ψ maps resulting from these two monomers (11).

In Figure 8, the smaller overlays on the allowed zones depict our guesses at the effects of restricting the range of linking bond angle or the ranges of Φ and Ψ. The effect of a narrowed range of linking bond angle is relatively small; most of the allowed zone is still found. Holding Φ and Ψ constant substantially reduces the range of allowed *n* and *h* values. Because of the range of monomeric geometries and linking bond angles, however, a number of different helices can still be formed. The computer program is also structured so that other tests, such as for the existence of a particular hydrogen bond, can be easily added. This approach enables the researcher to extrapolate the results of a single-crystal study of an oligomer to polymers.

Even if monomeric geometry and linking bond angle were invariant, *n*-*h* maps would still have some advantages over Φ-Ψ maps. One advantage is that *n*-*h* maps for different polymers are different (compare Figures 6 and 9), whereas Φ-Ψ maps have allowed zones that look similar. Another problem with Φ-Ψ maps is that small changes in Φ and Ψ sometimes result in large changes in *n* and *h*. Such a coarse (in terms of *n* and *h*) grid search might cause some feasible *n*-*h* combinations to be inadvertantly disallowed.

More importantly, *n*-*h* maps can represent the allowed shapes for polymers that are more complicated than the simple homopolymers discussed herein. That advantage of *n*-*h* maps arises because the chemically repeated unit is allowed to have various shapes. In the *n*-*h* mapping approach, a geometric change in the chemical repeating unit of a heteropolymer is, in principle, no different from an internal change in the simpler monomers of a homopolymer. The results from a stereochemical study of each can still be placed on *n*-*h* maps.

The *n*-*h* maps are also useful in themselves. Relative molecular flexibility can be assessed by comparing the allowed areas for two different polymers. We think that 1/*n* vs. *h* plots would have advantages for this use (3).

Once all allowed shapes of a given polymer are represented on a single map, some fundamental questions can be raised. Compare, in Figure 9, the large range of allowed shapes with the observed shapes of amylose, indicated by the dots in the drawing. (Both right-handed and left-handed models are shown for the observed polymorphs because the experimental evidence is usually not clearly in favor of either chairality.) the observed forms seem to be grouped. Why are some shapes preferred over others? This question is discussed in references 4 and 12.

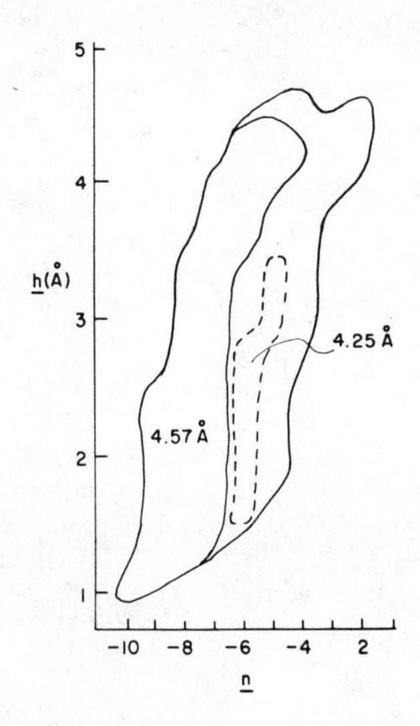

Figure 7. Map for left-handed amylose, showing the regions allowed by two different monomers having virtual bond (O(4)–O(1)) lengths of 4.57 Å and 4.25 Å. A similar experiment was reported in Ref. 11, except that the two zones were on different Φ-Ψ maps. Here, the two allowed zones are superimposed on the total allowed zone resulting from consideration of all residues available from single-crystal studies of maltose. Increments for n *and* h *were 1 and 0.2 Å, respectively.*

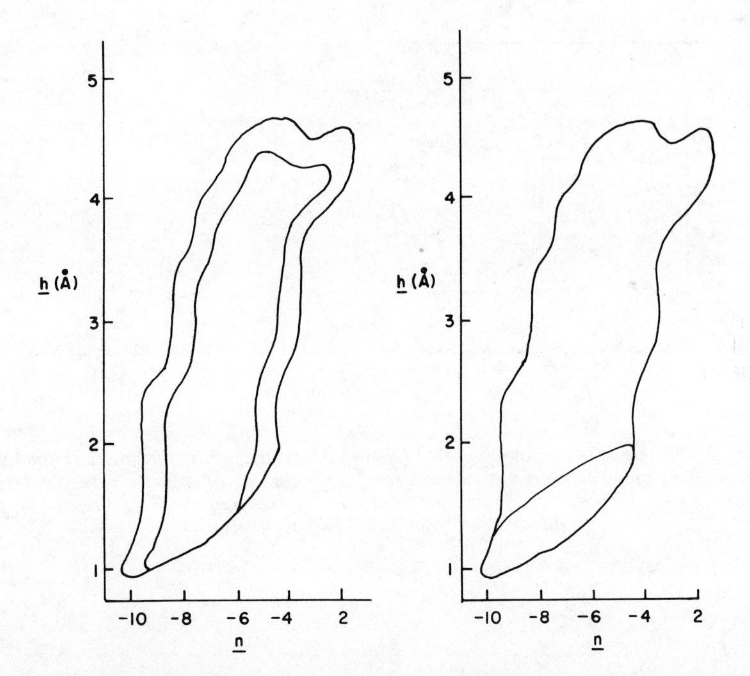

Figure 8. Restricted zones on n–h *maps of left-handed amylose resulting from limits on:* (left) *the linking bond angle: estimated region allowed by* $117° < \tau < 119°$ *compared with* $113° < \tau < 122°$*;* (right) *linking torsion angles: estimated region allowed when* $20 < \phi < 0$, $-10 < \psi < 0$ *compared to no restrictions.*

Both are superimposed on the total zone of Figure 7. The restricted zones in this figure only are estimates for the purpose of illustration. Because of the amount of computer time required with the old program and the deadline for publication of the proceedings, which prevented completion of the new program, this work has not been done.

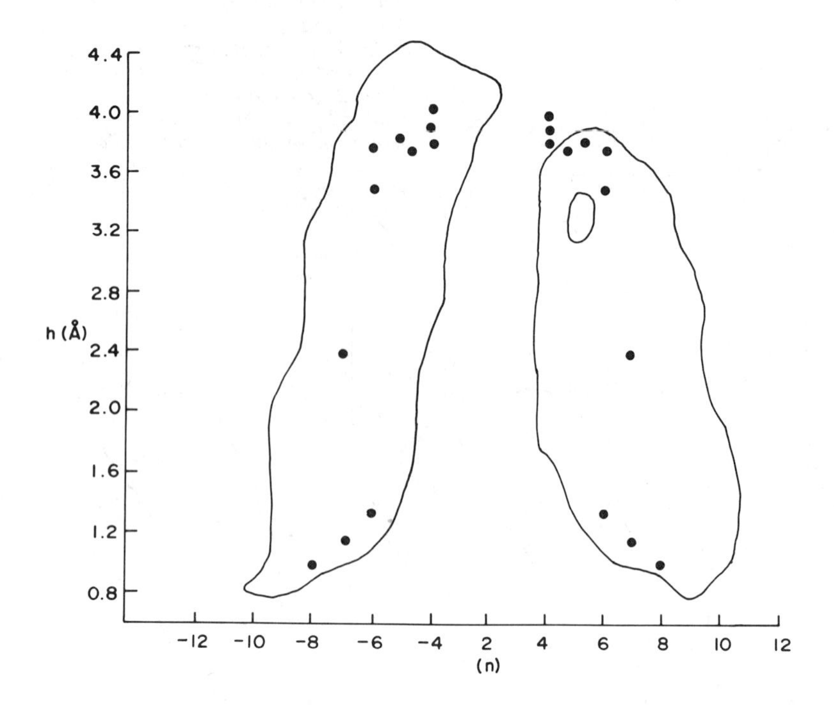

Figure 9. The n–h *map for amylose showing observed* n–h *values as dots on the total allowed zones. Note that there are two cluster of dots that correspond to efficient packing modes. For further information,* see *Ref.* 4. *Increments for* n *and* h *were 1 and 0.2 Å, respectively.*

As we come to understand the role of molecular shape in determining physical properties of polymers, we may wish to try to impose on a molecule a previously unobserved shape that was indicated in the allowed zones of an n-h map.

We see only two disadvantages in the n-h mapping approach. The finer structural details seem to be buried; no longer are Φ and Ψ available at a glance. Such details, however, may be called from computer memory for a particular model. Another problem with n-h mapping is that it requires more computer time than does a single geometry, Φ-Ψ map. With our original computer program and a slow computer, the last figure, which used 13 monomeric geometries, took about 72 hours. The new program, NHMAP, should be more than 10 times faster on the same computer. Other computers would be even quicker.

Acknowledgments

The authors thank Dexter French for helpful discussions. Some of our thoughts on the allowed and preferred shapes arose in discussions with D. A. Rees.

Literature Cited

1. Nata, G.; Corradini, P. Nuovo Cimento Suppl., 1960, 15, 9.

2. Ramachandran, G. N.: Ramakrishnan, C.; Sasisekharan, V. in Ramachandran, G. N., ed., "Aspects of protein Structure," Academic Press: London, 1963; p. 121.

3. French, A. D.; Murphy, V. G. Kainuma, K. J. Jpn. Soc. Starch Sci. (Denpun Kaga Ku), 1978, 25, 171.

4. French, A. D. Bakers Dig., 1979, 53, 39.

5. French, A. D.; Murphy, V. G. Polymer, 1977, 18, 489.

6. Zugenmaier, P.; Sarko, A. Biopolymers, 1976, 15, 2121.

7. Arnott, S.; Scott, W. E. J. Chem. Soc. Perkin Trans. II, 1972, 7, 324.

8. Pensak, D. A.; French, A. D. This Book.

9. Pensak, D. A.; French, A. D. Unpublished data.

10. French, A. D.; Murphy, V. G. Carbohydr. Res., 1973, 27, 391.

11. Goebel, C. F.; Dimpfl, W. L.; Brant, D. A. Macromolecules, 1970, 3, 644.

12. Rees, D. A. "Polysaccharide Shapes," Halsted Press: New York, 1977, p. 55.

RECEIVED May 21, 1980.

15

Theoretical Estimates of Helical Structure in Polynucleotides

WILMA K. OLSON

Department of Chemistry, Rutgers University, New Brunswick, NJ 08903

Potential energy studies over the last decade have supported the hypothesis that the principal conformations of the nucleic acids are implicit in their chemical architecture (1-10). Because the potentials are functions of molecular structure, these studies have helped to elucidate the interrelationship between the base, sugar, and phosphate moieties of the nucleotide repeating unit and the gross morphology of the polynucleotide chain. The energies usually reflect short-range nonbonded interactions in single-stranded nucleic acid fragments and, in principle, apply only to the treatment of ideal unperturbed randomly coiling chains (11). The data, however, provide useful starting conformations in constructing models of single-stranded polynucleotide helices (12, 13, 14, 15) and in analyzing X-ray crystallographic data (16).

In order to extend these conformational energy studies to the analysis of multi-stranded nucleic acid systems, it is necessary to devise a procedure to identify the arrangements of the polynucleotide backbone that can accommodate double, triple, and higher order helix formation. As a first step to this end, a computational scheme is offered here to identify the double helical structures compatible with given base pairing schemes. The feasibility of a duplex is estimated on the basis of semi-empirical energy estimates of base stacking and hydrogen bonding in a miniature double helix (i.e., complementary dinucleoside monophosphates) of specified conformation. The computed energies, however, are simply a measure of the geometric acceptability of the structure rather than a reliable measure of helix stability. The method is applicable to polynucleotide duplexes generated by any plausible base pairing arrangement and is thus useful in testing and comparing the various hydrogen bonding schemes that may stabilize a duplex.

A direct approach like this complements traditional DNA model building studies (17, 18, 19) that identify indirectly the combinations of backbone parameters that fit a particular set of helical parameters or structural constraints. The present method also offers a means to extrapolate theoretical and experimental

0-8412-0589-2/80/47-141-251$05.00/0

studies of low molecular weight nucleic acid analogs to polynucleotide systems. Since the local parameters can be continuously varied, this scheme further provides a means to study helix flexibility and to interpret macroscopic properties of the nucleic acid helices in solution.

Double Strand Formation

Double helical structures may be constructed from complementary single-stranded polynucleotide chains sharing a common helical axis according to the procedure outlined below. The two strands of the complex are assumed to be regular helices defined by a common set of backbone and glycosyl torsion angles. The data presented here are limited to model poly(dA)•poly(dT) double helices stabilized by Watson-Crick base pairs between antiparallel strands.

Helical parameters (n,h) and cylindrical atomic coordinates $\{r_i, \theta_i, z_i\}$ associated with each single strand of the polynucleotide complex may be obtained from the six backbone rotation angles (ψ', ϕ', ω', ω, ϕ, ψ) and the glycosyl rotation (χ) using the virtual bond scheme illustrated in Figure 1. The helical repeat n and step height h may be expressed as simple functions of the virtual bond length v and the angles θ^V and ψ^V that describe the orientation of adjacent chain residues:

$$\cos(\frac{n}{4\pi}) = -\cos(\frac{\theta^V}{2})\sin(\frac{\psi^V}{2}) \quad (1)$$

$$h \sin(\frac{n}{4\pi}) = v \cos(\frac{\theta^V}{2})\cos(\frac{\psi^V}{2}) \quad (2)$$

The virtual bond parameters may be obtained from the six fixed chemical bond lengths, the six fixed valence angles, and the six variable backbone rotation angles following procedures outlined elsewhere (12, 20). Atoms comprising each nucleotide repeating unit may also be transformed into a common virtual bond coordinate system following published methods (20). The final coordinate system assigned each single-stranded helix is chosen so that the z-axis coincides with the helix axis and so that the x-axis passes through the origin of the initial virtual bond. The orthogonal matrix $\underset{\sim}{T}(\theta,\theta^V,\psi^V)$ that effects the coordinate transformation from the frame of the initial virtual bond to that of the helix is obtained, following Shimanouchi and Mizushima (21), from the expression:

$$\underset{\sim}{T}(\theta,\theta^V,\psi^V)\ \underset{\sim}{T}(\theta^V,\psi^V) = \underset{\sim}{Z}(\theta)\ \underset{\sim}{T}(\theta,\theta^V,\psi^V) \quad (3)$$

where $\underset{\sim}{T}(\theta^V,\psi^V) = \underset{\sim}{X}(\psi^V-2\pi)\ \underset{\sim}{Z}(-\theta^V)$ is the matrix relating coordinate systems of adjacent virtual bonds with

$$\underset{\sim}{X}(\psi^V) = \begin{bmatrix} 1 & 0 & 0 \\ 0 & \cos\psi^V & -\sin\psi^V \\ 0 & \sin\psi^V & \cos\psi^V \end{bmatrix} \quad (4)$$

and

$$\underset{\sim}{Z}(\theta^V) = \begin{bmatrix} \cos\theta^V & -\sin\theta^V & 0 \\ \sin\theta^V & \cos\theta^V & 0 \\ 0 & 0 & 1 \end{bmatrix} \quad (5)$$

The parameter θ in Eq. 3 is the cylindrical repeating angle of the helix in radians and is given by $2\pi/n$.

The relative spatial locations of the two chains forming the theoretical duplex are determined on the basis of the geometry of a "perfect" Watson-Crick base pair. In this ideal situation the three atoms A-H•••B comprising each hydrogen bond are virtually collinear (as measured by $\cos\gamma \lesssim \sqrt{3}/2$ with $\gamma = \pi - \measuredangle$A-H-B) and the terminal heavy atoms A and B are separated by a distance D of approximately 2.9 Å. Cylindrical coordinates of the complementary poly(dT) chain, the backbone of which initially coincides with that of the poly(dA) strand, are obtained by inversion through the xy plane passing through z = 0 Å to alter the direction of the system followed by translation Δz and rotation $\Delta\theta$ about the helix axis to place the thymine bases in an optimal base pairing location (see Figure 2). The cylindrical coordinates (r_i, θ_i, z_i) of poly(dT) atom i expressed in the frame of the single-stranded helix are thus transformed to (r_i, $-\theta_i + \Delta\theta$, $-z_i + \Delta z$) in the double helix. The magnitudes of $\Delta\theta = \theta^I_{04} - (-\theta_{04})$ and $\Delta z = z^I_{04} - (-z_{04})$ are estimated from the ideal coordinates (r^I_{04}, θ^I_{04}, z^I_{04}) of the 04 thymine atom hydrogen bonded in a perfect orientation to N6 and H6 of adenine and the actual coordinates (r_{04}, θ_{04}, z_{04}) of the 04 atom in the poly(dT) single strand. If the magnitudes of r^I_{04} and r_{04} are similar, a double helix can be constructed with at least one linear hydrogen bond (N6(A)-H6(A)•••04(T)). If the r^I_{04} and r_{04} values differ appreciably, duplex formation is impossible. A Watson-Crick duplex is formed if both the N6(A)-H6(A)•••04(T) and the N1(A)•••H3(T)-N3(T) hydrogen bonds are of ideal geometry with the average $\langle D \rangle$ = 2.8-3.1 Å and $\langle\gamma\rangle \leq 30°$.

The double helices thus constructed are tested for geometric acceptability using semiempirical potential energy functions that reflect the extent of intrastrand base stacking and interstrand base pairing in a miniature dinucleoside monophosphate duplex of the same conformation. As described elsewhere (22), the base stacking potential V_S is chosen to display a minimum at a close to ideal base separation distance of 3.3 Å. Energies within

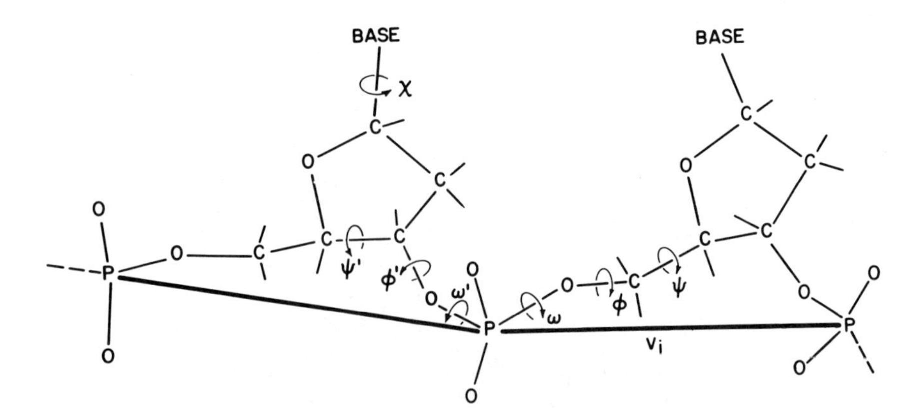

Figure 1. Schematic of a polydeoxyribonucleotide chain backbone showing the internal rotations (ψ′, φ′, ω′, ω, φ, ψ, and χ) and virtual bond vectors **v** *(heavy lines) characterizing each repeating unit*

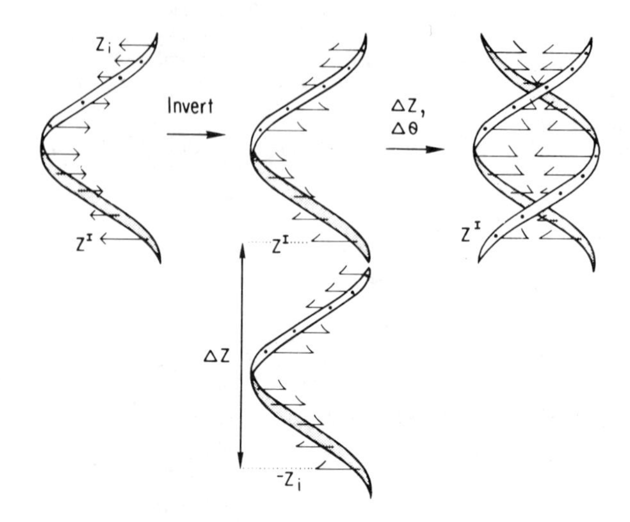

Figure 2. Schematic illustrating the theoretical construction of an ideally hydrogen-bonding polynucleotide helix from two complementary antiparallel strands of identical conformation (22)

4-5 kcal mole of this minimum are characteristic of partially overlapping base arrangements and 3-4 Å separation distances. The base pairing potential V_{HB} is minimized at optimum hydrogen bonding geometries. As the distance between the bases increases or the linearity parameter γ deviates from zero, V_{HB} is defined to merge smoothly with the ordinary van der Waals and Coulombic interactions between bases.

Potential Energy Surfaces

The above potential energy approach of building double helical polynucleotides is illustrated below for some hypothetical B-family poly(dA)·poly(dT) chains. For simplicity, the helices are generated as a function of the phosphodiester angles ω' and ω only. The remaining four backbone rotations and the glycosyl angle are set at constant values typical of B-DNA helices (see below). A fixed C2'-endo sugar pucker where $\psi' = -43°$ is chosen instead of the closely related C3'-exo conformation with $\psi' \sim -23°$ frequently reported in fiber diffraction studies (17, 23). According to potential energy estimates of pentose pseudorotation in model nucleosides (Olson, W.K., unpublished data), the C2'-endo state is lower in energy than the C3'-exo state. The C2'-endo pucker is also observed with greater frequency in the known X-ray crystal structures of low molecular weight nucleic acid analogs (24).

Details of the theoretically predicted double helical conformations are evident from the composite contour diagrams of base stacking and hydrogen bonding energies in Figure 3. The four fixed backbone rotation angles are here set at the following values defined with respect to trans = 0°: $\psi' = -43°$, $\phi' = -35°$, $\phi = 45°$, $\psi = -140°$. The glycosyl torsion χ is maintained at 75° with respect to trans arrangements of O1'-C1'-N9-C4 in the dA residues and of O1'-C1'-N1-C2 in the dT units. The energies are calculated at increments of 5° between 85° and 135° in both ω' and ω and the contours in the resulting two dimensional grid are located by interpolation. The dashed and solid contours in Figure 3 are indicative of helices with favorable base stacking and hydrogen bonding energies, respectively. These contours are located 4 kcal/mole above the minimum V_S value denoted by + and the minimum V_{HB} value located at ×. The surface is also divided into left- and right-handed helical fields by the dotted contour of h = 0 Å.

A noteworthy feature of Figure 3 is the large area of conformation space associated with favorable single-stranded base stacking. The 4 kcal/mole V_S contour coincides approximately with the set of geometrically acceptable base stacking conformations determined previously (25). The low energy base stacking occurs exclusively in right-handed structures for the choice of fixed parameters chosen here. This preference for right-handed helices persists upon minor variations in the B-DNA backbone (26) and occurs also in the A-DNA helices reported previously (22). The minimum base stacking state of the ω' ω pair in Figure 3 appears at (115°, 105°), corresponding to a relatively tightly wound helix with n = 9.5 and h = 3.2 Å. Adjacent bases in this conformation adopt an almost ideal orientation, the angle Λ between the overlapping base planes being 6° and the separation of base planes being 3.4 Å.

According to the V_{HB} contours in Figures 3, only a small proportion of polynucleotide conformations meet the somewhat rigorous criteria of acceptable hydrogen bonding in DNA double helices. The A•T bases on complementary strands associate in acceptable Watson-Crick base pairing arrangements when the backbone is confined to two narrow ranges centered along the ω' = 105° axis. Conformations in the smaller domain centered at (105°, 85°), however, introduce severe steric contacts between consecutive bases in each strand that rule out double helical structures. Both hydrogen bonding and base stacking achieve acceptable values in the larger area centered at (105°, 112°). The minimum energy double helix located at (105°, 115°) is an unusual 13-fold structure with h = 3.6 Å. A slight displacement of the bases away from the helix axis (denoted by × in Figure 4) introduces a small hole down the core of this complex. As a consequence of molecular size, consecutive A residues in the poly(dA) chain exhibit greater stacking overlaps than consecutive T bases of the complementary strand. The planes of complementary A and T bases describe an angle τ = 5 indicative of a slight propeller twisting of base pairs.

Minor variations of the backbone and glycosyl rotations from the fixed values used in the sample computations above produce a variety of theoretically acceptable double helices. As evident from the partial list of structures in Table I, these structures include several 10-fold duplexes similar to the B-DNA models from fiber diffraction studies as well as the larger 13-fold complex. Despite the large fluctuations in h from 1.7 to 4.3 Å, the bases associate at standard separation distances (3 Å ≤ <Z> ≤ 4 Å and 2.8 Å ≤ <D> ≤ 3.0 Å) and orientations (Λ ≤ 30° and <γ> ≤ 30°) in all cases. In order to avoid severe steric contacts at small values of h, the bases may tilt up to values of η = 45° with respect to the standard orientation (η = 90°) perpendicular to the helix axis.

Flexible Helices

The large number of theoretically acceptable duplexes described above suggests that DNA may adopt a variety of closely related helical structures in solution. A conformational blend of such helices within a single complex offers a more realistic interpretation of configuration-dependent properties of DNA in solution than the various rigid molecular models currently in use (27, 28). Chemical exchange studies (29, 30, 31, 32, 33) indicate, however, that the DNA duplexes do not remain perfectly

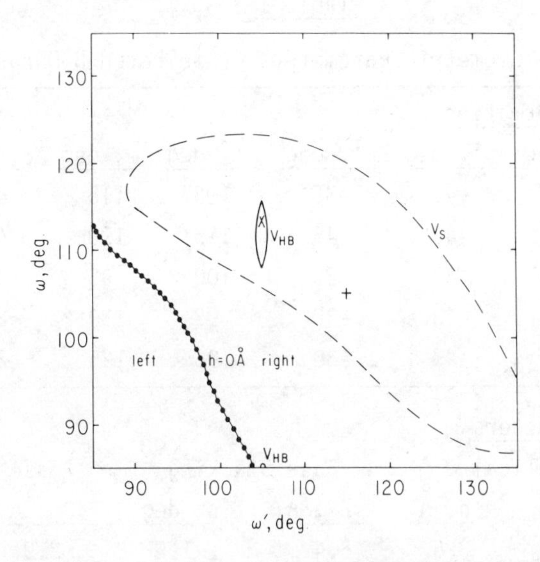

Figure 3. Contour diagram of the base stacking potential energy V_S of sequential adenine bases and the hydrogen bonding potential energy V_{HB} of the complementary A · T base pairs as a function of the phosphodiester rotation angles ω′ and ω. The energy contours enclose conformations within 4 kcal/mol of the minima marked by (+) for V_S and (×) for V_{HB}. The dotted contour of h = 0 Å *divides the space into fields according to chirality.*

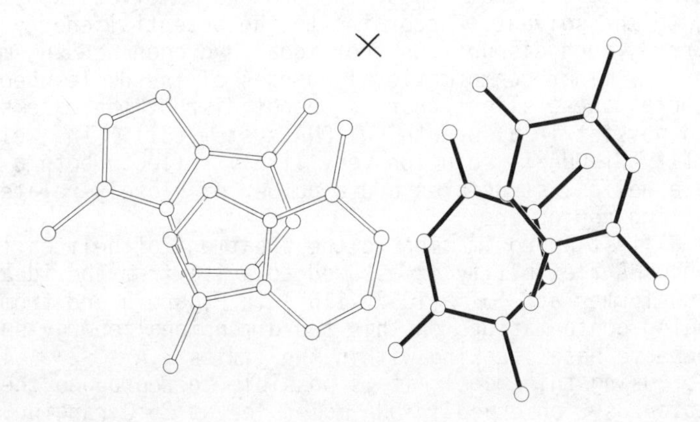

Figure 4. Detailed molecular representation of the theoretically predicted base stacking and base pairing of A (single lines) and T (solid lines) bases in the low energy ω′ω = 105°, 115° helix of Figure 3. The view is drawn perpendicular to the helix axis represented by (+).

Table 1

Comparative Geometric Parameters of Selected B-DNA Duplexes

Duplex	Conformation						
	χ, deg	ψ', deg	ϕ', deg	ω', deg	ω, deg	ϕ, deg	ψ, deg
I	75	-43	-35	105	115	45	-140
II	75	-43	-45	115	120	45	-140
III	95	-43	-25	100	95	45	-140
IV	70	-43	-30	95	130	29	-131
V	90	-43	-30	95	125	29	-131

Duplex	Parameters					
	Helical		Base Stacking		Base Pairing	
	n	h, Å	$\langle Z \rangle$, Å	Λ, deg	$\langle D \rangle$, Å	$\langle \gamma \rangle$, deg
I	13.0	3.6	3.4	1.5	2.9	4.5
II	10.3	3.4	3.0	13.2	2.9	9.4
III	10.5	1.7	3.7	26.2	2.9	7.5
IV	10.2	4.3	3.8	16.6	2.8	3.6
V	10.6	3.9	4.0	7.7	3.1	11.5

intact in solution. The polynucleotide backbone also undergoes "breathing" motions that expose the protons involved in hydrogen bonding to the solvent. According to the potential energy surface in Figure 3, such disruptions from ideal hydrogen bonding may involve only minor conformational changes of the duplex geometry. Furthermore, these slight changes do not disrupt the base stacking known to persist in a "breathing" DNA system (31). The well-known flexibility of DNA in solution very likely reflects both a variety of double helical structures and a number of closely-related non-base-pairing conformers.

As a first approximation to the treatment of helices in solution, DNA flexibility is assumed to arise from the ideal duplex conformer at $\omega'\omega$ = (105°, 115°) in Figure 3 and from the surrounding conformations on this two-dimensional energy surface that preserve base stacking within the limits 3 Å $\leq \langle Z \rangle \leq$ 4 Å and $\Lambda \leq 12°$. Using this model, it is possible to reproduce the radii of gyration $\langle s^2 \rangle$ observed in DNA molecules at 25°C ranging between 400 and 5000 base pairs in length (25). These limited motions are further consistent with the observed frequency of loop formation and cyclization in DNA of various lengths (34). The large proportion of non-base-paired states in the model (cf. Figure 6), however, is not consistent with the measured fraction (1/20) of open states of DNA at 0°C (33). An improved model that includes several ideal duplexes and a smaller number of related non-base-pairing structures is expected to reproduce the known proportion of breathing as well as the chain dimensions in DNA.

The flexible helix modeled here is best described by the entire array of conformations it can assume. A comprehensive picture of this array is provided by the three-dimensional spatial probability density function $W_0(\mathbf{r})$ of all possible end-to-end vectors (25, 35). This function is equal to the probability per unit volume in space that the flexible chain terminates at vector position $\mathbf{r}$ relative to the chain origin $\mathbf{0}$ as reference. An approximate picture of this distribution function is provided by the three flexible single-stranded B-DNA chains of 128 residues in Figure 5(a). The conformations of these molecules are chosen at random by Monte Carlo methods (35, 36) from the conformations accessible to the duplex model. The three molecules are drawn in a common coordinate system defined by the initial virtual bond of each strand. For clarity, the sugar and base moieties are omitted and the segments are represented by the virtual bonds connecting successive phosphorus atoms.

The three flexible chains in Figure 5(a) possess pseudohelical backbones and pseudohelical axes that change direction continuously. The individual turns of each pseudohelix vary considerably in size as a consequence of the range of local nucleotide motions. The gradual bending of the pseudohelical backbones describes trajectories with large radii of curvature. The end-to-end separations are less than that of the regular helix of the same length represented in Figure 5(b). The small group of random molecules crudely describe a spatial distribution function. The first 2-3 turns of each pseudohelix are roughly superimposable and also are only slightly altered from the regular helix. At these chain lengths, the molecules are rodlike and the distribution of r is confined to a small domain (34). Segments more removed from the chain origin, however, are found to deviate appreciably from the regular structure. The 128-segment structures are best described as wormlike chains. The distribution of end-to-end vectors of chains of this size is confined to an umbrella- or mushroom-shaped volume (25, 34). At longer chain lengths, the deviations between the various chain conformers become much more pronounced. Eventually at lengths of 8,000 to 10,000 nucleotides, the distribution function is an ideal Gaussian (25). The probability of occurrence of end-to-end vectors is completely random and all values of r are equally probable.

Flexible double helices are expected to follow approximately the trajectories of the DNA-B single strands outlined above. According to the measurements of DNA breathing, at least 95% of the motions of a double-stranded unit are limited to states that preserve Watson-Crick base pairing. The extent to which rotations of the $\omega'\omega$ pair about the 13-fold helix fulfill this criterion can

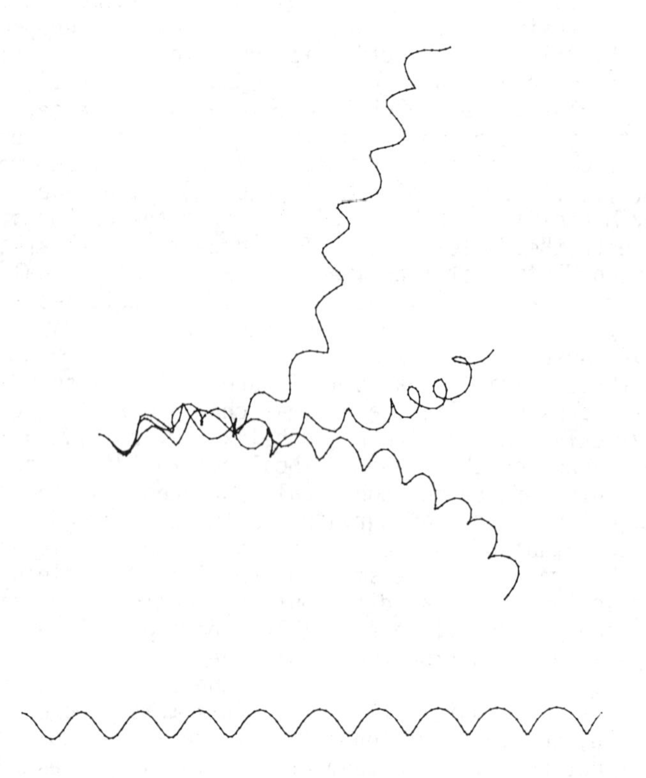

Figure 5. Computer generated perspective representation of single-stranded B-*DNA chains. The 128-residue chains are represented by the sequence of virtual bonds connecting successive phosphorus atoms. (a) Three representative flexible helices generated by Monte Carlo methods; (b) the regular $\omega'\omega = 105°, 115°$ helix predicted by potential energy methods.*

be estimated from the contour surface in Figure 6. Here the two strands of the duplex are not required to assume identical conformations. The solid lines in the figure are contours of the distance of separation d between C1' atoms on adjacent residues within different single-stranded helical units. Upon variation of ω' and ω over the range of phosphodiester rotations used above to describe flexible DNA single strands, the parameter d is found to vary between 4 and 5 Å. In contrast to the gradual changes in d, the dashed contours d' are more closely spaced and thus are more strongly dependent upon the ω' and ω rotations. The latter parameter is the C1' separation distance between two adjacent free bases that are hydrogen bonded with ideal Watson-Crick geometry to the single-stranded helical backbone. Duplex formation is possible if the complementary chain can assume a backbone arrangement that links these two free bases. In Figure 6 this complementary strand conformation is estimated as a state with a value of d identical to the d' value required by the state of the first strand. The single point denoted by × where d = d' = 4.8 Å in both strands is the ideal duplex predicted in Figure 3. The motions of a flexible duplex are thus limited in Figure 6 to the shaded area where d = d' = 4-5 Å.

Summary

The above potential energy method provides a convenient way to identify directly those conformations of the nucleic acid backbone and base that can participate in double helix formation. While these "energies" are necessarily approximate, they afford a basis for clear discrimination between sterically allowed and sterically forbidden structures. The "energy" approach also offers a means to extrapolate experimental studies (nmr, X-ray, etc.) on the conformation of small model compounds to the polynucleotide level and to test the relevance of the data in a helical complex. In addition, the method provides a starting point for a refined potential energy analysis of double helical conformation and stability.

The contour surfaces that relate local motions of the backbone to helical "energies" provide additional insight into the conformational nature of helices in solution. The continuous variation of rotation angles permits a direct molecular treatment of chain flexibility and of various solution properties of helices (i.e., light scattering dimensions, cyclization and loop closure probabilities, extent of "breathing", etc.). Earlier models of DNA in solution involve either specific regular helical structures (27, 28) or approximate artificial descriptions such as rigid rod, wormlike coil (37, 38), or Gaussian. The computations described here also clarify why the various descriptions of DNA apply at different chain lengths. The configuration-dependent properties of DNA of all sizes reflect very limited local molecular motions. With increase in chain length the motions of the DNA as a whole

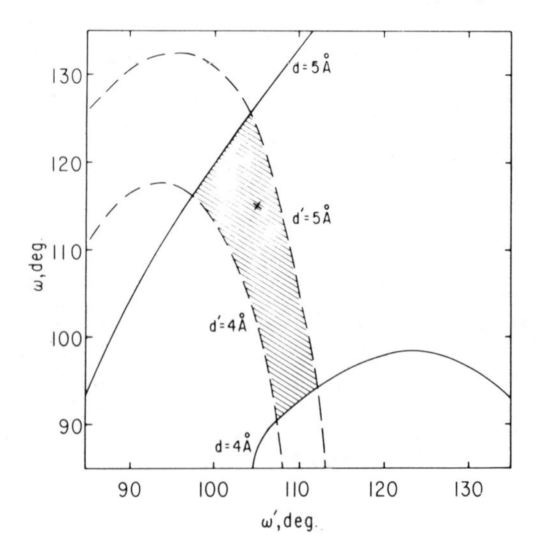

Figure 6. Contour diagram of the distances d *and* d′ *between successive Cl′ atoms in a DNA duplex as a function of the phosphodiester angles. The (×) denotes the conformation of the ideal theoretical helix of Figure 3 with* d = d′ = *4.8 Å. The shaded area on the diagram describes the local motions that preserve base stacking in a flexible duplex.* See *text for further explanation.*

magnify gradually and approach eventually the behavior of an ideal random coil.

List of Symbols

1. Mathematical Conventions

< >	The statistical mechanical average of the quantity enclosed, taken over all configuration of the chain.
$\underset{\sim}{T}$	The matrix $\underset{\sim}{T}$ of order 3 x 3.
$\underset{\sim}{r}$	The chain vector connecting the ends of the polynucleotide.
$\underset{\sim}{O}$	The origin of the chain.

2. Roman Letter Symbols

d	Distance between C1' atoms on adjacent residues in a particular helical unit.
d'	Distance between the C1' atoms that are attached to two free bases bound to a particular helical unit.
dA	Deoxyadenosine.
dT	Deoxythymidine.
D	Distance between the terminal heavy atoms of a hydrogen bond.
h	Helical step height.
n	Helical repeat = $2\pi/\theta$.
r_i	Cylindrical radius associated with atom i.
v	Length of the virtual bond connecting successive P atoms of the polynucleotide.
V_{HB}	Potential energy of hydrogen bonding.
V_S	Potential energy of base stacking.
$W_{\underset{\sim}{O}}(\underset{\sim}{r})$	Distribution function of the vector $\underset{\sim}{r}$ in three dimensions with respect to the chain origin as reference.
x	Cartesian axis.
y	Cartesian axis.
z	Cartesian axis.
z_i	Cylindrical vertical displacement associated with atom i.
Z	Separation of an atom in base i + 1 from the plane of base i.

3. Greek Letter Symbols

γ	Angle described by the three atoms comprising a hydrogen bond.
θ	The cylindrical repeating angle, in radians, of the polynucleotide helix.
θ_i	Cylindrical rotation angle associated with atom i.

θ^V	The complement to the pseudovalence angle formed by successive virtual bonds.
Λ	Angle described by the planes of adjacent bases in the same polynucleotide strand.
τ	Angle described by the planes of two bases arranged in a hydrogen-bonding geometry.
ϕ	Torsion angle about atoms P-O5'-C5'-C4'.
ϕ'	Torsion angle about atoms C4'-C3'-O3'-P.
χ	Glycosyl torsion angle between base and pentose.
ψ	Torsion angle about atoms O5'-C5'-C4'-C3'.
ψ'	Torsion angle about atoms C5'-C4'-C3'-O3'.
ψ^V	The pseudo rotation angle defined with respect to a 0° = trans arrangement of three successive virtual bonds.
ω	Torsion angle about atoms O3'-P-O5'-C5'.
ω'	Torsion angle about atoms C3'-O3'-P-O5'.

Acknowledgment

The author is grateful to the National Institutes of Health (USPHS Grant GM 20861) and the Charles and Johanna Busch Memorial Fund of Rutgers University for laboratory support and to the Center for Computer and Information Services of Rutgers University for computer time. A Career Development Award from the USPHS (GM 00155) and an A. P. Sloan Fellowship to W.K.O. are also gratefully acknowledged.

Literature Cited

1. Olson, W. K.; Flory, P. J. Biopolymers, 1972, 11, 25-56.
2. Sasisekharan, V., in "Conformation of Biological Molecules and Polymers"; Bergmann, E. D.; Pullman, B., Eds. Fifth Jerusalem Symposia on Quantum Chemistry and Biochemistry, 1973; pp. 247-260.
3. Yathindra, N.; Sundaralingam, S., in "Structure and Conformation of Nucleic Acids and Protein-Nucleic Acid Interactions"; Sundaralingam, M.; Rao, S. T., Eds. University Park Press: Baltimore, MD, 1975; pp. 649-676.
4. Thornton, J. M.; Bayley, P. M. Biochem. J., 1975, 149, 585-596.
5. Broyde, S. B.; Wartell, R. M.; Stellman, S. D.; Hingerty, B.; Langridge, R. Biopolymers, 1975, 14, 1597-1613.
6. Pullman, B.; Saran, A. Prog. Nucleic Acids Res. and Molec. Biol., 1976, 18, 216-326 and references cited therein.
7. Govil, G. Biopolymers, 1976, 15, 2302-2308.
8. Tosi, C.; Clementi, E.; Matsuoka, O. Biopolymers, 1978, 17, 67-84.
9. Thiyagarajan, P.; Ponnuswamy, P. K. Biopolymers, 1978, 17, 533-553 and 2143-2158.
10. Broch, H.; Vasilescu, D. Biopolymers, 1979, 18, 909-930.

11. Olson, W. K. Biopolymers, 1975, 14, 1775-1795.
12. Olson, W. K. Biopolymers, 1976, 15, 859-878.
13. Yathindra, N.; Sundaralingam, M. Nucleic Acids Res., 1976, 3, 729-747.
14. Fujii, S.; Tomita, K. Nucleic Acids Res., 1976, 3, 1973-1984.
15. Hingerty, B.; Broyde, S. Nucleic Acids Res., 1978, 5, 127-137.
16. Stellman, S. D.; Hingerty, B.; Broyde, S.; Subramanian, E.; Sato, T.; Langridge, R. Biopolymers, 1973, 12, 2731-2750.
17. Arnott, S.; Chandresekaran, R.; Selsing, E., in "Structure and Conformation of Nucleic Acids and Protein-Nucleic Acid Interactions"; Sundaralingam, M.; Rao, S. T., Eds. University Park Press: Baltimore, MD, 1975; pp. 577-596 and references cited therein.
18. Zhurkin, V. B.; Lysov, Yu. P.; Ivanov, V. I. Biopolymers, 1978, 17, 377-412.
19. Miller, K. J. Biopolymers, 1979, 18, 959-980.
20. Olson, W. K. Macromolecules, 1975, 8, 272-275.
21. Shimanouchi, T.; Mizushima, S. J. Chem. Phys., 1975, 23, 707-711.
22. Olson, W. K. Biopolymers, 1978, 17, 1015-1040.
23. Arnott, S.; Selsing, E. J. Mol. Biol., 1974, 88, 509-521.
24. Murray-Rust, P.; Motherwell, S. Acta Cryst., 1978, B34, 2534-2546.
25. Olson, W. K. Biopolymers, 1979, 18, 1213-1233.
26. The two-dimensional ω'ω energy surfaces of each of the theoretical duplexes listed in Table I exhibit the same general features of Figure 3.
27. Hogan, M.; Dattagupta, N.; Crothers, D. M. Proc. Natl. Acad. Sci. USA, 1978, 75, 195-199.
28. Levitt, M. Proc. Natl. Acad. Sci. USA, 1978, 75, 640-644.
29. Printz, M. P.; von Hippel, P. H. Proc. Natl. Acad. Sci. USA, 1965, 53, 363-370.
30. Frank-Kamenetskii, M. D.; Lazurkin, Yu. S. Ann. Rev. Biophys. Bioeng., 1974, 3, 127-150.
31. Teitelbaum, H.; Englander, S. W. J. Mol. Biol., 1975, 92, 55-78 and 79-92.
32. McGhee, J. C.; von Hippel, P. H. Biochem., 1975, 14, 1281-1296 and 1297-1303.
33. Kallenbah, N. R.; Mandal, C.; Englander, S. W., in "Stereodynamics of Molecular Systems"; Sarma, R. H., Ed. Pergamon Press: New York, 1979; pp. 271-282.
34. Olson, W. K., in "Stereodynamics of Molecular Systems"; Sarma, R. H., Ed. Pergamon Press: New York, 1979; pp. 297-314.
35. Yevich, R.; Olson, W. K. Biopolymers, 1979, 18, 113-145.
36. Jordan, R. C.; Brant, D. A.; Cesaro, A. Biopolymers, 1978, 17, 2617-2632.
37. Kratky, O.; Porod, G. Rec. Trav. Chim. Pays-Bas, 1949, 68, 1106-1122.
38. Yamakawa, H.; Shimada, J.; Fujii, M. J. Chem. Phys., 1978, 68, 2140-2150.

RECEIVED June 24, 1980.

Contribution of Electron Diffraction on Single Crystals to Polymer Structure Determination

FRANCOIS BRISSE and ROBERT H. MARCHESSAULT

Department of Chemistry, Universite de Montreal,
C.P. 6210, SUCC. A., Montreal, Quebec, Canada H3C 3V1

X-ray fiber diffraction provides the most important method for the structure determination of chemically regular crystalline polymers. Shortcomings of this approach are:

- ambiguities regarding choice of the space group,
- overlap of diffracted intensities and imperfect orientation of the fibers,
- limited number of diffraction data.

Recently, single crystal preparations from natural (1-3) and synthetic (4,5) polymers have become a well established art. This has allowed electron diffraction to complement X-ray diffraction information. This approach is valuable for:

- obtaining 2 to 3 times as many base plane reflections (hk0) allowing more precise unit cell dimensions,
- clearer observation of systematic absences in hk0 data,
- additional intensity data of the hk0 type.

The polymer single crystals could add significantly to structure determinations by using fiber diagrams and single crystal data in a simultaneous refinement. However, in electron diffraction analysis, the agreement between observed and calculated intensities is generally not as good as for X-ray diffraction (6). Although the structure factors for X-ray and electron diffraction hk0's are usually of comparable value, one occasionally finds very significant differences for a given reflection. Morphological aspects of single crystals are readily related to unit cell orientations in the electron diffractogram.

Experimental

Micrograph and electron diffraction patterns were obtained with a Philips EM300 electron microscope which was used at 80kV for imaging and 100 kV (λ = 0.03702 Å) for diffraction, in the selected-area diffraction mode. For d-spacing calibration, gold was evaporated on the specimen grid prior to deposition of the crystals in suspension form. The diffracted intensities recorded on Kodak Electron Image film #4463 were measured using a

0-8412-0589-2/80/47-141-267$05.00/0

Joyce-Loebel MK III C flat-bed microdensitometer. The structure factor magnitudes were derived from the integrated intensities according to $|F(hk\ell)|^2 = I(hk\ell)$. The largest errors in the intensity measurements were for the darkest spots. The scattering factors of C, O & H for the diffraction of electrons were taken from the International Tables for X-Ray Crystallography, vol. IV (7). No correction for multiple scattering and n-beam dynamical interactions were applied because the thickness of the crystals was of the order of 100 Å (6).

Electron Diffraction of Poly(n-methylene terephthalates), (nGT polymers).

Poly(trimethylene terephthalate) = 3GT $[C_6H_4\text{-CO-O-}(CH_2)_3\text{-O-CO}]_n$

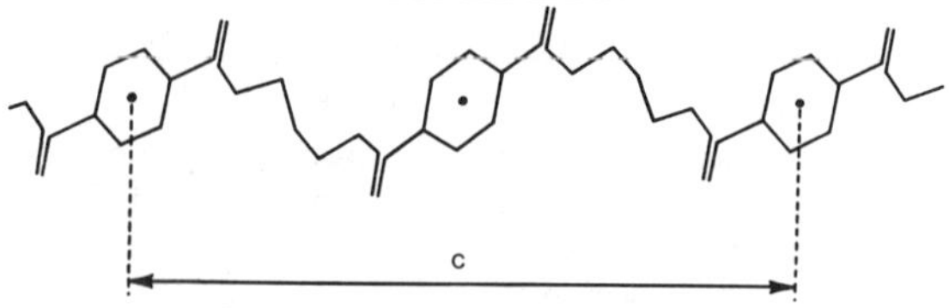

c

Single crystals for electron micrography and diffraction were obtained by dissolution of the polymer in nitrobenzene at 170° C followed by filtration. The solution was slowly cooled to 134° C and kept at this temperature for 12 h. The polymer that had crystallized was filtered and redissolved in nitrobenzene at 170° C, cooled to 139° C then very slowly to 136° C, kept at this temperature for 24 h and cooled to room temperature leaving the polymer single crystals in suspension in nitrobenzene. A total of 25 reflections could be observed on the electron diffractogram (hk0 section) of this material. This was an improvement over the X-ray fiber diagram since only five reflections had been recorded on the equator (see Table I).

The d-values of the electron diffractogram allowed us to fully index the X-ray fiber diagram using a triclinic unit cell. Without the electron diffraction reciprocal cell parameters (a^*, b^*, γ^*), the task of indexing the fiber diagram would have been more difficult because of the paucity of hk0 reflections on the equator.

On the electron diffractogram, the 010 reflection was by far the strongest and we had difficulty evaluating its intensity with respect to that of the other much weaker intensities. The structure amplitudes calculated for the structure of 3GT, arrived at by the model compound approach (8), are in very reasonable agreement with the corresponding values derived from the observed intensities considering the assumptions involved. These were a) the molecule has a two-fold axis of rotation passing through the central methylene group and b) the chain is rotated as a whole around the c-axis since we did not have the facility to refine the structure by varying the individual torsion angles.

TABLE I

hk0 reflections of 3GT observed on the X-ray fiber diagram (XR) and on the electron diffractogram (ED)

d	hkℓ	|Fo| XR	|Fo| ED	d	hkℓ	|Fo| XR	|Fo| ED
5.720	010	100	100	1.797	-230		10
4.301	100 }	26	20	1.543	130		3
4.289	-110 }		18	1.539	-140		5
2.950	110		12	1.532	-310		3
2.939	-120		13	1.530	-320		3
2.860	020		25	1.475	220		–
2.317	-210		5	1.469	-240		5
2.150	200 }		8	1.434	300 }		12
2.144	-220 }	25	11	1.430	-330 }	23	12
2.055	120		11	1.430	040 }		3
2.048	-130		12	1.283	310		2
1.907	030	54	24	1.278	-340		3
1.804	210		3	1.125	320		2

Poly(hexamethylene terephthalate) = 6GT $[C_6H_4-CO-O-(CH_2)_6-O-CO]_n$

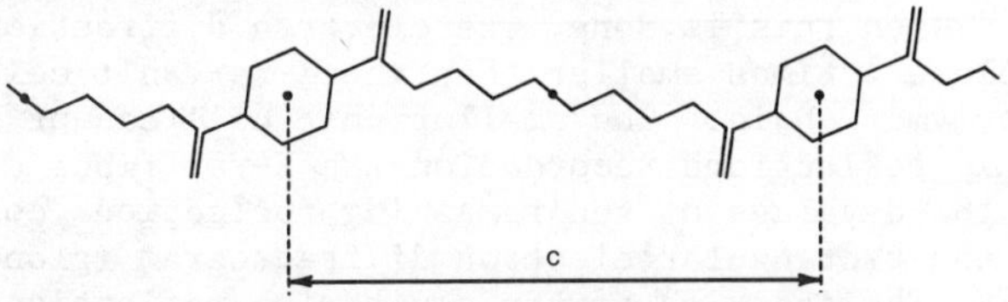

Depending on the choice of solvent, the temperature of dissolution and the rate of cooling, single crystals with at least two well differentiated morphologies could be obtained in a reproducible manner.

When nitrobenzene alone is used as a solvent (170° C) and with slow cooling, lamellar single crystals are obtained. Their electron diffractograms have an oblique cell (plane group p2). With a mixture of nitrobenzene/decane (65/35) or nitrobenzene/hexane the dissolution temperature is reduced to about 70° C. This gives "globular" single crystals whose electron diffractograms belong to the pgg (or pmg) plane group and the unit cell is rectangular. Occasionally both crystal forms were observed in the same preparation. These are shown on Figure 1.

The X-ray fiber diagram of 6GT could only be indexed (9) with a triclinic unit cell containing 4 chains per cell. This is quite unusual since there is only one chain per unit cell for all the other known nGT. The reciprocal lattice parameters resulting from the electron diffractograms are compared in Table II with the corresponding values calculated from published X-ray data. The a* and b* reciprocal unit cell parameters of the p2 cell (lamellar single crystals) are twice as large as those calculated using the published triclinic unit cell obtained from a fiber diagram while the γ* angle is identical.

Table II

Reciprocal unit cell dimensions of 6GT

	Electron diffractogram		X-ray fiber diagram
	This work		(9)
	pgg	p2	triclinic cell
a*	0.1106 $Å^{-1}$	0.1999 $Å^{-1}$	0.1029 $Å^{-1}$
b*	0.0726 $Å^{-1}$	0.2540 $Å^{-1}$	0.1231 $Å^{-1}$
γ*	90.0°	76.2°	80.0°

The direct space unit cell cannot be computed since only three of the six reciprocal unit cell dimensions may be obtained by electron diffraction. However since there is an obvious relationship between the electron diffraction (p2) and the X-ray fiber unit cells, the missing parameters (c*, α*, β*) may be assumed to be identical to those derived from the X-ray fiber diagrams. When this is done, the electron diffraction unit cell with a volume 4 times smaller than the X-ray unit cell contains but one polymer chain. The smaller unit cell cannot account for all the hk0 reflections recorded on the X-ray fiber diagram. However, the d-values of the remaining reflections coincide with those of the rectangular electron diffractogram (globular single crystals). The indexing of the equatorial reflections of the fiber diagram by means of the two unit cells is presented in Table III. This would seem to indicate that the two crystalline forms detected by electron diffraction co-exist in the fiber.

As far as the structure of 6GT is concerned, a fully extended planar chain fits well in the simple triclinic unit cell; this is confirmed by the good agreement between observed and calculated structure amplitudes for electron diffraction (10).

To the two-dimensional rectangular unit cell corresponds a three dimensional unit cell which may be either monoclinic or orthorhombic. This is what we are now trying to establish. Presumably, if the two cells (triclinic and monoclinic or orthorhombic) do coexist in the fiber, the fiber repeats should be identical. Consequently, a monoclinic unit cell with a large β angle would be preferable since the c directions in the two cells (triclinic and monoclinic) would coincide.

Table III

Indexing of the hk0 level of the X-ray fiber diagram using electron diffraction data for 6GT

Electron Diffraction Rectangular Cell		X-Ray Fiber Diagram	Electron Diffraction Oblique Cell (p2)	
d_o	hkℓ	d_o	d_o	hkℓ
		4.989	4.995	100
4.521	200	4.536		
4.096	130	4.110		
		3.772	3.946	010
3.445	040	3.494	3.533	$1\bar{1}0$
3.222	230	3.235		
2.520	330	2.561		
2.297	060	2.293		
		2.026	2.006	$\bar{1}20$
1.924	170	1.947	1.973	020
1.827	360	1.804	1.767	$2\bar{2}0$

Electron Diffraction of Polyalkanoates

Poly-β-Hydroxybutyrate. It has been shown (11,12,13) that poly-β-hydroxybutyrate (PHB) is a head-to-tail polyester based on the monomer D(−)-β-hydroxybutyric acid which leads to an optically active polymer having the formula:

$$\left[-\underset{CH_3}{\overset{H}{\underset{|}{\overset{|}{C}}}} - \underset{H}{\overset{H}{\underset{|}{\overset{|}{C}}}} - \underset{O}{\underset{\|}{C}} - O - \right]_n$$

The crystal structure has been reported by two independent groups: Cornibert and Marchessault (14) for the optically active natural polymer and Yokouchi *et al.* (15) for an optically neutral polymer made form a d,l monomer mixture. In spite of this, the two structures are in excellent agreement, the largest observed difference in the coordinates of the atoms is less than 0.18 Å.

Monolamellar single crystals of PHB were obtained by placing one milligram of active PHB of $\bar{M}_n$ = 42,000 in 4 ml of a 2/1 (v/v) mixture of ethanol and chloroform in a sealed tube and heating to 73° C. The solution was then cooled to 60° C and kept at this temperature for 24 hrs. The precipitate formed was redissolved by heating at 73° C just until the disappearance of turbidity and the tube was again placed in a 60° C bath for 24 hrs. The second dissolution led to a self-seeded crystallization and the large crystals shown in Fig. 2. The insert in Fig. 2 is the electron diffractogram recorded for this optically active systems (16).

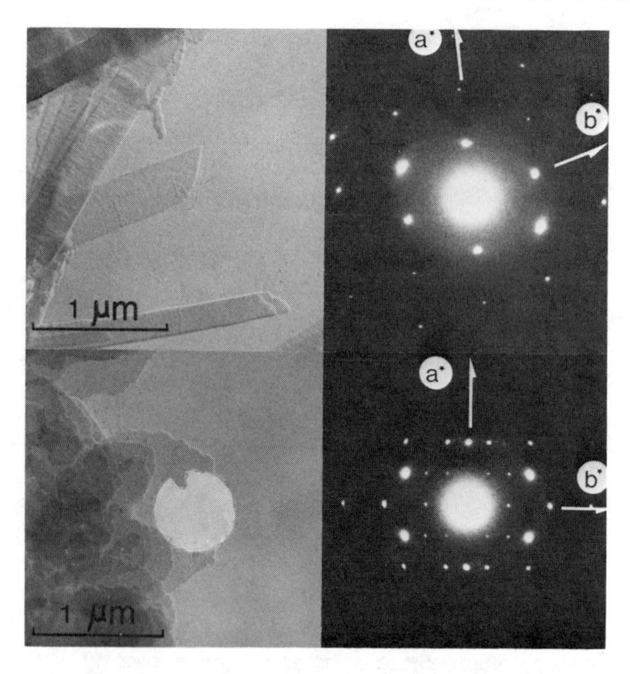

Figure 1. 6GT single crystals and corresponding electron diffractograms. (top) *Single crystals with an oblique cell;* (bottom) *single crystals with a rectangular cell.*

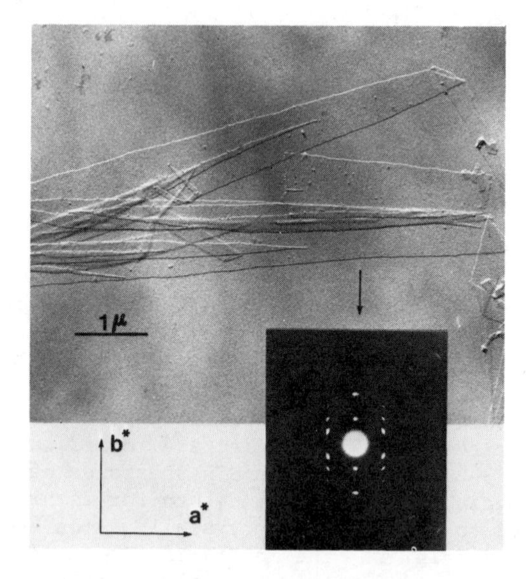

Figure 2. Single crystals of PHB and its electron diffractogram

The unit cell parameters a and b were 5.76 Å and 13.20 Å respectively which agrees well with published data (14,15) based on fiber diagrams. Clearly the diffractogram is an hk0 reciprocal lattice net as expected if the symmetry axis of the molecular helix is perpendicular to the crystal face. Since only even reflections are observed along the h00 and 0k0 directions the pgg base plane symmetry is confirmed in keeping with the proposed $P2_12_12_1$ space group for the orthorhombic unit cell (14,15).

The agreement index from the observed single crystal electron diffraction structure factors (28 separate reflections) compared to the calculated structure factors for the Yokouchi *et al.* model was 0.33. This is far from the values of 0.24 and 0.135 reported for the two published x-ray fiber analyses (14,15). This suggests that structure refinement based on x-ray diffraction data should simultaneously make use of base plane electron diffraction data rather than "after the fact" comparison as in this case.

A noteworthy fact is that for some reflections the Fo's from x-ray diffraction were significantly different from those derived from electron diffraction. This is illustrated in Table IV (16).

Table IV

Comparison of Some Observed (Fo) Structure Factors Derived from X-Ray Fiber Diffraction and Single Crystal Electron Diffraction on PHB.

hkℓ	Fo(X-ray)	Fo(Electron)
110	100.0	100.0
020	142.5	100.0
040	84.1	64.0
120	0	40.0
130	45.1	26.5

The calculated structure factors are in reasonable agreement for all of these reflections when due account is taken of the difference in scattering factors for x-ray and electron. Clearly, the marked difference in intensity of the 120 reflection for x-rays *vs* electron irradiation provides useful additional confirmation in checking the structure.

Poly(ε-Caprolactone). Two separate x-ray studies (17,18) have been reported for the crystal structure of poly(ε-caprolactone). Both studies involved excellent fiber diffraction data but the final chain conformations and packing differed somewhat. The two structures are compared in Fig. 3.

Space group and unit cell parameters are essentially the same although the small difference in c axis values led to the assumption of different chain conformations: planar zig-zag *vs* a non-planar chain. In the latter case, a shortening of the chain in the c direction was achieved by allowing the ester group to deviate from a planar *trans* arrangement. These differences in chain conformation were carried through the two structure refinements resulting in differences in chain orientation angle Θ and chain translation τ in the final packing.

Electron diffraction was considered as a possible approach for distinguishing between these two proposed structures. Accordingly, 5 ml of a 0.05% by weight solution of poly(ε-caprolactone) in toluene was dialysed against a mixture of 100 ml of propanol-toluene in a proportion of 9/1 by volume. The polymer precipitated in a few hours after room temperature dialysis was initiated. As seen in Fig. 4 the single crystals have the usual lamellar morphology and the large base platelets are covered with several lamellae which grow from screw dislocation.

The electron diffractogram in Fig. 4 shows the expected symmetry based on the known unit cell dimensions. The systematic absences for the $P2_12_12_1$ space group were noted and some 42 separate reflections could be observed which provided 16 distinct baseplane reflections. These were used to compare the observed structure factors for the two structures shown in Fig.3. The results are shown in Table V (19).

Table V

Comparison of Observed and Calculated Structure Factors for Electrons from Poly(ε-caprolactone) Single Crystals (19).

hkℓ	Fo	$\|Fc\|^*$(Ref. 17) planar zig-zag	$\|Fc\|^*$(Ref. 18) non-planar
110	113.0	151.6	147.0
200	106.5	104.1	140.9
210	31.2	25.6	35.5
020	52.6	69.0	42.7
120	19.2	10.5	13.5
310	53.4	28.1	45.7
220	38.5	40.2	28.4
400	37.1	16.1	27.8
320	16.0	11.0	15.3
410	8.0	13.7	6.8
130	25.3	25.2	11.6
230	8.0	5.1	3.1
420	11.8	6.9	12.2
510	15.2	4.8	11.0
330	8.0	13.2	2.4
520	7.0	7.6	7.0
$R = \Sigma\|\Delta F\|/\Sigma Fo =$		28.1	26.5

* $|Fc|$ calculated using the coordinates given in the references cited.

As can be judged form the small difference in the agreement index, it is difficult to conclude that one structure is to be preferred over the other. In fact, the exact value of the

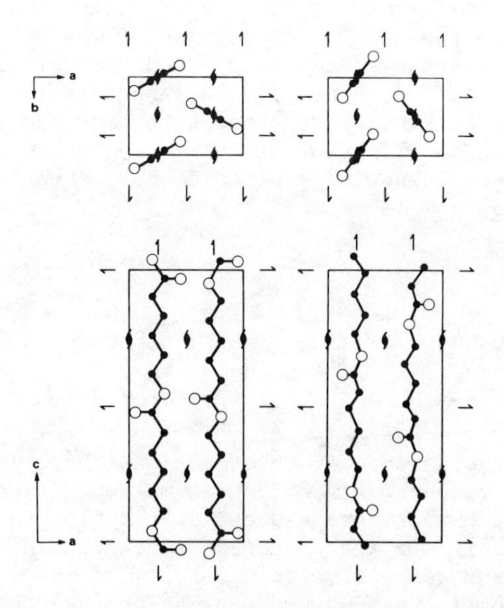

*Figure 3. Comparison of the proposed structures for poly (ε-caprolactone): (*left*) from Ref.* 17; *(*right*) from Ref.* 18

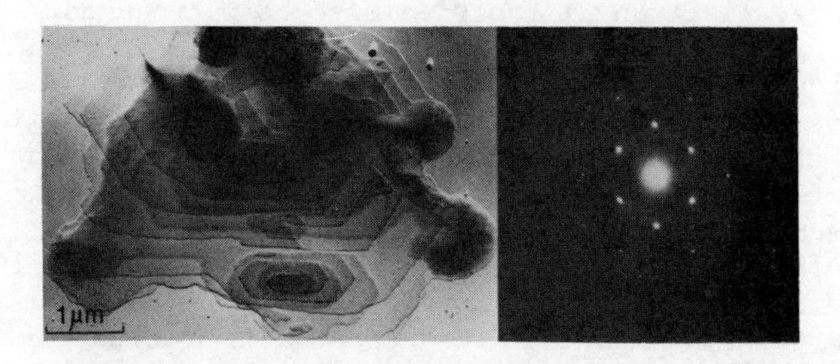

Figure 4. Single crystal of poly(ε-caprolactone) and its electron diffractogram

orientation angle in structures such as this have always been controversial. One needs only consider that there is still controversy (20,21) in the case of polyethylene which is a propotype for the aliphatic polyesters. Bunn's original proposal (20) for Θ in the polyethylene structure was 49° while a more recent analysis (21) gave 35°.

What is more disturbing, is the difference in chain conformation in the two cases. Distortion of the ester group from its planar *trans* arrangement is reported for certain non-polar aliphatic polyesters such as polyethylene adipate and suberate (22) but these are studies reported some time ago and could bear reexamination in the light of better x-ray methodology. In particular, present knowledge of torsional potentials for the ester group (23):

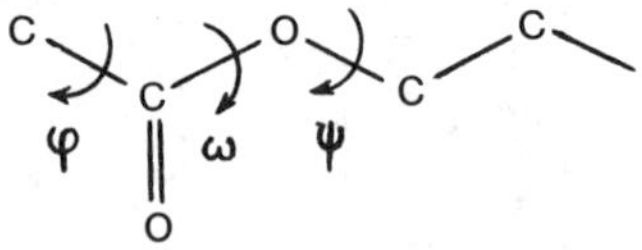

C O C
C C
φ ω ψ
O

leads to the conclusion that the rotational barriers for angles ω, ϕ and ψ are respectively 8.75 kcal/mole, 0.2 kcal/mole and 0 kcal/mole. One therefore expects deformation to be mainly in angles ϕ and ψ. In the proposed structure (18) for poly(ε-caprolactone) the respective values for ω, ϕ and ψ are: -7°, 5° -4° which is not in keeping with the values expected on the basis of torsional potentials. A statistical study of small molecule esters shows that the angle ω has a strong preference for values close to 0° (24).

To be useful in resolving the dilemma such as those encountered above, electron diffraction must be introduced into the structure refinement in parallel with X-ray diffraction analysis.

Although systematic absences are almost always more easily recognized in electron diffraction, occasionally dynamic diffraction effects lead to a weak diffraction for a supposedly absent reflection. Repeated observations on crystals of various thicknesses, allow a decision in such doubtful cases.

Finally, problems due to hydration or solvent inclusions, which are often observed with natural polymers, are easily accounted for by comparing X-ray and electron data since electron microscopy and diffraction usually take place in vacuum.

ACKNOWLEDGEMENTS

This work was supported by the National Research Council of Canada and the Ministère de l'Education du Québec.

LITERATURE CITED

1. TAYLOR, K.J., CHANZY, H. & MARCHESSAULT, R.H., J. Mol. Biol. 1975, 92, 165-167.
2. ROCHE, E., CHANZY, H., BOUDEULLE, M., MARCHESSAULT, R.H. & SUNDARARAJAN, P., Macromolecules, 1978, 11, 86-94.
3. GEIL, P.H. "Polymers Single Crystals" 1963. New York: Interscience.
   CLAFFEY, W., GARDNER, K., BLACKWELL, J., LANDO, J. & GEIL, P.H., Philos. Mag., 1974, 80, 1223-1232.
4. YAMASHITA, Y., J. Polym. Sci., 1965, A3, 81-92.
5. HACHIBOSHI, M., FUKUDA, T. & KOBAYASHI, S., J. Macromol. Sci., 1969, 33, 525-555.
6. DORSET, D.L., Acta Cryst., 1976, A32, 207-215.
7. International Tables for X-ray Crystallography 1974, Vol. IV. Birmingham: Kynoch Press.
8. BRISSE, F., PEREZ, S. & MARCHESSAULT, R.H., 1980, NATO Advanced Study Institute. Advances in the Preparation and Properties of Stereoregular Polymers. (Tirrenia, Italy October 1978). Dordrecht: Reidel Publishing Company.
9. JOLY, A.M., NEMOZ, G., DOUILLARD, A. & VALLET, G., Makromol. Chem., 1975, 176, 479-494.
10. BRISSE, F., POULIN-DANDURAND, S. & REVOL, J.F. To be published (1980).
11. LEMOINE, M., Ann. Inst. Pasteur, 1925, 39, 144.
12. OLSON, I., MERRICK, J.M. & GOLDSTEIN, I.J., Biochemistry, 1965, 4, 453.
13. MARCHESSAULT, R.H., OKAMURA, K. & SU, C.J., Macromolecules, 1970, 3, 735-740.
14. CORNIBERT, J. & MARCHESSAULT, R.H., J. Mol. Biol. 1972, 71, 735-756.
15. YOKOUCHI, M., CHATANI, Y., TADOKORO, H., TERANISHI, K. & TANI, H., Polymer, 1973, 14, 267.
16. MARCHESSAULT, R.H., COULOMBE, S., MORIKAWA, H. & REVOL, J.F. Can. J. Chem. (submitted).
17. BITTIGER, H., MARCHESSAULT, R.H., NIEGISCH, D., Acta Cryst., 1970, B26, 1923-1927.
18. CHATANI, Y., OKITA, Y., TADOKORO, H., YAMASHITA, Y., Polymer J., 1970, 1, (5), 555-563.
19. NOE, P., Mc.Sc. thesis, Chem. Dept. Université de Montréal, 1979.
20. BUNN, C.W., Trans. Far. Soc., 1939, 35, 482-491.
21. ZUGENMAIER, P., Doctoral dissertation, Freiburg University, 1969.
22. TURNER-JONES, A., BUNN, C.W., Acta Cryst., 1962, 15, 105-113.
23. SCOTT, R.A., & SCHERAGA, H.A., J. Chem. Phys., 1966, 45, 2091-2101.
24. CORNIBERT, J., HIEN, N.V., BRISSE, F. & MARCHESSAULT, R.H. 1974, 52, 3742-3747.

RECEIVED May 21, 1980.

17

Electron Diffraction and Dark Field on Ultrathin Sections of Textile Fibers

R. HAGEGE

Institut Textile de France, 35 Rue des Abondances, 92100 Boulogne S/Seine, France

X-Ray diffraction studies of textile fibers have led to the development of techniques to calculate or estimate the following fiber characteristics :

. percents crystallinity (% crystalline fraction)
. crystalline perfection (as characterized by various indexes)
. mean dimensions of crystallites
. long-period as estimated by small angle X-Ray scattering, etc...

However, electron diffraction, although frequently used for polymer single crystals studies, has seldom been applied to textile fibers, and particularly to ultrathin sections of those materiels. (1,2) the majority of published papers dealing with electron diffraction of fibers is concerned with isolated fibrils or fragments prepared by mechanical milling. As far as we know, apart from a preliminary report about mercerization of cotton fibers (3) and a recent paper about chemically etched PETP filaments (4), high modulus aramide fibers and carbon fibers are practically the only systems with detailed (recently) published work on electron diffraction and dark field of ultrathin sections of textile or paratextile fibers (5, 6, 7). Here we report similar studies on 2GT polyester and 66 polyamide fibers and define local order indexes.

0-8412-0589-2/80/47-141-279$05.75/0

Material and methods

Various kinds of multifilament coutinuous yarns of 2GT polyester (referred to as PETP in the present paper) were investigated :

. 1 FT textured continuous yarn ($PETP_1$)
. 1 multifilament from a geotextile spun bond membrane made of BIDIM (Rhone-Poulenc-Textile (RPT) - marque déposée) ($PETP_2$)
. 1 thermoset multifilament for tire-cord application ($PETP_3$)

Various 66 polyamide continuous yarns were also examined (referred to as, "type 2", "type 4" and "type 5") : these are experimental yarns prepared in different conditions and supplied by RPT.

Only transverse sections (i.e. sections cut, as precisely as possible, normal to the fiber axis) were examined. The production of electron microdiffraction patterns and dark field micrographs has to be done on very thin sections (if possible sections thinner than those used for ordinary bright field work). This is very difficult to do with this kind of material. Staining with Ag_2S following the usual method in our laboratory (8) often, if not always, is useful in improving the quality of the sections, without disturbing either the production of a diffraction pattern characteristic of the polymer species, or the photograph in dark-field conditions. Ag_2S staining procedure : the fibers are treated under 15 Atm at 20°C by gazeous H_2S. After mild rinsing with alcohol or acetone, they are immersed for 16 hours at 20°C into an aqueous 0,1 M solution of silver nitrate. Sorbed H_2S is thus transformed into Ag_2S and "in situ" precipitated into the accessible regions. The fibers are then rinsed by water and dried before embedding and cutting.

Electron Diffraction Micrographs

Production of the negative prints. In order to allow a quantitative treatment of the micrographs, the width of the electron beam was carefully controlled. The "irradiated area" is the area of the section illuminated by the electron beam, whereas the "selected area" is isolated from the irradiated area by means of a selection aperture which has a diameter of 0.5 or 1 μm, i.e., 1/10 to 1/5 of the latter. The section is searched under very mild conditions of irradiation so that the area of interest does not deteriorate, the filament being generally undersaturated and all other settings to the minimum.

In the present work, we have tried to examine the validity of the following method for quantitative purposes : first we photograph the "intact" diffraction pattern under defined conditions, then we photograph the amorphous pattern under the same conditions. The amorphous pattern is obtained by increasing the radiation rate until the crystalline diffraction disappears. In order to record the "crystalline" and "amorphous" patterns of the same selected area on the same photographic emulsion (to minimize the influence of photographic processing conditions), the following procedure was adopted :

a) visually identify the "crystalline" pattern in mild irradiation conditions (those conditions are inadequate for the photography)
b) introduce an objective aperture of 75 μm: this is of sufficient size so not to hide the main diffraction arcs. This aperture prevents the irradiation of the photographic emulsion far away from the center of the "intact" pattern, which allows for other prints to be taken on the same plate or film
c) center the "diffraction spot" (i.e. the impact of the unscattered beam) in a convenient position (e. g. near one corner of the photographic plate) and record the diffraction pattern with carefully defined conditions of irradiation and exposure time
d) "destroy" the crystalline order of the selected area and its surroundings (controlling it visually) by a strong increase of the radiation rate during adequate time
e) record the "destroyed" pattern in the same conditions as in c) with the center of the diffraction spot in a different part of the photographic plate

f) repeat d) and e) to get a second exposure of the "destroyed"pattern. This is to check the disappearance of the crystallinity
g) return to "image" conditions and expose the irradiated zone and its surroundings ; the beam should be focused (less than it was for the recording of diffraction pattern) and deflected and the preparation shifted so that this exposure is not superimposed on the preceeding ones.

Three diffraction patterns and one micrograph are thus recorded on the same plate or film. In some cases, we tried to evaluate the degradation of the crystalline patterns during one exposure ; this is analogous to an idea of a previous work of Dobb (9) who has studied the kinetics of the disappearance of the main interferences in the electron diffraction pattern of cellulose under constant irradiation rate by a method of "time-lapse series", Our aim was to try to determine a "true" crystallinity by an extrapolation to zero dose. In such a mode, the "crystalline" pattern of the same selected area is recorded three times (on the same emulsion) the irradiation rate corresponding to an exposure being set only during the exposure time.
On the same plate or film, 4 diffraction pattern (three "crystalline" patterns with a beginning of degradation and an "amorphous" one obtained as already explained) and one morphological view are obtained. Figure 1 shows an example of a slide with 5 exposures of a PA 66 filament (4 diffraction patterns and 1 morphologic view) obtained from a particular zone of the cross-section of an Ag_2S-stained fiber ; the 3 "crystalline" patterns show an increasing degradation (in the following way 1, 2, 3) and "A" is the amorphous pattern.

This process (using one of the modes previously described) was applied to a set of 3 to 12 different selected areas on the transverse section of one fiber. The photographic series is a completed record of the whole fiber cross-section at an adequate magnification. On the latter micrograph the irradiated areas can be seen because of their lesser electron density in bright field conditions. In some cases, these irradiated areas could also be seen under dark field conditions.

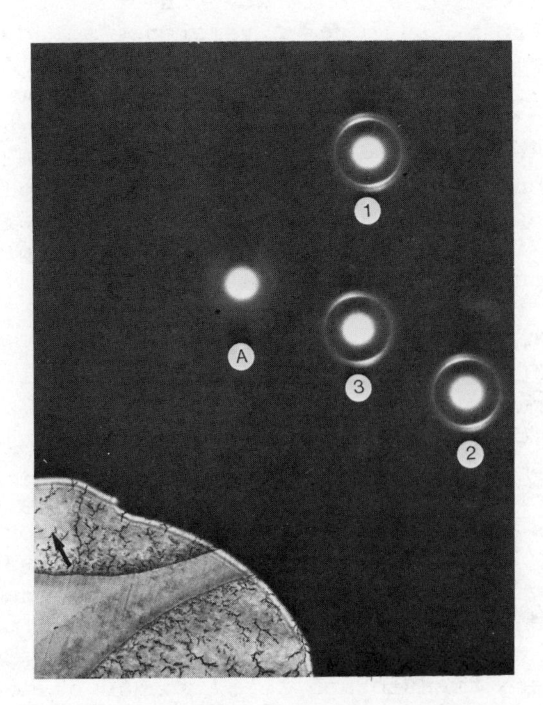

Figure 1. A slide with 5 exposures on the same emulsion.

In the lower left corner, zone (arrowed) of the cross section of an Ag_2S stained PA 66 fiber, which is responsible for crystalline pattern 1. Numbers 1, 2, and 3 are three successive exposures of the same pattern with an equal radiation rate—notice progressive fading of the outer, less intense (010) arc A: amorphous pattern (photographed in the same conditions) corresponding to the same area.

Quantitative Processing. Plates or film with the diffraction patterns were scanned with a Joyce-Loebl microdensitometer. Radial (2θ) densitometric plots of the crystalline pattern (eventually three successive exposures of the crystalline pattern are analyzed) and of the corresponding amorphous pattern were recorded on the same curve. In this way, the plot of the amorphous pattern was used as a reference standard. The densitometric recording began with the optical density of the non-irradiated emulsion ; this allowed the evaluation and normalization of the optical density of the diffraction pattern. When the analytical slit passed through the image of the border of the 75 μm objective aperture, the densitometric curve showed a sudden density raise "Δ d". (Fig. 6) The plots of the amorphous and crystalline patterns were thus normalized to the same reference "Δ d". Crystallinity was determined on the normalized curves by measuring the areas "C + A" and "A" under the crystalline and amorphous plots respectively.

In some cases, an "azimutal correction" was applied by the following method : an azimutal plot of the diffracted intensity was obtained using a polar coordinates table. The corrective factor for "C/A" was $\varepsilon/\varepsilon_0$ where is the area under the azimutal plot and ε_0 is the area of the rectangle having the same base and with the height equal to the maximum density of the azimutal plot. Corrected or uncorrected $\frac{C}{A}$ values were transformed to C% values by means of the formula :

$$C\% = \frac{100\,\frac{C}{A}}{1 + \frac{C}{A}}$$

which holds for C% values near but not equal to 100 % because we have no fully crystalline standard.

Dark Field Micrographs. Dark field micrographs were obtained by isolating a main diffraction arc on the microdiffraction pattern and, by tilting the electron beam at a correct angle, recording the image with the diffracted beam only. For the polyester, we have not been able as yet to obtain such micrographs of acceptable quality.

For 66 polyamide, we have obtained acceptable results by selecting the stronger reflection, that is in general the nearest one to the center of the pattern (in the transverse section).

Results

2GT polyester. Figure 2 is a montage showing the cross-section of a textured polygonal-shaped PETP strand stained with Ag_2S ; the metal is essentially peripheric with some radial inclusions. The diffraction patterns of 5 different selected areas are also present. Irradiated areas and selected areas can be resolved by their higher transmission. There is considerable variation in the azimutal width of the 3 main diffraction arcs while the orientation of the equator is only slightly variable. On this montage (as well as on the following ones), the diffraction patterns have been positioned in their true spatial position, after correction of the image rotation depending on magnification and on camera length.

Figure 3 is another example on Ag_2S stained PETP with 3 selected areas, the diffraction patterns of which have quasi-parallel equators.

Figure 4 corresponds to $PETP_3$ (Ag_2S stained) with 12 different selected areas and corresponding diffraction patterns. One can see that : 5 patterns have the same orientation, 3 are practically at right angles from the latter, 2 have an intermediate orientation and the last 2 are highly disoriented, with one Debye Scherrer type pattern.

Figure 5 is an example of a densitometric plot for $PETP_1$. The shoulder corresponding to the "amorphous halo" of the destroyed pattern is clearly visible. The normalization procedure is illustrated : in that specific case, the amorphous pattern is underexposed in comparison with the crystalline pattern. Point A corresponds to the optical density d_{oo} of the background of the slide (that is outside the image of the border of the objective aperture) ; point B corresponds to the "initial" density d_o of the crystalline pattern and B^1 to the "initial" density d^1_o of the amorphous pattern ; if the crystalline plot is taken as the reference, the amorphous plot density has to be multiplied by the normalization factore i.e.

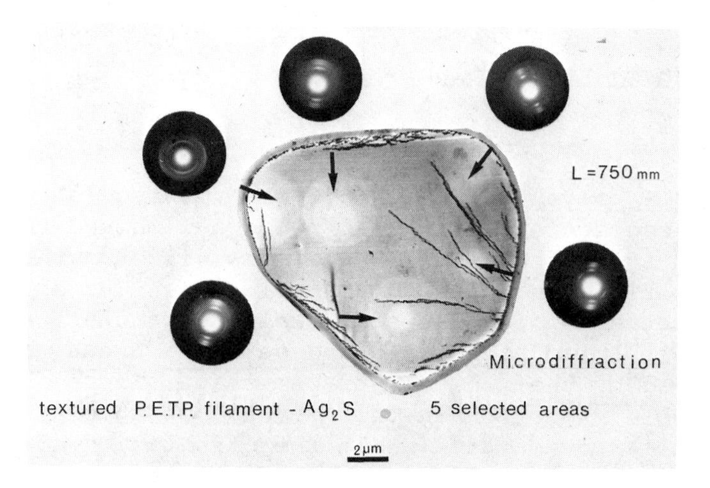

Figure 2. Cross section of a textured polygonal-shaped PETP strain stained with Ag_2S.

The photograph (positive print) of the cross section of the fiber has been cut around with scissors, leaving a small amount of embedding araldite around the fiber-end stuck on a "bristol" paper The diffraction patterns were stuck close to the corresponding areas (to which they are related by arrows) and their orientations are corrected according to the image/diagram rotation introduced by the microscope.

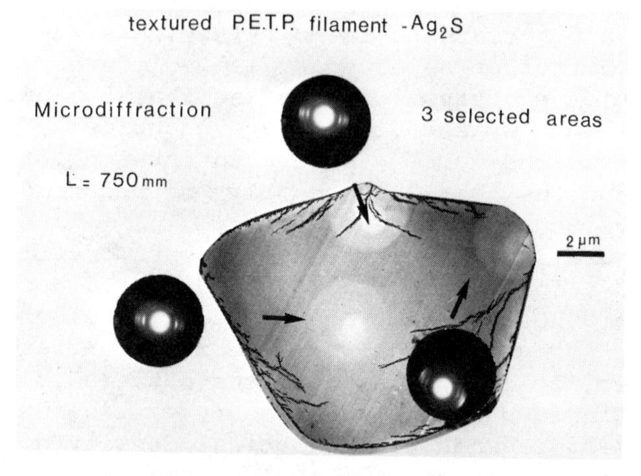

*Figure 3. Ag_2S-stained PETP with three selected areas, the diffraction patterns of which have quasi-parallel equators (*see *also Figure 2)*

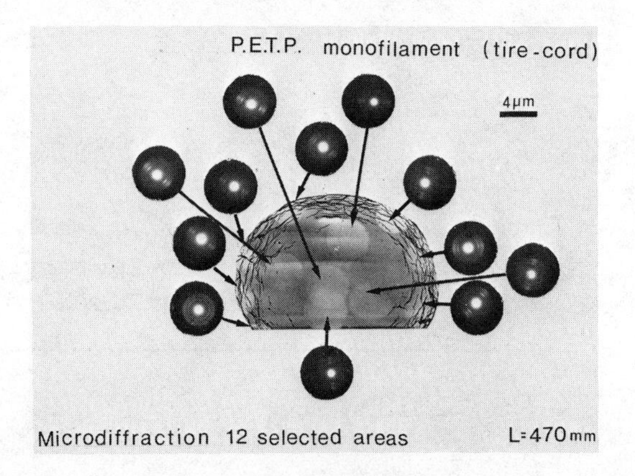

*Figure 4. Ag_2S-stained $PETP_3$ with 12 different selected areas and corresponding diffraction patterns (*see *also Figure 2)*

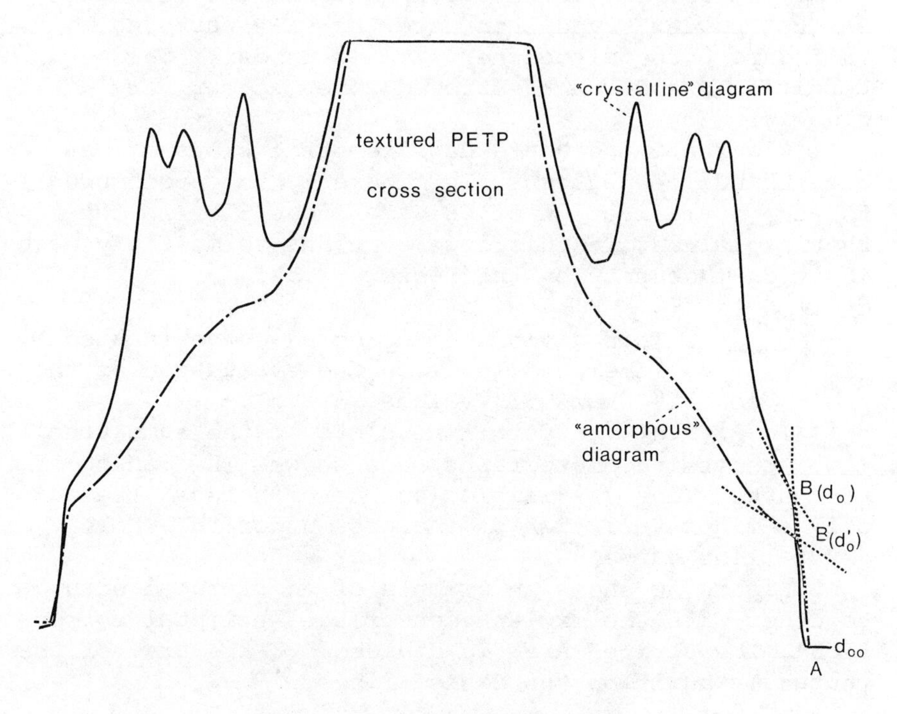

Figure 5. Densitometric plot for $PETP_1$

$$\frac{d_o - d_{oo}}{d_o^1 - d_{oo}}$$

Table I.

	Electron diffraction crystallinity	X-Ray crystallinity (RULAND's Method)
$PETP_1$	28,8 %	27,0 %
$PETP_2$	33,4 %	32,0 %

66 Polyamide. Figures 6 and 7 are montages of 66 polyamide. The diffraction patterns are much closer to "powder" type than were the PETP patterns. In figure 8 the altogether view is in dark field conditions and the 5 destroyed zones appear very clearly.

Figures 8 and 9 are densitometric plots of various kinds of patterns. In figure 8, the procedure for the normalization of the amorphous plot is further explained. In this case, one has to multiply each individual density by the factor $\frac{d_o - d_{oo}}{d^1_o - d_{oo}}$. This gives the corrected curve (dashed line) and leads to the evaluation of the "C" and "A" values.

In figure 9, crystalline plots of the same area (two successive photographs) are shown. The fading of the weaker line is visible, but nevertheless, there is only a minor variation of the area under the whole crystalline curve.

Figure 10 shows an example of an azimutal scan, together with the explanation of the "azimutal correction". The striped area is divided by the area of the rectangle built on the dashed lines.

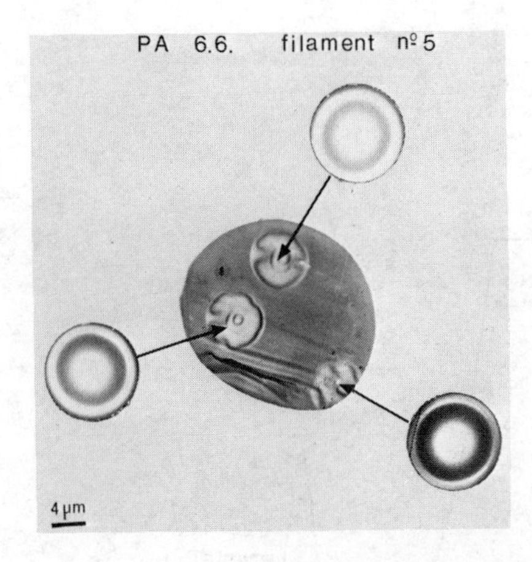

Figure 6. Montage of 66 polyamide (see also Figure 2)

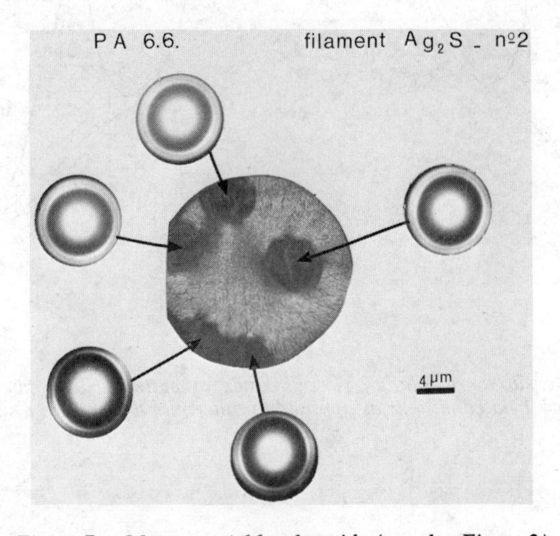

Figure 7. Montage of 66 polyamide (see also Figure 2)

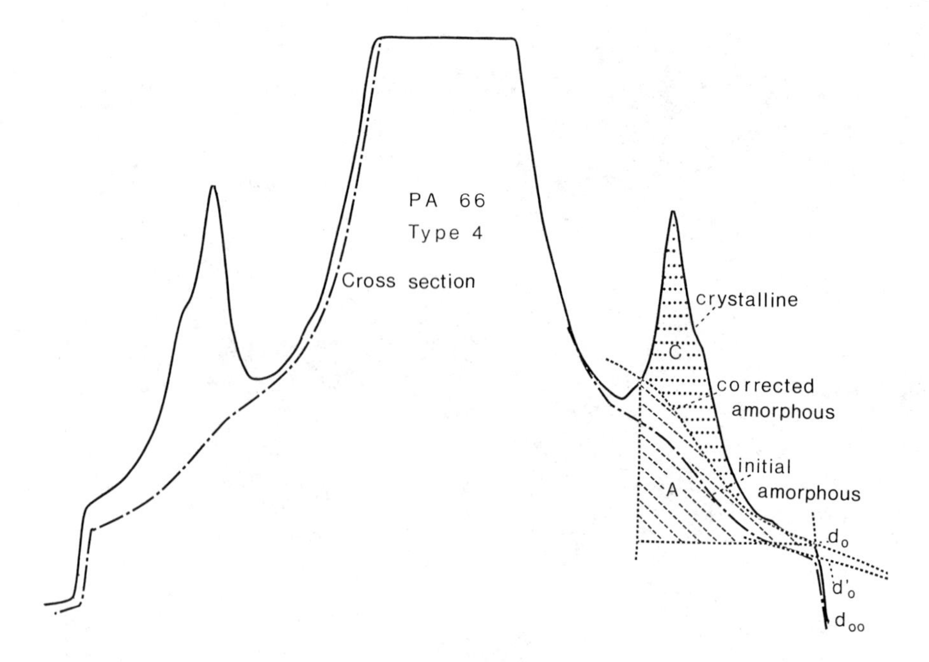

*Figure 8. Densitometric plot of various kinds of patterns. The fiber is photographed in dark field conditions as opposed to other montages. (*See *also Figure 2.)*

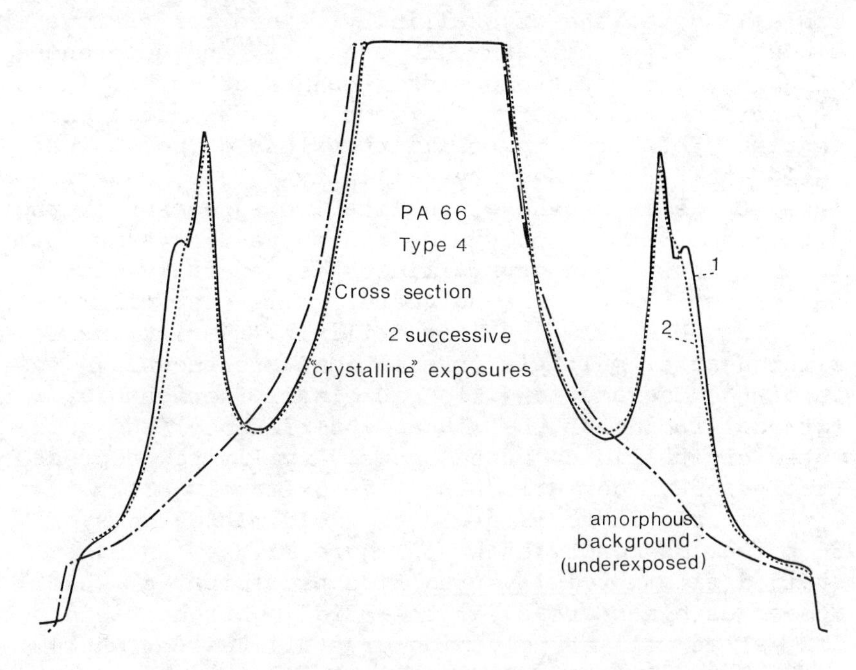

Figure 9. Densitometric plot of various kinds of patterns

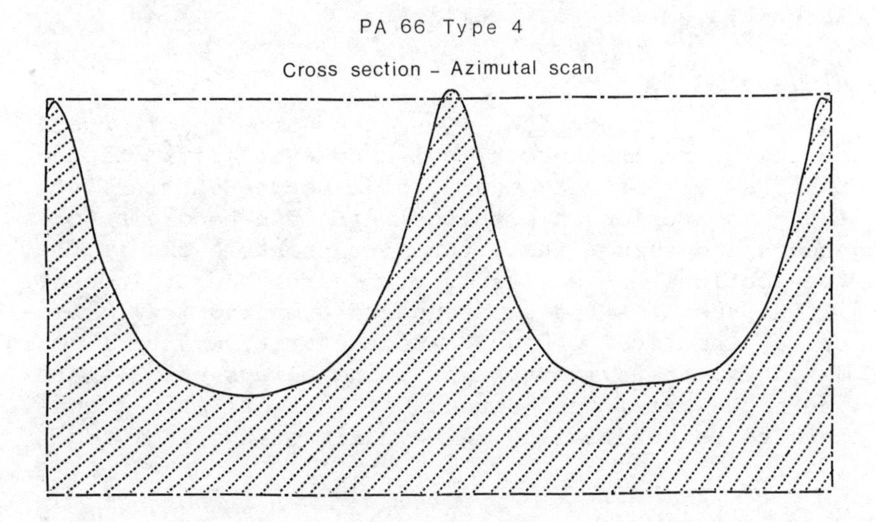

Figure 10. Example of an azimuthal scan

Table II gives the crystallinity values for various kinds of polyamide yarns. In one case, the occurence of a relatively amorphous skin can be detected. There is also an example of the effect of an azimutal correction. Absolute crystallinity values do not agree well with X-ray based crystallinity values. Nevertheless, there is a general qualitative agreement in the difference between polyamide and polyester yarns (it is well known that crystallinity is generally higher for polyamide than for polyester commercial filaments).

In table III, the "crystallinity" variations estimated after multiple irradiations are shown. In spite of the degradation which is clearly noticeable (gradual fading of the weaker interference) the estimated crystallinity often shows very limited decrease (the standard deviations are approximately 8 %).

Figures 11 to 13 are dark field micrographs of 66 polyamide monofilaments. Figure 11 show an Ag_2S stained filament. Silver sulfide precipitates, which appear as black areas (as they did in bright field images) as well as polyamide crystallites (bright spots) are visible. Figure 12 corresponds to a type 4 fiber (with skin-core morphology) where there is a lower density of crystallites in the skin region. Figure 13 corresponds to the case of type 5 fiber which has smaller crystallites.

Discussion

Local Orientation. The most striking observation of this work is that the selected area diffraction patterns are not in general of a Debye-Scherrer type. Among the various hypotheses which can be drawn to understand such a fact, the most probable one is that the sections are not truly transverse ones ; indeed, if one supposes the existence of a cylindrical symmetry at the level of each selected area, 0.5 to 1 µm in diameter (the symmetry axis being always parallel to the fiber axis) the "detectable" network main planes have to be parallel to the "c" axis of the individual

Table II. Crystallinity of PA 66-C%

Type 4		Type 4 (+Ag_2S)		Type 5	Type 2 (+Ag_2S)	
Core	Skin	Without azimutal corr.[on]	With azimutal corr.[on]			
%	%	%	%	%	%	%
38,5	32,3	51,6	36,4	44,1	57,6	Mean
6	6,2	8,2	4,4	9,7	7,8	Standard deviation

Table III. Type 2 Polyamide Degradation of the Diffraction Pattern

N° of the selected area	Cristallinity % after		
	the 1st irrad.[on]	the 2nd irrad.[on]	the 3rd irrad.[on]
1	65,0	68,9	66,3
2	56,2	55,3	53,3
3	63,4	61,5	62,7
4	63,8	65,4	63,5
5	61,3	58,8	55,5
6	46,7	45,4	45,1

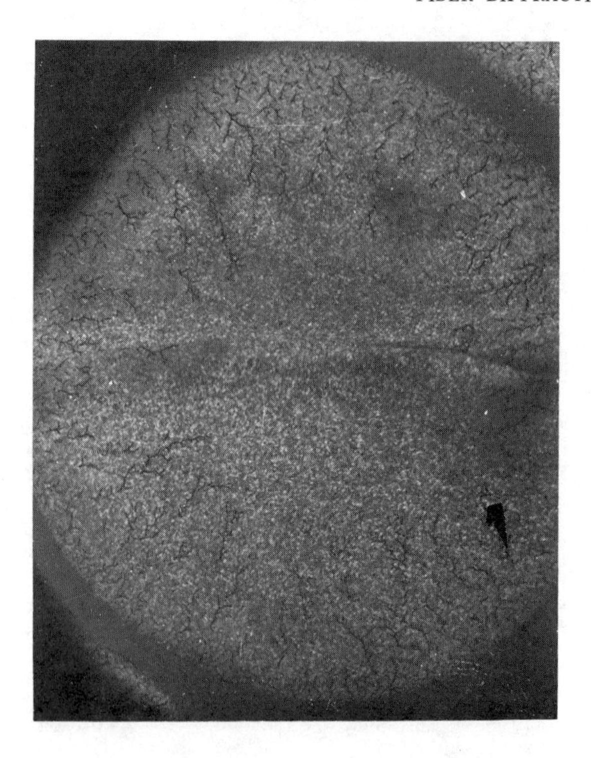

Figure 11. Dark field micrograph of 66 polyamide microfilaments

Figure 12. Dark-field micrograph of the cross section of an Ag_2S-stained Type 2 (see text) PA 66 fiber. Notice black deposits of silver sulfide in the periphery and white dots in the whole section, corresponding to crystallites in Bragg position.

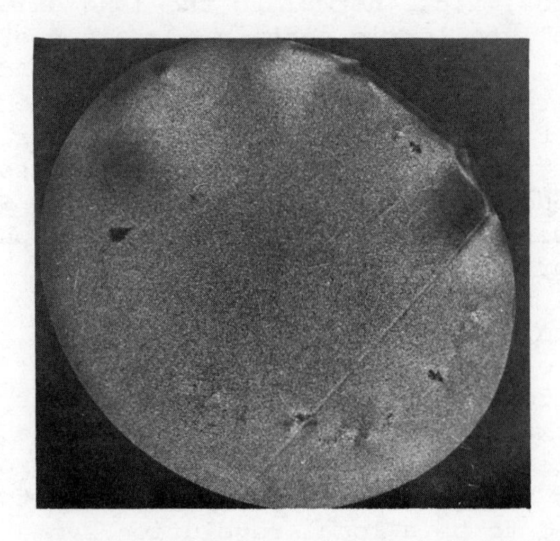

Figure 13. Dark-field micrograph of the cross section of a Type 4 PA 66 fiber. No silver sulfide deposits; notice the skin-core effect and the dimensions of the crystallites (white dots).

crystallites, and their angle with the incident beam has to be equal to the Bragg angle Θ. For a rough estimate, one can neglect this Bragg angle (this in fact is very small because of the low value of the wavelength associated to the electrons) and also the mean orientation angle of the c-axis of the individual crystallites (perfect orientation). In such a situation the "visible" network planes have a constant orientation in space ; they must be parallel to the incident beam and to the fiber-axis. The equator of the resulting diffraction pattern is normal to these planes, i.e., normal to the long axis of the theoretically elliptic section. As a support to this hypothesis, the equators of various local diffraction patterns are most frequently parallel to one another. This is not only true for the same cross section but also in different fibers which have been sectioned as an embedded bundle of parallel fibers. Nevertheless, we were not able to find any relationship between the orientation of the long axis of the pseudo elliptical section and that of the equator of the diffraction pattern.

Figure 14 is a scheme of the section (S) that was cut slightly oblique to the fiber axis (Δ_o) ; the beam axis (I) impinges normally in the center of the selected area (S) ; a cone of revolution (C) has been drawn round the axis (Δ), parallel to (Δ_o) and is supposed to be a local revolution symmetry axis. The angle Ø of this cone corresponds to the mean orientation of the c-axis of the individual crystallites ; another revolution cone (ε_o) is centered around (I) : its angle Θ_o is one of the main Bragg angles (Θ_o is generally much smaller than Ø). "Detectable" network planes show two main features : they are tangent to (ε_o) and they contain one generator of the cone (C). A plane normal to (I) intercepts on (C) as an ellipsa (E) and on (ε_o)as a small circle (γ_o).

In figure 15, examples of these two geometries are shown. Traces of the detectable network planes must have an orientation comprised between the two internal common tangents to (γ_o) and (E) ; it is thus possible to build geometrically (as the figure shows)

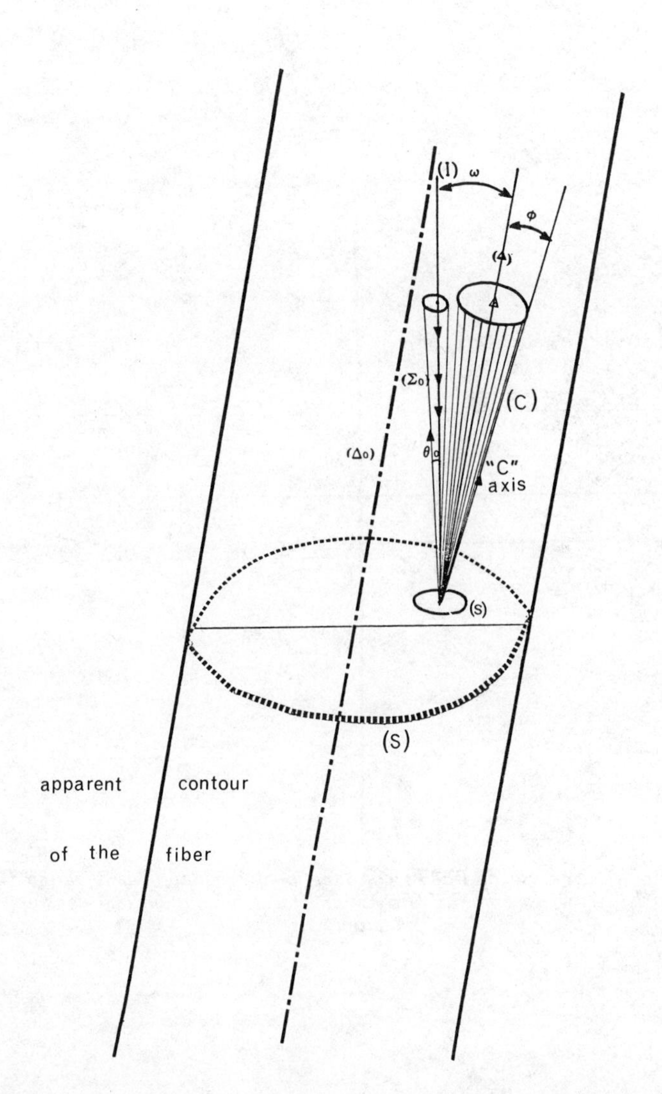

Figure 14. Similar to Figure 13, but with Type 5 PA 66 fiber (the cross sections have been thermally treated; notice very tiny, well-defined crystallites)

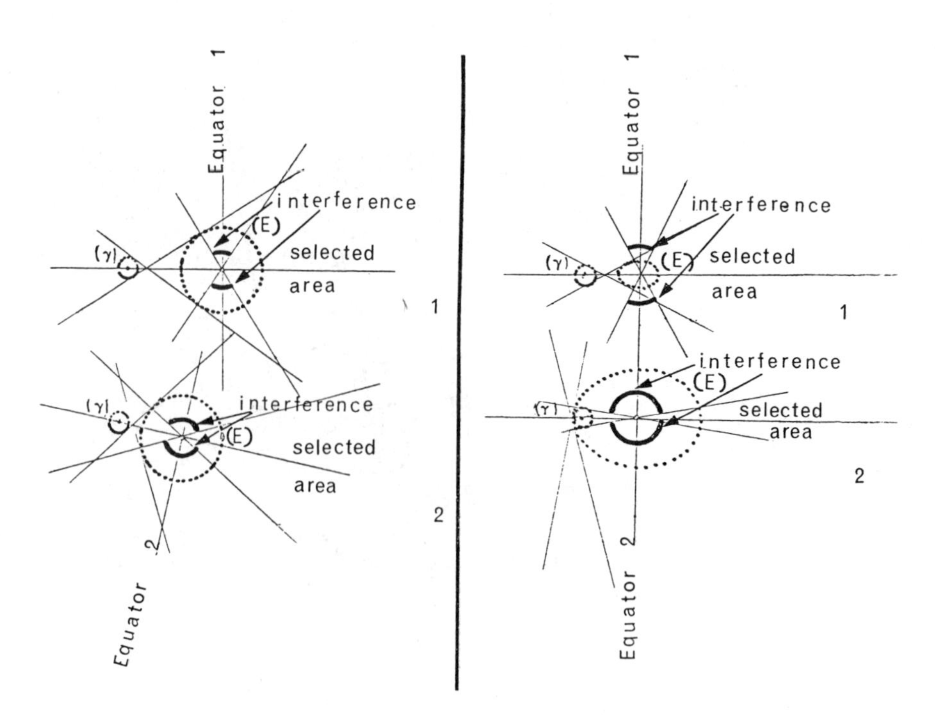

*Figure 15. (*Left*) Case of PETP: constant local orientation and variable local rotation axis; (*right*) case of PA 66: variable local orientation and constant local rotation axis*

the azimutal opening of the interference corresponding to the Bragg angle Θ_o.

There are two different ways to get local diffraction patterns "Debye-Scherrer" type (such patterns are obtained if (γ_o) is found within (E). In the first case (figure 15) the orientation of the local symmetry axis (Δ_o) is very close to that of (I) with the consequence that the whole cone (ε_o) is located within the cone (C) even if the local orientation is relatively good, i.e., angle Ø is small. In the second case, the local orientation is poorer, i.e., Ø is fairly large. As a result Ø and ω (angle between (Δ) and (I)) have equivalent values and again cone (C) surrounds cone (ε_o). The first case may correspond to PETP in which there would be a noticeable dispersion of the local symmetry axis around fiber axis. This could explain the relatively frequent variations in the orientation of the diffraction pattern in the specific case of PETP. The second case would correspond to 66 polyamide which has local variations of orientation and less overall orientation than in PETP. The two cases are presented in figure 15.

Local Crystallinity. The "azimutal correction" previously mentioned should be applied only in the second case discussed above, i.e., when the variation of the local orientation is the main factor in affecting the observed diffraction pattern. In this specific case, the diffracted intensity is spread over the whole interference. As a result, the correction should be made for 66 polyamide and not for PETP. This correction could have a large effect and would result in lowering the crystallinity differences between the two kinds of fibers. This azimutal correction gives a better account of the whole azimutal intensity than the normalization of the half-height width value as proposed in another paper (10).

Nevertheless, numerical values obtained in the present work are preliminary results and further experiments are underway. At the moment, we are unable to say whether section thickness can influence the crystallinity values obtained even through the use of "$\frac{C}{A}$" technique.

(This method partially eliminates the influence of the transmission of the sample "μt" factor). Moreover, it seems reasonable that the sections lack planarity and have a tilting artifact in at least part of the crystallites due to the action of the knife. These factors might have a large influence on the variations of orientation and intensity of the local diffraction patterns.

Conclusion

We have presented preliminary results on selected area electron diffraction of "pseudotransverse" ultrathin sections of PETP and PA66 filaments. These results suggest that the classical hypothesis of regular cylindrical symmetry in circular-shaped synthetic filaments is incorrect. Although the best way to study local variations of orientation is to work on longitudinal sections, our results strongly suggest that the cylindrical symmetry has to be considered at the level of a small area of less than 1 μm in diameter (and probably much less) and not at the level of the whole filament, as X-ray diffractometry has suggested, However, in PETP, a possible dispersion in the orientation of local symmetry axis, around the filament axis, is to be considered.

The crystallinity index, as obtained using the "amorphous" diffraction pattern (by destruction of local order under the electron beam) gives reliable information, i.e., in agreement with dark field images.

It should be emphasized that the search and photography of the diffraction pattern (using the procedures outlined above) does not bring about major degradation of the original crystallinity. This is possibly in part due to our use of a high vacuum electron microscope, in which the rate of contamination of the sample is small.

Acknowledgments

The author wishes to thank the workers of the laboratory for technical assistance and the "RHONE POULENC TEXTILE" Society for financial support of part of the present work.

<u>Literature cited</u>

1. HARRIS, P.H., 5th intern. cong. for Elec. Micr. BB-8 Acad. Press. N.Y. (1962)

2. SCOTT, R.G., Symposium on resinographic methods special technical publication N 60348 ASTM (1963)

3. BULEON, A., CHANZY, H., HAGEGE, R., Bull. Scien.ITF <u>4</u>, 229-233 (1975)

4. MURRAY,R., DAVIS, H.A., TUCKER, P., J. Appl. Polym. Sci., <u>33</u>, 177-196 (1978)

5. Brevet, Français, DU PONT, n° 2, 134, 582, Enregistrement n° 15,014 (1972)

6. DOBB, M.C., JOHNSON, D.J., SAVILLE, B.P., J. Polym. Sci. Polym. Phys., <u>15</u>, 2201 (1977)

7. DOBB, M.C., JOHNSON, D.J., SAVILLE, B.P., J. Polym. Sci. Polym. Symp., <u>58</u>, 237-251 (1977)

8. SOTTON, M., C. R. Acad. Sci. Paris 270B (1970)

9. DOBB, M.C., MURRAY, R., J. Micros. 101, 299-309 (1974)

10. SOTTON, M., ARNIAUD, A.M., RABOURDIN, C., J. Appl. Polym. Sci., <u>22</u>, 2585-2608 (1978).

RECEIVED May 21, 1980.

Structure of High Modulus Fibers of Poly-*p*-Phenylene Benzbisthiazole

ERIC J. ROCHE[1], TOSHISADA TAKAHASHI[2], and EDWIN L. THOMAS

Department of Polymer Science and Engineering, The University of Massachusetts, Amherst, MA 01003

Poly-p-phenylene benzbisthiazole, with repeat unit 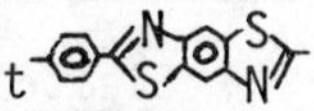 has recently been synthesized as part of the "Air Force Ordered Polymer Research Program" (1). This program, directed towards the preparation of very high strength, high temperature resistant polymers, has led to the synthesis of various rigid rod polymers, among which PBT shows the most promising properties. PBT fibers spun from a nematic dope of the polymer can exhibit a higher modulus than the well-known poly-p-phenylene terephthamide (PPT) fibers.

In previous work, the x-ray diffraction pattern of PBT fibers was interpreted as arising from a nematic arrangement of PBT molecules (2); this arrangement was derived on the assumption that PBT molecules can be considered as cylindrically symmetric, which in turn, suggested a hexagonal packing of such rods. Although the characteristic features of the x-ray pattern, i.e., broad equatorial reflections, meridional streaks and the absence of other (hkl) reflections were fully explained, this model does not agree with the observed density of the fibers. Electron microscopic data presented in this paper suggest another, more ordered model. This study also illustrates the capabilities of electron microscopy techniques, particularly electron diffraction coupled with dark field imaging, for the characterization of the structure of fibrous material.

Experimental

Material. A 9.85 weight percent nematic solution of PBT in methane sulfonic acid was dry jet-wet spun using a tapered glass capillary of 96 μm exit radius. The fiber was coagulated in a

[1]Permanent Address: CERMAV-CNRS, 53X, 38041 Grenoble-cedex,France
[2]Permanent Address: Faculty of Engineering, Fukui University, Bunkyo 3-9-1, Fukui 910, Japan

0-8412-0589-2/80/47-141-303$05.00/0

50/50 methanol/sulfolane bath at room temperature.

Electron Microscopy. A JEOL 100 CX electron microscope, operated at 100 kV, was used throughout this work. To prepare thin specimens, fibers immersed in water were repeatedly peeled into small fragments with the aid of sharp needles. The fibrillar fragments were then directly picked up on carbon coated grids. In some cases, the suspension was mildly sonicated to aid in dispersal of the fragments.

Darkfield (DF) imaging was performed with the tilted beam technique, the reflection selected by a 6×10^{-3} rad.objective aperture.

All pictures were recorded on Kodak SO-163 films, with maximum magnification (DF) of 10,000 X.

Results and Discussion

Electron Diffraction. Preparation of the fibers for electron microscopy is illustrated in Figure 1 which shows the image (scanning electron microscopy) of a partially peeled fiber. The internal fibrillar structure is quite apparent. Repeated splitting gives fragments suitable for transmission electron microscopy (TEM), as shown in Figure 2. Small fibrils of variable width, as small as a cross section of about 70 Å, are observed in some regions. An electron diffraction pattern, oriented as indicated in the Figure, could be easily recorded from the highly oriented fibrillar bundles. Three types of patterns were obtained, as shown in Figures 2, 3a and 3b respectively, indicating differing degrees of order for the same structure, encountered in different locations of a given fiber. Analysis of the most ordered patterns was most instructive. In addition to the very high number of equidistant meridional streaks (up to 20 orders being observable on the negatives), which correspond to a fiber repeat of 12.35 Å, the equatorial reflections are well resolved, and indicate the following spacings (the letters following each distance have their usual meaning for the relative observed intensities):

5.83 S	1.82 M-W
3.54 VS	(1.75) VW
3.16 M	1.71 W
(2.96) VW	

Faint, smeared (hkl) reflections are also observed, but are not resolved enough to be used for unit cell determination. Equatorial and meridional spacings indicate the following monoclinic cells:

Unit Cell I: $a = 5.83$, $b = 3.54$, $c = 12.35$
$\gamma = 96°$, $z = 1$

Unit Cell II: $a = 7.10$, $b = 6.65$, $c = 12.35$
$\gamma = 63°$, $z = 2$

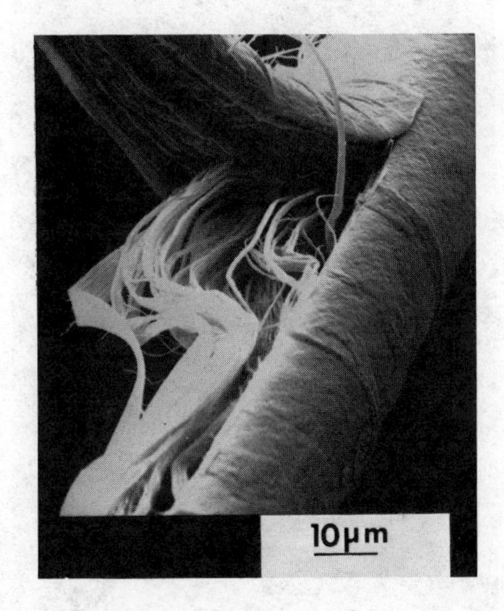

Figure 1. Scanning electron micrograph of a partially peeled PBT fiber

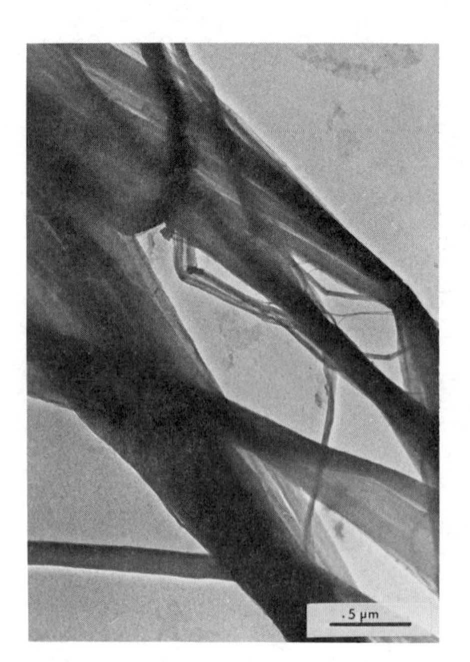

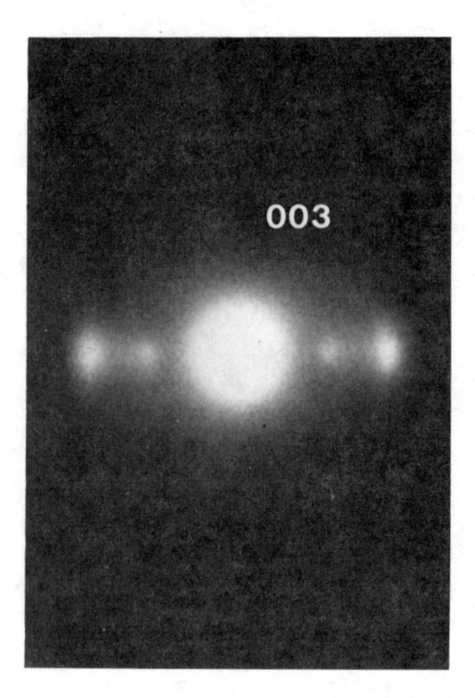

Figure 2. Bright field electron micrograph of a fibrillar fragment of a PBT fiber and corresponding electron diffraction pattern

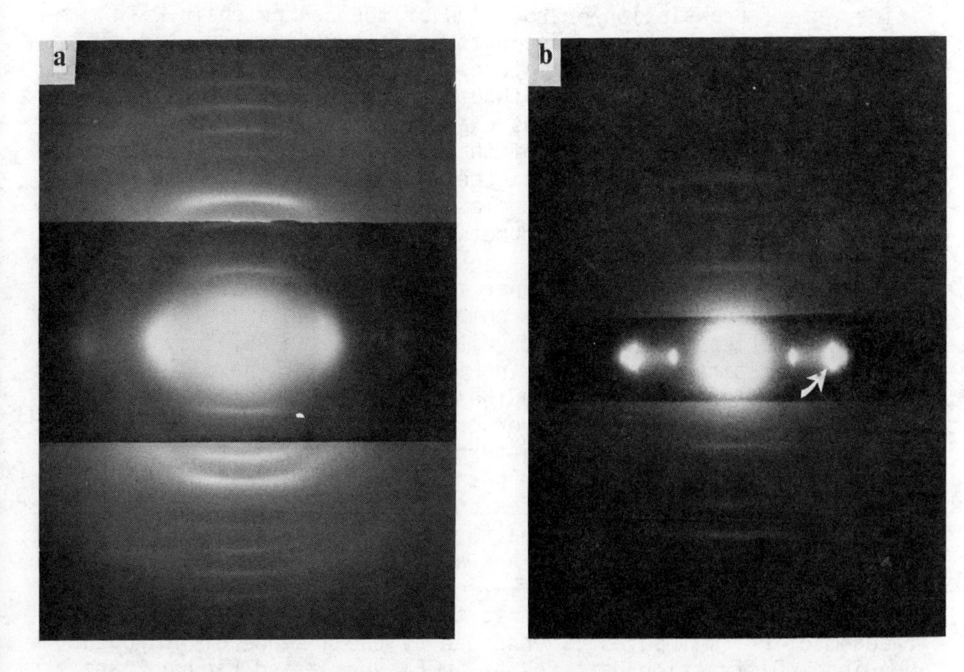

Figure 3. Electron diffraction patterns from fibrillar fragments of a PBT fiber exhibiting different degrees of order. The arrow points toward the reflection used for DF imaging. Note the splitting of this reflection, indicating a higher degree of order as compared to the same reflection in the pattern of Figure 2.

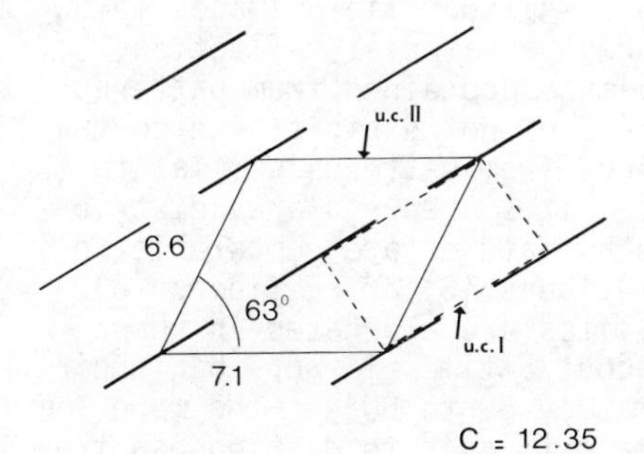

Figure 4. Possible unit cells of the PBT crystal structure (projection down the $\vec{c}$ axis)

Unit cell I corresponds to a very simple arrangement of parallel sheets, whereas Unit cell II (Figure 4) would allow more possibilities, i.e., translation or rotation of the center chain with respect to the corner chain. The unit cells have a calculated density of 1.69, which closely matches the observed density of approximately 1.6 for a fiber without macroscopic voids. The value of the fiber repeat corresponds exactly to the length of the repeat unit (3). Confirmation of the monoclinic packing, setting angle and choice between the two cells may be possible with more highly ordered fibers. Nevertheless, the present packing already allows a very reasonable interpretation of the different electron diffraction patterns.

If we assume a nearly planar molecular conformation (in the crystal structure of the model compound [3], the angle between the two moieties of the repeat unit is 23°), the PBT "crystallite" may be schematically shown as in Figure 5 (an all-parallel packing is taken for simplicity). The schematic emphasizes the two-dimensional character of this ordering, which explains essential features of the diffraction patterns. Small random translations of the chains along their axes causes the loss of all non-equatorial reflections and the appearance of reciprocal lattice discs as observed, for example, in pattern 3b. Such disorder is typical of rigid rod systems. It has already been observed from nematic fibers of poly-γ-benzyl-l-glutamate (4) and certain aramid fibers (5). In comparison to these cases, the translational freedom of PBT molecules is enhanced by the absence of hydrogen bonding. Diffraction patterns as in Figure 3a and Figure 2 are much more frequently observed than pattern 3b. These patterns can be considered to arise from crystallites of smaller lateral extent with more complete translational freedom along the chain axes. The number of meridional streaks is unaffected by the extent of this disorder, as it derives from the rigidity of the molecules.

Darkfield Electron Microscopy. All darkfield images below have been obtained from the strongest equatorial reflection, arrowed in Figure 3b. Darkfield images obtained from patterns similar to the one shown in Figure 2 do not exhibit high contrast. Figure 6 is such a DF image obtained from patterns similar to the pattern in Figure 3. Small crystallites are regularly distributed throughout the fragment. Such features are comparable to other observations on certain PPT fibers (5) or PE fibers (6). The strongest diffracting crystallites are elongated in shape with their average length being about 5 times larger than their width, which ranges from approximately 60 to 80 Å. The more numerous, smaller crystallites are difficult to distinguish from the background, due to inelastic scattering. Figure 7 shows corresponding brightfield (BF) and DF images of a peeled fragment. Although no peculiar contrast is observed in the BF image, a very marked banding transverse to the fiber direction appears in the

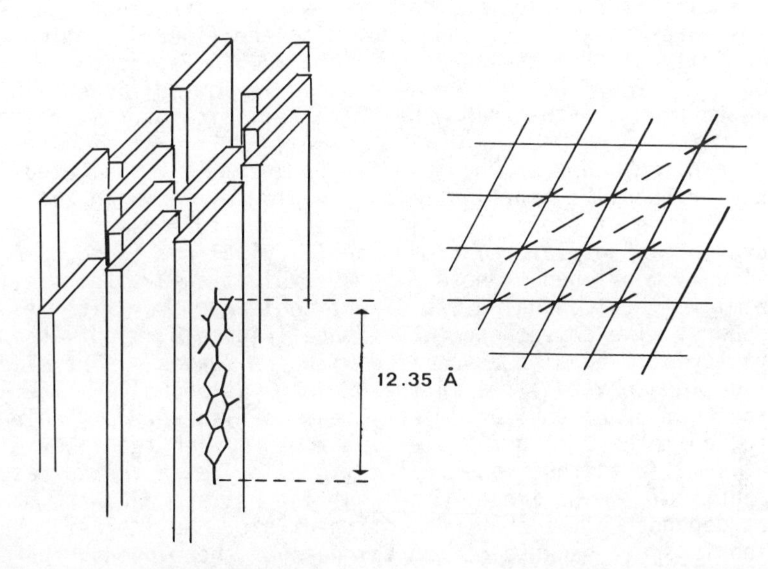

Figure 5. Schematic of the proposed arrangement of PBT molecules in a crystallite

Figure 6. Equatorial dark field image of a fragment of a PBT fiber

DF image. This observation, again, is comparable to what has been seen in PPT fibers (5). In the thinner part of the fragment the banding is regular, and each transverse dark band, ~ 200 Å in width, makes an angle of about 70° with the fiber direction. The periodicity of the banding is of about 1200 Å, as compared to 5,000 Å for PPT fibers. The resolution is not sufficient in this image to precisely describe the characteristics of the crystallites.

When fragments of the same fibers are mildly sonicated, thinner ribbonlike fibrils are observable, as shown in Figure 8 (left). Each of these flat ribbons appears to consist of smaller "microfibrils" of lateral dimensions varying from 50 to 80 Å as previously mentioned. These ribbons exhibit a wavy texture, each microfibril changing its direction in register with its neighbors. The corresponding DF image (Figure 8, right) shows a characteristic banding associated with the "waves". The size of well defined crystallites in the DF image is about 40 x 150 Å but a great number of smaller crystallites is also noted. Their long direction makes a slight angle (10° to 20°) with the fiber direction, similar to that made by the microfibrils in the corresponding parts of the BF image. The contrast between dark and bright zones depends on the sharpness of the kinks. As inelastic scattering is small because of the thinness of the ribbons, the grey background is mainly due, here, to diffuse scattering from less ordered regions.

This last observation suggests that orientation conditions, and not noncrystalline zones, are responsible for the observation of the dark zones. This is further demonstrated in Figure 9, which shows an "s-shaped" fiber. As the fiber progressively changes its orientation, the small zones (A) reverse their contrast from dark to bright, and conversely for the large zones (B). Therefore, textured ordered regions are present all along the ribbon, with neighboring crystallites in approximately the same orientation within a band. The banding period here varies from 1,000 to 2,000 Å. A double system of banding is also noted (see region C in Figure 9).

From the above observations, a schematic of the texture of these kinked ribbons is proposed in Figure 10. The ribbon is built up of closely packed microfibrils, well apparent in the kink zones. Each microfibril consists of a succession of narrow crystallites embedded in a somewhat less ordered matrix. The left hand portion of the schematic illustrates the banding observed in the DF image. Whether the bands appear due to the fragmentation of the fibers during sample preparation, or are characteristic of the as-spun fibers, is not known at present. The nonlinear stress-strain and the elongation at break (3%) suggest the bands may be shear bands.

Whatever their origin, the bands reflect the susceptibility of the fibrils to transverse kinking or buckling. Development of a skin-core morphology in the coagulation process may explain the

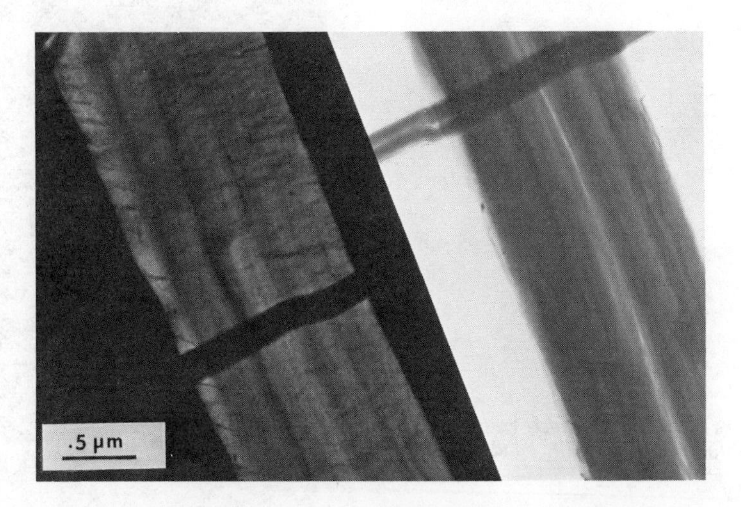

Figure 7. BF (right) *and DF* (left) *images of a fragment of a PBT fiber showing the band structure which appears in DF*

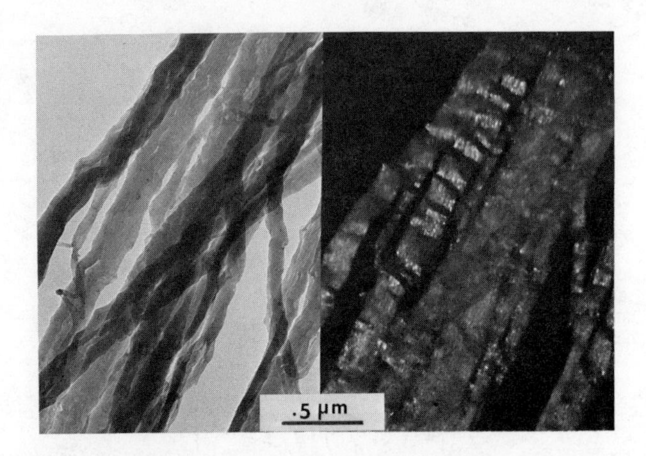

Figure 8. BF (left) *and DF* (right) *images of ribbonlike fragments of a PBT fiber showing the fibrillar texture of the ribbons*

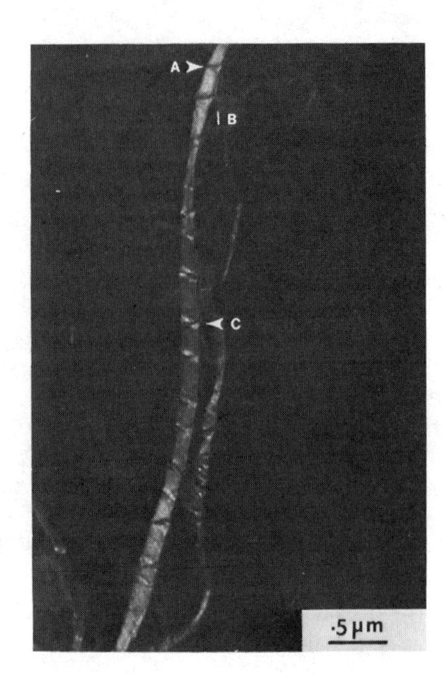

Figure 9. Electron micrograph showing orientation effects on the banded structure

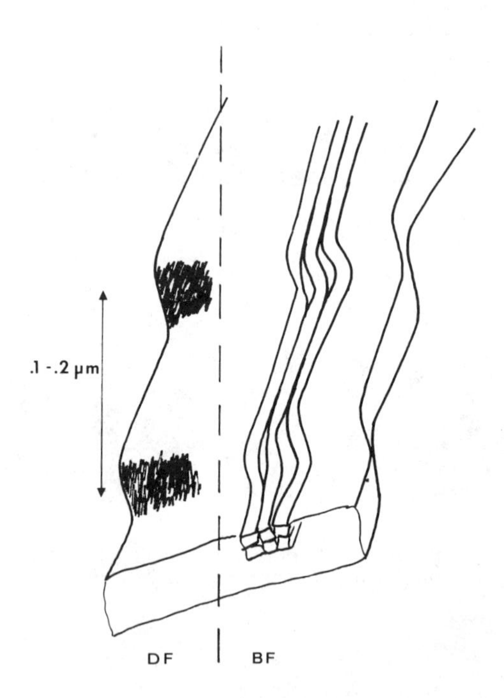

Figure 10. Schematic of the fibrillar structure of the ribbonlike fragments obtained after peeling and mold sonication of PBT fibers

discrepancy between darkfield images (banded or not), so they may also explain the different degrees of order encountered.

Conclusion

As already illustrated by a previous paper (5) on the structure of high modulus fibers, electron microscopy can be very successful when applied to these beam-resistant materials, and so constitutes an essential complement to x-ray studies. In the present work, electron diffraction coupled with BF and DF imaging has allowed detection of the best ordered zones within PBT fibers which illustrates the structure possibly obtainable by fiber processing refinement. The well ordered structures observed thus far compare rather well, with the exception of their fibrillar texture, to the structure of PPT high-modulus fibers. The two dimensional character of the crystallites is likely due to the freedom of axial translation of the molecules. Future work should determine if this feature is a direct consequence of the chemical structure of the PBT molecule or is simply the result of non-optimized processing conditions.

Acknowledgements

The fiber studied was kindly provided by Professor G. Berry of Carnegie-Mellon University. The authors thank Mr. S. Allen for furnishing the SEM picture of Figure 1. We also thank Dr. A. Kulshreshtha and Mr. W. Adams for helpful discussions throughout this work. Financial support was received from the U.S. Air Force through contract #F33615-78-C-5175 and the Materials Research Laboratory of the University of Massachusetts. One of us (EJR) is indebted to the CNRS for favoring his stay at the University of Massachusetts.

Literature Cited

1. Helminiak, T.E. 177th ACS Meeting, Hawaii, 1979, 675.
2. Adams, W.W.; Azaroff, L.V.; Kulshreshtha, A.K. Z. Kristal., in press.
3. Fratini, A.V.; Wiff, D.R.; Wellman, M.W.; Adams, W.W., to be published.
4. Samulski, E.T.; Tobolsky, A.V. Biopolym., 1971, 10, 1013.
5. Dobb, M.G.; Johnson, D.J.; Saville, B.P. J. Polym. Sci., 1977, 58, 217.
6. Gohil, R.M.; Petermann, J. J. Polym. Sci., Polym. Phys. Ed., 1979, 17, 525.

RECEIVED May 21, 1980.

19

Refinement of Cellulose and Chitin Structures

J. BLACKWELL, K. H. GARDNER[1], F. J. KOLPAK[2], R. MINKE[3], and W. B. CLAFFEY[4]

Department of Macromolecular Science, Case Western Reserve University, Cleveland, OH 44106

This symposium presented an unusual opportunity in that we discussed the methods used to determine polymer structures from fiber diffraction data, rather than concentrating on the actual structures derived and their possible implications. At Case Western Reserve University we have been involved in determination of the structures of cellulose and chitin. This paper describes our analyses (1-6) of the structures of cellulose I and II and α- and β-chitin, emphasizing the manner in which structural decisions were taken in each case. Efforts to determine these structures have a history of over 60 years, and it has only been with the advent of least squares techniques for the refinement of polymer structures (7) [notably the LALS method (8)], and the development of our present knowledge of polysaccharide stereochemistry, that solutions have become possible. In what follows we will look first at our methods for measuring intensities and thereafter will review the work on each of the four structures.

Intensity Measurements

As has been commented upon in other papers in this symposium, methods for measurement of intensities in fiber diagrams have received little attention in the last 15 years, during which time considerable progress has been made in handling the atomic coordinates for least squares refinement. Most intensity data has been obtained from linear densitometer traces through the x-ray reflections. We used these procedures for our work on cellulose I and β-chitin. For cellulose I, (1) the area under the peak on a radial scan was determined after subtraction of an estimated

Current addresses:
[1]Central Research and Development Department, E.I. du Pont de Nemours and Co., Experimental Station, Wilmington, DE 19898
[2]International Paper Corporation, Corporate Research Center, Tuxedo Park, NY 10987
[3]50 Yeelim Blvd., Apt. 13, Beersheva, Israel
[4]Research Division, Cleveland Clinic, Cleveland, OH 44106

0-8412-0589-2/80/47-141-315$05.00/0

background curve, and this was then corrected for fiber disorientation by the method of Cella et al. (9) The work for β-chitin (5) utilized intensities measured (10) by combination of radial and tangential scans, following the methods of Marvin et al. (11)

The inaccuracies inherent in the above approach are well known, especially the problem of assigning the background, which can lead to very large errors for weak reflections. In our later work on cellulose II (3,4) and α-chitin (6) we obtained an x,y map of the x-ray photograph in optical density units using an EDP scanning microscope. This instrument samples optical density at polar coordinates and is less than ideal in that the data must then be processed to give an x,y map analogous to that produced by the Optronics instrument. We normally process the data to give a 216x216 grid of optical density values, in which each data point is the arithmetic average of approximately ten measurements in the region close to the x,y point. Figure 1 shows typical data for two equatorial reflections in a cellulose x-ray pattern. The reflection is identified by the inner contour, which is drawn by hand. The background is obtained from the average value in a band one increment wide round the reflection. This is subtracted from each number within the reflection and the differences are summed to give the intensity. We have applied Lorentz and polarization corrections for the center of the peak to the summed intensity. (It would be more accurate to correct each x,y point before summation and this will be included in future work.)

The unit cell dimensions determined by least squares refinement of the observed and calculated d-spacings are given in Table I for the four structures considered below.

Table I

Unit Cells for Cellulose and Chitin Structures

	a(Å)	b(Å)	c(Å)	γ(°)	space group
Cellulose I					
Valonia (x-ray) 8-chain (1)	16.34	15.72	10.38	97.0	$P2_1$
Valonia (e.d.) 2-chain (2)	8.18	7.86	10.34	97.0	$P2_1$
Cellulose II					
Fortisan (3)	8.01	9.04	10.36	117.1	$P2_1$
Mercerized cotton (4)	8.02	8.99	10.36	116.6	$P2_1$
β-Chitin					
Pogonophore tube (5)	4.85	9.26	10.38	97.5	$P2_1$
α-Chitin					
Lobster tendon (6)	4.74	18.86	10.32	90	$P2_12_12_1$?

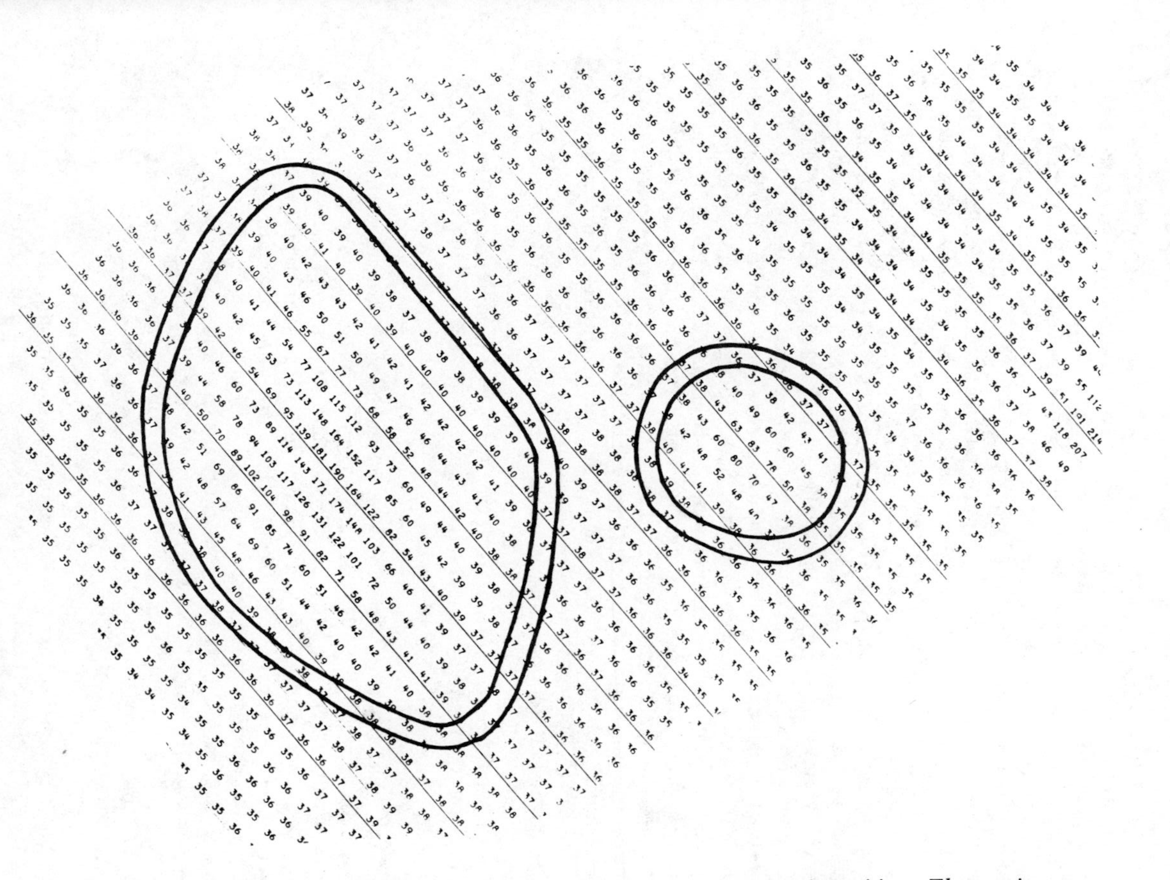

Figure 1. Typical x,y grid of optical density data used for measuring intensities. The sections shown contain two equatorial reflections for a cellulose x-ray pattern.

Cellulose I

The unit cell determined for Valonia cellulose I contains disaccharide units of eight chains, and is obtained from the two chain unit cell proposed by Meyer and Misch (12) by doubling the a and b axes. This doubling is necessitated by three weak reflections in the Valonia pattern. The other reflections with odd h and/or k are too weak to be detected and hence the differences between the four Meyer and Misch cells making up the eight chain cell must be very small. Thus we have taken the two chain unit cell as an adequate approximation to the structure. Odd order 00ℓ reflections are absent and thus the space group is $P2_1$. Examination of possible structures indicates that the 2_1 axes must be coincident with the chain axes, and hence the space group symmetry is compatible with a parallel or antiparallel arrangement of the chains. Our procedure was to construct both parallel and antiparallel chain models, and to refine them using the LALS programs to select the best model. As has been observed by many other groups, the intensity distribution, notably the very weak 002 and strong 004 reflections, necessitates an approximate quarter stagger of the chains. Thus we needed to consider four models defined as follows:

p_1 - parallel "up" chains with the center chain staggered by $+c/4$;
p_2 - parallel "down" chains with the center chain staggered by $+c/4$;
a_1 - antiparallel chains with an "up" chain at the origin and a "down" chain at the center staggered by $-c/4$;
a_2 - antiparallel chains as in a_1 but with the center chain staggered by $+c/4$.

An "up" chain has $Z(O5) > Z(C5)$; in each case the glycosidic oxygen O1 is at (0,0,0) for the corner chain and the stagger defines the c displacement of O1 for the center chain. It should be noted that the different chain polarities in models p_1 and p_2 lead to different models, regardless of the chain stagger; change of the stagger from $+c/4$ to $-c/4$ leads to the same model with a change of origin. In contrast a_1 and a_2 are variants of the same basic model, and differ only in the chain stagger. However, models a_1 and a_2 were found to occur in separate minima during the x-ray refinement, and it was convenient to treat them separately.

Each of these models was refined against the x-ray intensity data, consisting of 36 observed and 40 unobserved (non-meridional) reflections. The atomic coordinates for the cellulose chain were derived from standard bond lengths and angles, and the chain was constrained to have a 2_1 screw axis repeating in 10.38Å, and to form an intramolecular O3-H···O5' hydrogen bond. This chain is rigid except for the rotational freedom of the $-CH_2OH$ group. The models were defined in terms of the following refinable parameters:

$\phi 1$ and $\phi 2$, the orientations of the corner and center chains about their axes;
X1 and X2, the torsion angles defining the $-CH_2OH$ orientations on the center and corner chains;
S, the chain stagger;
K, a scale factor and B, an isotropic temperature factor.
In the initial work X1 was set equal to X2; later work showed that refinement of X1 and X2 as independent variables did not lead to significantly different values for the two chains.

The four models were refined against the observed data only, leading to structures with the following residuals: $R_{p1} = 0.179$, $R_{p2} = 0.202$, $R_{a1} = 0.207$, and $R_{a2} = 0.249$. Statistical tests show that model a_2 can be rejected in favor of a_1. The choice between the parallel models p_1 and p_2 is more difficult but examination of model p_2 shows that this model cannot be fully hydrogen bonded, and hence p_2 is rejected in favor of p_1, which also gives better x-ray agreement. Thus models p_1 and a_1 were taken as the most likely parallel and antiparallel models for further refinement. At this point the unobserved data were included, calculating weighted R' and R" where $w = 1$ for observed and $w = 1/2$ for unobserved reflections. F(hkl) for an unobserved reflection was set at two thirds an assigned threshold and was included only if the calculated structure amplitude exceeded the threshold. The final residuals for the two models were $R'_{p1} = 0.233$, $R'_{a1} = 0.299$, and $R''_{p1} = 0.215$, $R''_{a1} = 0.270$. Application of the Hamilton statistical test (13) to these data indicates that the a_1 model can be rejected at the 99.5% level.

Thus a parallel chain model is proposed for cellulose I, as shown in Fig. 2. The $-CH_2OH$ group is oriented close to the tg conformation (14) and hence each glucose residue forms two intramolecular hydrogen bonds: O3-H···O5' and O2'-H···O6; and also an intermolecular hydrogen bond O6-H···O3 to the next chain along the a axis in the 020 plane. The structure can be seen as an array of staggered hydrogen bonded sheets. It is interesting that the a_1 model is very similar to this: the same type of hydrogen bonded sheets are present but successive sheets have opposite polarity. The a_1 model is stereochemically acceptable and the decision in favor of the parallel chain structure is made on the basis of x-ray data alone. It should be noted that in separate work Sarko and Muggli (15) also proposed a parallel chain model for cellulose I based on x-ray data for *Valonia*. Calculations by French and Murphy (16) based on our intensity data also showed a preference for the parallel chain model.

Electron Diffraction

High quality electron diffraction patterns of *Valonia* cellulose fibrils were first obtained by Honjo and Watanabe.(17) These patterns contain a large number of reflections and this technique promises a significant increase in the possible

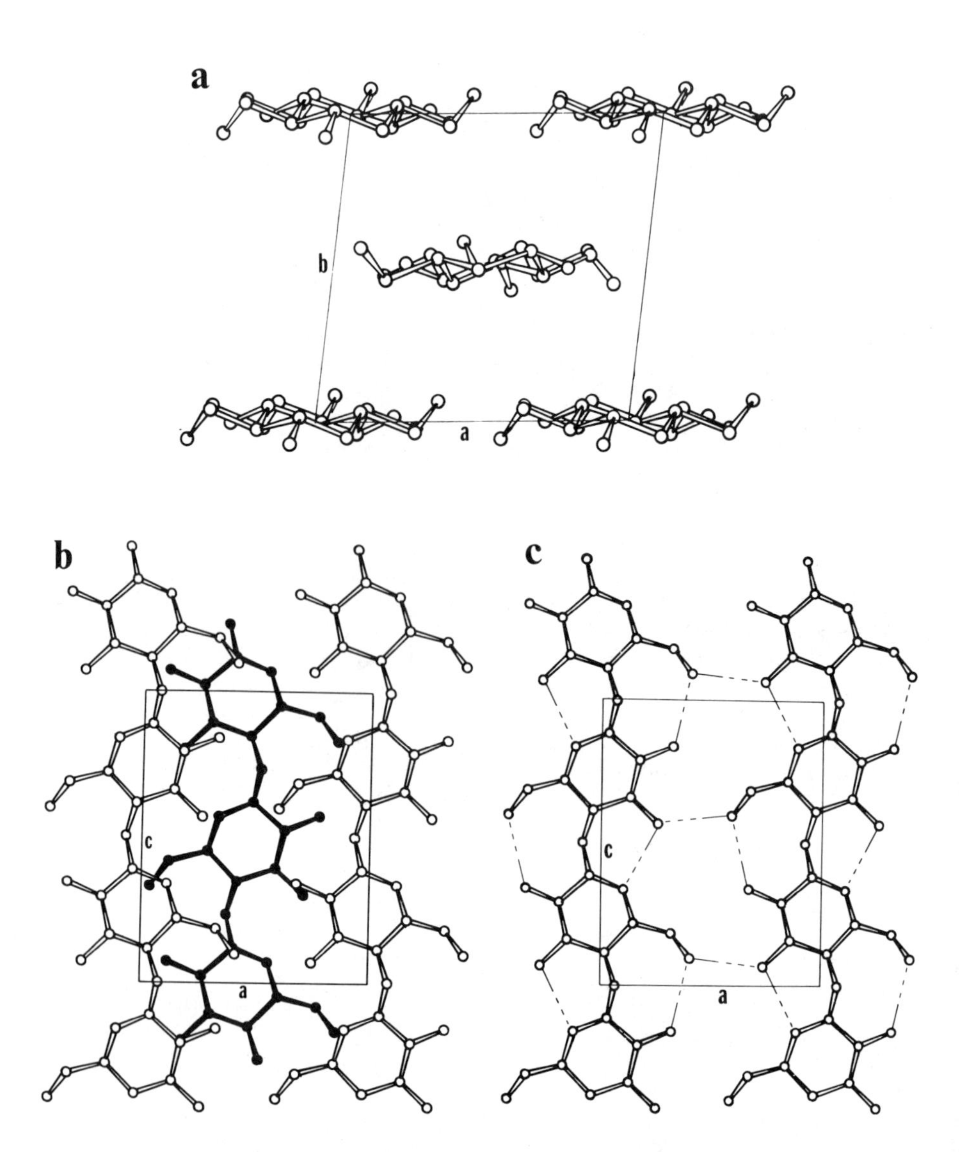

Figure 2. Structure of Cellulose I: (a) ab *projection; (b)* ac *projection; (c) hydrogen bonding network in the 020 plane*

resolution of polymer structures based on diffraction data. Electron diffraction has been used in this laboratory (18) to refine a synthetic polymer structure.

Previous electron diffraction work on *Valonia* cellulose (17,19) concentrated on determination of the unit cell and space group, and no efforts were made to make quantitative use of the intensities. Electron diffraction patterns were obtained from thin sheets stripped from the *Valonia* cell wall and had the appearance of fiber diagrams. However, it has been shown (20) that the microfibrils in such specimens tend to be oriented with the 110 planes perpendicular to the surface, and thus these planes have the highest probability of satisfying the Bragg condition. Examination of the diffraction pattern shows that the ratio of the 110 and $1\bar{1}0$ intensities is very much higher than for the x-ray pattern of specimens with fiber disorientation. We assumed a Gaussian distribution of microfibril rotations about the preferred orientation, and applied the following correction factor to the calculated structure amplitude

$$F\ corr(hkl) = F\ calc(hkl)(1/\sigma)\ \exp\left[-(\Psi-\Psi_o)^2/2\sigma^2\right]$$

where σ is the width of the Gaussian distribution, Ψ is the angle between the hkl reciprocal lattice vector and the h0l reciprocal lattice plane, and Ψ_o is the value of Ψ which defines the midpoint of the distribution.

The refinement (2) proceeded in the same way as for the x-ray work, except that σ and Ψ_o were refined as additional variables. We assumed that the scattering was kinematic. The cross sectional dimensions of the microfibrils are 200x100Å and our previous work on synthetic polymer single crystals showed that the kinematic approximation was adequate for such small crystallites. Intensity measurement presented considerable difficulty in that multiple film exposures could not be obtained. Sequential exposures of the same area of the specimen led to problems of beam damage, and patterns from different areas were not comparable due to differences in the preferred orientation. As a result, only the 28 strongest non-meridional intensities could be measured. These were all for reflections which could be indexed by the Meyer and Misch unit cell, and thus the two chain unit cell was used for the refinement.

Refinement of the four quarter staggered models led to the following residuals: $R_{p1} = 0.255$, $R_{p2} = 0.247$, $R_{a1} = 0.254$ and $R_{a2} = 0.294$. All of these values are relatively close. It is significant however that model p_1 is very close to that refined by x-ray work, whereas the other three differ significantly from their x-ray counterparts. The agreement for the meridional reflections is unacceptable for both the a_1 and a_2 models: in particular a strong 002 intensity is predicted, compared to the very weak observed reflection. In addition, the two chains in the unit cell have different orientations about their axes such

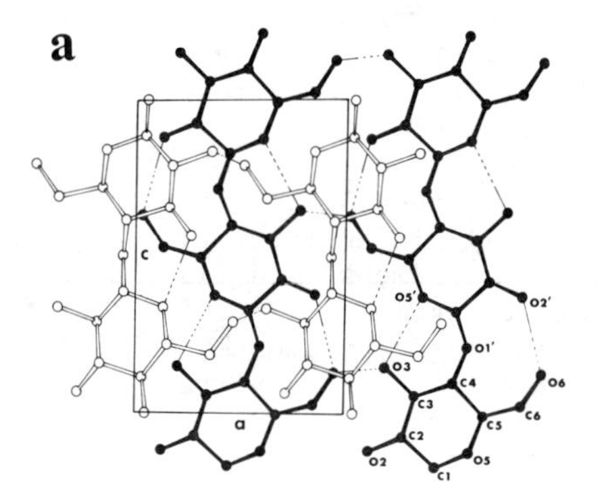

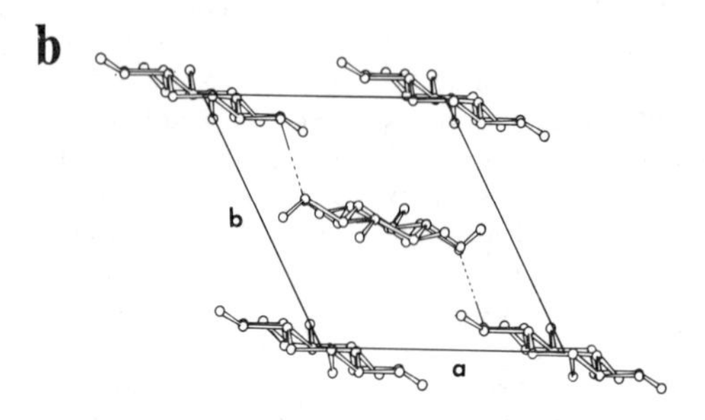

Figure 3. Structure of Cellulose II: (top) *(a)* ab *projection; (b)* ac *projection.* (Right) *(c) hydrogen bonding network for the center "up" chains; (d) hydrogen bonding network for the corner "down" chains; (e) hydrogen bonding in the 110 plane.*

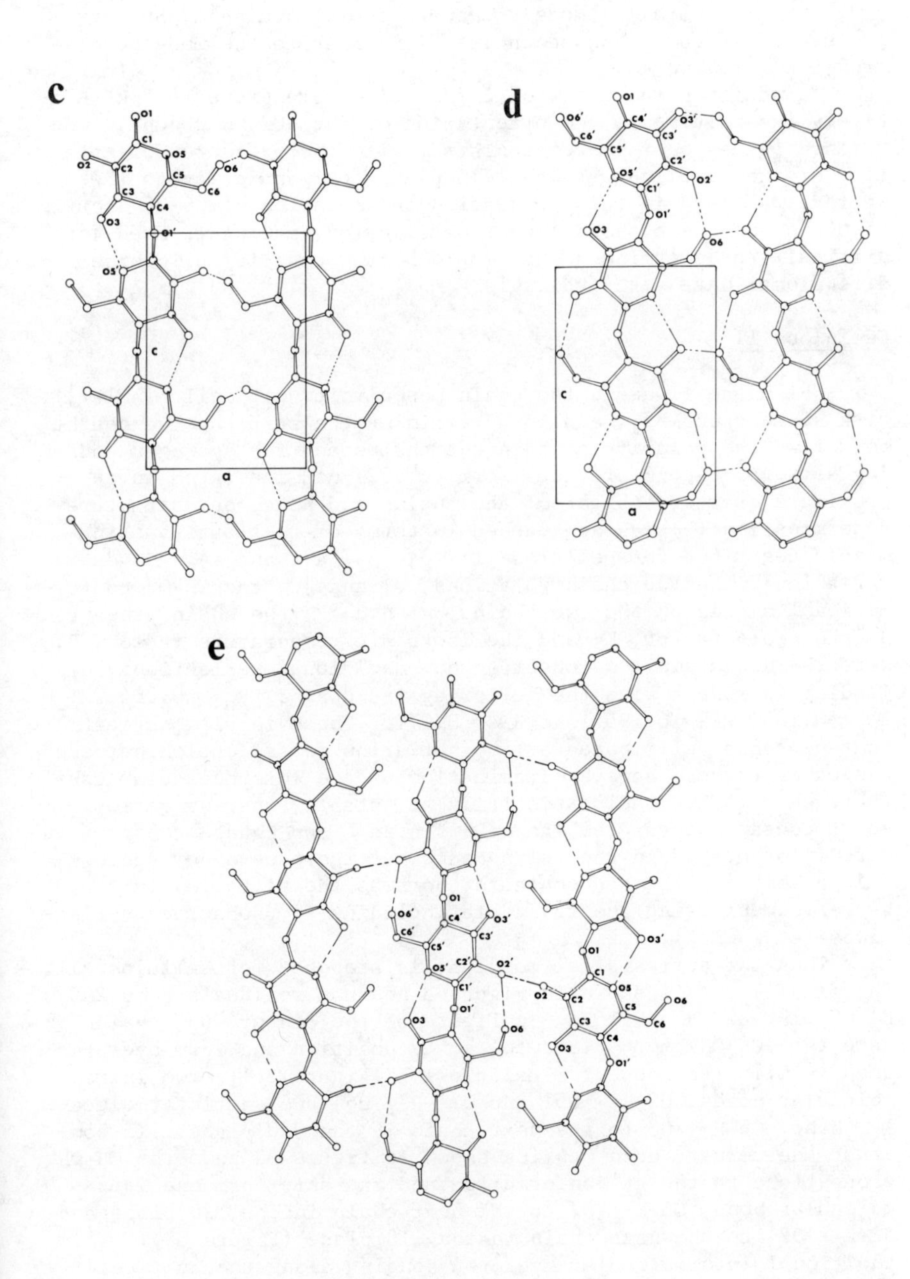
c
O1
C1
O2
C2
O5
C5
C3
C4
O6
C6
O3
O1'
O5'
c
a
d
O1
O6'
C4'
O3'
C6'
C3'
C5'
C2'
O5'
O2'
C1'
O1'
O3
O6
c
a
e
O1
O6'
C4'
O3'
C6'
C3'
C5'
C2'
O2'
O5'
C1'
O1'
O3
O6
O3'
O1
C1
O2
C2
O5
C5
O6
C6
C3
C4
O3
O1'

that absent equatorials are predicted to have appreciable intensities. On these grounds the antiparallel models can be rejected. The two parallel models cannot be distinguished, but our preference is for p_1 based on its similarity to the model derived from the x-ray work.

In addition the electron diffraction data proved to be relatively insensitive to the orientation of the $-CH_2OH$ groups. The refined values of Ψ_o were consistent with the preferred orientation of the lamellae with the 110 planes perpendicular to the surface. Viewed in this projection the rotation of C6-06 about C5-C6 is seen as a short linear oscillation, making it much more difficult to determine χ than would be the case if full three dimensional data were available.

Cellulose II

Cellulose II has a two chain monoclinic unit cell (Table 1) with space group $P2_1$, and in determining the structure we sought to define the polarity of adjacent chains and the hydrogen bonding network. Again previous work (21,22) had indicated an approximate quarter stagger of the chains and hence our structure determination (3) was approached in terms of the four starting models described for cellulose I: p_1, p_2, a_1, and a_2. Seven variables: the two chain rotations, $\phi 1$ and $\phi 2$, the CH_2OH conformational angles on the two chains, $\chi 1$ and $\chi 2$, the chain stagger, S, the scale factor, K, and the isotropic temperature factor, B, were refined against 44 observed non-meridional reflections, leading to models with the following residuals: $R_{p1} = 0.254$, $R_{p2} = 0.188$, $R_{a1} = 0.195$, and $R_{a2} = 0.171$. These results suggest that p_1 can be eliminated but do not allow for selection between the other three. However examination of the intermolecular contacts shows only model a_2 is fully acceptable, whereas serious short contacts occur for the other three. Non-bonded constraints were incorporated in the refinements but these were not successful in removing the bad contacts; nor was the situation improved by refinement using the full data including 41 unobserved reflections (at $w = 1/2$).

Thus the antiparallel model a_2 is proposed for cellulose II. The final model is shown in Figure 3 and had residuals $R = 0.235$, $R' = 0.219$, and $R'' = 0.167$. In Figure 3c the center "up" chains have the $-CH_2OH$ group close to the *tg* position and form hydrogen bonds similar to those for cellulose I (Figure 2c): two intramolecular bonds, 03-H···05' and 02'-H···06, and one intermolecular bond, 06-H···03 to the next chain in the 020 plane. In contrast the corner "down" chains shown in Figure 3d have the CH_2OH group close to the *gt* conformation and can only form one intramolecular bond, 06-H···02 to the next chain in the 020 plane and 02-H···02' to the next chain in the 110 plane (Figure 3c). This additional intermolecular hydrogen bonding along the unit cell diagonal in cellulose II is an important difference from cellulose I (in addition to the chain polarity) and may account for the higher stability of form II. Work by Stipanovic and Sarko (23) has also indicated an antiparallel arrangement of chains.

Determination of the cellulose II structure required application of stereochemical restrictions taking account of the packing of chains in the crystal lattice. The nature of these restrictions is apparent in the R-map shown in Figure 4. Holding all the other variables constant, the $-CH_2OH$ conformations on the two chains, X1 and X2, were incremented and R" was calculated for each position.(4) Contours of equal R" are plotted against X1 and X2 in Figure 4. It is interesting that a number of (shaded) minima are obtained, and that these correspond to permutations of the staggered positions for the CH_2OH group: gg, gt, and tg. Of these minima only the (gt, tg) combination is stereochemically acceptable in the cellulose II unit cell. A number of minima which are indistinguishable from (gt, tg) on the basis of the R" value are therefore rejected because of bad contacts. When regenerated cellulose is examined in the electron microscope (24) it is seen to consist of small fibrils with widths in the range 20-40Å. An estimated 20-30% of the CH_2OH groups of the entire specimen would thus occur on the surface of these crystallites, where the packing restrictions are relaxed. The existence of these multiple minima does not *prove* that other conformations are definitely present, but this is likely to be the case since the tg conformation occurs only rarely in carbohydrate structures,(14,25) and probably occurs in cellulose due to the constraints inherent in packing the polymer chains. In the absence of these constraints, the CH_2OH groups will probably adopt the gt and gg conformations on the microfibrillar surface.

Chitin

Chitin is the 2-acetamido derivative of cellulose and serves as the fibrous component of skeletal tissues in many lower animals. At least two polymorphic forms of chitin have been recognized,(26) of which the α- and β-forms are the best characterized. The unit cells and space groups of α- and β-chitins are given in Table 1. Both have approximately the same fiber repeat as cellulose, and apparently have the same 2_1 helical conformation.

β-Chitin

Highly crystalline β-chitin is obtained from pogonophore tubes and the spines of certain diatoms, and is analogous in both crystallinity and morphology to the cellulose obtained from *Valonia* cell walls. Intensity data (10) were obtained for 61 observed non-meridional reflections for a specimen of dispersed (sonicated) crystallites of pogonophore tube (*Oligobrachia*

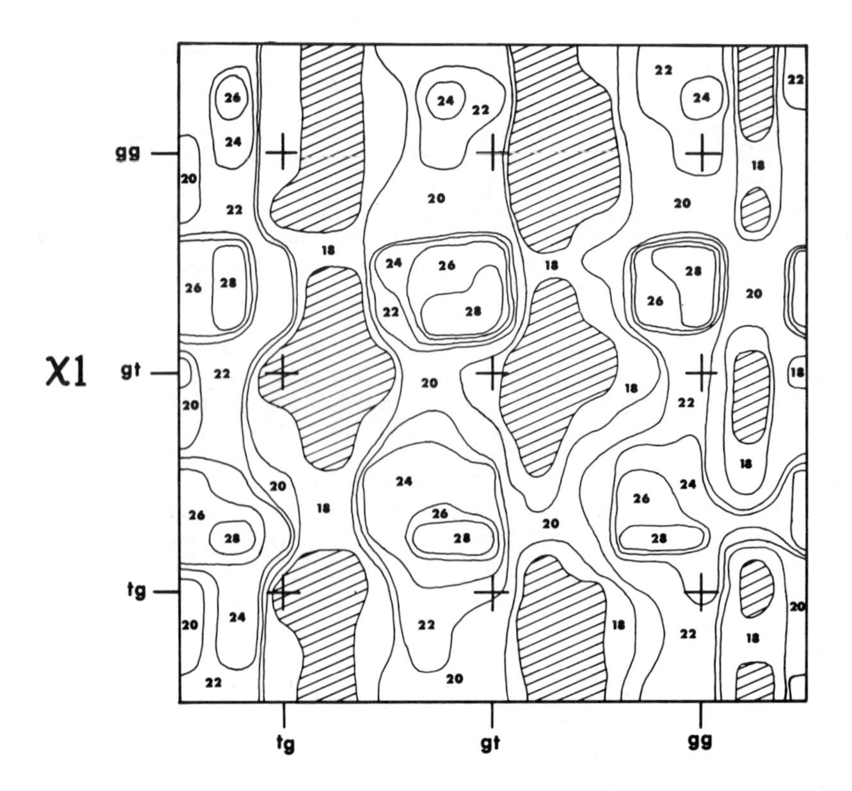

Figure 4. R-map for Cellulose II: variations of R″ with χ1 and χ2, the CH_2OH orientations on neighboring chains

ivanovi) oriented in a poorly crystalline protein matrix. Five observed reflections in the range of the observed data were set at zero. The atomic coordinates used were the same as for cellulose, except that O2-H is replaced by a planar $-NHCOCH_3$. The unit cell contains a disaccharide unit of a single chain, and hence the structure is an array of parallel chains. The structure was refined (5) in terms of the following parameters:

ϕ, the orientation of the chain about its axis;
χ, the rotation of the CH_2OH group;
χ', the rotation of the planar $-NHCOCH_3$ group about C2-N2;
K, the scale factor, and B, the isotropic temperature factor.

The refined structure had residuals R = 0.244, R" = 0.274. The amide groups are approximately perpendicular to the chain axis and form hydrogen bonds N-H···O = C, as indicated by the polarized infrared spectrum. The O6-H group is close to the gt conformation but is not hydrogen bonded and has a short O6···O5 contact. Elimination of the short contact with a non-bonded constraint increases the residuals to R = 0.250, R" = 0.288, an insignificant change. However the O6-H groups remain unbonded, contrary to infrared indications.

A search for possible hydrogen bonding networks reveals two possibilities. A change of ~20° in χ would allow for a hydrogen bond O6-H···O = C to the carbonyl in the next chain in the 100 plane. The chains are linked in sheets by N-H···O = C hydrogen bonds, and hence this arrangement of the O6-H group is described as an intrasheet hydrogen bond. Alternatively, a shift of ~180° in χ would allow for an O6-H···O = C bond to the carbonyl in the next chain in the $1\bar{1}0$ plane, described as an intersheet hydrogen bond. Constraints were included to require formation of each bond in turn, and the resultant models had the following residuals intrasheet bond: R = 0.267, R" = 0.302; intersheet bond: R = 0.342, R" = 0.367. The intersheet model can be rejected in favor of the intrasheet model at the 99.5% level. The non-bonded and intrasheet models cannot be distinguished and hence the intrasheet model must be selected as this is fully hydrogen bonded. The final model for β-chitin is shown in Figure 5.

α-Chitin

A $P2_12_12_1$ space group has been proposed for α-chitin, and a structure was published by Carlstrom,(27) based on refinements using optical diffraction masks. This structure gave reasonable agreement between the observed and calculated intensities.(28) However, the evidence for the $P2_12_12_1$ space group is not convincing. Only the absences of odd order 0k0 reflections can be demonstrated. Fortuitously, the b dimension of the unit cell is an approximate multiple of a($b \simeq 4a$) and hence it is not possible to determine absences for the h00 reflections. In addition a

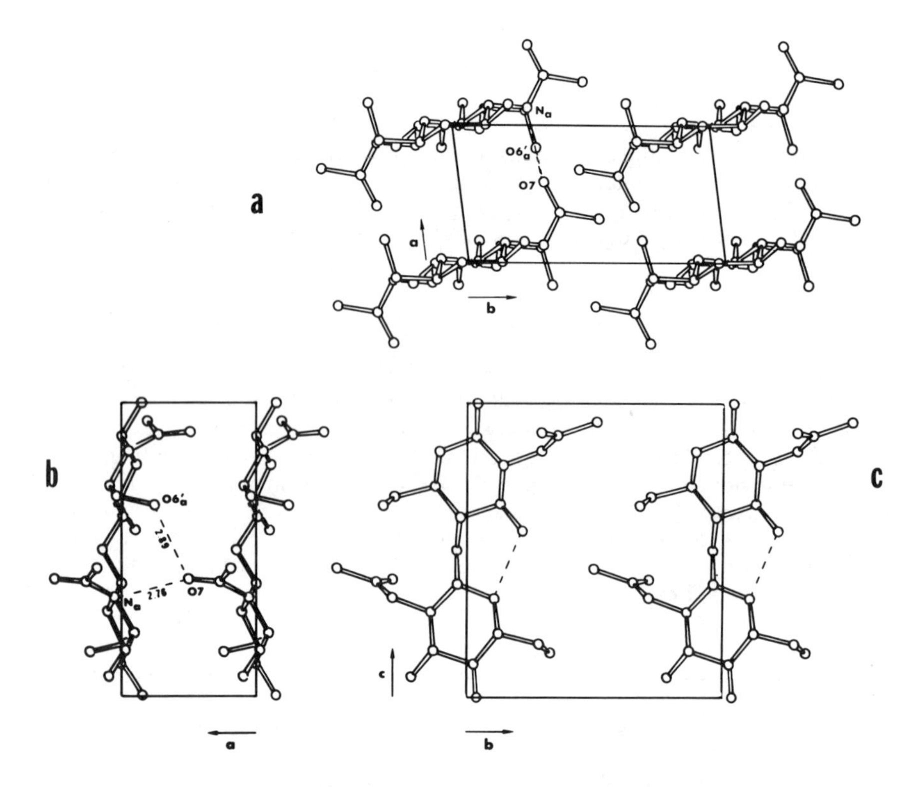

Figure 5. Structure of β-chitin: (a) ab *projection showing C═O · · · H—N bonds; (b)* ac *projection showing C═O · · · H—N and intrasheet O6—H · · · O ═C bond; (c)* ac *projection*

weak to medium intensity is seen in the x-ray pattern for the 001 reflection. Recent electron diffraction work on highly crystalline α-chitin in Saggita spines show odd order h00 and 001 reflections.(29) However, the intensity agreement for Carlstrom's structure show that the structure must be at least close to the $P2_12_12_1$ space group.

In addition to the problems with the space group, the structure can be criticized on two other grounds. Firstly, the CH_2OH are not hydrogen bonded, although the infrared spectrum indicates that there are no unbonded hydroxyls. Intermolecular hydrogen bonding between the sheets of chains would be expected since otherwise α-chitin should swell on hydration, as is observed for β-chitin.(10) Secondly, the amide I band in the infrared spectrum is a doublet at 1656 and 1621 cm^{-1} whereas only a singlet band is expected for the structure as proposed by Carlstrom.

In reexamining the structure (6) we utilized intensity data for 45 observed reflections in deproteinized lobster tendon (Homarus americanus). The unit cell contains disaccharide sections of two chains, and the model was refined in terms of the following parameters:

$\phi1$ and $\phi2$, the rotations of the two chains about their axes;
$\chi1$ and $\chi2$, the CH_2OH rotational angles;
$\chi'1$ and $\chi'2$, the $-NHCOCH_3$ rotations;
S, the chain stagger;
K, the scale factor, and B, the isotropic temperature factor.

Initially we considered a $P2_12_12_1$ model, for which the ϕ, χ, and χ' angles are the same for the two chains. This led to a model with residuals $R = 0.184$, $R'' = 0.161$ (observed data). This is judged to be very good agreement, but the model is similar to that proposed by Carlstrom and is subject to the same criticisms.

The defects in the model could not be overcome with the $P2_12_12_1$ space group and it was necessary to consider the lower symmetry, $P2_1$. Refinement was attempted for both parallel and antiparallel chain models. At this point it was found that the data was not sufficiently sensitive to allow simultaneous refinement of more than six variables, and it was necessary first to refine the $\phi1$, $\phi2$, S, K, and B, and then consider the other variables. The best parallel chain model obtained had $R'' = 0.25$, which could easily be rejected in favor of an antiparallel model at $R'' = 0.16$. The $\phi1$, $\phi2$, and S parameters refined for the antiparallel $P2_1$ model were not significantly different from those for the $P2_12_12_1$ model.

To consider possible hydrogen bonding schemes, an R" map was plotted, varying the $-CH_2OH$ rotations, $\chi1$ and $\chi2$, on the two chains; this map is shown in Figure 6. The $P2_12_12_1$ structures lie on the dashed diagonal for $\chi1 = \chi2$. The refined $P2_12_12_1$ model lies at the minimum at (149,149). This minimum is elongated, stretching from (120,180) to (180,120). At the two extremities of this minimum it is found that an acceptable hydrogen bonding network can be formed: the CH_2OH group on one chain forms an intramolecular O6-H···O=C hydrogen bond, and this allows for the $-CH_2OH$ group on the other chain to form an intermolecular (intersheet) O6-H···O6 hydrogen bond. This model has a residual R" = 0.166, indistinguishable from the non-bonded model. However, the model has a bad contact between the intramolecularly bonded O6 and the methyl group of the amide side chain. This contact could probably be eliminated by flexing the chain coordinates, with little effect on the residual, but we lacked the programming capabilities to do this, and overcame the defect by a constraint which produced a small rotation of the amide group, resulting in an increase in the residual to R" = 0.190. The model is not fully refined and it seems very reasonable that R" could be reduced by adjustment of the model, and hence the bonded model is judged to be acceptable.

The symmetry of Figure 6 indicates that two possible models exist: an intramolecular bond on chain 1 with an intermolecular bond on chain 2, or vice versa. On looking at molecular models, selection of one option for a particular site does not appear to dictate the selection at neighboring sites. Thus a random mixture of the two options was considered by placing half oxygens at the two positions for each residue. This statistical model is shown in Figure 7, and had a residual of R" = 0.188, not significantly different from the non-statistical model.

Placing half oxygens at the two O6 positions restores the $P2_12_12_1$ symmetry. The recent electron diffraction data is against this symmetry. This would argue that selection of hydrogen bonding scheme at a particular site does indeed affect the choice at neighboring sites, and this cascades through the crystalline domain. However, regardless of this problem, the model overcomes the other defects of the Carlstrom model. The hydrogen bonding scheme involves all the OH groups and does link the sheets through intermolecular bonds. In addition, there are two types of amide groups, those which accept one hydrogen bond and those which accept two bonds. This simple difference should be sufficient to explain the splitting of the amide I bond in the infrared spectrum.

This research has been supported by the N.S.F. Polymer Program, most recently through grant No. DMR-76-82768.

Abstract

This paper is a review of x-ray diffraction work in the authors' laboratory to refine the structures of cellulose I and II, and α- and β-chitin, concentrating on the methods used to select between alternate models. Cellulose I is shown to consist of an array of parallel chains, and this conclusion is supported by a separate refinement based on electron diffraction data. In the case of cellulose II, both parallel and antiparallel chain

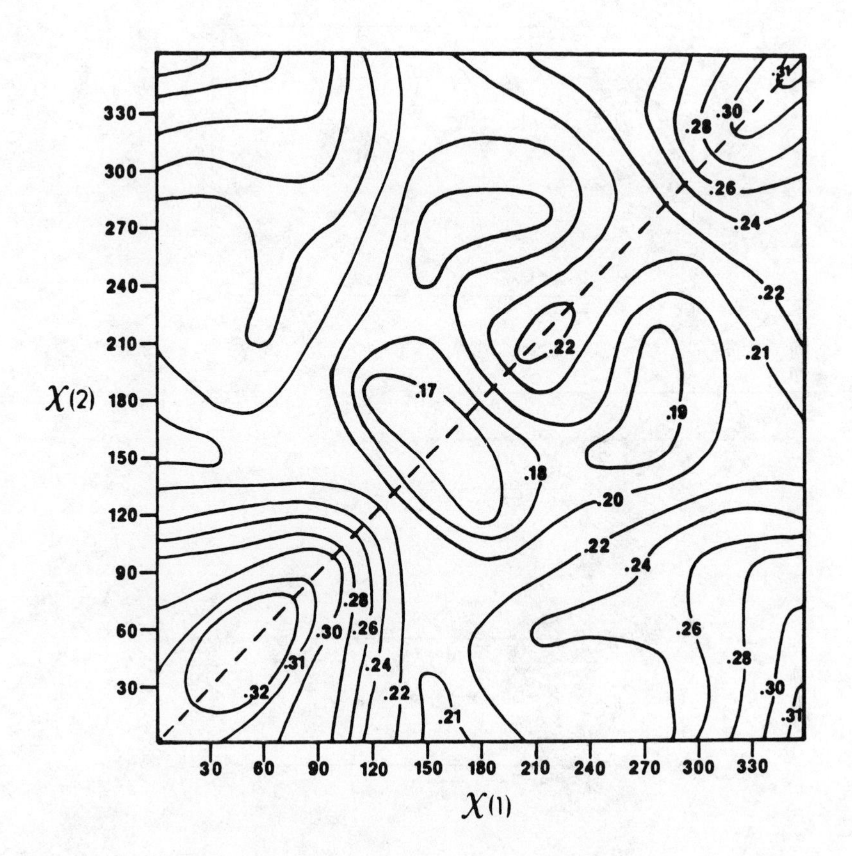

Figure 6. R-map for α-chitin: variation of R″ with χ1 and χ2, the CH_2OH orientations on neighboring chains. The dashed line corresponds to $P2_12_12_1$ models, in which χ1 = χ2.

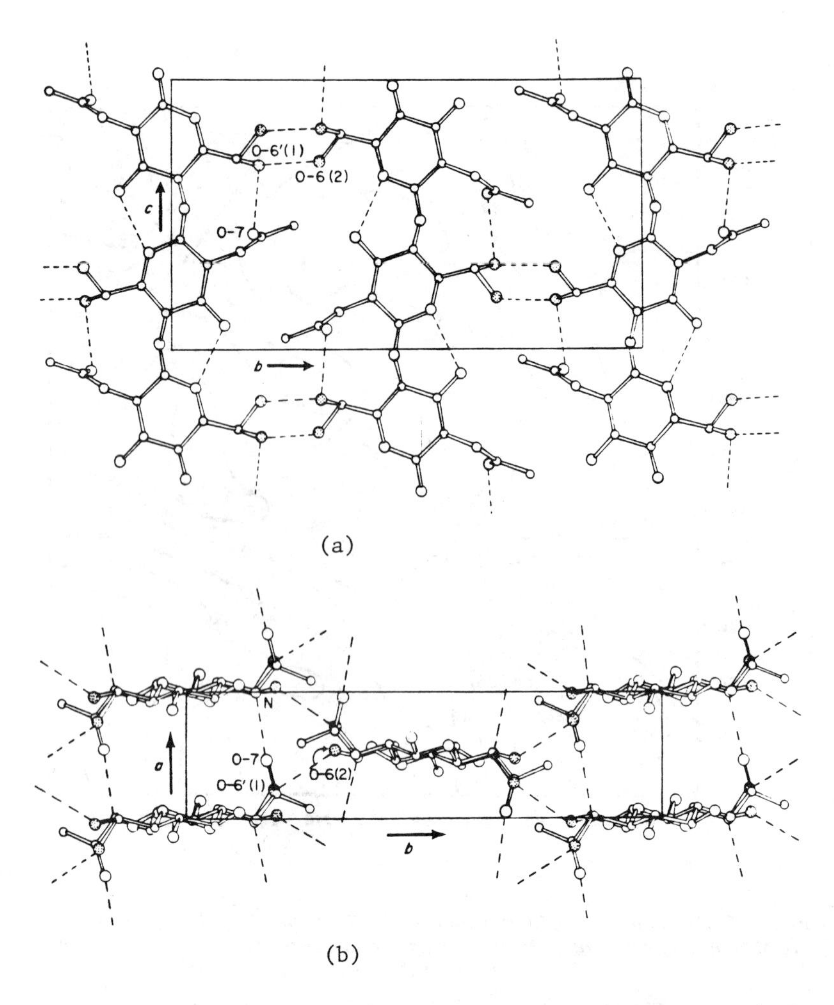

Figure 7. Structures of α-chitin with statistical hydrogen bonding network: the shaded atoms represent half (O6) oxygens oriented to form intra- and intermolecular hydrogen bonds ((a) ac *projection; (b)* ab *projection)*

models are consistent with the x-ray intensity data, but only the antiparallel model is stereochemically acceptable. The initial refinements of both α- and β-chitin lead to models in which the $-CH_2OH$ groups are not hydrogen bonded, but insertion of constraints leads to bonded models which are not significantly different in terms of the x-ray agreement. The refined models for all four structures are described in each case.

Literature Cited

1. Gardner, K.H. and Blackwell, J. Biopolymers, 1974, 13, 1975-2001.
2. Claffey, W.B. and Blackwell, J. Biopolymers, 1976, 15, 1903-1915.
3. Kolpak, F.J. and Blackwell, J. Macromolecules, 1976, 9, 273-278.
4. Kolpak, F.J.; Weih, M.; Blackwell, J. Polymer, 1978, 19, 123-131.
5. Gardner, K.H. and Blackwell, J. Biopolymers, 1975, 14, 1581-1595.
6. Minke, R. and Blackwell, J. J. Mol. Biol., 1978, 120, 167-181.
7. Arnott, S. and Wonacott, A.J. Polymer, 1966, 7, 157-166.
8. Smith, P.J.C. and Arnott, S. Acta Cryst., 1978, A34, 3-11.
9. Cella, R.J.; Lee, B.; Hughes, R.E. Acta Cryst., 1970, A26, 118-124.
10. Blackwell, J. Biopolymers, 1969, 7, 281-298.
11. Marvin, D.A.; Spencer, M.; Wilkins, M.H.F.; Hamilton, L.D. J. Molec Biol., 1961, 3, 547.
12. Meyer, K.H. and Misch, L. Helv. Chim. Acta, 1937, 20, 232-244.
13. Hamilton, W.C. Acta Cryst., 1965, 18, 502-510.
14. Sundaralingam, M. Biopolymers, 1966, 6, 189-213.
15. Sarko, A. and Muggli, R. Macromolecules, 1974, 7, 486.
16. French, A.D. and Murphy, V.G. Adv. Chem. Symp., 1977, 48, 12.
17. Honjo, G. and Watanabe, M. Nature, 1958, 326-328.
18. Claffey, W.B.; Gardner, K.H.; Blackwell, J.; Lando, J.B.; Geil, P.H. Phil. Mag., 1974, 30, 1223-1232.
19. Fisher, D. and Mann, J. J. Polymer Sci., 1960, 42, 189-194.
20. Frey-Wyssling, A. Biochim. Biophys. Acta, 1955, 18, 166-168.
21. Jones, D.W. J. Polymer Sci., 1958, 32, 371-394.
22. Jones, D.W. J. Polymer Sci., 1960, 42, 173-188.
23. Stipanovic, A.J. and Sarko, A. Macromolecules, 1976, 9, 851.
24. Kolpak, F.J. and Blackwell, J. Text. Res. J., 1978, 48, 458-
25. French, A.D. and Jeffrey, G.A. Molecular Structure by Diffraction Methods (publ. The Chemical Society, London), 1978, vol. 6, p. 183-223.
26. Rudall, K.M. Adv. Insect Physiol., 1963, 1, 257-313.
27. Carlstrom, D. J. Biophys. Biochem. Cytol., 1957, 3, 669-683.

28. Marchessault, R.H. and Sarko, A. Adv. Carbohydrate Res., 1967, 22, 421-482.
29. Atkins, E.D.T.; Dlugosz, J.; Ford, S. Int. J. Biolog. Macromolecules, 1979, 1, 29-32.

RECEIVED August 18, 1980.

Accuracy of Polymer Structure Determination

A Comparison of Published Structures of Poly(tetramethylene terephthalate)

I. H. HALL

Department of Pure and Applied Physics, The University of Manchester,
Institute of Science and Technology, P.O. Box 88, Manchester M6O 1QD U.K.

Although the crystalline structures of many polymers have been determined, it is unusual for these to include any estimate of the probable error in the various parameters. But unless these are known, deductions which are sometimes made from the values of the parameters (e.g. the assignment of infra-red absorption band frequencies) can be misleading.

This is illustrated by the case of poly (tetramethylene terephthalate) (4GT). Three independent determinations have been made of the crystalline structure of the α-phase of this material. (1,2,3). The conformation angles are given in Table I (see Figure 1 for key) from which it will be seen that they all

Table I

Proposed Conformations of the α-phase of 4GT

	Mencik(1)	Yokouchi et al (2)	Hall & Pass (3)
τ_1	174.8°	173.8°	179.4°
τ_1^1	-3.5°	1.8°	-0.6°
τ_2	177.5°	178°	-177.9°
τ_3	-90.6°	-88°	94.3°
τ_4	-88.4°	-68°	-79.3°

describe similar structures although there are differences between them, some quite large. Stambaugh et al (4) compare these structures and conclude that the one published by Mencik is best. Table I shows that in this the conformation angle of one of the methylene bonds deviates from a gauche value of -60° by -28.4°. They subsequently (5) use this deviation in assigning infra-red absorption bands, and in a discussion of conformational energy. However, until information on the accuracy with which the

0-8412-0589-2/80/47-141-335$05.00/0

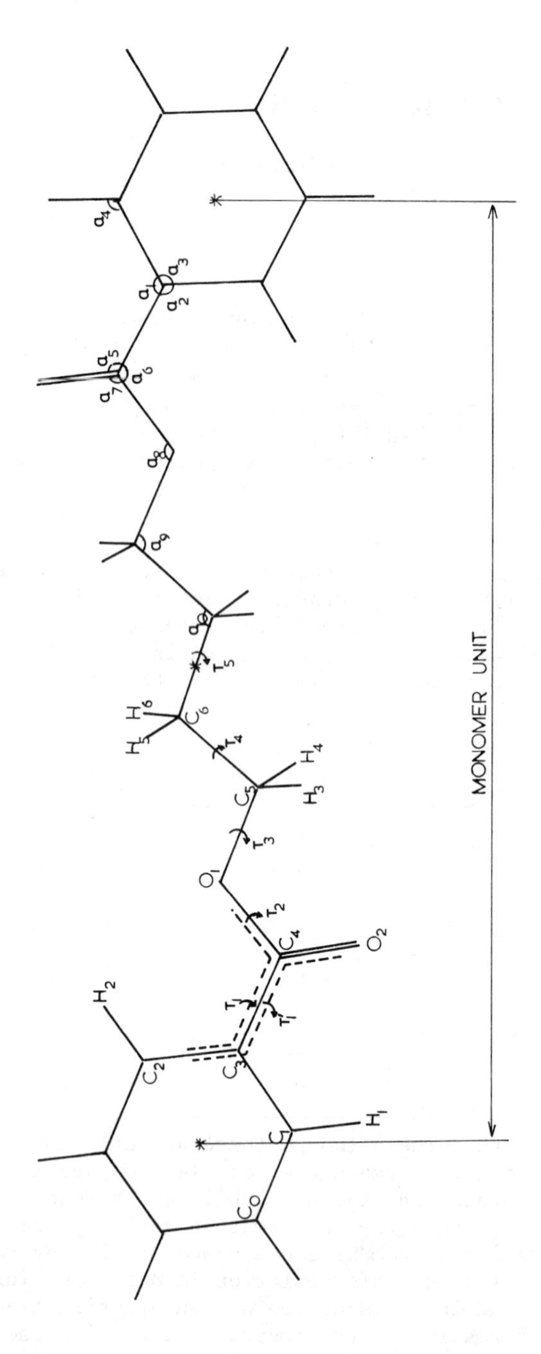

Figure 1. Monomer of 4GT ((*) centre of symmetry)

parameters have been determined becomes available it is impossible to know whether the structures differ significantly, or whether the deviation of this particular angle from a gauche conformation is significant.

In the present paper the raw data used by the various authors in their structural determinations of 4GT are used to investigate whether their structures differ significantly, and to produce the model, complete with error estimates, which best satisfies all their data.

Method of Investigation. All three determinations of the structure follow the same sequence of steps.

(a) The unit cell is determined.
(b) A set of X-ray intensity data is obtained.
(c) Values are assumed for the bond parameters.
(d) A model is proposed.
(e) This model is refined against the intensity data.

It is apparent from this that there are four sources of error (to be called factors) - the unit cell parameters, the intensity data, the bond parameters, and the starting model. Three independent choices are available for each of these four factors from the published structual determinations. In the present investigation it is first of all assumed that each of the three choices is equally good. Any set of factors comprising any combination of choices is then a valid starting point of refinement. Thus refinements can be performed on 81 different combinations of factors. Those factors which make the greatest contribution to the variability of the parameters describing the final structure are identified by applying the techniques of factorial analysis of variance, and Hamilton's test (6) is used to test the assumption that all choices of these factors are equally good. If any choice proves to be significantly worse than the others, it is taken out of the analysis and the remaining sets used to determine an average model and the uncertainty in its parameters.

The choices of each of the factors are discussed below.

Unit Cell. The unit cells published by each group of authors are shown in Table II. In this table Mencik's values for a and b and for α and β have been interchanged to agree with the conventions of other authors. Also, opportunity has been taken to correct misprints in the values of α and γ in ref.3. If the locations of reflections are calculated from these cells and compared with measured values, no preference is indicated for any particular one (7). The differences between them are too small to cause changes in the indexing of reflections.

Structure Factor Data. Only Yokouchi et al include a structure factor set in their published paper; Mencik offered sets by application to the author. Because of the large amount of data, values will not be reproduced here. The set of Hall and

Pass may be had on application to the present author.

Table II

Unit Cell Parameters of the α-phase of 4GT

	Mencik (1)	Yokouchi et al (2)	Hall & Pass (3)
a(Å)	4.83	4.83	4.89
b(Å)	5.96	5.94	5.95
c(Å)	11.62	11.59	11.67
α^{o}	99.9	99.7	98.9
β^{o}	115.2	115.2	116.6
γ^{o}	111.3	110.8	110.9
Vol ($Å^3$)	260.0	260.4	262.8

Mencik measured the intensity of strong spots using a Joyce double-beam microdensitometer, and assessed weak spots visually. Angular and geometric corrections were applied and an unspecified scaling technique was used. This was applied individually to layer lines up to the fourth, fifth and higher layer lines were treated as a group. Yokouchi et al again use a microphotometer, correcting for the Lorentz-polarisation factor and obliquity effect. Thus, apart from scaling, these two groups of authors would appear to use essentially similar techniques, involving, a single intensity scan through each reflection.

Hall and Pass used a rather different method (8), whereby the intensity was measured on a 25μm lattice of points covering the entire area of a reflection. From this an intensity contour map of the reflection was created and its total intensity determined by measuring the area within each contour line, multiplying this area by the intensity difference between adjacent contours, and summing these products for all contour areas within the boundary of the reflection. The measured intensity was corrected by application of the value at the centre of the reflection of the Lorentz-polarization factor.

Each author includes reflections that the others omit, and Mencik includes many weak, high-order, reflections ignored by the others. It is possible to identify a set of 42 reflections common to all authors, and to compare these common sets they have been scaled to a common average. From the subsequent values it was clear that for several reflections there was good agreement between two of the authors, the third disagreeing badly,suggesting that he had made some error in the measurement of this particular reflection. However, it did not appear that any one author had more disagreements than any other.

"R-factors" between pairs of these scaled common structure factor sets were calculated using the formula

$$R_{12} = \frac{\left| |F_1| - |F_2| \right|}{|F_1|}$$

and are shown in Table III.

Table III

"R-factors" Calculated between Pairs of Observed Structure Factor Sets Determined by Different Investigators

Investigators		R-factor
Hall and Pass (3)	Yokouchi et al (2)	18.7%
Hall and Pass	Mencik (1)	22.1%
Mencik	Yokouchi et al	15.6%

The discrepancies revealed by this table are not unusually large; similar, or larger, differences have been seen in data sets from other polymers. If different laboratories differ by these amounts in the measurement of structure factors, then disagreements between them in structures which give R-factors below about 20% are unlikely to be significant, since two equally good sets of structure factors can differ by this amount. That statement does, however, presuppose that all structure factor sets are equally good. If one laboratory has a superior technique to the other two then a structure based on that laboratory's data having an R-factor less than about 20% could be significantly better than the others. To prove that, though, other considerations must be introduced to demonstrate the superiority of technique. Methods of quantifying error in intensity measurements are urgently required if accuracy of polymer structure determination is to be improved.

Except for the comparisions discussed above, the full structure-factor set supplied by the author has been used in all cases.

A further complication arises with this material in that the diffraction pattern is not a true fibre pattern. The molecular chain axis in the crystallites is tilted with respect to the fibre axis (2,3) and this causes reflections to be displaced above and below the mean layer line position. Thus it is possible that some reflections which would overlap in a true fibre pattern to such an extent that their intensities could not be determined separately, might be resolved with the present material. When this happens for a reflection of observable intensity, then it will be measured as a discrete reflection and the resolution of the final structure will be higher than if the material had true fibre symmetry. However, Stambaugh et al (4) claim that for some of the reflections which have been assigned to overlapping groups, the displacement due to this tilt is such that if they were of observable intensity they would be seen as discrete reflections. They must therefore be below the threshold of observable intensity and should be omitted from the group and treated as discrete, unobserved, reflections.

This correction has not been made. Since the wrongly assigned reflection is unobserved, it does not contribute to the intensity of the group, whose measured intensity is thus the same whether or not it is included. Hence its inclusion in the calculated intensity causes no error. However the full improvement of resolution which could be achieved by taking advantage of the tilted crystal orientation has not been realised, and so the uncertainties in the final structure might be slightly greater (and similar to those in a true fibre) than they otherwise would be.

By retaining the groupings of overlapping reflections assumed by each author, their individual differences of judgement are retained and made to contribute to the uncertainties in the final structural model. The determination of the uncertainties caused by such differences of judgement is the main purpose of the present investigation

Bond Parameters. The chemical structure is shown in Figure I. All authors assumed that the conformation was centrosymmetric about the points indicated, and all treated the benzene ring as a rigid unit. They are less specific concerning their treatment of the carbonyl unit. (i.e. that part of the molecule comprising the three bonds, C_3C_4, C_4O_2,C_4O_1 in Figure 1); it will be clear from the values of τ_1 and $\tau_1{}^1$ in Table 1 that for the structures of both Mencik and Yokouchi et al these three bonds are not co-planar. However, they are closely co-planar in all published structures of low molecular weight analogues of this substance and so were forced to be so in the present work. The plane of this unit was, however, allowed to change with respect to that of the benzene ring; this was the practice of all authors in their original work. (An assertion by Stambaugh et al (4) that Desborough and Hall (7) force these two planes to remain parallel is untrue and must be based on a misunderstanding of our paper).

Since the structures were originally determined Pérez and Brisse (9-13) have published the structures of a series of closely related low molecular weight compounds. These probably provide the most reliable values of bond parameters that are available and since they were not available for the original work, choices have been modified slightly where there are large discrepancies with these values. These alterations are shown in Tables IV and V. Where there is a value in brackets in these tables it is the modified value which has been used in the present work. The bond lengths and angles may be identified by reference to Figure 1.

Table 1V

Bond Lengths Assumed for 4GT

Bond	Range from Pérez Structures (Å) (9-13)	Mencik (Å) (1)	Yokouchi et al (Å) (2)	Hall and Pass (Å) (3)
C_oC_1	1.37 - 1.38	1.40	1.395	1.38
C_1C_3	1.38 - 1.40	1.40	1.395	1.38
C_2C_3	1.39 - 1.40	1.40	1.395	1.39
C_4C_3	1.48 - 1.48	1.49	1.49	1.48
O_2C_4	1.20 - 1.21	1.23	1.23	1.21
O_1C_4	1.33 - 1.34	1.26(1.33)	1.36	1.34
C_5O_1	1.44 - 1.49	1.41	1.43	1.44
C_6C_5	1.49 - 1.50	1.53	1.54	1.50
C_7C_6	1.49 - 1.52	1.53	1.54	1.50

Table V

Bond Angles Assumed for 4GT

Bond	Range from Pérez Structures (9-13)	Mencik (1)	Yokouchi et al (2)	Hall and Pass (3)
α_1	118° - 120°	119°	120°	119°
α_2	121° - 123°	119° (120°)	120°	121°
α_3	119° - 123°	121°	120°	120°
α_5	124° - 125°	121° (124°)	123°	125°
α_6	113° - 113°	119° (114°)	114°	113°
α_7	122° - 123°	120° (122°)	123°	122°
α_8	116° - 117°	119°	111° (116°)	119° (117°)
α_9	105° - 108°	104°	110° (108°)	105°
α_{10}	112° - 115°	105° (112°)	110° (112°)	113°

Starting Model Both Mencik (1) and Hall and Pass (3) used a similar scheme to construct a starting model. The molecule from the benzene ring to the carboxyl group was treated as a rigid, planar unit and the orientation within the unit cell found which gave the best R-factor. The CH_2 units were than added to complete the model.

Yokouchi et al (2) used conformational energy calculations, revealing six low-energy conformations. Each of these was placed in the unit cell with the chain axis coincident with the c-axis and rotated about this axis to minimise the R-factor. One of them gave a much lower R-factor than the others, and this was chosen as the starting model.

Insufficient information is given in the original papers to enable these models to be re-created, and in order to investigate the effect of starting model each authors published structure was used as a starting point for the refinements performed here.

Refinement Method. As well as the factors discussed above, the refinement method can also influence the final structure. It might do this through the following features:

(a) the treatment of unobserved reflections,
(b) the weighting of structure factors,
(c) the type of refinement calculation.

The effects of these were not investigated in the present work; instead, a common procedure was adopted throughout.

Unobserved reflections (up to the smallest d-spacing at which a reflection was observed on a given layer line) were included in the refinement if their calculated structure factor was greater than the threshold value, otherwise they were omitted. Since all of the authors disregarded unobserved reflections in their original work, it was necessary to estimate threshold values.

The photographs from which our original data set was obtained were used to do this, taking as a guide the intensities of weak reflections superimposed upon a background similar in intensity to that where the unobserved reflection would be expected. Having established values for our own data set, it was assumed that threshold intensities of unobserved reflections in other sets would be in a similar ratio to the intensities of nearby observed reflections.

In adopting this procedure we have reduced the independence of the original data sets. However, it has been our recent experience that omission of unobserved reflections can lead to bad choices of final refined models. A good model (i.e. one for which the calculated intensity of most unobserved reflections is low) will only be slightly affected by the precise choice of threshold, since only those few reflections whose calculated intensity is greater than this will contribute to the refinement.

Unless the structure factors are weighted in some way, the

refinement will be biased to give good agreement for the strongest reflections. This will tend to give a low R-factor - since the strongest reflections also make the dominant contributions to the value of R - with a refinement effectively based on fewer reflections than are present in the list. For this reason, it is normal practice in the crystallography of low-molecular weight materials to weight the structure factors so that the weaker reflections make a greater contribution to the refinement. A scheme commonly used is due to Cruickshank (14), and with this all reflections, on average, contribute equally to the sum of residuals.

Stambaugh et al (4) have criticised the use of this scheme with polymeric fibres on the grounds that the weak reflections are, in general, those at high diffraction angle which are therefore diffuse, probably overlapping groups, and therefore measured less accurately than the stronger ones. They should therefore be underweighted. There is some truth in this argument. However some weak reflections are unambiguously indexed and can be measured with similar accuracy to strong ones. Also, it is our experience that unless some weighting is used the final structure is very sensitive to quite small changes in the structure factors of a few strong reflections, and almost totally insensitive to large changes in the weaker ones. Thus it will be less satisfactory to use no weighting at all than the Cruickshank scheme.

There is need for careful consideration of the use of weighting schemes for polymeric work, but this must wait until methods of intensity measurement are available which allow error estimates to be made. A satisfactory scheme can then be evolved. For the present it was decided to use the Cruickshank method.

For all refinements, the "LALS"(15) computer program was used.

Results and Discussion. The conformation is defined by a set of bond rotation angles defined in Figure 1 but these are insufficient to describe the structure; the location of the molecule in the unit cell must also be given. It is placed so that the centre of the benzene ring lies at the cell origin and the chain axis coincides with the c-axis. However the rotation of the chain about its axis is still undefined. This will be specified by the angle θ. The molecule is placed in the cell so that the benzene ring normal lies in the (010) plane. θ is the rotation about the axis required to bring the molecule to its correct orientation and is positive if the rotation is anticlockwise looking in the negative c-direction.

Though strictly they are not necessary, two other angles are given to help visualise the orientation of the terephthalate residue. ψ is the angle between the benzene ring normal and the c-axis (the nearer this angle is to 90° the nearer the plane of the residue comes to containing the c-axis), and ϕ is the angle the bond C_3C_4 makes with the c-axis (this gives the rotation of the axis of the residue in its own plane away from the c-axis).

Factorial analysis of variance determines the contributions

to the overall variability caused by the various factors in the experiment. In the present case a value for each structural parameter will be obtained from each of 81 different refinements and the variability in the value of a given parameter will be described by its variance. The factorial analysis will determine the contribution to this variance caused by the differences in unit cells, structure factor sets, assumed values of bond parameters and starting models. However as well as determining these contributions, it will also determine the contributions caused by interactions between factors. For example, if the variability caused by using different structure factor sets depended upon which unit cell was being considered, the variance due to the structure factor-cell interaction would be significantly greater than the residual (i.e. that which remains when all contributions due to main factors and interactions between them has been considered). In the present work none of the interactions differed significantly from any other, so it was assumed that they were all non-significant and their variance has been averaged and included as the residual.

The remaining results of the analysis are given in Table VI. The variances are expressed as standard deviations to make their interpretation easier. All the main factors are included, although their contribution to the variability of some of the parameters is not significant.

Table Vl

Standard Deviation of Structural Parameters Caused by Main Factors in Structure Analysis

Factor	θ	ψ	ϕ	τ_1	τ_2	τ_3	τ_4
Structure Factors	8.0°	4.2°	4.5°	12.7°	15.9°	3.5°	22.4°
Unit Cell	3.3°	0.5°	1.2°	1.4°	4.2°	1.9°	4.3°
Starting Model	0.3°	0.6°	0.3°	2.2°	2.3°	3.7°	4.7°
Bond Parameters	1.7°	0.8°	1.6°	4.6°	6.0°	2.5°	34.0°
Residual	0.9°	0.3°	0.2°	1.7°	1.5°	1.7°	3.0°

The overwhelming contribution to the variability arises from differences between the structure factor sets. τ_4 is the only exception and in this case the major contribution is caused by variability in choices of bond parameters. These make a contribution which increases in importance with the distance along the molecule chain from the origin of co-ordinates.

A surprising feature is the significant contribution to the variability of some parameters caused by differences in the starting models. This might be expected if they differed widely, but in this case the published structure of each author was used and all three were fairly close (see Table I). The point at which a refinement is terminated is arbitrary and it could be that had further cycles of refinement been allowed the final structures from different starting models would have converged. This possibility was investigated and it was clear that the structures were not assymtotically approaching a common form.

It is not clear whether the above conclusions are peculiar to the application of the LALS refinement system to this material or whether they are of general applicability. For a polymeric fibre, the diffraction pattern contained a good number of well-defined reflections. Difficulties in intensity measurements are likely to be less severe than in many cases, and so disagreement between structure factor-sets are unlikely to be excessive. The data-parameter ratio was in all cases better than 10:1, which is again quite good for a polymeric fibre. Some preliminary experiments with a different material having a worse ratio have indicated a much greater sensitivity to small changes in the starting model.

For the present material, the bulk of the scattering power is in the terephthaloyl unit and so the structure factors would be expected to play an important role in locating this. However, once it is fixed geometrical factors become important to meet the symmetry constraints, and so the bond parameters would be expected to play an increasingly important role in defining the conformation as the centre of symmetry is approached. Also there is some uncertainty in the values which should be given to the lengths of the methylene bond C_6C_5 and C_7C_6. In structures of low-molecular weight analogues determined by Pérez and Brisse these bond-lengths have always been about 1.50Å, whereas a methylene bond-length is more usually about 1.54Å. The true value in the polymer is uncertain and this is reflected in the values chosen for these bonds.

Thus it would seem that there are unusual features in this system which might make it particularly sensitive to the choices of some bond parameters, and that there are difficulties in assigning reasonable values to some bonds. Thus the variability caused by these features might be peculiar to this system. In all other respects, however, there is no reason to expect excessive variability; the converse is a more likely expectation.

Since the structure factor sets provide the greatest contribution to variability, the assumption that all are equally good is worth re-examining. The R-factor was calculated for each refined structure using only the reflections in the set common to all authors. (The refinement was performed on the particular authors full set). This procedure was adopted because one author might include a reflection omitted by the others because of difficulties in measuring its intensity. Had such a reflection been included in the R-factor calculation it would unfairly bias the value. From these calculations, it was clear that the R-factors obtained using the structure factors of Yokouchi et al, were always lower than those for the same combination of factors except that another author's structure factors were used.

Furthermore, using Hamilton's test, all the differences were significant at the 5% level of probability. Thus it was concluded that for all combinations of other factors, the structure factor set of Yokouchi et al gave the best description of the final model. Further considerations were therefore restricted to this set.

Within this group of structures, none of the different combinations of factors gave an R-value significantly worse - at the 5% level of probability - than the best. Thus there is no justification for choosing any one in preference to any other, and so the mean was taken as the best structure. The standard deviations of the parameters of this model were calculated from the overall variance within a structure factor set. That is, the contribution to uncertainty due to errors in the structure factor set have been omitted. Thus these standard deviations are almost certainly underestimates, for whilst the structure factors of Yokouchi et al might be significantly better than those of the other authors, they are certain to contain some error.

The structural parameters of this model, together with two standard deviations, are given in Table V11. If this is compared with Table 1, and the reasonable assumption made that similar uncertainties apply there, then it is seen that the differences between the original models are not significant, and that they do not differ significantly from this new one.

Table V11

Mean Structural Parameters of the α-phase of 4GT

θ	$3.6^{o} \pm 2.1^{o}$
ψ	$73.3^{o} \pm 0.7^{o}$
ϕ	$25.8^{o} \pm 0.8^{o}$
τ_1	$177.2^{o} \pm 3.6^{o}$
τ_2	$-177.5^{o} \pm 3.7^{o}$
τ_3	$-88.0^{o} \pm 3.7^{o}$
τ_4	$-75.0^{o} \pm 12.5^{o}$

Also, although τ_4 is almost certainly greater than the -G value of -60°, because of its large uncertainty it is dangerous to use arguments which assume that it is accurately known.

Conclusions. "R-factors" calculated between pairs of structure-factor sets determined by three different laboratories varied between 16 and 22%, and these large inaccuracies were the dominant contributor to errors in the parameters of the crystalline structure. Thus if the accuracy with which the structure of polymeric fibres is determined is to be improved. it is essential both to increase the accuracy of intensity measurement, and to quantify its errors.

However, even if the structure factors were known with complete accuracy, uncertainties in the range of 4° to 13° would still remain in the conformation angles, uncertainties in the values of the bond parameters being the major contributor to this error.

Abstract. Three independent determinations have been made of the crystalline structure of the α-phase of poly (tetramethylene terephthalate). The data on which these determinations have been based are used to asses the contributions to the uncertainties of the structural parameters caused by errors in the unit cell parameters, structure factors, and bond parameters. The effects of differences in the model from which refinement is started are also assessed. The major contribution to uncertainty arises from errors in the structure factors (the "R-factor" between structure factor sets from two different laboratories can be greater than 20%) but errors in bond parameters also make a sizeable contribution. Hamilton's test indicates that one of the structure factor sets used in this study is less inaccurate than the other two and using this the best model satisfying all the other data is estimated together with the uncertainties in its parameters.

Literature Cited.

1. Mencik, Z. J.Polym.Sci.(Polym. Phys. Edn.), 1975, 13, 2173
2. Yokouchi, M.; Sakakibara, Y.; Chatani, Y.; Tadokoro, H.; Tanaka, T.; Yoda, K. Macromolecules, 1976, 9, 266.
3. Hall, I.H.; Pass, M.G. Polymer, 1976, 17, 807.
4. Stambaugh, B.; Koenig, J.L.; Lando, J.B. J.Polym. Sci. (Polym. Phys. Edn.), 1979, 17, 1053.
5. Stambaugh, B.; Lando, J.B.; Koenig, J.L. J.Polym. Sci. (Polym. Phys. Edn.), 1979, 17, 1063.
6. Hamilton, W.C. Acta Crystallogr., 1965, 18, 502.
7. Desborough, I. J.; Hall, I. H. Polymer, 1977, 18, 825
8. Hall, I. H.; Pass, M. G. J. Appl. Crystallogr., 1975, 8, 60.
9. Pérez, S.; Brisse, F. Acta Crystallogr., 1976, B32, 470.
10. Pérez, S.; Brisse, F. Can. J. Chem., 1975, 53, 3551.
11. Pérez, S.; Brisse, F. Acta Crystallogr., 1976, B32, 1518
12. Pérez, S.; Brisse, F. Acta Crystallogr., 1976, B32, 2110.
13. Pérez, S.; Brisse, F. Acta Crystallogr., 1977, B33, 1673.
14. Pilling, D.E.; Cruickshank, D. W. J.; Bujosa, A.; Lovell, F.M.; Truter, M.R. "Computing Methods and the Phase Problem in X-ray Crystal Analysis", Pergamon: Oxford; 1961. p. 32.
15 Smith, P. J. C.; Arnott, S. Acta Crystallogr., 1978, A34, 3.

RECEIVED February 19, 1980.

21

Technique and Sample Preparation for Plant Tissue

D. R. KREGER

Department of Plant Physiology, University of Groningen,
Postbus 14, 9750 AA Haren (Gr.), The Netherlands

Methods were discussed to reveal texture in plant materials *in situ* by X-ray diffraction. A uniform texture in plant materials is generally confined to areas of microscopic dimensions, and the oriented materials are normally of poor crystallinity. Special methods are therefore needed to obtain fibre diffraction from such areas. They may be distinguished as microbeam techniques, artificial orientation of micro-areas, chemical methods improving crystallinity in oriented but poorly crystalline materials, and combinations of these possibilities.

The examples of application presented were taken from the authors personal experience. They concerned a microbeam technique in which the collimators and camera are adapted respectively to produce X-ray beams down to about 10 microns diameter and registration of the diffraction on flat film 10 mm diameter at distances down to 1 mm from the specimen (1, 2). This enabled fibre patterns to be obtained from wax coatings on plants (1) and a single starch granule (3), in both instances leading to new insights about the ultrastructure of these objects.

An example of artificial orientation was the orientation attained in cell walls of cork tissue by compression, which gives rise to orientation of the walls in a plane perpendicular to the direction of pressure. Irradiation with the beam in the plane of orientation resulted in a fibre pattern from which the orientation in the walls of the alicyclic wax component (friedelin) could be determined (4).

Improvement of fibre diffraction by chemical means was exampled by the results obtained with hydrolytic and other extractions of fungal tissue, in particular with reference to the (1-3)-β-D-glucan component in the fibrous mycelial ropes (rhizomorphs) of *Armillaria mellea*. The rhizomorphs in the native state produce only one, oriented but diffuse and weak spacing attributable to (1-3)-β-D-glucan. After the adequate chemical extractions they yielded a complete fibre

0-8412-0589-2/80/47-141-349$05.00/0

pattern of the glucan. This led to the derivation of a triple helical structure for this glucan chain (5) in both the hydrated and dry polymorph (5, 6), essentially similar to the structure of (1-3)-β-D-xylan (7).

Literature Cited

1. Kreger, D. R.: An X-ray study of waxy coatings from plants. Rec. Trav. bot. néerl., 1948, 41, 603-736.
2. Kreger, D. R.: In Bouman, J., Ed. "Selected Topics in X-Ray Crystallography"; North-Holland Publishing Company: Amsterdam, 1951, p. 340-352.
3. Kreger, D. R.: The configuration and packing of the chain molecules of native starch as derived from X-ray diffraction of part of a single starch grain. Biochim. Biophys. Acta,1951, 6, 406-425.
4. Kreger, D. R.: X-ray diffraction of stopper cork. J. Ultrastructure Res., 1958, 1, 247-248.
5. Jelsma, J.; Kreger, D. R.: Ultrastructural observations on (1-3)-β-D-glucan from fungal cell walls. Carbohydrate Res., 1975, 43, 200-203.
6. Jelsma, J.: "Ultrastructure of Glucans in Fungal Cell Walls"; Thesis, University of Groningen, 1979.
7. Atkins, E. D. T.; Parker, K. D.: The helical structure of a β-D-1,3-xylan. J. Polymer Sci.: C, 1969, 28, 69-81.

RECEIVED May 28, 1980.

22

Fiber Diffraction and Structure of (1-3)-α-D-Glucan in Fungal Cell Walls

J. JELSMA and D. R. KREGER

Department of Plant Physiology, University of Groningen, Postbus 14, 9750 AA Haren (Gr.), The Netherlands

X-ray powder patterns have shown that (1-3)-α-D-glucan occurring in fungal cell walls may crystallize in three polymorphs, a native form (polymorph I), a precipitated, hydrated form (II) and a dry form (III) (1). A fibre orientation was found in fungal tissue, in particular the trama of Piptoporus betulinus, in which the hyphae run parallel. Fibre patterns of the three polymorphs of the glucan were obtained from this tissue after removal of part of the other cell wall constituents by adequate chemical extractions and enzymatic treatments. The orientation in the native polymorph (I) was retained after transformation into III by drying, and this polymorph was transformed into II by remoistening the specimen.

Evaluation of the fibre patterns showed that the glucan chains have a stretched form with twofold screw symmetry and a fibre repeat period of 0.835 nm, as predicted by model building and conformational analysis. The differences between the polymers exist therefore in their chain packing. A monoclinic unit cell with a=0.581 nm; b=1.00 nm; $\gamma=96^{\circ}$ was derived for polymorph I and orthorhombic cells for II and III respectively with the base plane axes a=0.502 nm; b=0.963 nm for II and a=0.457 nm; b=0.865 nm for III. These cells contain 4 glucose residues, and on account of spatial considerations the ribbon-like chains in projection on the base plane were supposed to be oriented with the longer axis parallel to the b-axis (i. e. with the pyranose rings near parallel to the bc-plane) in both polymorph II and III, whereas in polymorph I a diagonal position appeared more favourable.

In each of the three polymorphs some weak reflections could be indexed only by doubling the a-axis, while some meridional reflections did not fit in the structure anyway. The structure of polymorph II was confirmed by fibre patterns from stacked films of the precipitated glucan. Electron micrographs showed short microfibrils with a beaded

0-8412-0589-2/80/47-141-351$05.00/0

appearance in the native glucan (polymorph I), while the precipitated glucan (polymorph II) showed flat platelets (2).

Literature Cited

1. Jelsma, J.; Kreger, D. R.: Polymorphism in crystalline (1-3)-α-D-glucan from fungal cell-walls. Carbohydrate Res., 1979, 71, 51-64.
2. Jelsma, J.: Ultrastructure of Glucans in Fungal Cell Walls. Thesis, University of Groningen, 1979.

RECEIVED May 28, 1980.

Crystal Structure of (1→3)-α-D-Glucan

KOZO OGAWA—Radiation Center of Osaka Prefecture, Sakai, Osaka 593 Japan

KEIZO OKAMURA—Department of Wood Science and Technology, Faculty of Agriculture, Kyoto University, Kyoto 606 Japan

SACHIKO OKA and AKIRA MISAKI—Department of Food and Nutrition, Faculty of Science and Living, Osaka City University, Osaka 558 Japan

Both in theory and in practice there exist eight glucopyranose homopolymers, and some of the molecular conformations of three of these, i.e. cellulose and amylose (1,2,3,4), and (1→3)-β-D-glucan (5,6,7) have been established by x-ray analysis. Although (1→3)-α-D-glucan is among the five homopolymers previously unsolved by x-ray diffraction, possible chain conformations were predicted with computers to be an extended ribbon (8,9), a single helix (9), or a double or triple helix (10).

(1→3)-α-D-Glucan is mainly found in micro-organisms or in the cell walls of many kinds of fungi and yeast (11,12,13). X-ray powder patterns of the fungal glucans have been extensively studied by Jelsma and Kreger (14,15). They reported that the glucan crystallized as three polymorphs (I-III); polymorph I was observed only in native tissue, II (hydrate form) appeared on precipitation of the glucan from alkaline solution, and III (dehydrated form) arose after drying. Bacteria found in human saliva also produce (1→3)-α-D-glucans that are interesting in connection with dental caries (16). Specifically, *Streptococcus mutans* and *Streptococcus salivarius* synthesize from sucrose α-glucans that are very sticky and insoluble in water. The glucans have a (1→3)-α-D-glucan backbone with short side chains of (1→6)-α- (and also (1→4)-α- for *S. salivarius*) linked glucose residues. These glucans form dental plaques and consequently contribute to dental caries.

The purpose of our study was to obtain a well-defined x-ray fiber pattern of a bacterial (1→3)-α-D-glucan, and, based on the pattern, to determine the conformation of the glucan. We believe that knowledge of molecular conformation is important for understanding plant physiology and the dental significance of (1→3)-α-D-glucan. It would also provide basic knowledge useful in the study of more complicated heteropolysaccharides containing (1→3)-linked α-D-glucopyranose units. X-ray powder patterns of our bacterial samples are also reported for comparison with those of fungal (1→3)-α-D-glucans reported by Jelsma and Kreger (14).

0-8412-0589-2/80/47-141-353$05.00/0

Experimental

Sample of salivarius. As previously reported (17), a linear glucan consisting solely of (1→3)-α-D-glucosidic linkages was prepared by mild Smith degradation of an α-glucan produced by S. salivarius TTL-LP1. Viscosity measurement of the glucan gave [η] = 0.52 dl/g in 0.5N NaOH solution at 25°C (sample A, powder form).

Since attempts to prepare a continuous film or fiber from solutions of sample A in aqueous alkali, hydrazine hydrate, or N-methylmorpholine N-oxide-DMSO mixture were not successful, the glucan was acetylated and a continuous transparent film of the glucan acetate was obtained from a 5% solution in chloroform. A well-oriented glucan film was obtained by stretching the glucan acetate film by 6.5 times its original length in glycerine at 150°C and deacetylating in 2M sodium methylate-methyl alcohol solution. The low crystallinity of this film was substantially improved by annealing in water at 140°C, under tension, in a sealed bomb (sample B, oriented film).

Another oriented film made from sample A was prepared by the same procedure as for sample B, except that sample A was pretreated with hydrazine hydrate. The viscosity was [η] = 0.50 dl/g in 0.5N NaOH (sample C, oriented film).

Sample of mutans. A (1→3)-α-D-glucan supplied by Dr. Sudo was prepared by dextranase degradation of an α-glucan produced by S. mutans OMZ 176 in a sucrose medium (16). The number average degree of polymerization of the glucan, determined by the modified Somogyi-Nelson method (18) with laminaribiose as a standard, was 43.8. The DP of sample A could not be determined by this method indicating that the glucan produced by S. salivarius glucan had a higher molecular weight. Mutan was soluble in DMSO but did not produce any continuous film from DMSO solution or any other solvent mentioned above. Sample D was therefore mutan powder. Its viscosity was [η] = 0.33 dl/g in 0.5N NaOH.

Methods. The wide and small angle x-ray patterns were recorded in a flat film camera at controlled relative humidity using Ni-filtered CuKα radiation generated at 40 kV and 15 mA. The density of the glucan films were measured by flotation in a carbon tetrachloride-m-xylene mixture.

Results

Fiber patterns. In vacuum, the stretched and annealed glucan film made from S. salivarius (sample B) gave a clear x-ray pattern as shown in Figure 1 (17). All 29 reflections (Figure 2) can be indexed with an orthorhombic unit cell (Tables I and II). In the previous report (17), the unit cell was reported to be

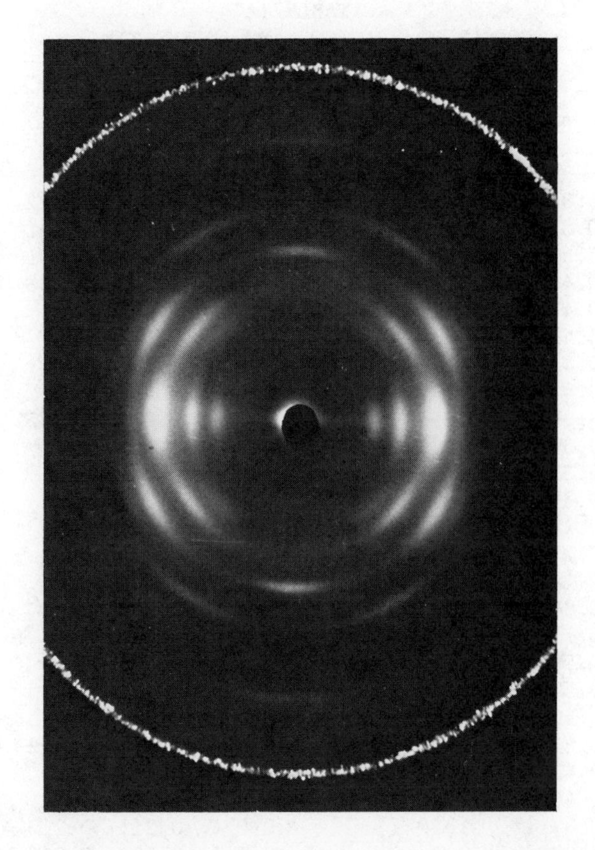

Carbohydrate Research

Figure 1. X-ray fiber diagram of Sample B in vacuum. The fiber axis is vertical (17).

TABLE I
OBSERVED SPACINGS AND INTENSITIES
FOR THE FIBER PATTERN IN VACUUM

h k l	Spacing (nm) calc.	Spacing (nm) obs.	Int.* (obs.)
2 0 0	0.823	0.821	S
1 1 0	0.826		
2 1 0	0.623	0.628	S
0 2 0	0.477	0.476	VS
3 1 0	0.476		
1 2 0	0.459		
2 2 0	0.413	0.410	M
4 0 0	0.412		
4 1 0	0.378	0.373	W
1 3 0	0.313	0.311	W
4 2 0	0.312		
5 1 0	0.311		
3 3 0	0.275	0.275	VW
6 0 0	0.274		
0 0 1	0.844	0.844	VW
1 0 1	0.751	0.758	VW
1 1 1	0.590	0.587	W
2 0 1	0.589		
2 1 1	0.501	0.503	S
0 2 1	0.416	0.414	VS
3 1 1	0.414		
1 2 1	0.403		
2 2 1	0.371	0.371	M
4 0 1	0.370		
4 1 1	0.345	0.338	W
3 2 1	0.331		
5 0 1	0.307	0.299	W
0 3 1	0.298		
6 1 1	0.252	0.253	W
0 0 2	0.422	0.422	S
1 1 2	0.376	0.377	VW
2 0 2	0.375		
2 1 2	0.349	0.348	S
0 2 2	0.316	0.314	M
3 1 2	0.316		
2 2 2	0.295	0.295	VW
4 0 2	0.295		
3 2 2	0.274	0.275	W
0 3 2	0.254	0.254	W
1 3 2	0.251		
4 2 2	0.251		
5 1 2	0.251		
1 0 3	0.277	0.276	S
0 1 3	0.270		
1 1 3	0.266	0.265	W
2 0 3	0.266		
2 1 3	0.256	0.255	VW
3 0 3	0.250	0.242	W
0 2 3	0.242		
3 1 3	0.242		
0 0 4	0.211	0.210	VW
2 1 4	0.200	0.200	W

*VS, very strong; S, strong; M, medium; W, weak; VW, very weak

TABLE II
CRYSTAL DATA FOR THE FIBER PATTERN IN VACUUM

Crystal system	Orthorhombic
Space group	$P2_12_12_1$
$\underline{a}$ (nm)*	1.646±0.005
$\underline{b}$ (nm)*	0.955±0.003
$\underline{c}$(fiber axis)(nm)*	0.844±0.001
Volume (nm^3)	1.327
Number of glucose residue	8
ρ(calc.)(g/cm^3)	1.62
ρ(obs.)(g/cm^3)	1.54
Number of chain	4
Helix parameters	
$\underline{n}$	2
$\underline{h}$ (nm)	0.422

*Lattice parameters were obtained by a least-squares refinement program.

monoclinic with $\underline{a}$ = 0.823 nm, $\underline{b}$ = 0.955 nm, $\underline{c}$ (fiber axis) = 0.844 nm, and γ = 90°. However, by increasing the exposure time, a very weak new reflection appeared at 0.758 nm on the first layer line. The $\underline{a}$ axis was doubled in order to accomodate the new reflection which indexed as the (101) reflection (Table I). $P2_12_12_1$ seems to be the best space group (Table II).

As the relative humidity was increased (76% and 98%), the glucan film (sample B) showed an additional reflection on the equator having medium intensity and a $\underline{d}$ value of 0.971 nm, corresponding to the (010) reflection (Table I). The appearance and disappearance of the (010) reflection is reversible with changing relative humidity. Sample C, which was prepared by pretreating sample A with hydrazine hydrate, gave a fiber pattern with spacings identical to that of sample B. The reflection spots of sample C are somewhat sharper and more arcked than those of sample B, indicating higher crystallinity and less orientation. However, the fiber pattern of sample C did not show the (010) reflection at higher humidity.

The x-ray fiber pattern of sample B before annealing did not differ essentially from that of the annealed film, though the crystallinity was considerably lower.

Small angle x-ray scattering of the oriented films (samples B and C) showed a discrete interference along the fiber axis corresponding to a long period of 14.0 nm (Figure 3). The stretched film of sample D prepared with the same regeneration procedure used for sample B showed no orientation. This would be

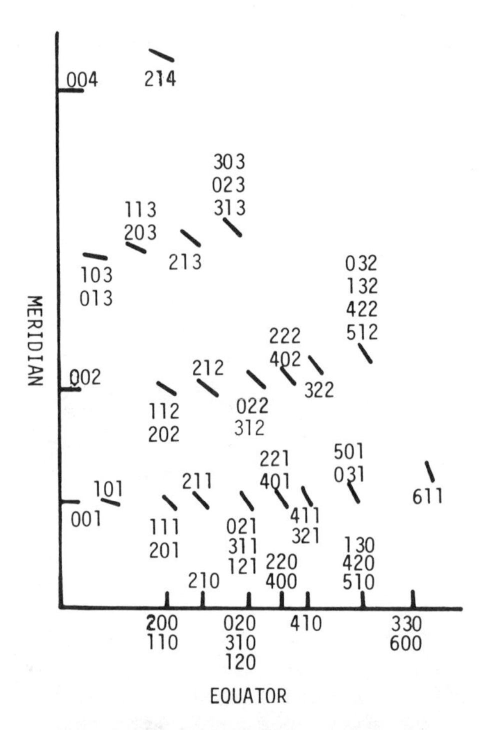

Figure 2. Indexed schematic display of the x-ray pattern corresponding to that in Figure 1

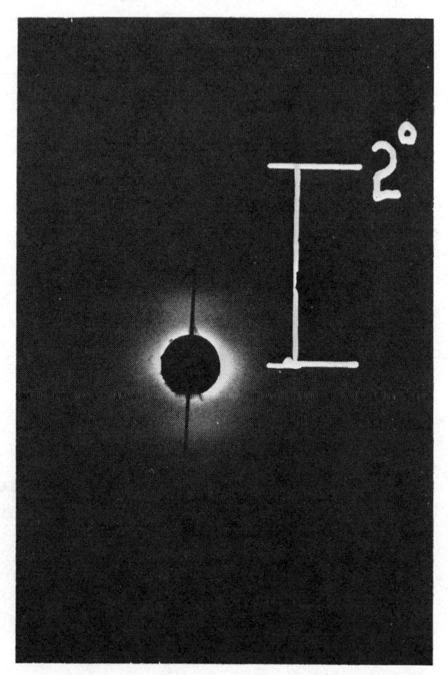

Figure 3. The small-angle x-ray scattering of Sample B in vacuum

due to its lower molecular weight.

Powder patterns. Sample A was highly crystalline and showed two interconvertible powder patterns: one at high relative humidity (76% - 98%), and the other *in vacuo* (Table III). The *d* spacings of these patterns differ from those of fiber patterns of either sample B or sample C (Table I). Mutan (sample D) also showed a clear pattern similar to that of sample A *in vacuo*. However, this pattern did not change by varying the relative humidity (Table III).

TABLE III
OBSERVED SPACINGS AND INTENSITIES FOR POWDER PATTERNS

S. salivarius (sample A)		*S. mutans* (sample D)
at 76% and 98% r.h.	in vacuum	at all r.h. studied
0.966 (VS)	0.870 (S)	0.868 (S)
0.788 (VW)		
0.639 (VW)	0.627 (S)	0.624 (S)
0.542 (VW)	0.518 (M)	0.514 (M)
0.498 (VS)	0.459 (VS)	0.457 (VS)
0.423 (VS)	0.415 (VS)	0.412 (VS)
0.395 (W)	0.383 (W)	0.382 (W)
0.362 (W)	0.348 (VW)	0.351 (VW)
0.323 (M)	0.315 (W)	0.316 (W)
0.304 (VW)	0.281 (W)	0.278 (W)
0.273 (M)		
0.261 (VW)	0.259 (W)	0.256 (W)
0.212 (W)	0.215 (VW)	0.216 (VW)
0.202 (VW)	0.204 (VW)	0.198 (VW)

Discussion

Fiber patterns. Despite the presence of a (001) reflection (Figure 1), a twofold screw axis along the fiber axis has been incorporated into the (1→3)-α-D-glucan backbone conformation, because tilting the oriented film causes the (002) and (004) reflections to become strong, but not the (001). Therefore, the presence of a very weak (001) reflection is likely due to a slight disorder of O(6) along the glucan chain. The value of *h*, the advance per residue along the helix axis, 0.422 nm (Table II), is very close to the virtual bond length of 0.428 nm determined for α-D-glucose (19). This indicates that the (1→3)-α-D-glucan molecule is almost entirely extended along the fiber axis as predicted by the conformational calculations (8,9). Conformation and packing refinement of the (1→3)-α-D-glucan molecule are underway in collaboration with Professor Anatole Sarko, and these results will be available soon. Preliminary indications are that the adjacent chains have opposite sence along the *b* axis and the molecules are extensively hydrogen bonded. As with cellulose, this undoubtedly accounts for the insolubility of this glucan.

The discrete interference in small angle x-ray scattering (Figure 3) corresponds to a period of 14 nm. This spacing may be attributable to chain folding of the glucan molecule along the fiber axis. This interpretation is in keeping with an antiparallel packing of the glucan chain, as required by the symmetry of the $P2_12_12_1$ space group. This type of discrete diffraction is also noticeable in Fortisan fiber (20). In that case, cellulose acetate fiber was saponified to regenerated cellulose, then treated with dilute hydrochloric acid. The long spacing of (1→3)-α-D-glucan is therefore another similarity to a microstructural feature of cellulose.

Sample B, which is less crystalline than sample C, showed an additional spot on the equator corresponding to a (010) reflection at high relative humidity. We suggest that this spot is due to a slight disorder of the molecule along the *b* axis. This disorder would be caused by the addition of a water molecule to the glucan chain. Therefore, the $P2_12_12_1$ symmetry is lost.

Powder patterns. All the powder patterns of our bacterial glucans were similar to those of fungal (1→3)-α-D-glucans (14). The crystal structure of sample A (glucan from *S. salivarius*) at high relative humidity corresponded to Kreger's polymorph II (hydrated form), and the powder patterns of sample A *in vacuo* as well as sample D (mutan) at all humidities corresponded to his polymorph III (dehydrated form). It is noteworthy that the crystal structure of low molecular weight (1→3)-α-D-glucan (sample D) is not affected by the presence of water. Similar findings have been reported in the case of (1→3)-β-D-glucan (paramylon or curdlan) (21).

During our study, Jelsma and Kreger obtained three fiber patterns from a "native" trama tissue of *Piptoporus betulinus*, which they classified as polymorphs I III of (1→3)-α-D-glucan. Their fiber patterns are less oriented than our patterns from regenerated glucan, and the fungal (1→3)-α-D-glucans are contaminated by (1→3)-β-D-glucan (15). Since our fiber patterns are from "regenerated" material, our crystal structure is different from any of Kreger's polymorphs and it might be named polymorph IV in extension of Kreger's classification. Interestingly, all of their patterns have identical *n* (number of glucose residues per one turn of the helix) and similar *h* values, close to our values (Table II). However, these four polymorphs have different crystal parameters for dimensions other than the fiber axis. These results suggest that the typical chain conformation of (1→3)-α-D-glucan is a 2_1 ribbon, maximally extended along the fiber axis. This is another similarity to cellulose for which four polymorphs (cellulose I-IV) have the same fiber period, 1.03 nm.

The change of the crystal structure of (1→3)-α-D-glucan should be compared to the case of (1→3)-β-D-glucan (curdlan). Although curdlan changes its polymorphic form by annealing (6), neither the fiber patterns (samples B and C) nor the powder patterns (samples A and D) underwent polymorphic transformations due to annealing.

Kreger and Jelsma suggested that polymorph II was energetically more stable than polymorph III (14). However, we do not agree, since sample D (mutan) showed the powder pattern of polymorph III at all relative humidities, we believe the dehydrated form is more stable than the hydrated form. Furthermore, polymorph IV (regenerated form) seems to be even more stable. Its fiber pattern did not change significantly with varying relative humidity.

Acknowledgements

We thank Dr. R. H. Marchessault of Xerox Research Centre of Canada and Professor A. Sarko of State University of New York, College of Environmental Science and Forestry for their useful suggestions on analyzing x-ray fiber patterns. The sample of *Streptococcus mutans* was kindly given by Dr. M. Sudo of Kanonji Institute, Research Foundation for Microbial Diseases, Osaka University, to whom we are indebted.

Abstract

A linear (1→3)-α-D-glucan (sample A) was prepared by mild Smith degradation from a bacterial α-glucan elaborated by a strain of *S. salivarius* TTL-LP1. The oriented glucan film (sample B) was obtained by acetylating the glucan, followed by stretching in glycerine at 150°C and deacetylation. Annealing the oriented film in water at 140°C, under tension in a sealed bomb, improved the crystallinity. All the 29 reflections on the equator and four layer lines were indexed with an orthorhombic unit cell with *a* = 1.646 nm, *b* = 0.955 nm, and *c* (fiber axis) = 0.844 nm. $P2_12_12_1$ seemed to be the best space group. The crystal structure of sample A and another bacterial (1→3)-α-D-glucan (sample D) prepared from *S. mutans* fell in two of three polymorphs I-III of fungal (1→3)-α-D-glucan reported by Jelsma and Kreger; the crystal structure of sample B might be named as polymorph IV. All polymorphs of (1→3)-α-D-glucan observed to date have almost identical helix parameters *n* = 2 and *h* = 0.422

nm. These h values are close to the virtual bond length of 0.428 nm determined for α-D-glucose, suggesting that the typical chain conformation is an extended, ribbon-like structure. The similarities of (1→3)-α-D-glucan to cellulose are discussed.

Literature Cited

1. Marchessault, R. H.; Sarko, A. Advan. Carbohydr. Chem., 1967, 22, 421-482.
2. Marchessault, R. H.; Sundararajan, P. R. Advan. Carbohydr. Chem. Biochem., 1976, 33, 387-404.
3. Sundararajan, P. R.; Marchessault, R. H. Advan. Carbohydr. Chem. Biochem., 1978, 35, 377-385.
4. French, A. D. Bakers Digest, 1979, 53, 39-46,54.
5. Jelsma, J.; Kreger, D. R. Carbohydr. Res.,1975, 43, 200-203.
6. Marchessault, R. H.; Deslandes, Y.; Ogawa, K.;Sundararajan, P. R. Can. J. Chem., 1977, 55, 300-303.
7. Bluhm, T. L.; Sarko, A. Can. J. Chem., 1977, 55, 293-299.
8. Rees, D. A.; Scott, W. E. J. Chem. Soc., 1971, B, 469-479.
9. Sathyanarayana, B. K.; Rao, V. S. R. Biopolymers, 1972, 11, 1379-1394.
10. Bluhm, T. L.; Sarko, A. Carbohydr. Res., 1977, 54, 125-138.
11. Kreger, D. R. Biochim. Biophys. Acta, 1954, 13, 1-10.
12. Johnston, I. R. Biochem. J., 1965, 96, 659-664.
13. Bacon, J. S. D.; Jones, D.; Farmer, V. C.; Webley, D. M. Biochim. Biophys. Acta, 1968, 158, 313-315.
14. Jelsma, J.; Kreger, D. R. Carbohydr. Res., 1979, 71, 51-64.
15. Jelsma, J. Doctoral dissertation, 1979, University of Groningen.
16. Ebisu, S.; Misaki, A.; Kato, K.; Kotani, S. Carbohydr. Res., 1974, 38, 374-381.
17. Ogawa, K.; Misaki, A.; Oka, S.; Okamura, K. Carbohydr. Res., 1979, 75, C13-C16.
18. Hiromi, K.; Takasaki, Y.; Ono, S. Bull. Chem. Soc. Jpn., 1963, 36, 563-569.
19. Brown, G. M.; Levy, H. A. Science, 1965, 147, 1038-1039.
20. Schulz, J.; John, K. Cellulose Chem. Technol., 1975, 9, 493-501.
21. Marchessault, R. H.; Deslandes, Y. Carbohydr. Res., 1979, 75, 231-242.

RECEIVED May 21, 1980.

24

Ultrastructure of Curdlan

NOBUTAMI KASAI
Department of Applied Chemistry, Faculty of Engineering, Osaka University, Yamadakami, Suita, Osaka 565 JAPAN

TOKUYA HARADA
Institute of Scientific and Industrial Research, Osaka University, Yamadakami, Suita, Osaka 565 JAPAN

1. Introduction

During the studies on *Alcaligenes faecalis* var. *myxogenes* 10C3, Harada and co-workers found that a spontaneous mutant, 10C3K produces an insoluble exocellular polysaccharide(1). This polysaccharide is entirely composed of D-glucosyl residues which are connected almost exclusively by β-(1→3)-linkages. This and the similar glucan formed by some strains of *Agrobacterium* are named Curdlan because they form irreversibly, resilient gel when heated in water.

Because of interesting and specific properties of the gel, curdlan has many potential uses not only in food industry but also as a film, a fiber, a support for immobilizing enzymes and a binding agent in tobacco product etc.(2). Besides these, an antitumor activity(3) has been reported.

In connection with these uses the conformation of curdlan molecule in solution and in gel and the ultrastructure of aggregates of curdlan molecules are considered very important. This paper deals with conformational studies on curdlan molecules by means of light scattering, ^{13}C NMR, X-ray diffraction, and morphological studies using an electron microscope.

0-8412-0589-2/80/47-141-363$05.25/0

2. Conformation of Curdlan Molecule in Solution

1). Conformation in a cadoxen-water mixture(4)

For the light scattering study, Hirano *et al.* divided a curdlan sample into nine fractions by fractional precipitation at 25°C with cadoxen* as a solvent and a propanol-water(3:1 by volume) mixture as a precipitant. The light scattering measurement was carried out in the 1:1 water-diluted cadoxen at 25°C.

From the measured scattering intensities Hirano *et al.* obtained the angular dependence of the infinite-dilution reduced scattering intensity $(K_c/R_\theta)^{\frac{1}{2}}_{c\to 0}$ and the concentration dependence of the zero-angle reduced scattering intensity $(K_c/R_\theta)^{\frac{1}{2}}_{\theta\to 0}$. They then determined the values of weight-average molecular weight $\overline{M}_w$, z-average mean-square radius of gyration $\langle S^2\rangle_z^{\frac{1}{2}}$, and second virial coefficient A_2 of the fraction by the usual method. The results are presented in Table I.

The values obtained for $\overline{M}_w$ are about one order of magnitude larger than the reported number-average molecular weights determined by the method of Manners *et al.* (5)

Figure 1 gives the molecular weight dependece of $\langle S^2\rangle_z^{\frac{1}{2}}$. The data points may be fitted by a straight line as shown in Fig. 1.

$$\langle S^2\rangle_z^{\frac{1}{2}} = 3.2 \times 10^{-2}\overline{M}_w^{0.53} \qquad \text{(in nm)}.$$

Figure 1 includes the data from cellulose in the same solvent(7). At fixed $\overline{M}_w$, the values of $\langle S^2\rangle_z^{\frac{1}{2}}$ for curdlan are about one half those for cellulose, indicating that the curdlan molecule assumes a more coiled form than does the cellulose molecule in 1:1 water-diluted cadoxen.

The double logarithmic plots of intrinsic viscosity $[\eta]$ for curdlan(4) and cellulose(6,7) against $\overline{M}_w$ suggest that the curdlan and cellulose molecules are not very flexible.

---

* Tri(ethylene diamine) hydroxide, $[Cd(en)_3](OH)_2$

Table I. Numerical results from light scattering measurements on curdlan fractions in the 1:1 water-diluted cadoxen at 25°C (4)

Fraction Number	$\overline{M}_w$ X10^{-5}	$\langle S^2\rangle_z$X10^{11}/cm^2	A_2X10^4/cm^3 mol g^{-2}
F-II	6.83	1.62	3.81
F-III	5.85	1.49	3.54
F-IV	3.48	0.751	3.85
F-V	2.60	0.534	4.38
F-VI	1.78	0.350	3.71
F-VII	1.24	0.262	6.58
F-VIII	0.888	0.159	6.11
F-12-2*	0.659	0.156	8.13

*Sample fractionated at Takeda Chemical Industries. $\overline{DPn}$ 131 by the method of Manners *et al.* (5).

Polymer Journal

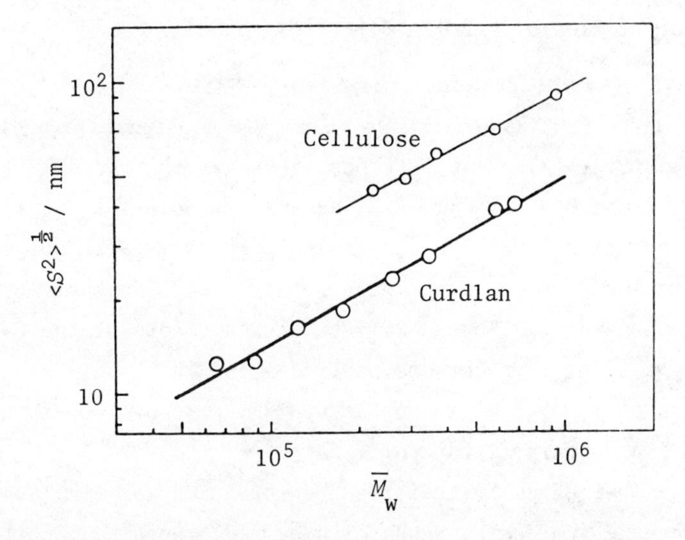

Polymer Journal

Figure 1. Molecular weight dependence of $\langle S^2\rangle_z^{\frac{1}{2}}$ *for curdlan in the 1:1 water-diluted cadoxen at 25°C (4). Data for cellulose (small circles) (7) are also plotted.*

A consideration of the molecular conformation using the worm-like chain model suggests that the curdlan molecule may contain helical portions but, as a whole, takes a random-coil conformation (4).

2). Conformation in an alkaline solution

Figure 2 shows ^{13}C NMR spectra of curdlan($\overline{DPn}$ 540) of an water suspension and alkaline solution reported by Saitô *et al.*(8). It is noteworthy that at a concentration of 0.22M NaOH the broad ^{13}C NMR signals decrease to less than one-tenth of those of the aqueous suspension. This sharp spectrum is characteristic of random-coil conformation. This is consistent with the results of Ogawa *et al.* (9). They determined the dependence of optical rotation, intrinsic viscosity, and the extinction angle of curdlan solution on concentration of NaOH and found the transition from ordered form to random-coil in about 0.2M.

3. Conformation of Curdlan Molecules in Gel

1). Conformation in gel.

Saitô *et al.*(8) have shown that ^{13}C NMR signals of gel can be observed (Fig. 3B). ^{13}C peak positions are shifted downfield by 2.8, 3.2, and 0.9 for C-1, C-3, and C-4 respectively, compared with those of water-soluble degraded polymer in a random-coil conformation. On the other hand, other C-2, C-5, and C-6 signals remain unchanged. These downfield displacements of the C-1 and C-3 signals occur at carbons participating in the glucosidic linkages of (1→3)-ß-D-glucan, and this fact could be explained in terms of a certain fixed conformation.

The energetically preferred conformations of (1→3)-ß-D-glucan have been determined by Sundaralingham(10), Rees and Scott(11), and Sathyanarayana and Rao(12). According to these results, wide and extended helical conformations are most probable. It has been shwon that ^{13}C peaks of molecules in the rigid double- or triple-helical conformation are completely lost(13-15). However, rather narrow ^{13}C peaks have been reported for single helical molecules (16,17). These facts suggest that the ^{13}C NMR visible portion should be ascribed to the single-helical region.

The line widths of C-1--C-5 ^{13}C resonance peaks in the gel are about 150 Hz, which are rather broad in comparison with those of alkaline solution (0.22M NaOH)(Fig. 2D) and also with those of the lower molecular weight fraction (Fig. 3A). By comparing the integrated peak intensities (with nuclear Overhauser effect surpressed) with those of the corresponding peaks in the 0.22M NaOH solution, in which the curdlan molecules are considered to have random-coil conformation, it was estimated that these observed ^{13}C peaks account for only 20-30 and 60% for the C-1--C-5 and C-6 carbons of the total gel, respectively. Hence, the remaining invisible portion of the ^{13}C should be ascribed to the junction zones of the gel net work and to residues located closely to these junctions.

These results lead to a conclusion that the visible ^{13}C NMR peaks are due to the single helical portion. The NMR-invisible portion of the gel is present as multi-helical junction-zones.

2). Conformation in and ultrastructure of an oriented gel.

As the X-ray diffraction diagrams of an oriented gel (Fig. 6) are poorly crystalline, it has been useful to conduct X-ray crystal structure analyses of oligomer of β-(1→3)-D-glucan. The molecular structures of laminarabiose and its acetyl derivative have been determined.

(1) Laminarabiose(18)

As shown in Fig. 4, the molecules of laminarabiose have a conformation, different from the fully-extended one, which is stabilized by an intramolecular hydrogen bond between O-4'--H and O-5[2.786(7) Å] (*Cf.* intramolecular O-2····O-2'=3.920(6) Å). The ring-to-ring confromation can be described as (Φ , Ψ) = (27.9°, -37.5°) according to the definition of Sathyanarayana and Rao(12),

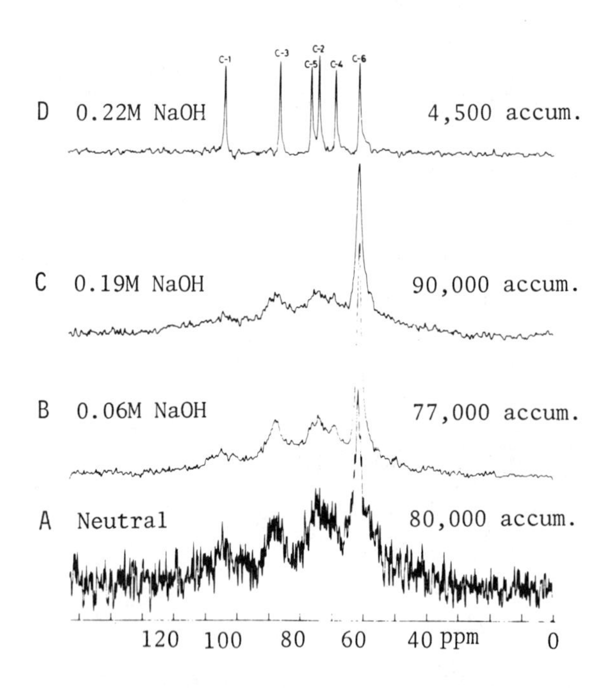

Biochemistry

Figure 2. ^{13}C NMR spectra of curdlan($\overline{DPn}$ 540) in water suspension (A) and alkaline solution (90° pulse, repetition time 0.6 s) (8)

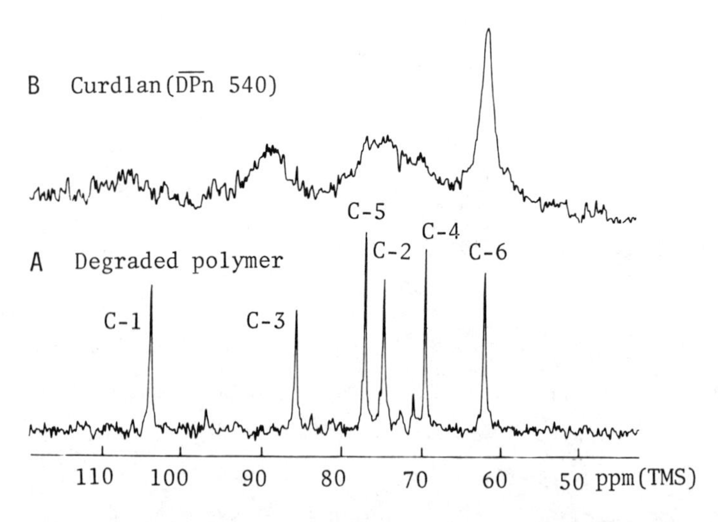

Biochemistry

Figure 3. ^{13}C NMR spectra of curdlan and of a degraded polymer ($\overline{DPn}$ 13) in solution (90° pulse, repetition times 1 s for A and 0.6 s for B) (8)

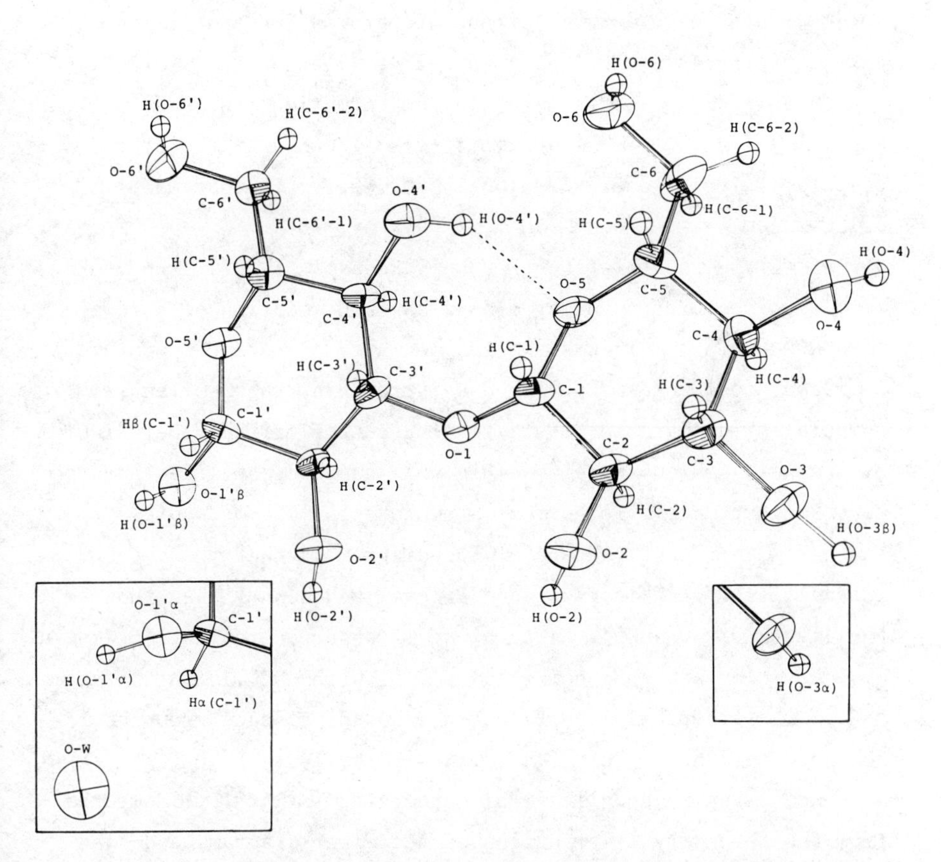

Figure 4. Molecular structure of laminarabiose (18)

The C and O atoms are shown as thermal ellipsoids with 50% probability level, and the H atoms are represented as spheres with B = *0.75 Å². Atoms that are related to the α-anomer molecule (~38%) and the water of crystallization (~19%) associated with the α-anomer molecules are shown in squares.*

and it is located in the comparatively low energy region of the energy contour diagram of β-(1→3)-glucan(12).

(2) Methyl 2,3,4,6,2',4',6'-hepta-*O*-acetyl-β-D-laminarabioside(19).

Because hydrogen bonds are prohibited by the acetyl groups the molecules have a fully-extended conformation and the inter-oxygen distances are O-4'····O-5=3.224(14) and O-2'····O-2=3.604(14) Å (Fig. 5). The ring-to-ring conformation can be described as (Φ,Ψ)=(42.5°, 4.7°), and it is also located in the other comparatively low energy region in the energy contour map(12).

(3) Curdlan

a. Room temperature structure Takeda *et al.*(20) prepared a film specimen, regenerating from a 1N NaOH solution of curdlan ($\overline{DPn}$=400). This film was swollen in water at room temperature. A fibrous specimen was obtained by stretching the swollen film uniaxially (draw ratio 3.5 : 1). The center of the fiber diagram of the wet, fibrous specimen (Fig. 6A) has a cross-like appearance that suggests a simple helical structure for the molecule. Another significant feature of the diagram is that the average intensities of the layers 1 and 2 were much weaker than those of the layers 0, 3, and 4. A total of 17 observed reflections could be indexed by an orthorhombic unit cell with a=26.4, b=16.4 and c=22.6$_5$ Å (fiber axis); but the corresponding space group could not be determined.

Referring to a cylindrical Patterson function(21) computed from the intensity distribution of the fiber diagram, and also considering the standard geometry of the D-glucose ring(22) and conformations about the β-(1→3) linkage obtained by the single crystal structure analyses of laminarabiose(18) and its acetyl derivative(19), reasonable models of 7/1 and 6/1 single helices with fiber repeat period of 22.6$_5$ Å(fully-extended model) could be constructed. The Fourier transform of each helix was calculated using the equation given by Klug *et al.*(23). The results for the relative magnitude of the intensities of the 7/1

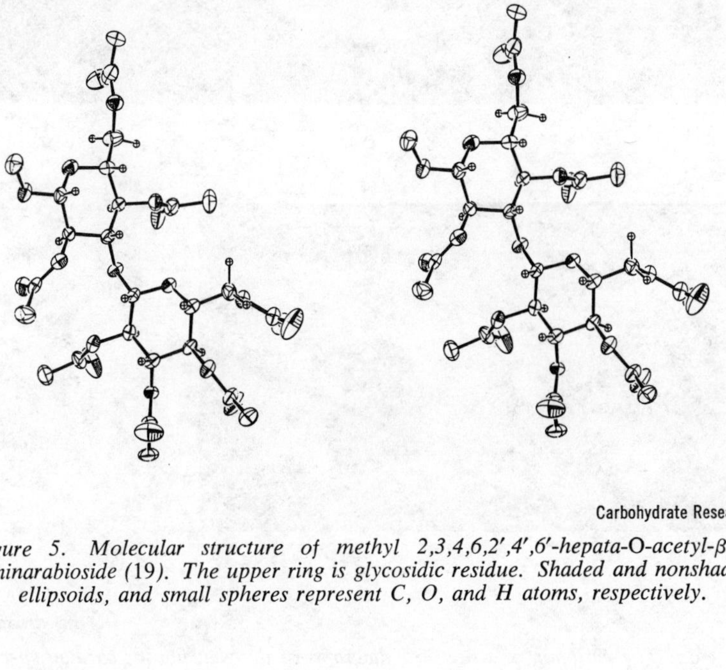

Carbohydrate Research

*Figure 5. Molecular structure of methyl 2,3,4,6,2′,4′,6′-hepata-*O*-acetyl-β-*D*-laminarabioside (*19*). The upper ring is glycosidic residue. Shaded and nonshaded ellipsoids, and small spheres represent C, O, and H atoms, respectively.*

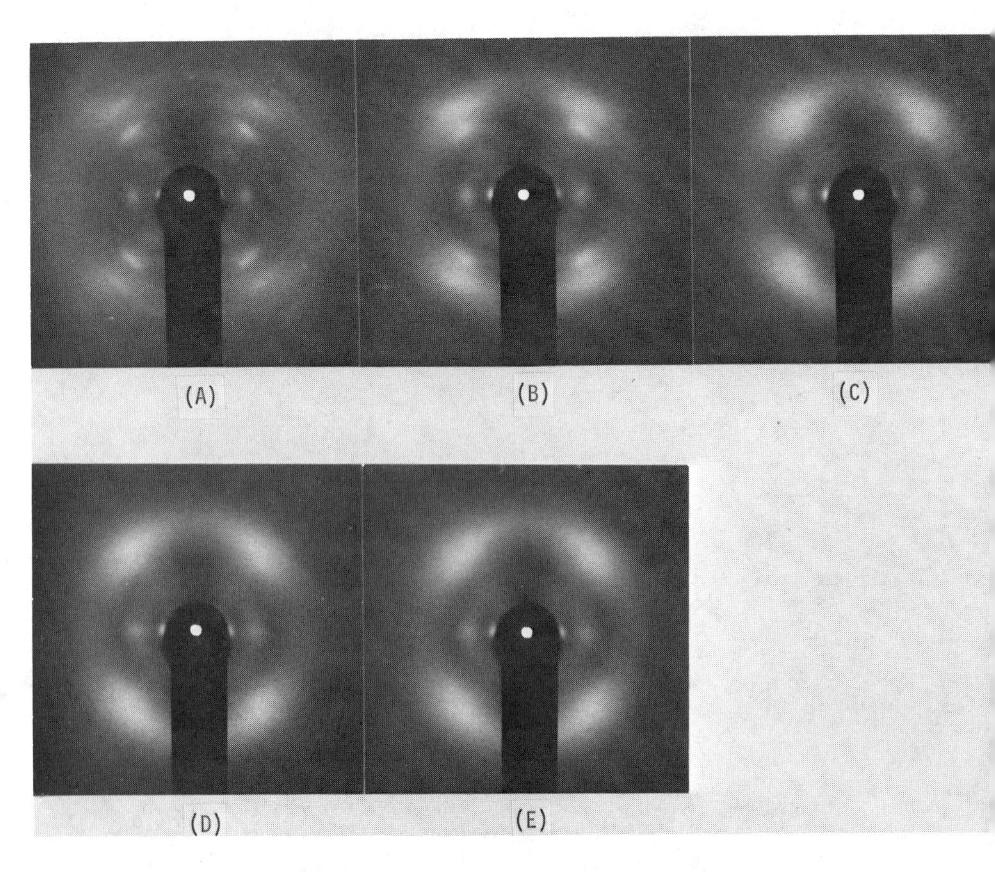

Polymer Journal

Figure 6. The variation of x-ray fiber diagrams of the wet, fibrous curdlan specimen ($\overline{DPn}$ 400) dried in the air (20). Starting from the original gel specimen (A), photographs from (B) to (E) were taken at regular intervals (1 hr).

single helix, either left- or right-handed, agree with the observed peak positions of the layers 0, 1, 2, and 3 better than those of 6/1 single helices. However, the calculated intensities of layers 1 and 2 are greater than the observed intensities. The Fourier transforms of the double- and triple-stranded 7/1 helices were also calculated using the equation given by Fraser *et al.* (24). Among them, that of a triple-helix showed a rather good fit with the observed intensity. However, the calculated intensities of layers 1 and 2 are very weak, being about one-hundredth of that at the equator. These facts suggest that the most of the (1→3)-β-D-glucan molecule in the wet fibrous specimen has a 7/1 single helical structure and the rest has a triple-stranded helical structure (Fig. 8). These results are consistent with the proposed structure based on the ^{13}C NMR studies mentioned before.

The X-ray small-angle scattering photograph of the same specimen shows only a diffuse streak along the equator (Fig. 7). Assuming that the inhomogeneties in the specimen consists of highly dispersed, uniform, long cylindrical micelles which are arranged parallel to the stretching direction, the diameter of the micelles was estimated to be about 80 Å by the Guinier plot(25) of the intensity distribution along the equator. The $\varepsilon^2 I$-ε plot(26) gives a measure of intermicellar distance in the equator as about 120 Å.

A variation of fiber diagram, taken at regular intervals, of a wet fibrous specimen dried in the air at room temperature, is also given in Fig. 6(20). During the course of drying, the broadening of each pattern destroyed more and more the details of the fiber diagrams, and parallel to this change the small-angle scattering weakened and finally disappeared. These facts suggest that the micelles lose intramicellar water(or crystalline water) by drying, which disturbs the regular molecular arrangement inside the micelle, and at the same time, the evaporation of intermicellar water causes adjacent micelles to contact directly, which

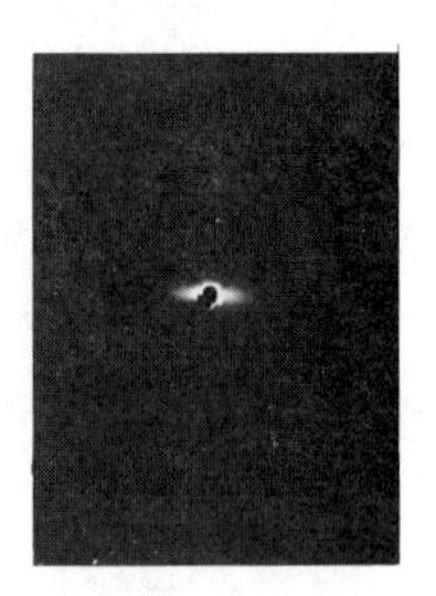

Polymer Journal

Figure 7. X-ray small-angle scattering diagram of the wet, fibrous curdlan specimen ($\overline{DPn}$ 400) (20)

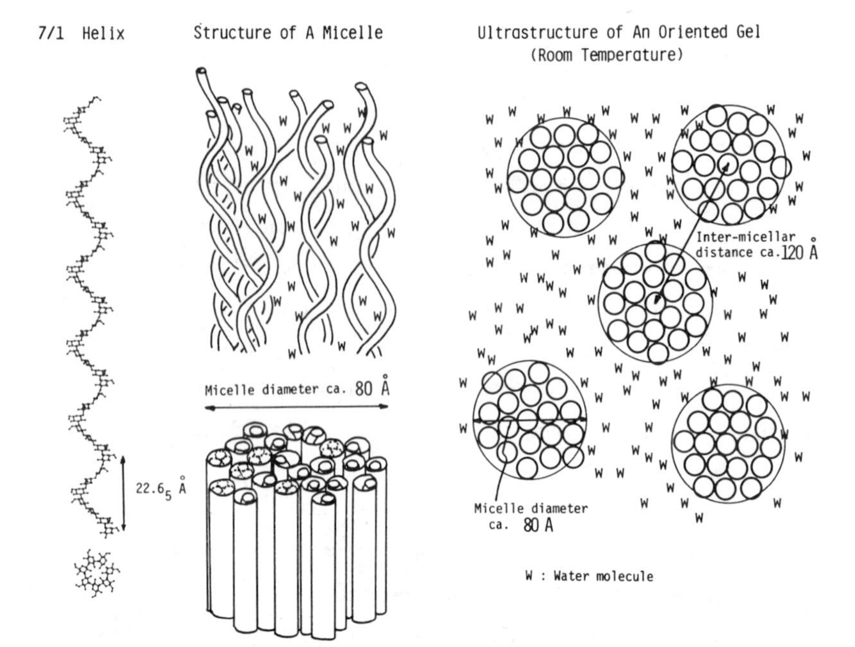

Figure 8. A model for the ultrastructure of curdlan gel at room temperature.

The oriented gel consists of micelles, of which the average diameter is about 80 Å. Micelle–micelle distance is about 120 Å. Area between micelles is filled with water molecules. Micelle interior is packed mostly by 7/1 single helical molecules that are hydrogen-bonded to one another by water molecules. The single helical molecule is also probably hydrated. Some parts of the micelle are occupied by molcules of triple-stranded helix, which are also hydrated to some extent.

does away with the inhomogeneity in the specimen (Fig. 8).

b. High temperature structure

Marchessault *et al.* (27) obtained weakly crystalline and oriented fibers(A) by extruding a 10% DMSO solution of crudlan into CH_3OH at room temperature and washing it in water. They found that the same fiber annealed in water, under tension, at 140°C, in a sealed bomb, is of higher crystallinity and occured as two reversible crystalline polymorphs: one at high relative humidity(B) and the other at humidities less than 20%(C).

The fiber period of 5.79 Å for (C) together with the results of a conformational study and the values of its intensity pointed to a triple-stranded 6/1 helical structure with $P6_3$ symmetry

Takeda *et al.* (19) heat-treated the wet fibrous specimen in a sealed bomb in the presence of water. At temperatures above 120°C the annealed specimen showed a remarkably different X-ray fiber diagram (Fig. 9): the first layer disappered, the fiber

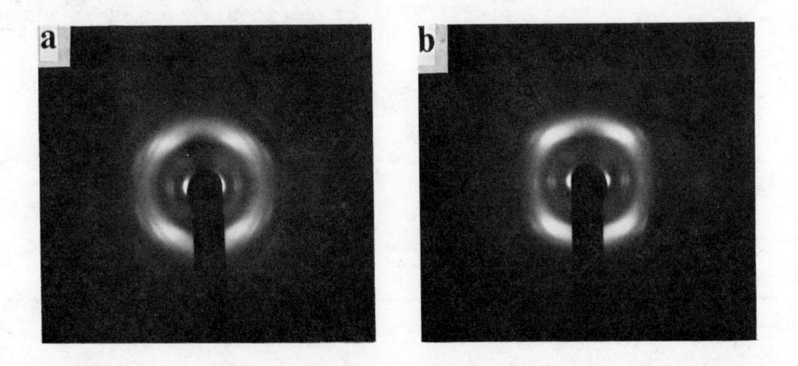

Polymer Journal

Figure 9. X-ray fiber diagrams of heat-treated (>120°C) curdlan specimen ($\overline{DP}n$ 400) (20): (A) dried in air and (B) dried in vacuum

period decreased to 18.1_2 Å. The annealed specimen dried in vacuum gave another X-ray diagram, having a fiber period of 5.7_2 Å (Table II). These two kinds of fiber diagrams respectively correspond to those of (B) and (C) obtained by Marchessault *et al.* (27).

Triple-stranded helical structure was also proposed for β-(1→3)-D-glucan by Atkins and Parker(28). Blum and Sarko(29) studied the structures of lentinan, a branched (1→3)-β-D-glucan and of pachyman, a (1→3)-β-D-glucan containing a few β-1,6-glucosidic lingakes, and proposed the existnece of a triple-strnaded helix in their structures.

Crystal data and a brief account for the structure of three forms of the curdlan are given in Table II. A schematic represen-

Table II. A brief account for the structure of three forms of Curdlan

		Room temperature or Heat-treated at temp.<120°C	Heat-treated at temp.>120°C in a sealed bomb in the presence of water, and	
			dried in air	dried in vacuum
Unit Cell	Takeda *et al.* (20)	a=26.4 Å b=16.4 c=22.6_5	a=b=15.4_1 Å c= 18.1_2	a=b=14.1_9 Å c= 5.7_2
		(A)	(B)	(C)
	Marchessault *et al.* (27)	a=b=17 Å c= 22	a=b=15.71 Å c= 18.82	a=b=14.38 Å c= 5.79
Structure		Single helix + Triple helix	Triple helix	Triple helix $P6_3$

tation of the structural change in the three forms are shown in Fig. 10.

4. Morphological study on the Curdlan(30,31)

An alkaline solution of curdlan($\overline{DPn}$ 455, 0.5%(w/v) in 0.1N NaOH) was neutralized with 0.1N HCl, and the resultant suspension of curdlan in a cellophane bag was dialyzed against water(renatured curdlan). Its suspension was heated at 60°C for 0.5 h and then portions were heated further at 90°C for 0.5 h or at 120°C (autoclaving) for 0.5 h or 4 h. A part of the curdlan suspension after dialysis was sonicated for 0.5 h at 20 Hz in an ultrasonic oscillator in ice water. These suspensions were stained negatively with 2% uranyl acetate. A Hitachi HU-11DS electron microscope was used with an accerelating voltage of 100KV. Original magnification of the electron microscope is 2 x 10^4.

Electron micrographs of the renatured and heated preparations are given in Fig. 11. The renatured curdlan has a fibrillar structure, microfibrils of which are composed of spindle-shaped component, about 100-200 Å wide and 1000-1500 Å long (Fig. 11A). This width of microfibrils seems to correspond roughly the inhomogeneities of about 80 Å across the microfibrils determined by X-ray small-angle scattering.

Rather similar micrographs were observed for the preparation set at 60°C(Fig. 11B). In the set at 90°C, which is reported(1) to be most resilient, the electron micrographs showed a change from fibrillar structure to a complicated arrangement of microfibrils connected to each other (Fig. 11C).

Heating at 120°C for 4 h resulted in the other fibrillar structure of the larger microfibrils with the dimensions of about 300-400 Å wide and 1000-1500 Å long (Fig. 11D), in which striations are not so clear as in the renatured and 60°C-set preparations. X-ray studies of the oriented curdlan gels showed that the heat treatment at temperatures above 120°C(autoclaving) caused a

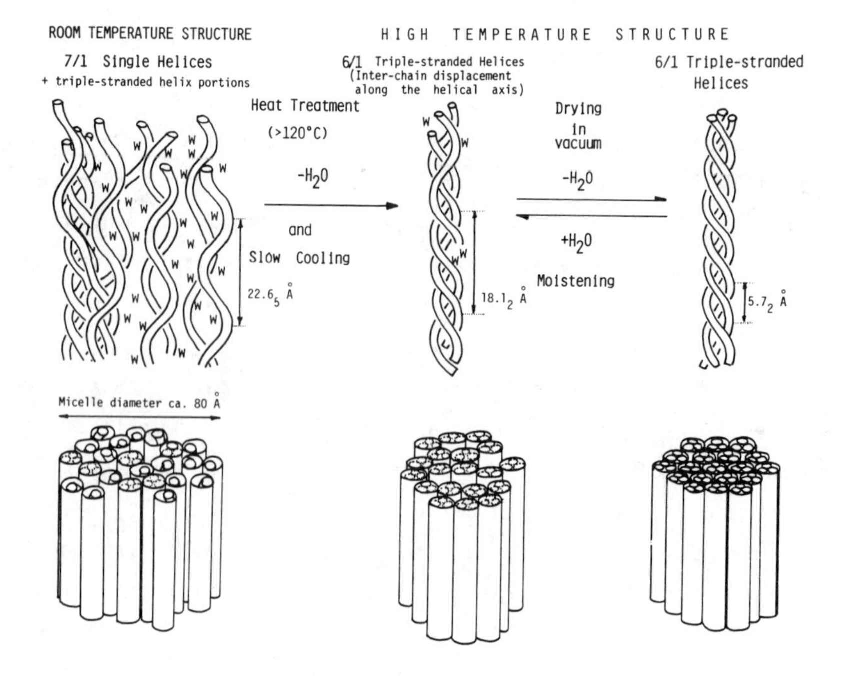

Figure 10. Schematic representation of structural change between three forms of curdlan.

(left) *Room-temperature structure (for details* see *Figure 8);* (middle) *high-temperature structure at high humidity. By heat treatment in a sealed bomb in the presence of water, curdlan molecules change their conformation to 6/1 triple-stranded helices. In a triple helix, component helices are displaced and are not in phase with each other along the helical axis (fiber period: 18.1_2 Å).* (right) *High-temperature structure at low humidity. Triple-stranded helical structure with* $P6_3$ *symmetry (fiber period: 5.7_2 Å.)*

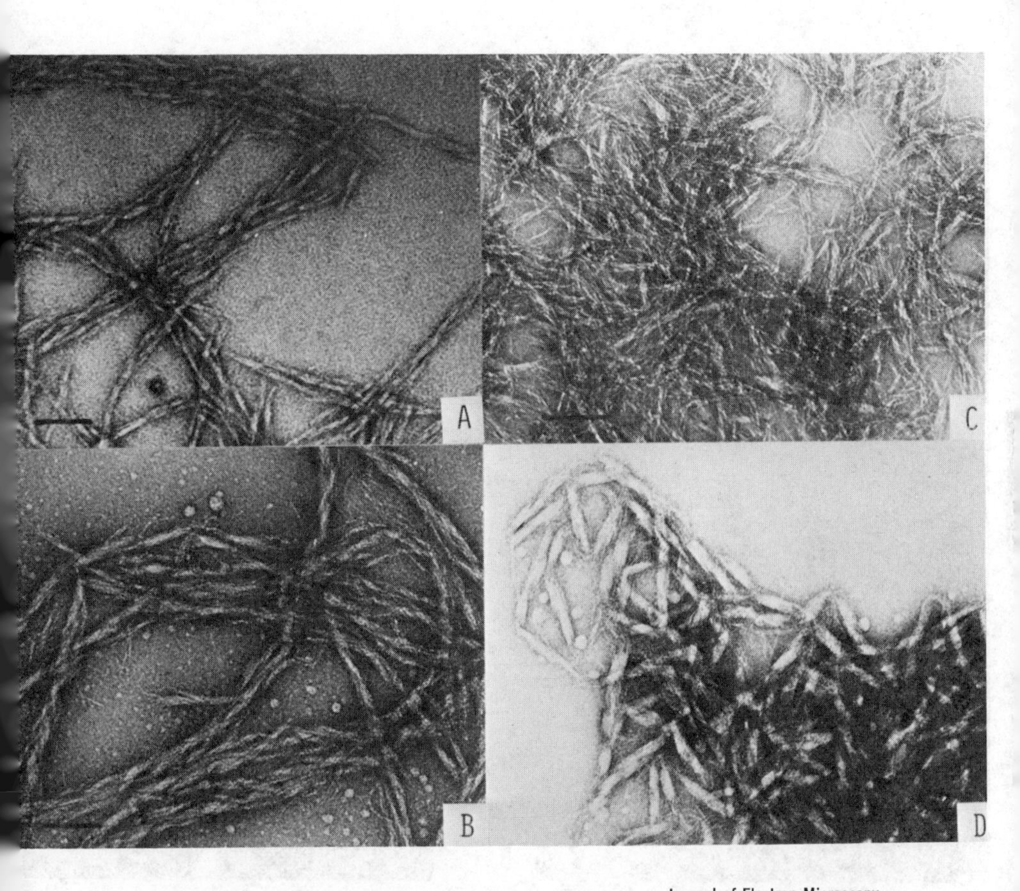

Journal of Electron Microscopy

Figure 11. Electron micrographs of the curdlan ($\overline{DPn}$ 455) (31): (A) regenerated specimen; (B) heated at 60°C for 0.5 hr; (C) heated at 90°C for 0.5 hr after the heat treatment at 60°C; (D) heated at 120°C for 4 hr after the heat treatment at 60°C. The bar marker represents 0.1 μm.

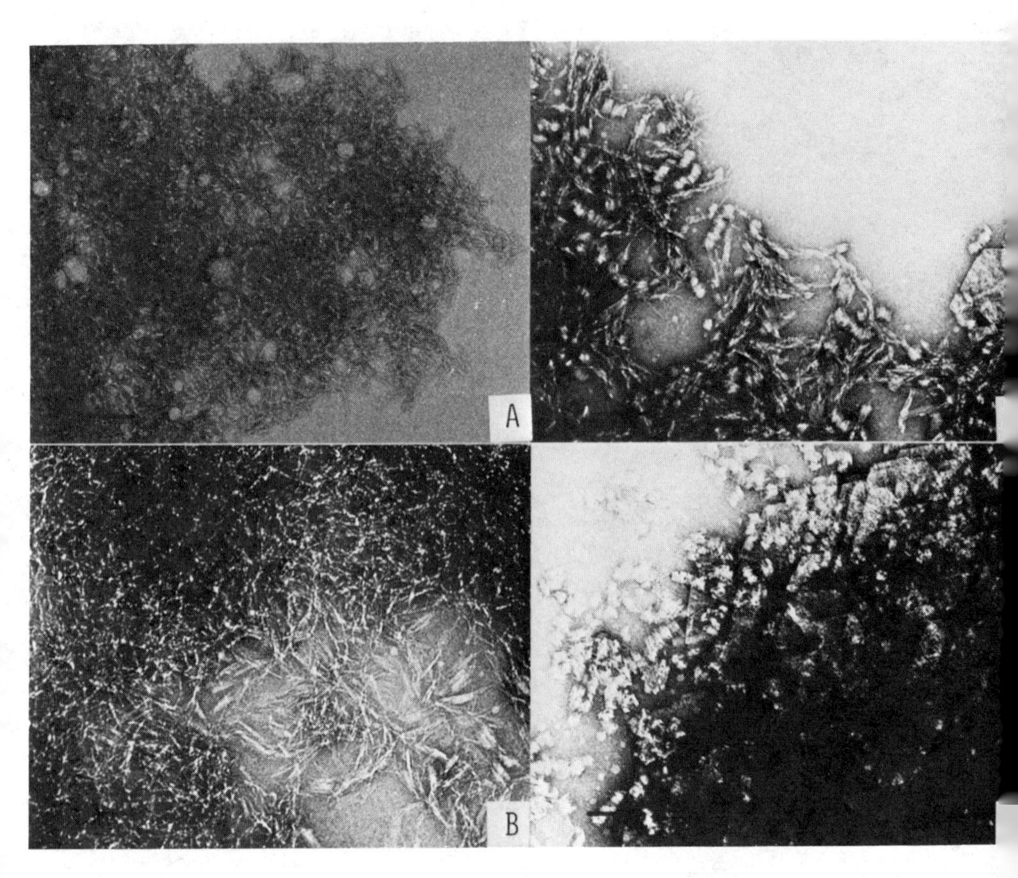

Journal of Electron Microscopy

Figure 12. Electron micrographs of β-(1 → 3)-D-glucan ($\overline{DPn}$ 49) (31). A, B, C, and D were treated as in Figure 11. The bar marker represents 0.1 μm.

change of molecular structure of the curdlan chains from the single to triple-stranded helices.

From samples of lower molecular weight ($\overline{DPn}$ 131 and 49), specimens were prepared in a similar way to those of $\overline{DPn}$ 455. Approximately similar electron micrographs were obtained with specimens of $\overline{DPn}$ 131. Figure 12 shows electron micrographs of $\overline{DPn}$ 49. The renatured and 60°C-set preparations gave much shorter and narrower microfibrils than those with higher $\overline{DPn}$. The 90°C-set preparation (Fig. 12C) contains many pseudo-lamellar crystals of about 150 Å wide and 150-1000 Å long, with clear cross-wise striations. Many lamellar crystals were also observed in this specimen. The preparation set at 120°C(Fig. 12D) gave hexagonal, lamellar crystals with different sizes. The thickness of these lamellar crystals were estimated about 80-90 Å from the average length of their shadows in the micrograph (Fig. 13)(31). Similar hexagonal lamellar crystals have been observed with esparto grass β-(1→4)-xylan(32-33).

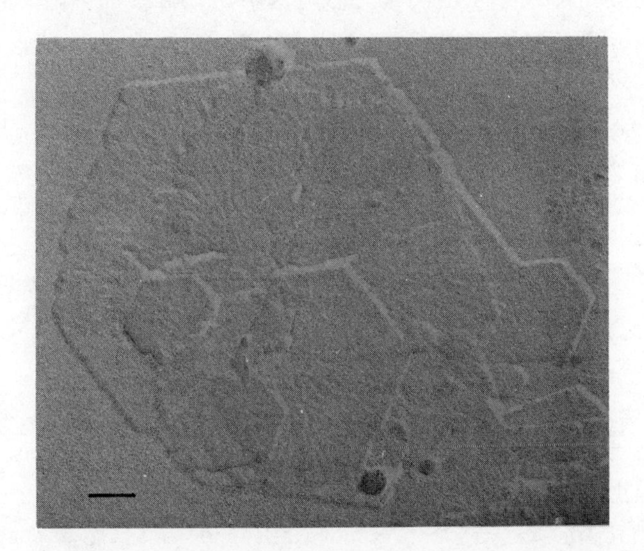

Journal of Electron Microscopy

Figure 13. Lamellar crystals of β-(1 → 3)-D-glucan ($\overline{DPn}$ 49) (31) shaded with platinum at an angle of 30°. The bar marker represents 0.1 μm.

Literature cited

1. Harada, T., Masada, K., Fujimori, K., and Maeda, I., *Agr. Biochem.* (1966) 30, 196.
2. Harada, T., *ACS Symposium Series* (1977) No. 45, 265.
3. Sasaki, T., Abiko, N., Sugiyama, Y., and Nitta, K., *Cancer Res.* (1978) 38, 379.
4. Hirano, I., Einaga, Y., and Fujita, H., *Polymèr J.* (1979) 11, 901.
5. Manners, D. J., Masson, A. J., and Strugeon, R. J., *Carbohyd. Res.* (1974) 32, 47.
6. Henley, D. *Arkiv. Kemi.* (1961) 18, 327.
7. Brown, W. and Wilkinson, R. *Eur. Polymer J.* (1965) 1, 1.
8. Saitô, H., Ohki, T. and Sasaki, T., *Biochem.* (1977) 16, 908.
9. Ogawa, K., Watanabe, T., Tsurugi, J., and Ono, S., *Carbohyd. Res.* (1972) 23, 399.
10. Sundaralingham, M., *Biopolymers* (1968) 6, 189.
11. Rees, D. A. and Scott, W. E., *Chem. Commun.* (1969) 1037.
12. Sathyanarayana, B. K. and Rao, V. S. R., *Biopolymers* (1971) 10, 1605.
13. Bryce, T. A., MaKinnon, A. A., Morris, E. R., Rees, D. A., and Thom, D., *Discuss. Faraday Soc.* (1974) 57, 221.
14. Chien, J. C. W. and Wise, E. B., *Biochem.* (1975) 14, 2475.
15. Smith, I. C. P., Jennings, H. J. and Deslauriers, R., *Acc. Chem. Res.* (1975) 8, 306.
16. Saitô, H. and Smith, I. C. P., *Arch. Biochem. Biophys.* (1973) 158, 154.
17. Allerhand, A. and Oldfield, E., *Biochem.* (1973) 12, 3428.
18. Takeda, H., Yasuoka, N. and Kasai, N., *Carbohyd. Res.* (1977) 53, 137.
19. Takeda, H., Kaiya, T., Yasuoka, N. and Kasai, N., *Carbohyd. Res.* (1978) 62, 27.
20. Takeda, H., Yasuoka, N. and Kasai, N., *Polymer J.* (1978) 10, 365.

21. Norman, N., *Acta Crystallogr.* (1954) 7, 462.

22. Arnott, S. and Scott, W. E., *J. Chem. Soc. Perkin II.* (1972) 324.

23. Klug, A. and Franklin, R., *Acta Crystallogr.* (1955) 8, 777.

24. Fraser, R. D. B., MacRae, T. P. and Miller, A., *Acta Crystallogr.* (1964) 17, 769.

25. Guinier, A., *Ann. Phys.* (1939) 12, 161.

26. Kratky, O. and Porod, G., *Acta Phys. Austriaca* (1948) 2, 133.

27. Marchessault, R. H., Deslandes, Y., Ogawa, K. and Sundarajan, P. R., *Canad. J. Chem.* (1977) 55, 300.

28. Atkins, E. D. T. and Parker, K. D., *J. Polymer Sci.*, Part C, (1969) No. 28, 69.

29. Bluhm, T. L. and Sarko, A., *Canad. J. Chem.* (1977) 55, 293.

30. Koreeda, A., Harada, T., Ogawa, K., Sato, S. and Kasai, N., *Carbohyd. Res.* (1974) 33, 396.

31. Harada, T., Koreeda, A., Sato, S. and Kasai, N., *J. Electron Microsc.* (1979) 28, 147.

32. Atkins, E. D. T., Booy, F. P. and Chanzy, H., Developments in Electron Microscopy and Analysis (1975) Academic Press, London, p. 319.

33. Marchessault, R. H. and Liang, C. Y., *J. Polymer Sci.* (1960) 43, 71.

RECEIVED May 21, 1980.

25

The Gelling Mechanism and Relationship to Molecular Structure of Microbial Polysaccharide Curdlan

W. S. FULTON and E. D. T. ATKINS

H. H. Wills Physics Laboratory, University of Bristol, Bristol BS8 1TL U.K.

The extracellular microbial polysaccharide curdlan is essentially a linear homopolymer of β-(1-3)-D-glucan as illustrated in Figure 1. It is produced from the mutant strain of the bacteria _Alcaligenis faecalis_ and was first isolated and investigated by Harada and coworkers (1-3) who coined the name curdlan to describe the gelling behaviour of the polysaccharide at elevated temperatures.

When an aqueous suspension (2%w/v) of curdlan is heated, a number of distinct phases can be observed. First the suspension clears forming a sol between 50-55°C. Then gelation begins at approximately 65°C forming a low set gel. This is a weak and easily disrupted gel. If the temperature is raised further to 95°C, a translucent, resilient gel (Fig. 2) is produced, which has been termed the high set gel. Both the low set and high set gels are thermally irreversible. As regards the morphology of these two types of gel, only that of the high set gel is reported in this work.

A number of X-ray diffraction patterns of oriented curdlan gels have been reported (2,4). They may be separated into two families. The first group bears a strong resemblance to the previously reported X-ray pattern for β-(1-3)-D-xylan and which was interpreted in terms of a six-fold triple-stranded molecule (5,6). Similar patterns have also been obtained from β-(1-3)-D-glucan in fungal tissue by Jelsma and Kreger (7) and from lentinan by Bluhm and Sarko (8). The second type of X-ray pattern, which exhibits layer lines of spacing 2.70nm, has been suggested by Takeda _et al._ (2) to correspond to a majority component of seven fold single helices.

We have examined the X-ray results from various oriented curdlan gels and report our findings and calculations on a number of models. In addition we have investigated the behaviour of curdlan gels in a series of deuterium exchange experiments monitored with infra-red spectroscopy.

0-8412-0589-2/80/47-141-385$06.50/0

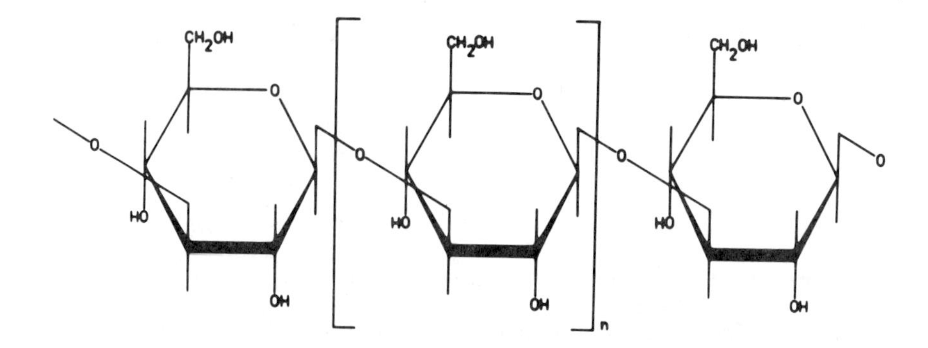

β-(1-3)-D-GLUCAN: CURDLAN

Figure 1. Chemical structure of curdlan

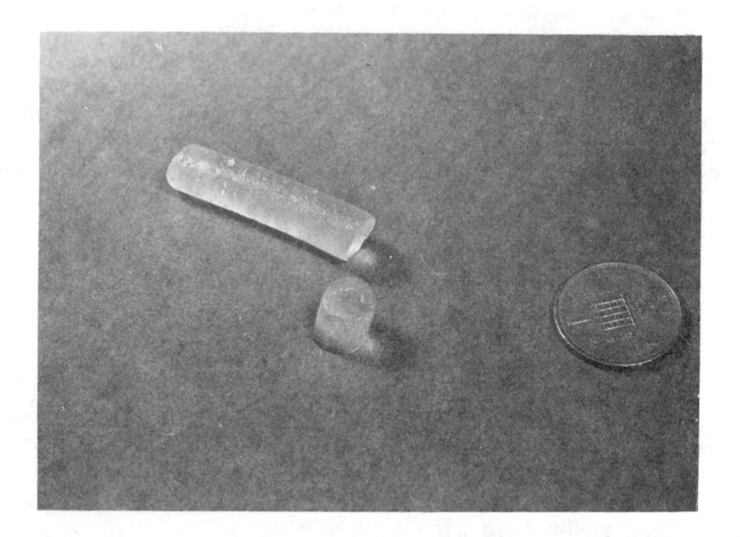

Figure 2. Resilient gel obtained from curdlan at temperatures in excess of 95°C

Materials and Methods

(a) Fibre Formation and X-ray Diffraction. The suspension of curdlan (2% w/v) was poured into a convoluted glass tube of internal diameter 1-2mm. and heated in a water bath to 95°C for 10 minutes to form a high set gel. The tube was removed from the bath, broken and gel fibres were pulled out of the tube and subsequently annealed under tension in an autoclave at 120°C for an hour before drying over silica gel. A typical X-ray diffraction pattern from this gel fibre is shown in Figure 3.

An alternative fibre forming technique involved the dissolution of curdlan in 1M NaOH to give a 5%-7% w/v solution, which was cast onto a teflon block. After drying, the resulting film was washed in water and simultaneously stretched to produce a thin gel fibre. This was examined by X-ray diffraction under the following conditions:

i) at 100% relative humidity (Fig. 4a)
ii) after drying over silica gel (Fig. 4b)
iii) after annealing in an autoclave and drying over silica gel (Fig. 4c).

X-ray diffraction photographs were recorded on flat film using a camera with pinhole collination (250 μm diam.) and Ni-filtered Cu Kα radiation. Calcite (characteristic spacing 0.3035 nm) was used as an internal diffraction calibration. Interplanar spacings were measured off the original photographs and the unit cell dimensions were refined, where possible, by a least-squares procedure.

(b) Film Formation and Infra-red Spectroscopy. All deuteration exchange experiments and drying of deuterated films were undertaken in a dry-box. Water vapour was excluded from the internal environment by molecular sieve and silica gel, which functioned as a dessicant and also gave an indication of the water content in the box. Liquid phase deuteration was carried out on the polysaccharide by completely immersing the clamped film in heavy water for at least 24 hours. The film was removed, dried and placed in an infra-red cell with sodium chloride windows and the spectrum recorded using the Unicam S.P. 100 infra-red spectrometer.

Gel Film. A gel film was produced by placing a suspension (2% w/v) of curdlan in water on a clean glass plate. A ground glass ring was laid on the plate and the suspension evenly distributed over the enclosed area. The plate was held over boiling water and heated to approximately 100°C for 10 minutes with minimal evaporation of water from the sample. After the film had gelled it was removed from the glass plate, laid on a teflon block and allowed to dry over the drying agent, phosphorous pentoxide.

Alkali Gel Film. Curdlan was dissolved in 1M NaOH (5%-7% w/v), spread into a thin film on a clean glass plate and allowed to dry. The film, when immersed in distilled water, swelled and became transparent as the excess alkali was washed out. The gel

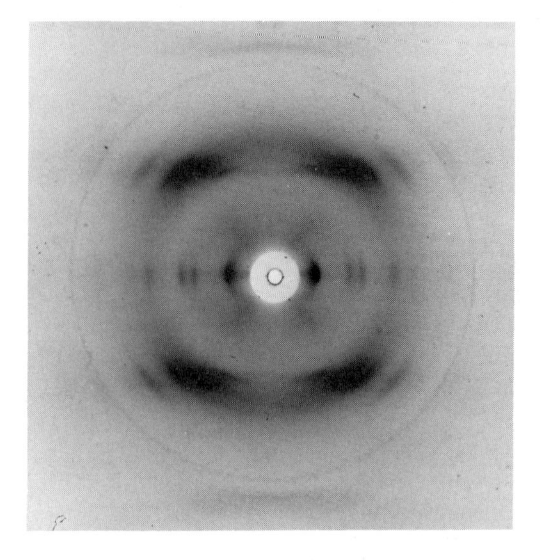

Figure 3. X-ray fiber diffraction photograph obtained from a dry, annealed-oriented fiber of curdlan gel (fiber axis vertical)

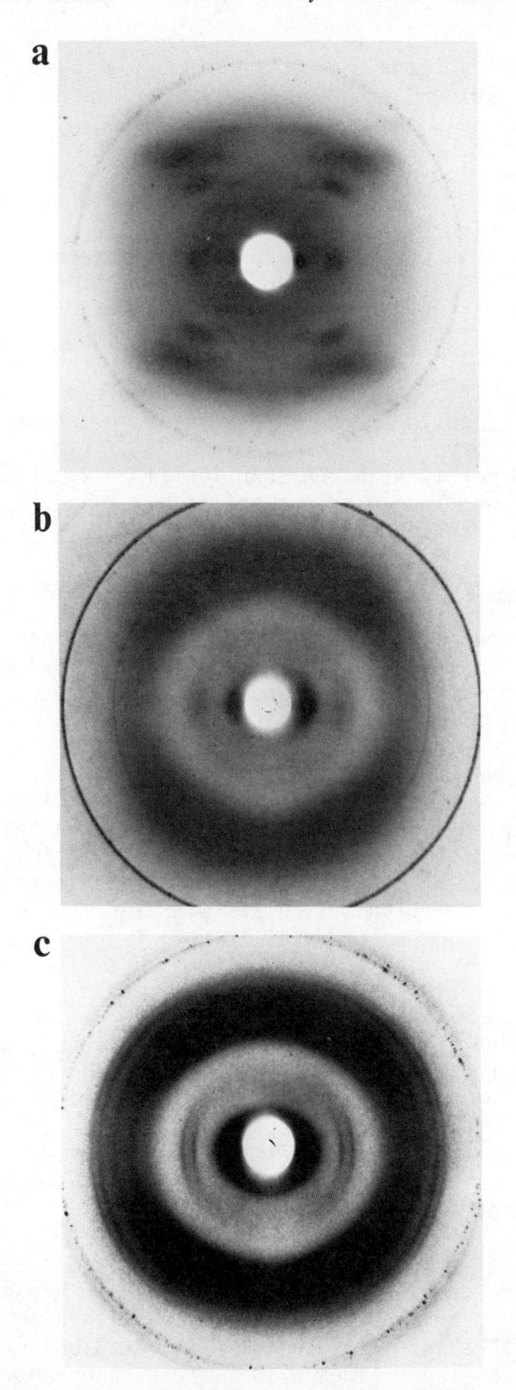

Figure 4. X-ray diffraction photographs of curdlan fibers obtained from alkali solution: (a) at 100% relative humidity; (b) after drying over silica gel; (c) after annealing and drying

film was carefully removed from the glass plate and allowed to dry on a teflon block.

Deuterated gel. Dried curdlan powder was mixed into a 2% w/v suspension in heavy water. This suspension was spread on a glass plate and placed in an airtight container, shown schematically in Figure 5. The container was heated to approximately 100°C, thus gelling the polymer. The film was washed in D_2O, to stop the rapid loss of D_2O from the hot film, and then carefully removed and dried. (All operations except that of heating were carried out in the dry-box.)

Deuterated alkali gel. NaOD was dissolved in D_2O to form a 1M NaOD solution. Curdlan was dissolved in this solution (5% w/v) and spread into a thin film on a teflon block. After drying, the film was washed in D_2O to remove any excess heavy alkali and then allowed to dry.

Film from Formic Acid. A solution (1% w/v) of the curdlan in formic acid was spread onto a teflon block and allowed to dry, forming a thin transparent film.

(c) Molecular Model Building. Molecular models were generated by using a linked-atom procedure (9). The positions of the atoms were defined in terms of the bond lengths, bond angles and torsion angles of the molecule. Stereochemical parameters for the β-D-glucose rings in the standard $^4C_1$ chair were based on the average set given by Arnott and Scott (10). Bond lengths and bond angles were held constant, including the angles for the glycosidic bridges which were assigned a value of 116.5°. This defines the glucose residues as rigid bodies and leaves the glycosidic torsion angles and also the torsion angle that defines the orientation of the hydroxmethyl group as explicit variables. These parameters were varied to produce a model with the appropriate helical symmetry and pitch and with the minimal amount of non-bonded steric compression (11,12).

(d) Cylindrically Averaged Fourier Transform. The cylindrically averaged Fourier transforms of the various structures were calculated using the equations of Franklin and Klug (13) and atomic scattering factors computed using the analytical expression given in International Tables of Crystallography (14). These values were modified to approximate the effect of the disordered water in the voids between chains (15).

X-Ray Diffraction

The X-ray diffraction pattern obtained from the dry, annealed fibre (see Materials and Methods) is shown in Figure 3. The reflections index on a hexagonal unit cell with dimensions $a = b =$ 1.438 nm, c (fibre axis) = 0.582 nm with meridional reflections occurring only on layer lines with $\ell = 2n$, where n is an integer. These constraints on the Miller indices hkl are in accordance with

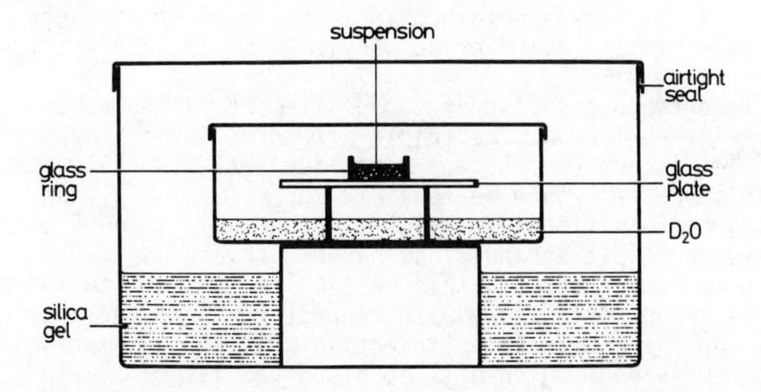

Figure 5. Diagrammatic representation of the equipment used to obtain curdlan gels in heavy water

space group assignment $P6_3$. The essential features of the X-ray pattern are similar to that obtained from β-(1-3)-D-xylan (6). (The dimensions of the xylan unit cell are slightly different: a = b = 1.340 nm, c (fibre axis) = 0.598 nm.) Atkins and Parker (6) were able to interpret such a diffraction pattern in terms of a triple-stranded structure. Three chains, of the same polarity, intertwine about a common axis to form a triple-strand molecular rope. The individual polysaccharide chains trace out a helix with six saccharide units per turn and are related to their neighbours by azimuthal rotations of $2\pi/3$ and $4\pi/3$ respectively, with zero relative translation. A similar model for curdlan is illustrated in Figure 6. Examinations of this model shows that each chain repeats at a distance 3 x 0.582 = 1.746 nm. Thus if for any reason the precise symmetrical arrangement between chains (or with their associated water of crystallization) is disrupted, we would expect reflections to occur on layer lines which are orders of 1.746 nm. Indeed such additional reflections have been observed via patterns obtained from specimens at different relative humidity (4) offering confirmation for the triple-stranded model.

Since the 5-position of successive saccharide units occur on the outside surface of the triple-strand molecule it was evident after the discovery of such a model for β-(1-3)-D-xylan that a similar structure would be feasible for β-(1-3)-D-glucan. Detailed conformational analyses by Bluhm and Sarko (8) have supported a triple-strand right-handed structure.

It is convenient, in this particular context, to index the observed Bragg reflections with respect to the longer 1.746 nm repeat for the c axis in anticipation of the conformational changes envisaged, which will be discussed later. Table I provides a list of the observed reflections together with the interplanar spacings. The measured density of 1.52 g/cm^3 is compatible with one triple-stranded molecule running through the unit cell.

The X-ray diffraction patterns obtained for the alkali prepared gel (see Materials and Methods) is shown in Figure 4a. It is similar to the pattern reported by Takeda et al. (5) who indexed the reflections on an orthorhombic unit cell with dimensions a = 1.64 nm, b = 2.64 nm and c (fibre axis) = 2.265 nm. It was proposed that the unit cell contained two single-strand seven-fold helices together with an estimated 250 water molecules.

The patterns we obtained show discrete equatorial reflections and streaked layer lines indicating longitudinal disorder in the specimen. The equatorial reflections index on a hexagonal net with dimensions a = b = 1.701 nm and the layer line spacing c (fibre axis) = 2.270 nm. The axial advance per saccharide residue in the triple-strand model discussed previously (Figs. 3 and 6) is 1.746/6 = 0.291 nm. It would appear logical therefore to associate the larger layer line spacing of 2.270 nm with seven or eight saccharide units although there is no

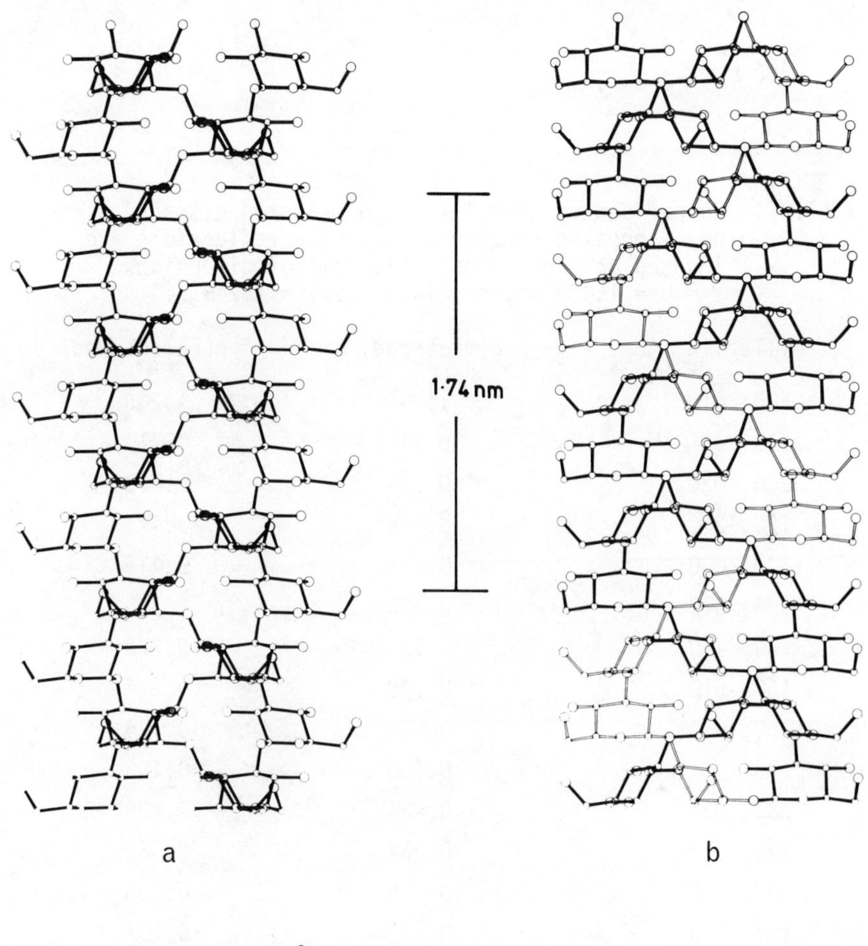

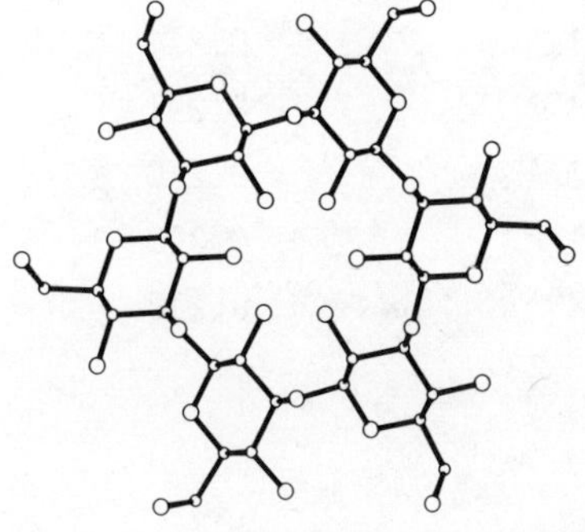

Figure 6. Computer generator ORTEP projections of the triple-strand structure corresponding to the x-ray diffraction pattern shown in Figure 3. The pitch of 1.746 nm for each six-fold helix is reduced to 1.746/3 = 0.581 nm by the symmetry related chains. (a) and (b) are different projections normal to the molecular axis and (c) is a projection parallel to the molecular axis.

Table I. Comparison of calculated and measured d-spacings for dry, annealed curdlan fibre. The reflections are indexed on a hexagonal unit cell of dimensions $\underline{a} = \underline{b} = 1.438$ nm, $\underline{c}$ (fibre axis) = 1.746 nm.

Miller indices	d(measured) nm	d(calculated) nm
100, 010	1.241	1.245
110	0.719	0.719
200, 020	0.621	0.623
120, 210	0.464	0.471
300, 030	0.411	0.415
130, 310	0.336	0.345
140, 410	0.263	0.272
150, 510	0.220	0.224
103, 013	0.530	0.524
113	0.455	0.453
203, 023	0.424	0.425
123, 213	0.364	0.366
303, 033,	0.335	0.338
133, 313	0.297	0.297
403, 043	0.271	0.274
006	0.284	0.291
116	0.270	0.270
126, 216	0.251	0.248
306, 036	0.238	0.238

particular need to have an integer number of saccharide residues per repeat. It is straightforward to envisage the six-fold triple-stranded structure untwisting slightly, destroying the perfect $P6_3$ symmetry relationship, and therefore allowing layer lines of spacing ∿2 nm to appear. The measured density is 1.24 g/cm^3 which would allow one triple-strand molecule per unit cell together with approximately 2 water molecules per glucose residue.

When the sample is dehydrated the X-ray diffraction pattern obtained is of poorer quality (Fig. 4b) and is similar to that reported for lentinan (8). We suggest that the removal of water causes a twisting of the chains back toward the six-fold triple-stranded model. On annealing, the sample completes this transition (Fig. 4c) by exhibiting a pattern similar to Figure 3. The reflections observed in Figure 4c index on a hexagonal unit cell with dimensions $a = b = 1.530$ nm, c (fibre axis) = 1.76 nm and the measured density is 1.52 g/cm^3.

A model for the seven-fold triple-stranded structure is shown in Figure 7.

Infra Red Spectra

A detailed band assignment is not presented in this work, but a summary is given in Table II. Throughout all the deuteration and rehydrogenation procedures there are three main features to note in the spectra.

i) The band at 3300 cm^{-1}, due to the stretching mode of the hydrogen bonded hydroxl group, persists after deuteration with only a small decrease in intensity which is estimated at 20%.

ii) The band at 1645 cm^{-1}, due to the in-plane deformation of the water molecule. This water is the strongly bonded water of crystallization which is inaccessible to exchange as it is still present in the deuterated films.

iii) The band at 2500 cm^{-1}, due to OD stretch on deuteration which is removed completely when the sample is rehydrogenated.

Gel-film (Fig. 8). Deuteration (B) shows that ROD-D_2O (cryst) is present with a slight reduction in H_2O (cryst). Therefore only a small proportion of the water of crystallization is exchangeable. Differences in the deuterated spectrum lead to the assumption that some of the OH groups undergo an exchange reaction to form OD. Rehydrogenation proves that this is a completely reversible reaction and that the D_2O (cryst) is exchangeable, contrary to the "buried" H_2O (cryst).

Fourier transform infra-red techniques were performed on a block of hydrated gel, similar in form to that shown in Figure 2. The resulting spectra are shown in Figure 9. Spectrum B is after the water background had been subtracted. A comparison of this spectrum with that of A in Figure 8 indicates no real structural difference between the bulk gel and a dried gel film.

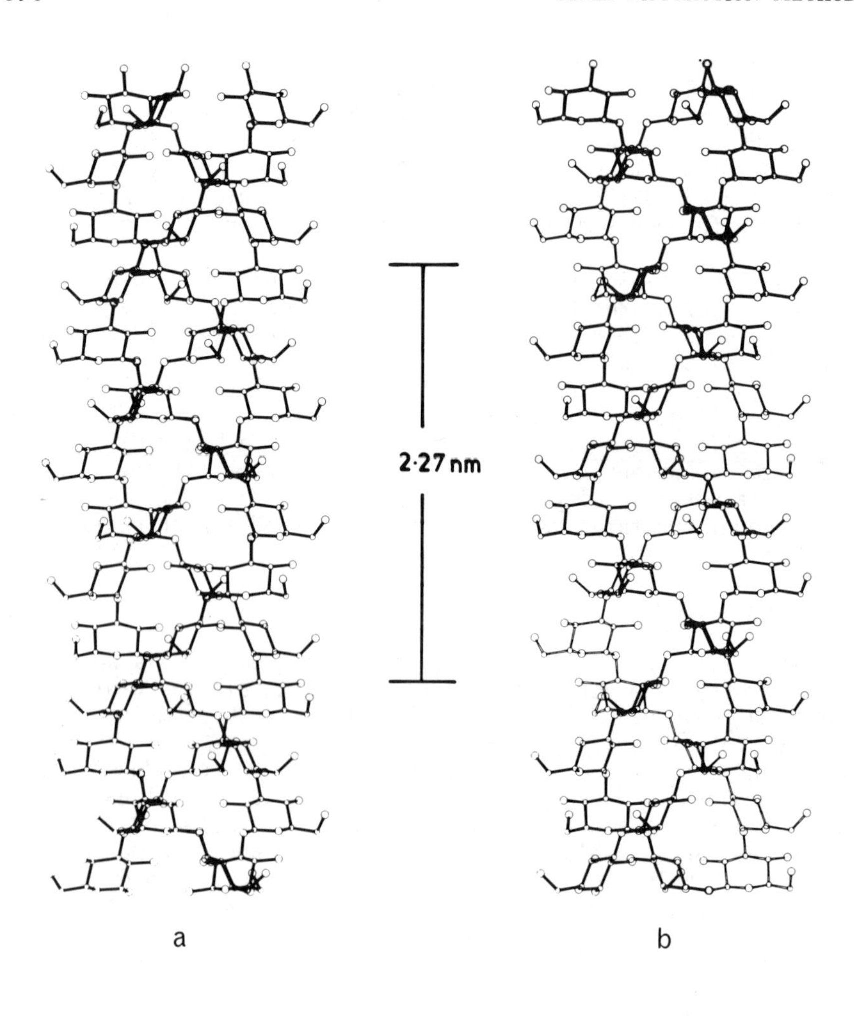

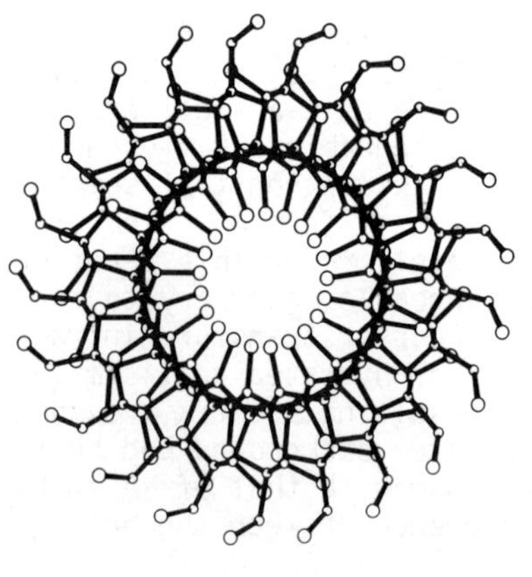

Figure 7. Proposed triple-strand structure incorporating seven-fold helical chains. Each chain related azimuthally by $2\pi/3$ and $4\pi/3$, respectively, and with zero relative translation and similar polarity. (a) and (b) are different projections normal to the helix axis; (c) is a projection looking along the helix axis.

Table II. The band assignment for the infra-red absorption spectrum of a curdlan gel film (1A).

Band cm^{-1}	Assignment
3370	OH stretch (hydrogen bonded)
2917 2894	CH, CH_2 stretch
1647	H-O-H bend
1460	CO-H in-plane bend (hydrogen bonded)
1423	CH_2 deformation
1374	CH bend
1317	CH_2 wag
1258	C-OH in-plane bend
1234	C-OH in-plane bend or CH in-plane bend
1162	C_1-O-C_3 antisymmetric stretch
1107	Antisymmetric in-phase ring stretch or C-O stretch
1082	C-O stretch (secondary OH groups)
1050	C-O stretch or C-C stretch
1007	C-O stretch
930	C-O stretch (primary alcohol)
894	C_1 group frequency (common to linkage)

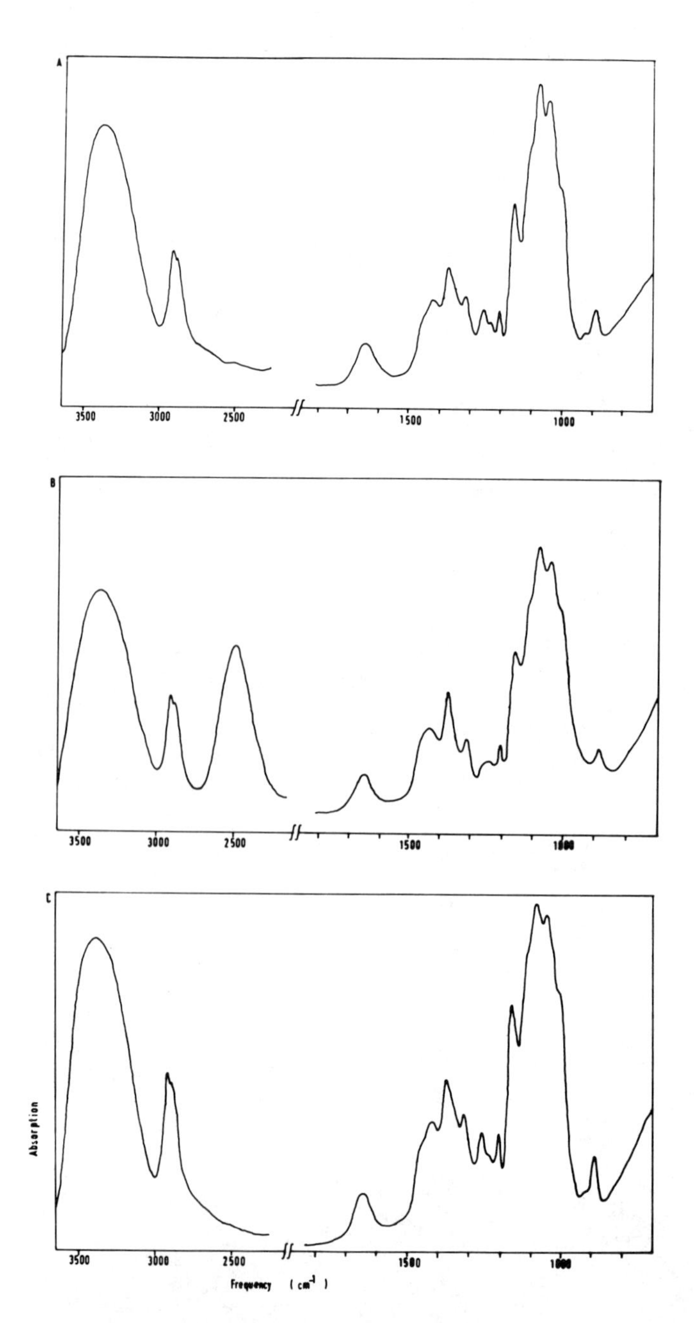

Figure 8. IR spectra obtained from curdlan sample gelled in water: (a) original film; (b) film after deuteration; (c) rehydrogenated film

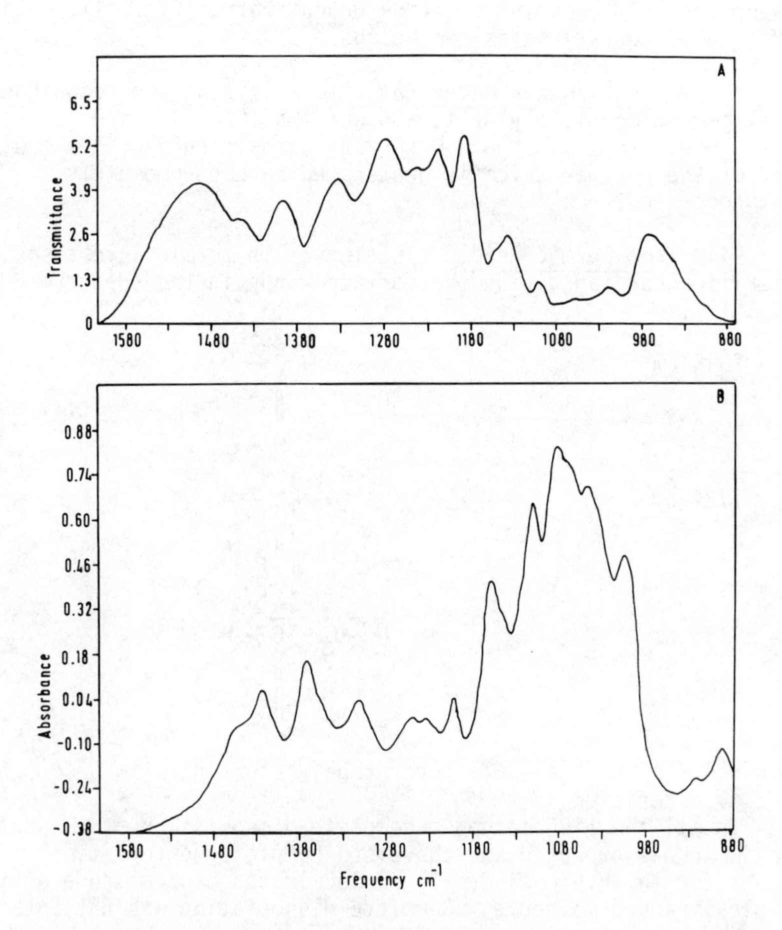

Figure 9. FTIR spectra of: (a) the hydrated gel block (see *Figure 2); (b) after water background was substracted*

Alkali Gel Film. The exchange reaction is only partial i.e. ROH and H_2O (cryst) bands remain on deuteration (Fig. 10). Apart from the ROD band there is no apparent change in the spectrum, implying no apparent structural change.

Gel in D_2O. Gelation in heavy water still leaves a high proportion of the polysaccharide undeuterated (Fig. 11), both ROH and H_2O (cryst) bands remaining.

Heavy Alkali/Heavy Water Gel Film. This strong technique still cannot completely deuterate curdlan (Fig. 12) indicated by the persistent ROH and H_2O (cryst) bands. Furthermore that part of the polymer which is deuterated can be completely rehydrogenated.

Film from Formic Acid. Dissolution in formic acid also esterifies curdlan giving the formate bands in the spectrum (Fig. 13).

2848 cm^{-1} —O—C—H (C=O)

1724 cm^{-1} —O—C—H (C=O)

1166 cm^{-1} —O—C—H (C=O)

Exchange for OD is a small percentage, occurring with groups not already esterified.

In all the five methods, complete dissociation of the polymer chains was expected and therefore complete deuteriation except for the hydroxyl groups buried in the central core of the triple-stranded molecule. Complete dissociation was not obtained even when using increasingly severe deuteration techniques.

Discussion

The X-ray evidence favours a model for the curdlan gel based on triple-stranded molecules. This structure is present in both the gel prepared from aqueous suspension and from alkali solution. The alkali gel contains the proposed seven-fold triple-stranded model (Figs. 7 and 14a) which converts to the more symmetric six-fold triple helix by annealing (Figs. 6 and 14b). This

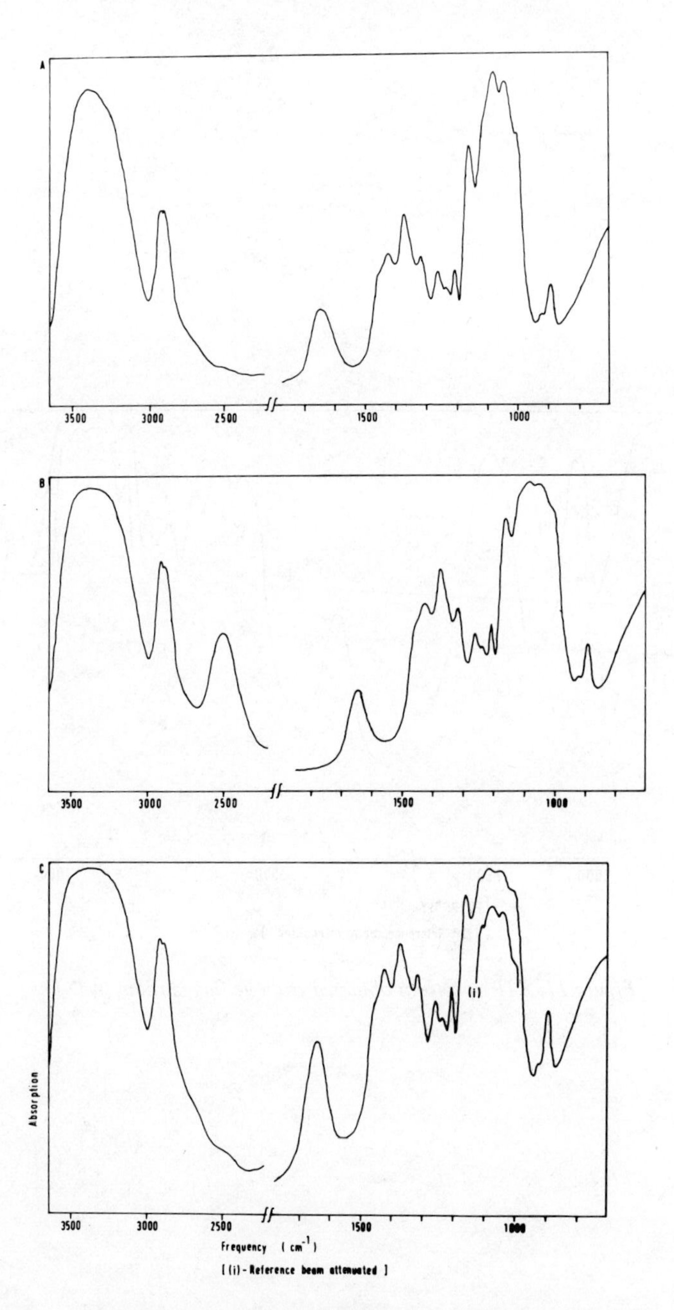

Figure 10. IR spectra obtained from curdlan after gelling from alkali: (a) original film; (b) after deuteration; (c) rehydrogenated film

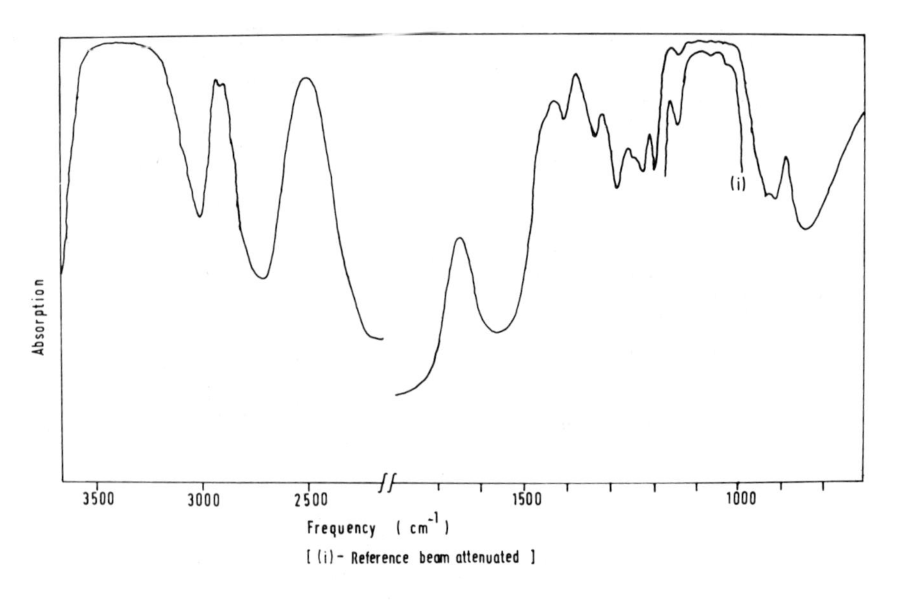

Figure 11. IR spectrum obtained from gelling curdlan in D_2O

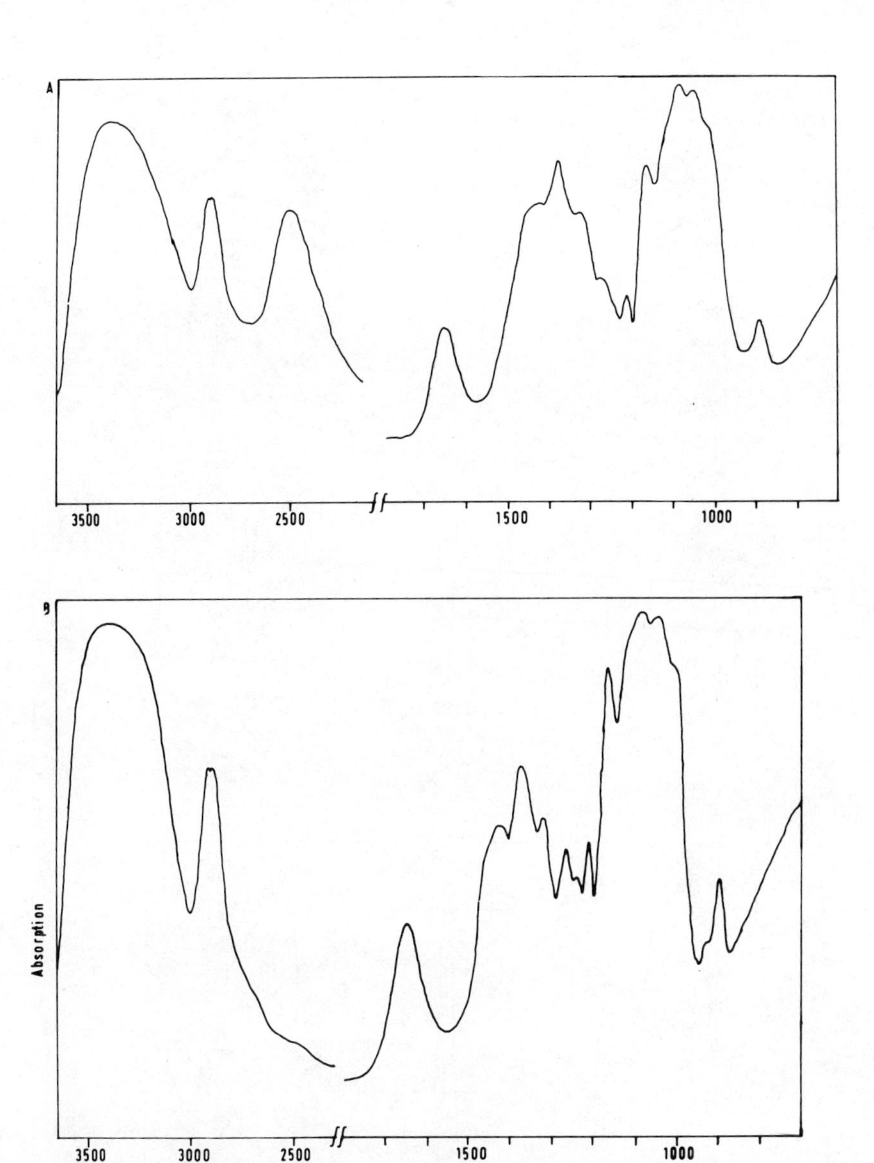

Figure 12. IR spectra of curdlan after gelling in NaOD and D_2O: (a) deuterated film; (b) rehydrogenated film

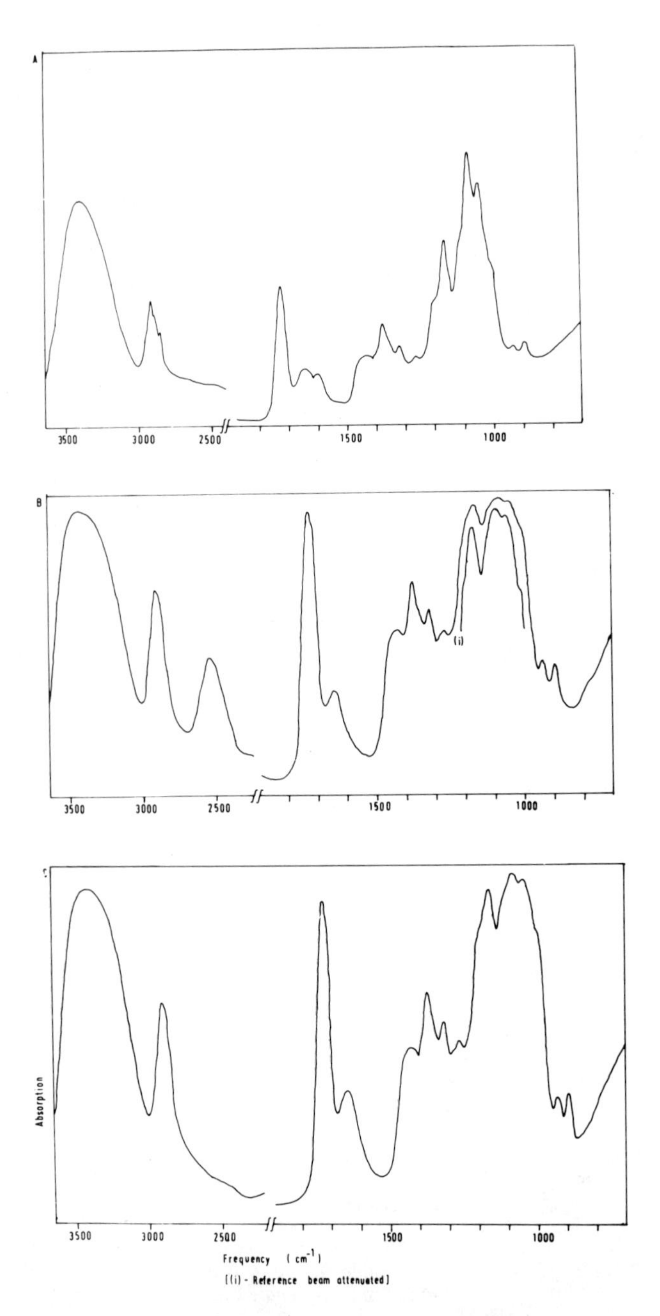

Figure 13. IR spectra obtained after dissolving curdlan in formic acid: (a) original film; (b) deuterated film; (c) rehydrogenated film

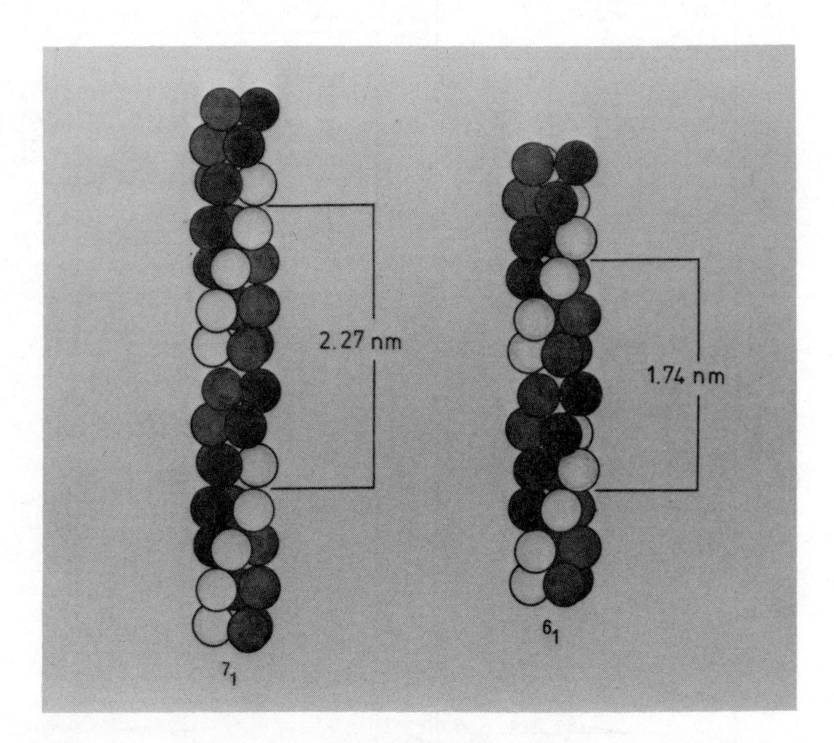

Figure 14. Comparison of triple-stranded models involve: (a) seven-fold helices; and (b) six-fold helices

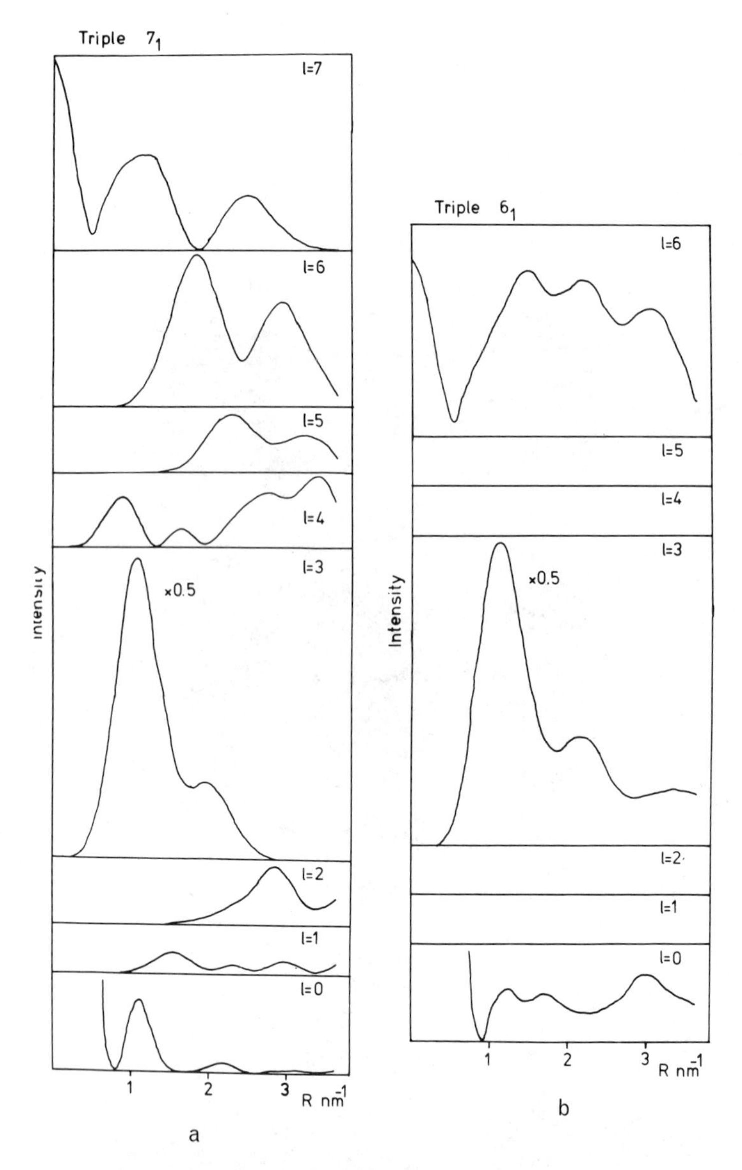

Figure 15. Cylindrically averaged transforms of the two models illustrated in Figures 6 and 7: (a) intensity distribution on successive layer lines for the triple-stranded 7_1 helix; (b) intensity distribution on successive layer lines for the triple-stranded 6_1 helix. Note layer lines with index l = *1, 2, 4, and 5 are absent.*

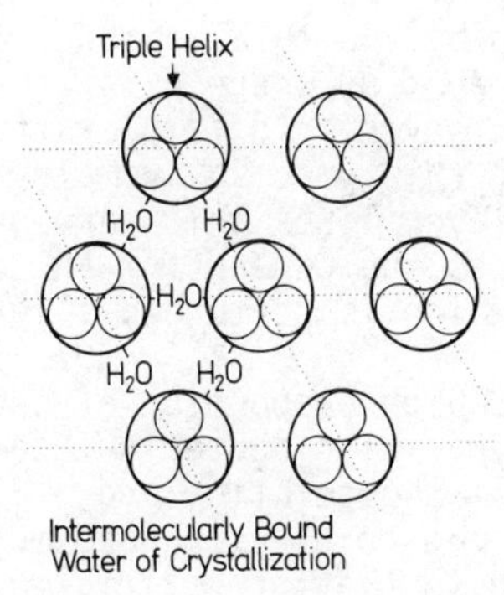

Figure 16. Cross section of an associated curdlan micelle showing the triple helices hydrogen bonded to the inaccessible water of crystallization

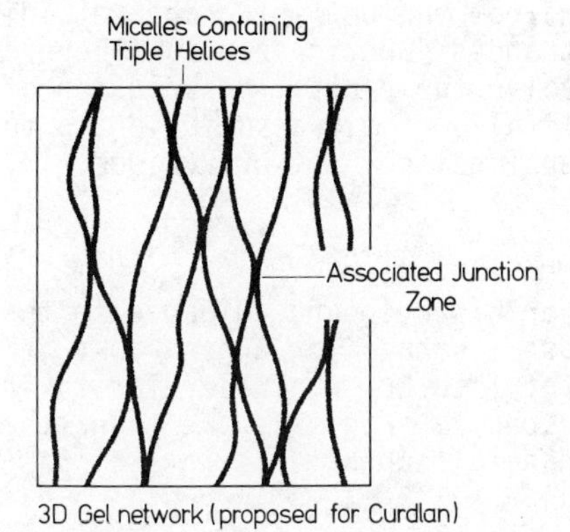

Figure 17. The 3D gel network proposed for curdlan. The micelles are associated at particular points forming the junction zones of the gel.

twisting of the helices is stereochemically feasible and does not involve disruption of the internal stack of interchain hydrogen bonded triads which stabilize the structure (16,17). This transition is illustrated schematically in Figure 14. The conversion from seven-fold to six-fold results in only a small reduction in the axial advance per saccharide from 0.324 nm to 0.291 nm.

The cylindrically averaged Fourier transform of the seven-fold and six-fold triple-stranded structures are shown in Figure 15. The Fourier transform of the six-fold triple-stranded model illustrates the symmetry of the system by the total absence of intensity on layer lines with index $\ell \neq 3n$, where n is an integer. The Fourier transform of the seven-fold triple-stranded structure shows that in destroying this precise symmetry relationship intensity occurs on all layer lines which are orders of the 2.27 nm spacing. This reinforces the concept of an indigenous triple-stranded structure which is perturbed slightly by the interaction of solvent.

The infra-red results show that a high proportion of the various gels are inaccessible to exchange with deuterium atoms. This phenomenon persists throughout all the gelation and dissolution processes. Consequently, the triple-stranded molecules must be bound by hydrogen bonding to the interstitial water of crystallization, thus forming a micellar domain. A cross section of the proposed model is shown in Figure 16. Thus we envisage the gelling mechanism of curdlan to involve the interaction of these micelles and *not* the untwining and retwining of single helices into triple-stranded junction zones. Therefore the gelation mechanism would involve the partial dissolution of the micelles, which in turn would allow their association to form the junction zones of the 3-dimensional gel network (Figure 17).

Acknowledgements

We thank Mr. C.G. Cannon, for his help and guidance in the infra-red analysis, and Professor T. Harada for a generous supply of material. We are grateful to Dr. G. Fraser for taking the Fourier transform infra-red spectra (Fig. 9). We thank the Science Research Council for financial support.

ABSTRACT

The molecular structure and gelling mechanism of the bacterial polysaccharide curdlan was investigated using X-ray diffraction and infra-red spectroscopy. Gels were obtained from heating an aqueous suspension to 95°C and also from treatment of the alkaline solution. X-ray diffraction patterns of the oriented gel, obtained from aqueous suspension were indexed on a hexagonal unit cell with dimensions $\underline{a} = \underline{b} = 1.438$ nm and $\underline{c}$ (fibre axis) = 1.746 nm and interpreted in terms of a triple-stranded structure with space group symmetry $P6_3$, by analogy with the previously published structure for β-(1-3)-D-xylan. The X-ray patterns of the gel formed from alkaline solution gave equatorial reflections which indexed on a hexagonal lattice $\underline{a} = \underline{b} = 1.701$ nm and a layer line spacing $\underline{c}$ (fibre axis) = 2.270 nm. This is interpreted as a perturbed triple-stranded structure which undergoes a transition on annealing to form the six-fold triple helix. Infra-red spectroscopy and deuteration of the polysaccharide gel films indicated the presence of inexchangeable hydroxyl groups and the interstitial water of crystallization. This interstitial water forms a hydrogen bonded network with the triple helices, binding them into a micellar structure; it is the association of these micelles which form the gel network.

Literature Cited

1. Harada, T. in "Extracellular Microbial Polysaccharides, ed. Sanford and Laskin, ACS Symposium Series 45, 1977, 265-283.

2. Takeda, H.; Yasuoka, N.; Kasai, N.; Harada, T. Polymer J., 1978, 10, 365-368.

3. Harada, T. in "Polysaccharides in Food", ed. Blanshard and Mitchell, Butterworths, London, 1978, 18-68.

4. Marchessault, R.H.; Deslandes, Y.; Ogawa, K.; Sundarajan, P. R. Can. J. Chem., 1977, 55, 300-303.

5. Atkins, E.D.T.; Parker, K.D.; Preston, R.D. Proc. Roy. Soc. B, 1969, 173, 205-221.

6. Atkins, E.D.T; Parker, K.D., J. Poly. Sci., 1969, C28, 69-81.

7. Jelsma, J; Kreger, D.R. Carbonhyd. Res. 1975, 43, 200-203.

8. Bluhm, T.L; Sarko, A. Can. J. Chem., 1977, 55, 293-300.

9. Arnott, S.; Wonacott, J.A. Polymer, 1966, 7, 157-166.

10. Arnott, S.; Scott, W.E. J. Chem. Soc., (Perkin Transactions II), 1972, 324-335.

11. Gardner, K.H.; Magill, J.M.; Atkins, E.D.T. Polymer, 1978 19, 370-378.

12. Smith, P.J.C.; Arnott, S. Acta Cryst. A, 1978, 34, 3-11.

13. Franklin, R.E; Klug, A. Acta Crystallogr., 1955, 8, 777-780.

14. International Tables for X-ray Crystallography, Kynoch Press, Birmingham, 1974.

15. Fraser, R.D.B.; MacRae, T.D.; Suzuki, E. J. Appl. Cryst., 1978, 11, 693-694.

16. Atkins, E.D.T.; Parker, K.D. Nature, 1968, 220, 784-785.

17. Fulton, W.S.; Atkins, E.D.T.; Cannon, C.G. (in preparation).

RECEIVED February 19, 1980.

Fiber Diffraction Studies of Bacterial Polysaccharides

K. OKUYAMA, S. ARNOTT, R. MOORHOUSE, and M. D. WALKINSHAW
Department of Biological Sciences, Purdue University, West Lafayette, IN 47907
E. D. T. ATKINS and CH. WOLF–ULLISH
H. H. Wills Physics Laboratory, University of Bristol, Bristol, BS8 1TL U.K.

Recently, the detailed primary structures of several bacterial polysaccharides have been determined (1,2,3). These polysaccharides are usually heteropolymers with large chemical repeating units. The complexity of their primary structures together with the limited information from their X-ray fiber diffraction patterns makes detailed structural analysis very difficult. The pioneering work in this field is the analysis of *Escherichia coli* K29 capsular polysaccharide (4). The chemical repeat of the *E. coli* K29 polymer contains six sugar residues, four in the backbone and two in the side-chain. This complex polysaccharide, nevertheless, can be spun into oriented and polycrystalline fibers providing 48 independent Bragg reflections down to 0.3nm resolution. A linked-atom least-squares analysis (4) yielded a detailed structure with a low crystallographic residual (R=0.26) and opened the possibility of similarly thorough analyses of other complex bacterial polysaccharides. In this paper we shall consider the molecular structures of two bacterial polysaccharides, one from *Xanthomonas campestris* and the other from *Klebsiella* serotype K8. In earlier X-ray studies of the former we concluded that a multistranded helical structure was unlikely (5). The

0-8412-0589-2/80/47-141-411$05.00/0

results of other physical probes (6,7) have prompted rescrutiny of that conclusion.

The primary structure of the extracellular polysaccharide from *X. campestris* was recently reinvestigated by Jansson *et al.* (2), and by Melton *et al.* (8). The chemical repeat unit is a pentasaccharide. The backbone is a β-1,4 glucan like cellulose and chitin (9,10) with a trisaccharide side-chain on each cellobiose unit (I).

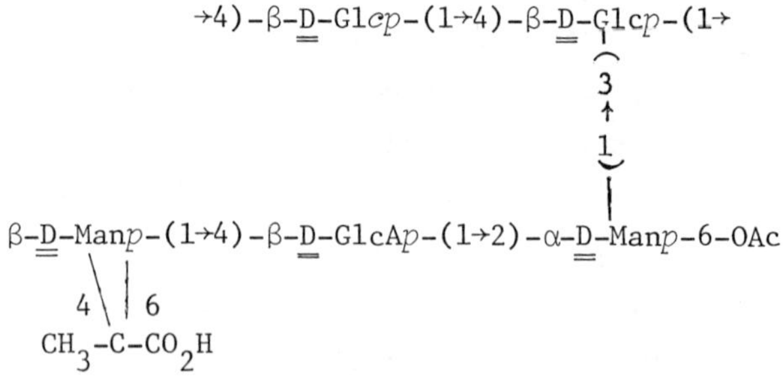

```
        →4)-β-D-Glcp-(1→4)-β-D-Glcp-(1→
                                |
                                3
                                ↑
                                1                    (I)
                                |
β-D-Manp-(1→4)-β-D-GlcAp-(1→2)-α-D-Manp-6-OAc
    4\ | 6
  CH3-C-CO2H
```

The primary structure of the *Klebsiella* K8 capsular polysaccharide was established by Sutherland (3). It is a poly(tetrasaccharide) with three sugar residues in the backbone and a monosaccharide side-chain (II).

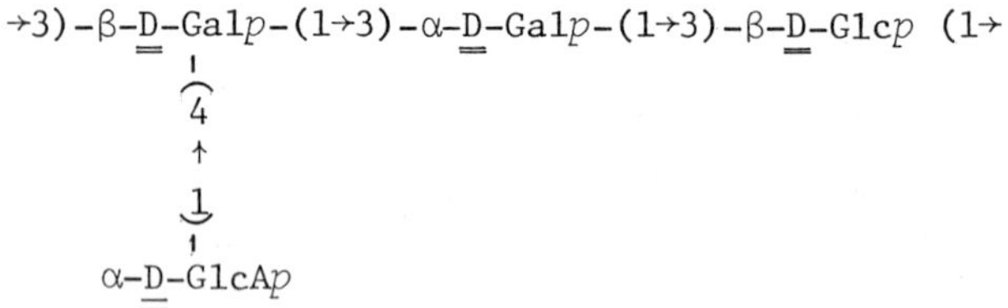

```
→3)-β-D-Galp-(1→3)-α-D-Galp-(1→3)-β-D-Glcp (1→
          |
          4                                         (II)
          ↑
          1
          |
  α-D-GlcAp
```

Molecular Structure of Xanthan

The xanthan X-ray diffraction pattern (Figure 1) indicates a system with well-oriented molecules but with limited lateral ordering. The Bragg reflections in the inner part of the pattern can be indexed on the basis of a rectangular unit cell with a=2.90nm, b=2.49nm and c=4.70nm. Reflections with $h+k+l\neq 2n$ are systematically absent. Apparently the xanthan molecules are packed together in a body-centered lattice. The more diffuse, continuous intensity on the layer lines suggests that the cyrstalline domains extend laterally only a few unit cell widths so that the xanthan fibres may be considered to be made up of fibrils as depicted in Figure 2.

The layer line spacing 4.70nm and the nodes of intensity every 5th layer line suggest a molecule with 5-fold screw symmetry and an axial translation per (cellobiose) repeat (h) of 0.94nm. This is notably shorter than the corresponding distance (1.03nm) in cellulose (9) and α-chitin (10). 5N-fold ($N\geq 2$) helix models are also possible involving N parallel, coaxial chains with special relationships between the chains. Various possibilities are given in Table I.

The body-centered lattice implies that each unit cell contains 2N chains (N = 1,2,3,...). For unit cells containing only carbohydrate residues (and no water) the calculated densities for successive values of N are 0.46, 0.92, 1.26, 1.83,... . Since the measured value is only 1.44 then clearly values of N>3 need not be considered. Even for N=3 there would be only 0.5H_2O per monosaccharide which is unlikely since the spacing between molecules shrinks 15% when the specimens are dried over silica gel. It seems that N=2 is the most likely possibility since there would be about 5 water molecules per monosaccharide as was observed for the *E. coli* capsular material (4). For N=1 the number of water molecules per monosaccharide would be 22.

Electron microscopy (6) suggests double or triple-stranded helices and hydrodynamic experiments (7) suggest that the mass per unit length of the xanthan molecule is twice that which

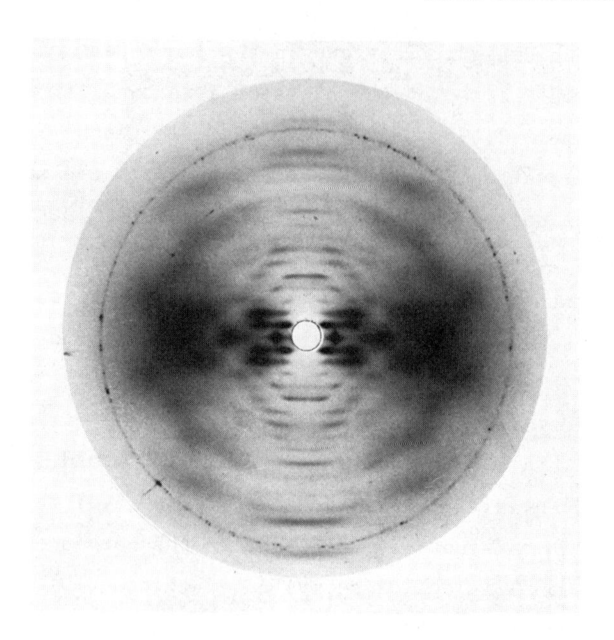

Figure 1. *Diffraction pattern from xanthan at 92% relative humidity. The pattern shows five-fold helical symmetry and fiber repeating unit of 4.7nm.*

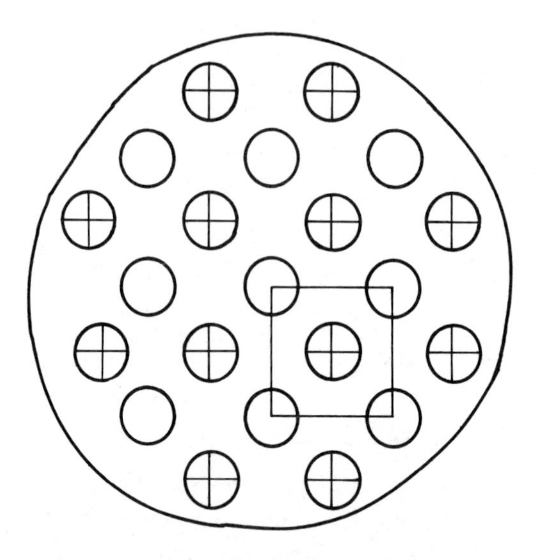

Figure 2. *Schematic of xanthan fibril. Small circle represents xanthan molecule or its assembly. The rectangle indicates the unit cell of the body-centered lattice in which molecules are packed.*

Table I. Various Possible Structures of Xanthan

Pitch	Helix symmetry	Possible structures
4.7 nm	5_1, 5_2, 5_3, 5_4	Single helix Parallel double helix no condition between chains Antiparallel double helix no condition between chains
9.4 nm	10_1, 10_3, 10_7, 10_9	Parallel double helix $\Delta w=0.0$ $\Delta\mu=180^\circ$
14.1 nm	15_1, 15_2, 15_4, 15_7, 15_8, 15_{11}, 15_{13}, 15_{14}	Parallel triple helix $\Delta w=0.0$ $\Delta\mu=\pm120^\circ$

would be expected from single helical chains of the kind we have described (5). Double helical molecules packed in body-centered lattices which in turn are organised in microcrystalline fibrils would appear to be the solution that best reconciles the X-ray diffraction data with other physical measurements. The possible double helical models are listed in Table II.

The backbone conformation of xanthan is defined by four dihedral angles at two glycosidic linkages if the pyranose rings are considered to have the fixed standard geometry (Figure 3). However, the known helical symmetry and axial rise of the disaccharide backbone repeat allow the number of degrees of freedom in the backbone to be reduced to two (11). Therefore, if the sterically allowed region for the two angles of one glycosidic bridge is divided into grid points at 5° intervals, it is possible to calculate the appropriate values of the other two bridge angles of the asymmetric unit and to determine whether they also fall in the allowed region. For the grid chosen there are 713 points. The number of these which lead to possible helical structures are shown in Table II.

No 5_2 helical model could be constructed on any of these grid points. In the case of other 5-fold helices, the single chain helices which could be constructed at several grid points were refined to minimize overshort non-bonded contacts. With the 5_3 helix, unacceptably short contacts persisted and therefore, we excluded this model from further considerations. The 5_4 chain had one short contact d(H4A···O2)=0.173nm.

The isolated 5_1 helices were free of any objectionable contacts. Since there is no special relationship between two chains in the 5-fold double helix, the relative orientation parameters, $\Delta\mu$ and ΔW, were obtained from the contacts map for each conformation. For both 5_1 and 5_4 double helices, conformations without fatal short contacts could be constructed. These double helices were subjected to a packing analysis in the

Table II. Exploration of Possible Double Helical Models for Xanthan

Double helical Models	Number of Grid Points	Non-bonded Contacts within double helix	Non-bonded Contacts between double helices
5_1 Parallel	255	good	good
5_1 Antiparallel	255	good	good
5_2 Parallel	0		
5_2 Antiparallel	0		
5_3 Parallel	76	bad	
5_3 Antiparallel	76	bad	
5_4 Parallel	450	good	bad
5_4 Antiparallel	450	good	bad
10_1 Parallel	276	good	bad
10_3 Parallel	60	bad	
10_7 Parallel	271	bad	
10_9 Parallel	552	bad	

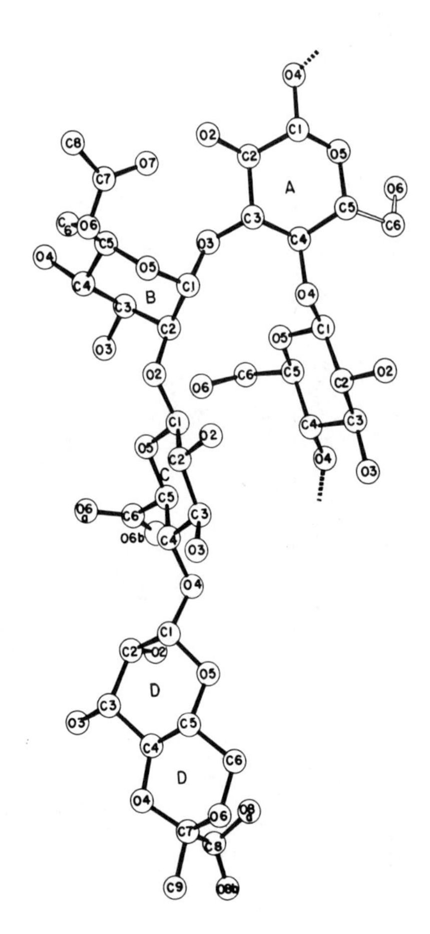

Figure 3. The pentasaccharide repeating unit of xanthan showing the atom labeling. The unlettered residue and residue A are β-D-glucose; B is α-D-mannose; C is β-D-glucuronate; and D is 4,6-O-(1-carboxyethyl)-β-D-mannose.

observed unit cell. For both parallel and antiparallel 5_4 helices there were numerous short contacts between molecules because of the large radius of the chains. Minimization did not relieve these contacts at all.

In the case of 10-fold helices, the two chains in a double helix would be related by 2-fold symmetry coincident with the helix axis. Starting from several grid points, the intra- and inter-molecular short contacts within a double helix were minimized. Only the double helix model with 10_1 chains could be constructed without any fatal contacts. However, we could not pack this model successfully in the unit cell because of the large helical radii for several atoms. The results of these calculations are summarized in Table II.

Thus, the parallel and antiparallel 5_1 double helices were the only models to survive. The conformations of these two models are illustrated in Figure 4. It is interesting that in both the parallel and antiparallel models the molecular conformations are very similar. In both cases the following intra-molecular hydrogen bonds were found: O5A···O3, O6A···O3, O3B···O5C, and O7B···O2A. One hydrogen bond O6A···O6A was found in the antiparallel structure as an inter-chain hydrogen bond within a double helix. The parallel structure also had two short contacts between double helices, d(C6D···H8B) = 0.186nm and d(H6D···H8B) = 0.11nm. It is possible that these short contacts together with the short intra-molecular hydrogen bond O5A···O3 (0.234nm) in both structures could be improved. For the moment the antiparallel 5_1 double helix appears to be the more successful model.

Structure of *Klebsiella* serotype K8 Capsular Polysaccharides

The X-ray diffraction pattern from the sodium salt of the *Klebsiella* serotype K8 capsule is shown in Figure 5. The

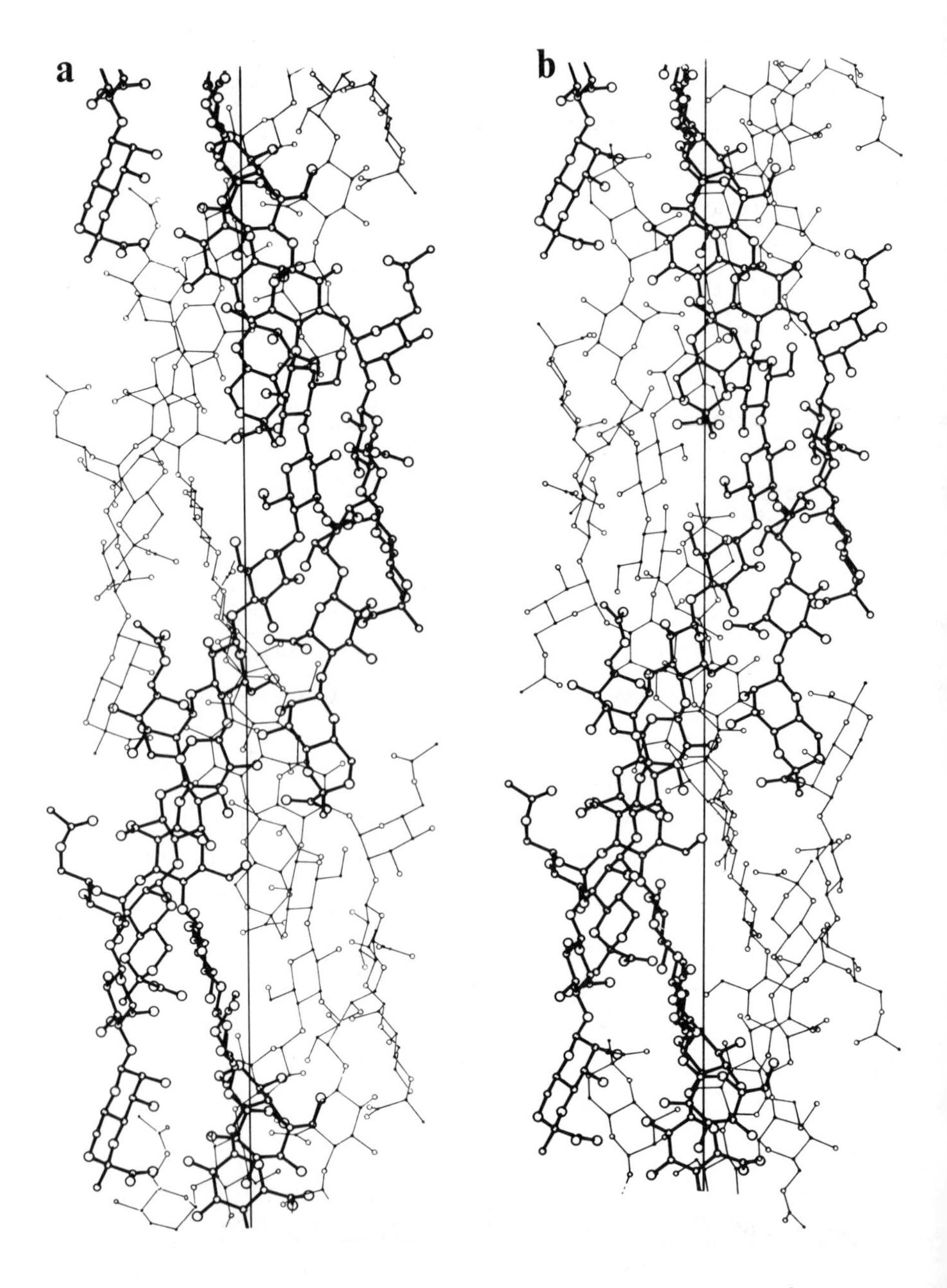

Figure 4. The 5_1 (a) parallel and (b) antiparallel double helices viewed perpendicular to the helix axis

reflections were indexed on the basis of an orthorhombic cell with a=1.473nm, b=1.254nm, c(fiber axis)=5.085nm. Even order meridional reflections ($00l$) suggest a two-fold screw axis along the fiber. The corresponding axial rise of 2.543nm is much too large to be attributed to one chemical asymmetric unit but has the value expected for two of them.

From the absence of odd reflections on ($h00$), ($0k0$) and ($00l$), the possible space groups are P222, $P222_1$, $P2_12_12$ and $P2_12_12_1$. In each case the number of chains in the unit cell would be a multiple of 4 if the molecules were in the general positions, but could be 2 when the molecular 2_1 axis coincided with a 2_1 space group axis. Since the number of chains calculated from the unit cell dimensions and the observed density (1.52 g/ml) is certainly no more than 3, the number of chains in the unit cell must be 2. Three space groups $P222_1$, $P2_12_12$ and $P2_12_12_1$ are possible. The packing schemes of $P222_1$ and $P2_12_12$, however, cannot explain the observed strong (110) intensity. Therefore we adopted $P2_12_12_1$ as the most probable space group. The two molecules in the unit cell are then related by symmetry, consequently only two packing parameters (μ and w) defining the orientation and height of one molecule in the unit cell are needed.

The *Klebsiella* K8 polysaccharide has another modification, in which the molecule has a four-fold screw symmetry and packs tetragonally. Since the intensity distribution in the diffraction pattern of the tetragonal form is very similar to that in the orthorhombic form, we assumed (as a first approximation) that the molecule has a four-fold helical (4_1 or 4_3) symmetry, the same in both crystal forms.

The main chain conformation of this polysaccharide is defined by six dihedral angles at three glycosidic linkages (Figure 6). In order to get allowed regions for these dihedral angles, disaccharide contact maps for each glycosidic linkage

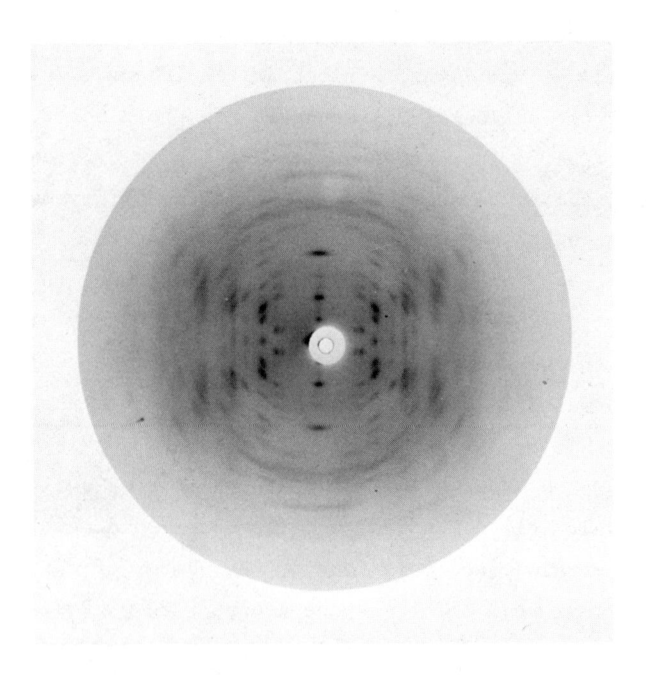

Figure 5. Diffraction pattern from the capsular polysaccharide of Klebsiella *serotype K8 at 90% relative humidity (provided by Dr. E. D. T. Atkins, University of Bristol, England)*

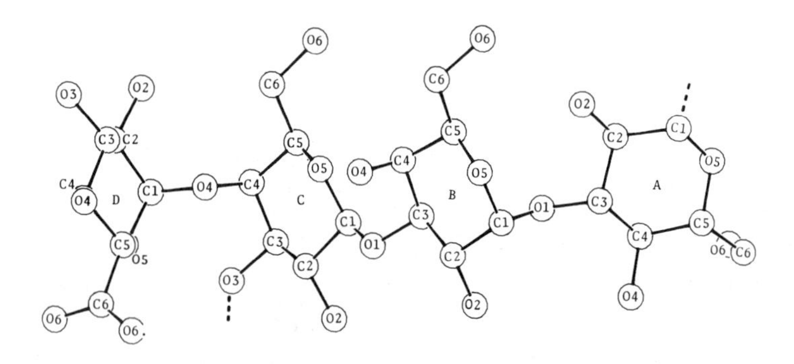

Figure 6. The tetrasaccharide repeating unit of Klebsiella *K8 polysaccharide showing the atom labeling. The residue A is β-D-glucose; B is α-D-galactose; C is β-D-galactose; and D is α-D-glucuronate.*

were calculated. Giving the middle values of allowed regions as a starting set, molecular models for 4_1 and 4_3 helices were constructed by LALS (12). Dihedral angles for these models are listed in Table III together with intra-molecular hydrogen bonds. Model 1 has the 4_1 helical symmetry. Models 2 and 3 have the 4_3 helical symmetry. All three glycosidic linkages of the backbone in Model 1, two in Model 3 and one in Model 2 are reinforced by hydrogen bonds acorss the linkages. By the criterion of maximum number of hydrogen bonds, Model 1 seems to be the best among them. There is no hydrogen bond possible between α-D-glucuronate and β-D-galactose.

Changing packing parameters μ and w stepwise, the inter-molecular contacts were examined for each model and crystallographic *R* factors calculated. Compared with Model 2 and 3, *R* for Model 1 was very high (more than 60%) at all the positions which had no fatal inter-molecular contacts. Further, according to the conformational calculations mentioned in the previous section, the allowed conformational space for 4_1 helices is extremely narrow compared with that for 4_3 helices. For these reasons, we discarded the 4_1 model at this stage. Model 2 gave a low value of *R* (37%) only at position $\mu=75^o$ and w=0.20 within the allowed region. *R* for Model 3 was above 45% at all positions in the allowed region. During these calculations, an intra-molecular hydrogen bond O4B···O5C had been imposed. When this was release, *R* decreased from 46% to 42% at $\mu=15^o$ and w=0.35. This structure was examined further by Fourier analysis and plausible positions of sodium and water molecules were found, reducing *R* to 28%. The molecular structure is shown in Figure 7.

Fourier analysis of Model 2 is still in progress. However, from the above experiments it is clear that the intra-molecular hydrogen bond scheme in *Klebsiella* K8 polysaccharide does not have hydrogen bonds strengthening every glycosidic linkage.

Table III. Molecular Models of *Klebsiella* K8

Parameters	4_1 helix	4_3 helix	
	Model 1	Model 2	Model 3
θ(C2C-C3C-O3A-C1A)	-84.9	-133.0	-113.5
θ(C3C-O3A-C1A-C2A)	176.2	171.7	170.7
θ(C2A-C3A-O1B-C1B)	-74.1	-95.9	-91.0
θ(C3A-O1B-C1B-C2B)	-138.1	-155.7	-151.1
θ(C2B-C3B-O1C-C1C)	-74.5	-149.2	-176.8
θ(C3B-O1C-C1C-C2C)	-139.9	158.2	157.8
θ(C5C-C4C-O4C-C1D)	156.8	146.1	150.4
θ(C4C-O4C-C1D-C2D)	-108.4	-124.3	-110.1
Intra-molecular hydrogen bonds	O2A···O5B	O2A···O5B	O2A···O5B
	O4A···O6C	O2A···O6B	O2A···O6B
	O2B···O5C		O4B···O5C

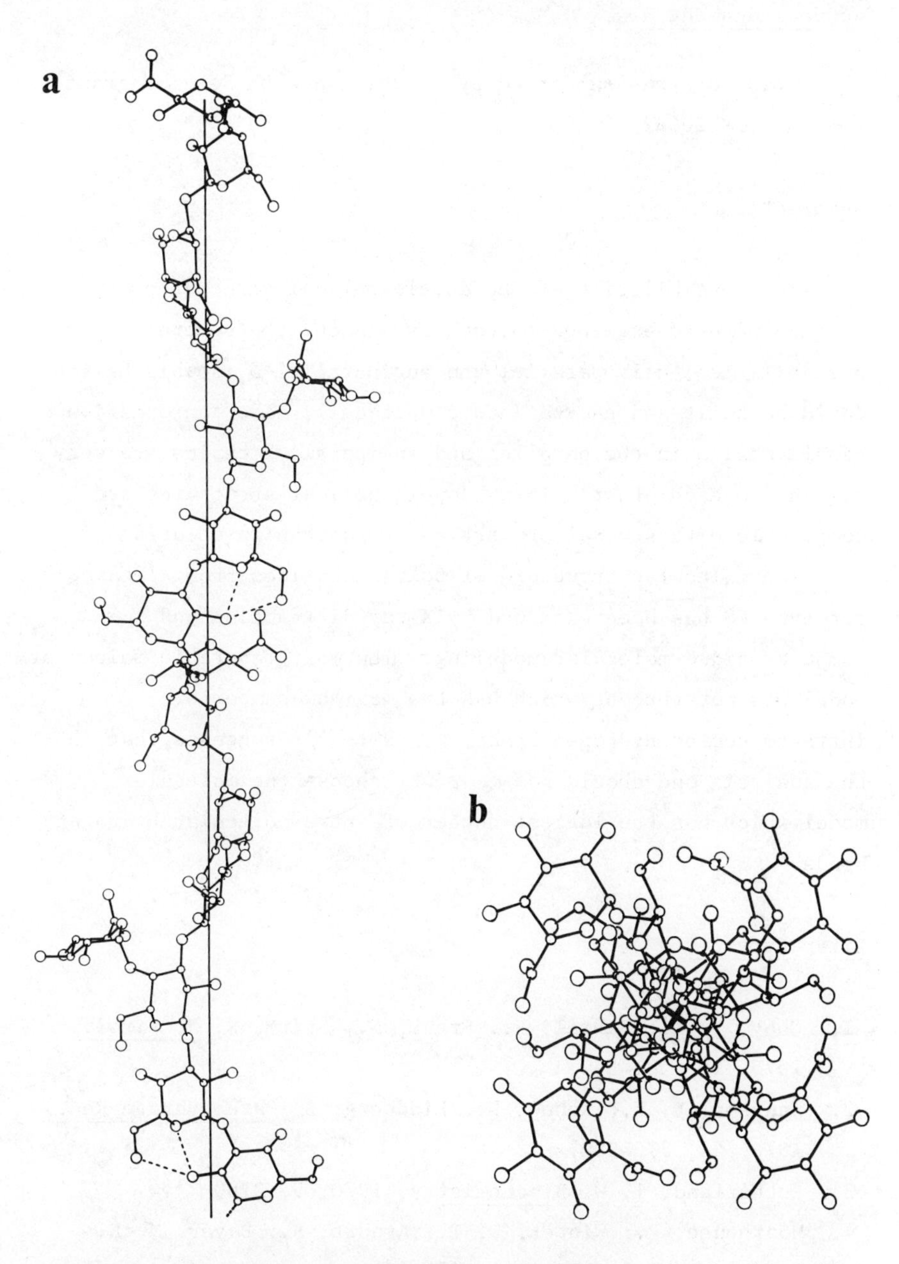

Figure 7. The 4_3 helix of Klebsiella *K8 capsular polysaccharide viewed (a) perpendicular to and (b) down the helix axis*

Acknowledgement

This work was supported by a Public Health Service grant to S.A. (GM 20682).

Abstract

The possibilities of the double helical structures of xanthan were re-examined thoroughly. Among the several possibilities, only parallel and antiparallel 5_1 double helices could be built and packed into a unit cell. The conformation of single chains in the parallel and antiparallel chains are very similar to each other. These double helical structures are compatible with several properties of xanthan in solution.

The molecular structure of polysaccharides from *Klebsiella* serotype K8 has been examined by X-ray diffraction and computer-aided molecular modeling. The most favorable molecular model was not the one which had the maximum number of intra-molecular hydrogen bonds. This result suggests that in the analysis one should not *a priori* choose the molecular model which has the largest number of intra-molecular hydrogen bonds.

Literature Cited

1. Choy, Y. M.; Fehmel, F.; Frank, N.; Stirm, S. J. Virol., 1975, 16, 581-590.
2. Jansson, P. E.; Kenne, L.; Lindberg, B. Carbohydrate Res., 1976, 45, 275-282.
3. Sutherland, I. W. Biochemistry, 1970, 9, 2180-2185.
4. Moorhouse, R.; Winter, W. T.; Arnott, S.; Bayer, M. E. J. Mol. Biol., 1977, 109, 373-391.
5. Moorhouse, R.; Walkinshaw, M. D.; Winter, W. T.; Arnott, S. "Cellulose Chemistry and Technology"; Arthur, J. C., Ed. American Chemical Society: Washington, D.C., 1977, p. 133.

6. Holzwarth, G.; Prestridge, F. G. Science, 1977, 197, 757-759.

7. Holzwarth, G. Carbohydrate Res., 1978, 66, 173-186.

8. Melton, L. D.; Mindt, L.; Rees, D. A.; Sanderson, G. R. Carbohydrate Res., 1976, 46, 245-257.

9. Marchessault, R. H.; Sarko, A. Advan. Carbohydrate Chem., 1967, 22, 421-482.

10. Carlström, D. J. Biophys. Biochem. Cytol., 1957, 3, 669-683.

11. Go, N.; Okuyama, K. Macromolecules, 1976, 9, 867-868.

12. Smith, P.J.C.; Arnott, S. Acta Cryst., 1978, A34, 3-11.

RECEIVED February 19, 1980.

Review of the Structures of *Klebsiella* Polysaccharides by X-ray Diffraction

H. F. ELLOWAY[1], D. H. ISAAC[2], and E. D. T. ATKINS

H. H. Wills Physics Laboratory, University of Bristol, Bristol, BS8 1TL U.K.

The polysaccharides which will be considered in this paper are all found in the extracellular capsular material which surrounds the cell wall of bacteria belonging to the Genus *Klebsiella*. The carbohydrate nature of these capsules was first demonstrated in 1914 by Toenniessen (1) and in 1926 Julianelle (2) was the first person to report on the presence of sharply defined and high specific types of polysaccharides in the different serotypes.

To date 83 different serotypes have been delineated and a complete investigation of their carbohydrate composition has been undertaken by Nimmich, (3, 4). He found that all consist of a regularly repeating unit of between three and seven saccharide residues which are chemically different in each strain. He further found that all are acidic, the predominant charged groups being the uronic carboxyl groups, several having, in addition, pyruvate groups. Figure 1 shows the variety of different structural patterns that have emerged from those *Klebsiella* polysaccharides whose chemical structure has been fully elucidated. This figure has been adapted from a table originally proposed by Dr. G.G.S. Dutton (personal communication). It should be noted that pyruvate groups have only been included in the case of those serotypes that contain no uronic acid residues. In fact only four serotypes without uronic acid have so far been found, and of these, three contain pyruvate groups and one, *Klebsiella* K38 is believed to be unique in the polysaccharide field, in containing 3-deoxy-L-*glycero*-pentulosonic acid residues (5). From this table the varying degrees of

[1]current address: Department of Physics, University of Keele, Keele, Staffs, ST5 5BG, U.K.

[2]current address: Department of Metallurgy and Materials Technology, University College of Swansea, Swansea, SA2 8PP, U.K.

0-8412-0589-2/80/47-141-429$07.50/0

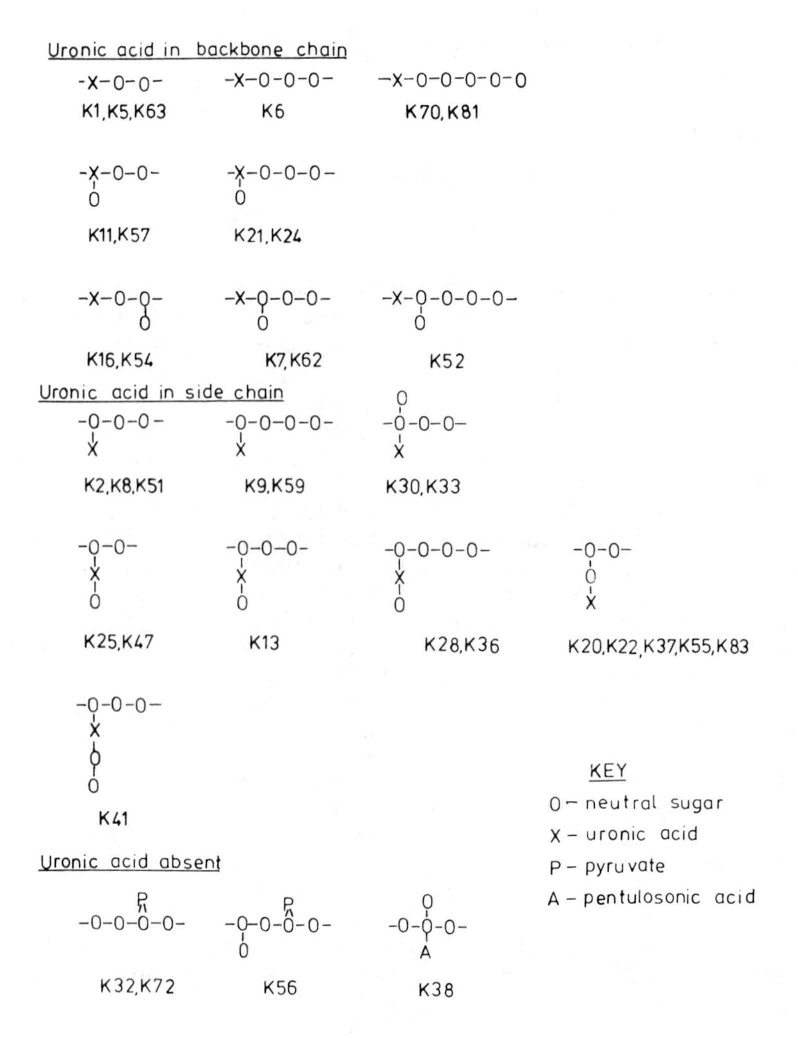

Figure 1. Variety of structural patterns found among Klebsiella *bacterial polysaccharides*

complexity found amongst the different serotypes is clearly seen, some consisting of linear chains, others having mono-, di-, or tri-saccharide side chains and a few having two side chains attached to the same saccharide. Within each group of serotypes which have the same basic structural pattern, there is a variety in the saccharide components and substituents and types of glycosidic linkage found. Indeed some novel types of glycosidic linkage geometry have emerged which will be considered individually later. Figure 2 shows both the different mono-saccharide residues, and also non-carbohydrate components, that have so far been recognised as constituents of these hetero-polysaccharides.

The examination of the possible molecular conformations exhibited by these polysaccharides is particularly pertinent and interesting in view of the fact that some of their biological functions are believed to be highly dependent on their individual conformations. One particular example of this concerns the specific antigenic properties exhibited by the capsular material. It is believed that the specificity of the antigens is dominated by the uronic acid residues present and that the different antigenic properties of the various polysaccharides are due to differences in the steric arrangements of these residues, resulting from their differing compositions (6). A second interesting area concerns bacteriophage attack. It is known that certin Klebsiella bacteriophage depolymerise specific Klebsiella serotypes (7) and it is hoped that detailed investigation of the three-dimensional structure of these polysaccharides and, in particular the spatial distribution of particular groups in relation to the glycosidic linkages being cleaved, will lead to a greater understanding of this bacteriophage attack.

Over the last few years a large number of different Klebsiella polysaccharides have been investigated using the techniques of X-ray diffraction followed, where appropriate, by computational analysis. In this paper we have endeavoured both to mention individually some of these polysaccharides that have particularly interesting conformational features and also to summarise some of the more general trends that have emerged from a study of this group of bacterial polysaccharides. It is pertinent to point out at this stage that each model shown in this paper is not necessarily the 'best' model for the particular polysaccharide, rather it is one of a number of energetically favourable models. Details of both the model building criteria used and computational techniques employed are given elsewhere (8).

Klebsiella K5 and Klebsiella K63

The chemical structure of Klebsiella K5, shown in Figure 3, was established by Dutton and Mo-Tai Yang (9). They found it

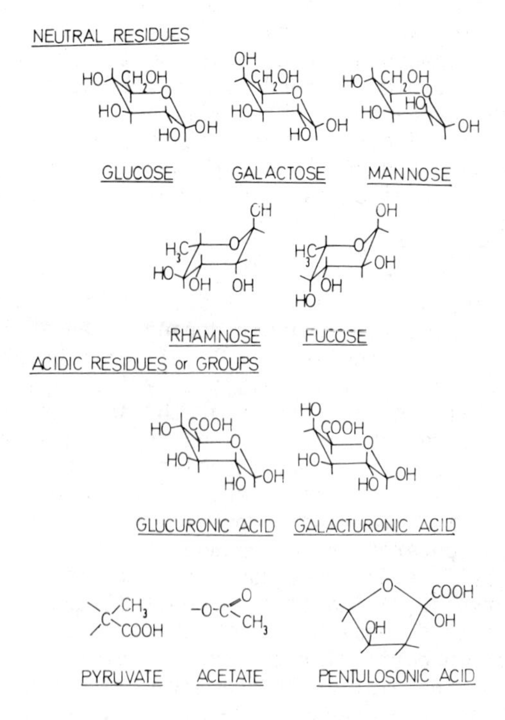

Figure 2. Monosaccharide constituents and noncarbohydrate components (acetate, pyruvate and pentulosonic acid) of Klebsiella *bacterial polysaccharides*

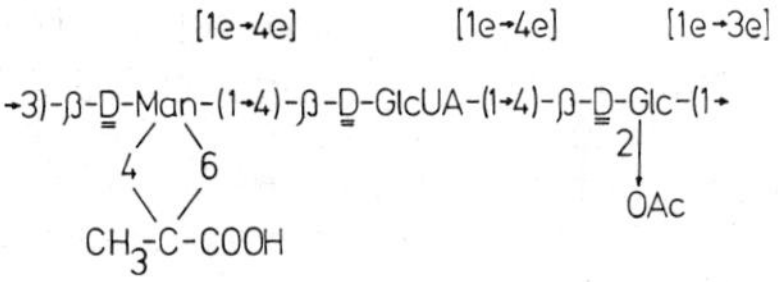

[1e→4e] [1e→4e] [1e→3e]

→3)-β-D-Man-(1→4)-β-D-GlcUA-(1→4)-β-D-Glc-(1→

4, 6: CH_3-C-COOH

2: OAc

Figure 3. Repeating chemical structure for Klebsiella *K5*

consisted of a linear trisaccharide repeating unit which contained in addition both pyruvate and O-acetyl side appendages. X-ray diffraction studies showed that the polysaccharide formed an extended 2-fold helix and three projections of the most favourable model determined from model building calculations are shown in Figure 4 (10). This conformation is both stereochemically acceptable and also allows the incorporation of three stabilising intrachain hydrogen bonds, one across each glycosidic linkage. It was found that the torsional angles at the 1, 3 linkage disposed the charged pyruvate group on the radial periphery of the molecule and in close proximity to the O-acetyl group, as is clearly seen in the second projection along the helix axis in Figure 4. If, as has been suggested, the non-carbohydrate constituents such as O-acetyl and pyruvate groups function as antigenic determinants, the close association of these groups may represent a determinant site.

One further point of interest concerning *Klebsiella* K5 is that the O-acetyl groups appear to have *little effect* on the helix stability. X-ray diffraction patterns taken of deacetylated material showed that the pitch and type of helix were both identical to those for the acetylated material and further, model building of a deacetylated chain gave rise to the same backbone conformation and system of hydrogen bonds (10).

It is interesting to compare the structure of *Klebsiella* K5 with that of *Klebsiella* K63 which, as can be seen *from figure* 1, is the only *other linear* trisaccharide so far found in the *Klebsiella* group. The chemical structure of K63 as determined *by J-P. Joseleau*, Grenoble (personal communcation) is shown in Figure 5 and Figure 6 shows an X-ray diffraction pattern obtained (in collaboration with Dr. H. Chanzy, Grenoble) from this material. This pattern showed that, like K5 polysaccharide, K63 polysaccharide also forms an extended 2-fold helix. Figure 7 shows three projections of the most favourable model found for K63 (11). From the projection down the axis the position of the carboxyl group on the periphery of the helix is clearly seen, occupying a similar position to that of the pyruvate group in *Klebsiella* K5 (Figure 4). A further feature of this model of *K63 is its* similarity to the α 1, 3 glucans, since all its saccharide residues are also α 1, 3 linked (Figure 5). In particular the occurence of both O(5)••H-O(2) and O(6)••H••O(2) hydrogen bonds between the adjacent galacturonic acid and galactose residues in K63 and consecutive galactose residues in the model for the α 1, 3 glucan polymorph I proposed by Jelsma (12) is notable.

Klebsiella K16 and *Klebsiella* K54

An interesting comparison is also possible between *Klebsiella* K16 nad *Klebsiella* K54. The chemical structure of

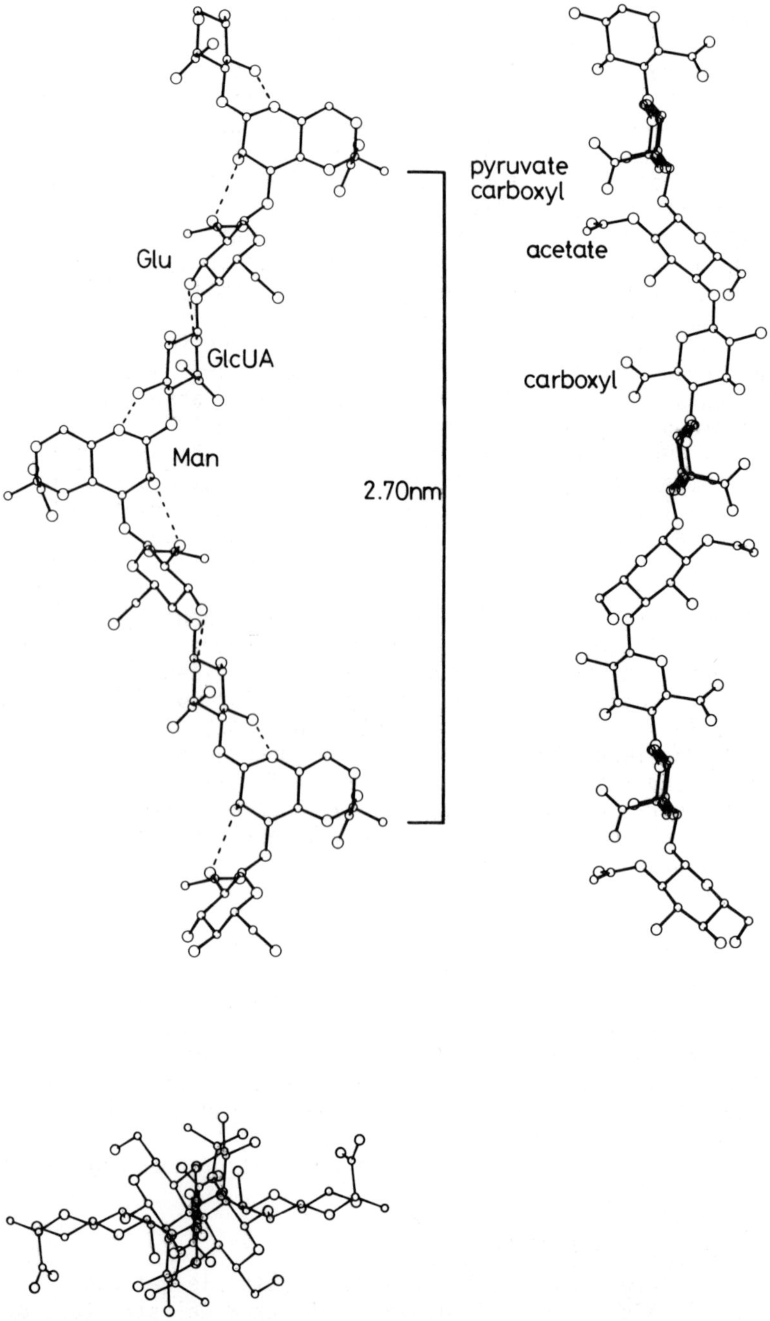

pyruvate carboxyl
acetate
Glu
GlcUA
carboxyl
Man
2.70nm

Figure 4. Computer-generated projections of a two-fold helix of Klebsiella *K5*

[1a→3e] [1a→3e] [1a→3e]

→3)-α-D-Gal-(1→3)-α-D-GalUA-(1→3)-α-L-Fuc-(1→

Figure 5. Repeating chemical structure for Klebsiella *K63*

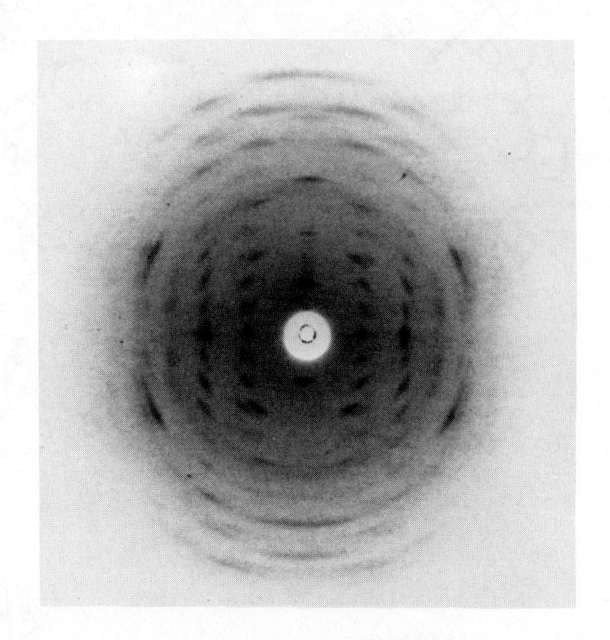

Figure 6. X-ray fiber diffraction pattern obtained (in conjunction with Dr. H. Chanzy) from an oriented film of the sodium salt of Klebsiella *K63*

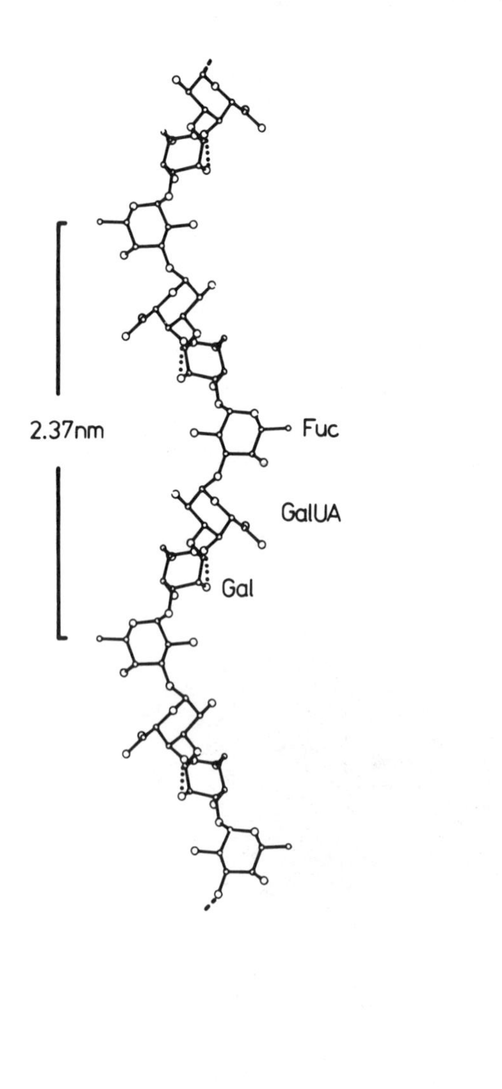

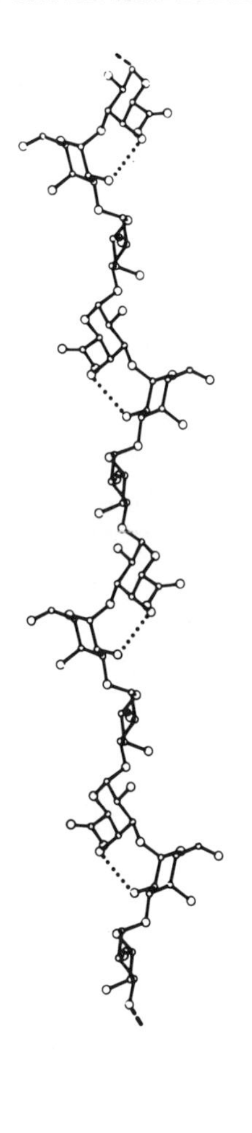

Figure 7. Computer-generated projections of a two-fold helical conformation for Klebsiella *K63*

K16 (Figure 8) was established by Chakraborty and Niemann, Freiburg (Stirm, private communication) and that of K54 (Figure 9) was determined by Conrad et al (13). From Figure 1 it is evident that both have the same structural pattern consisting of a tri-saccharide backbone, which includes a uronic acid residue, and a monosaccharide side chain. X-ray diffraction studies showed that both formed extended 3-fold helical structures (14) Model building calculations indicated that in K16 left-handed helices were preferred and three projections of one such model are shown in Figure 10. From the projection down the helix axis it is particularly significant to note the compact nature of the central core. In the case of K54, on the other hand, right-handed helices were found to be preferable. An examination of the projection down the axis of one such model shown in Figure 11 shows that in this instance the molecule has a hollow central core into which α-L-fucose O(2) atoms protrude. It is likely that small cations and/or water molecules will pack into this region. A further feature of K54 is the incorporation of a 1, 6 linkage in its backbone geometry, which thus introduces a third variable torsion angle at this linkage. In model building calculations the additional flexibility was reduced by restricting the O(6) conformation to be one of three acceptable positions; gauche-gauche, gauche-trans and trans-gauche. It was found to be impossible to build stereochemically acceptable left-handed helices in any of these three O(6) conformations and right-handed helices could only be built if the O(6) atom was placed in the gauche-gauche conformation (11). This conformation is similar to those found in single crystal structures of the 1, 6 disaccharide, α-melibiose (15) and the 1, 6 trisaccharide raffinose (16).

Klebsiella K38

K38 is another Klebsiella polysaccharide that incorporates a 1, 6 linkage in its backbone and in addition has the distinctive feature of incorporating a 3-deoxy-L-glycero-pentulosonic acid residue in its asymmetric unit (5) (Figure 12). Figure 13 is a typical X-ray diffraction pattern obtained from this material and shows that the molecule forms an extended 2-fold helical structure. Model building calculations showed that in the case of this polysaccharide the gauche-trans conformation was preferred for the O(6) atom of β-D-glucose at the 1, 6 linkage. Three projections of the most stereochemically favourable model found in these calculations for K38 are shown in Figure 14.

Klebseilla K8

Klebsiella K8 provides an example of a 4-fold helical structure as is apparent from the X-ray diffraction pattern

[1a→4e] [1e→4a] [1a→3e]

→3)-α-D-Glc-(1→4)-β-D-GlcUA-(1→4)-α-L-Fuc-(1→
4↑1
β-D-Gal

Figure 8. Repeating chemical structure for Klebsiella *K16*

[1e→4e] [1a→3e] [1a→6]

→6)-β-D-Glc-(1→4)-α-D-GlcUA-(1→3)-α-L-Fuc-(1→
4↑1
β-D-Glc

Figure 9. Repeating chemical structure for Klebsiella *K54*

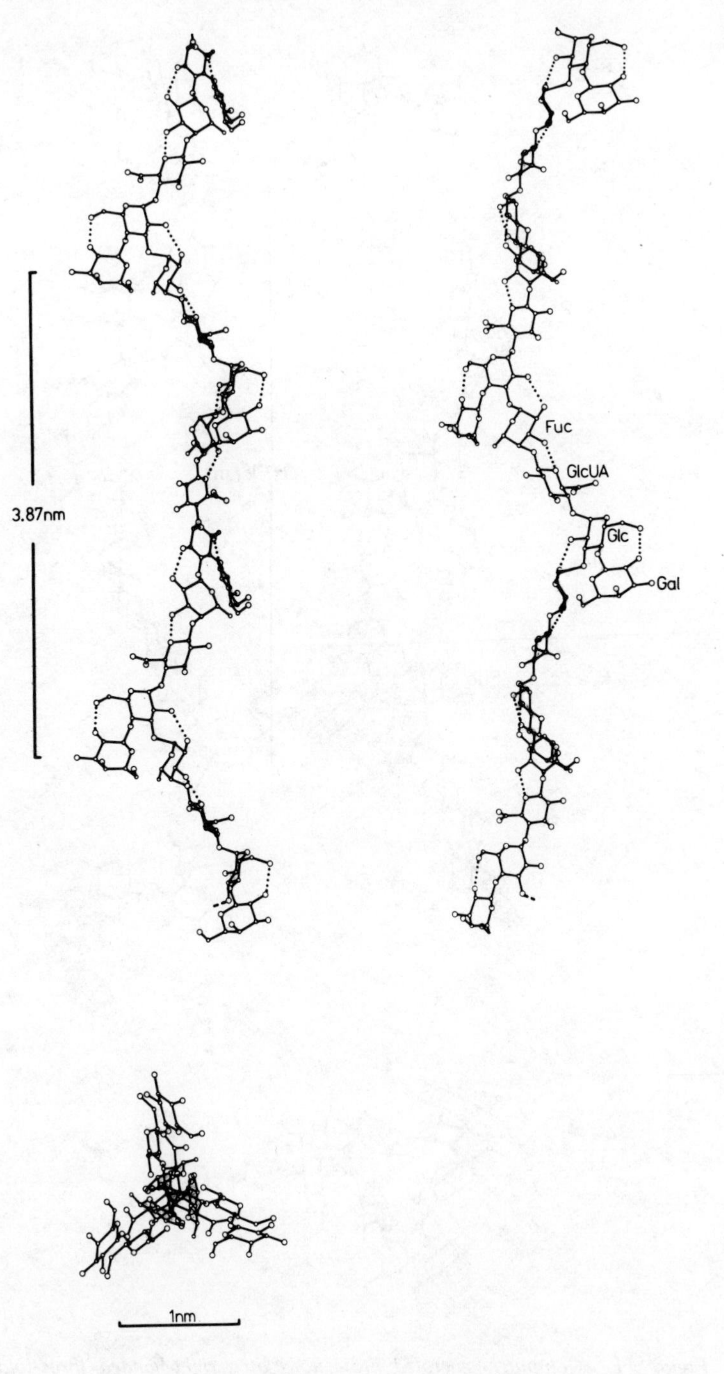

Figure 10. Computer-generated projections of a left-handed three-fold helical conformation for Klebsiella *K16*

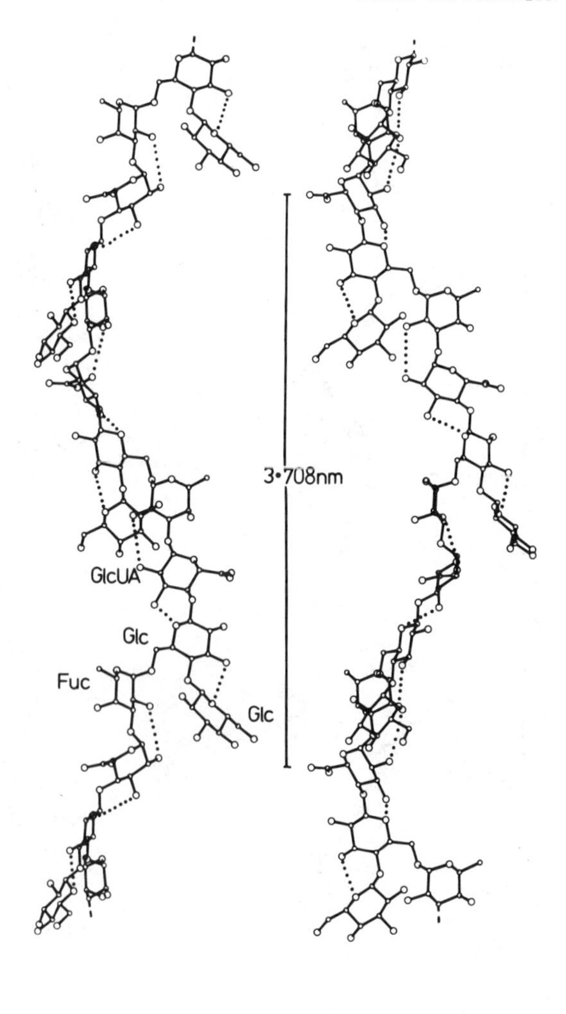

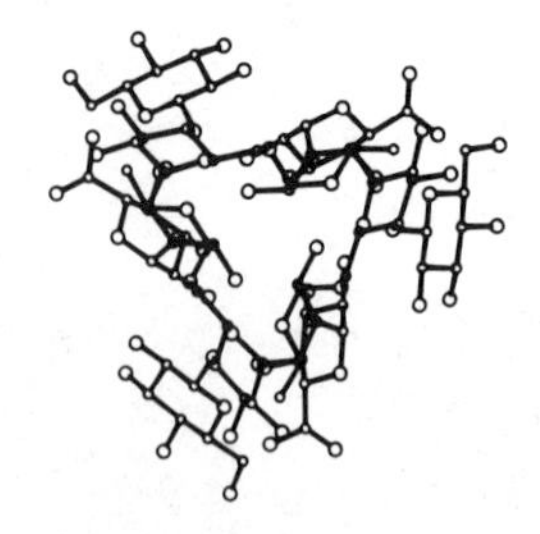

Figure 11. Computer-generated projections of a right-handed three-fold helical conformation for Klebsiella *K54*

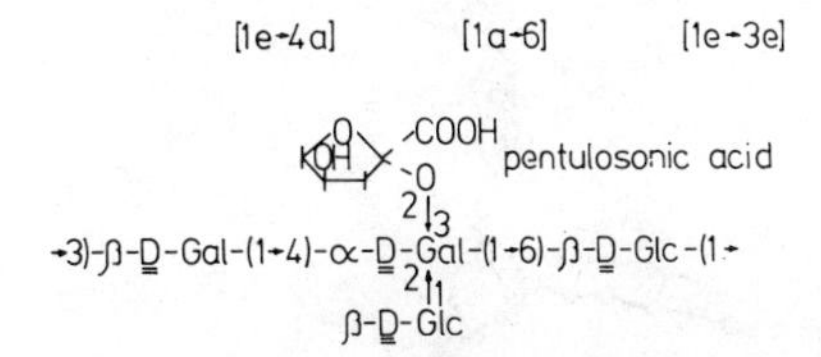

Figure 12. Repeating chemical structure for Klebsiella *K38*

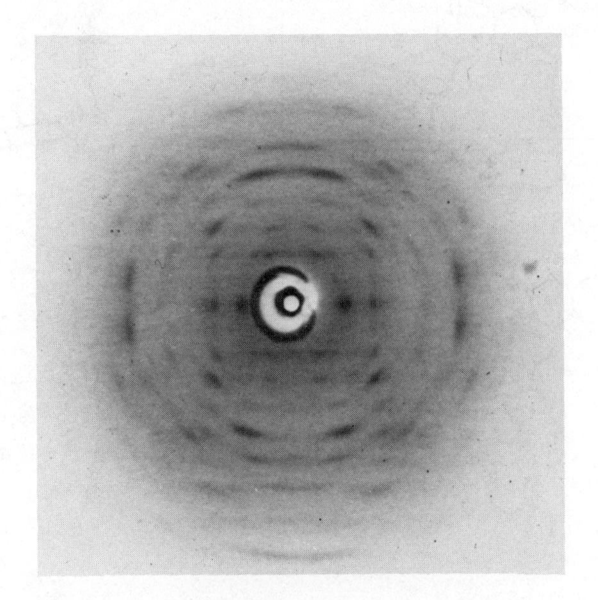

Figure 13. X-ray fiber diffraction pattern obtained from an oriented film of the sodium salt of Klebsiella *K38*

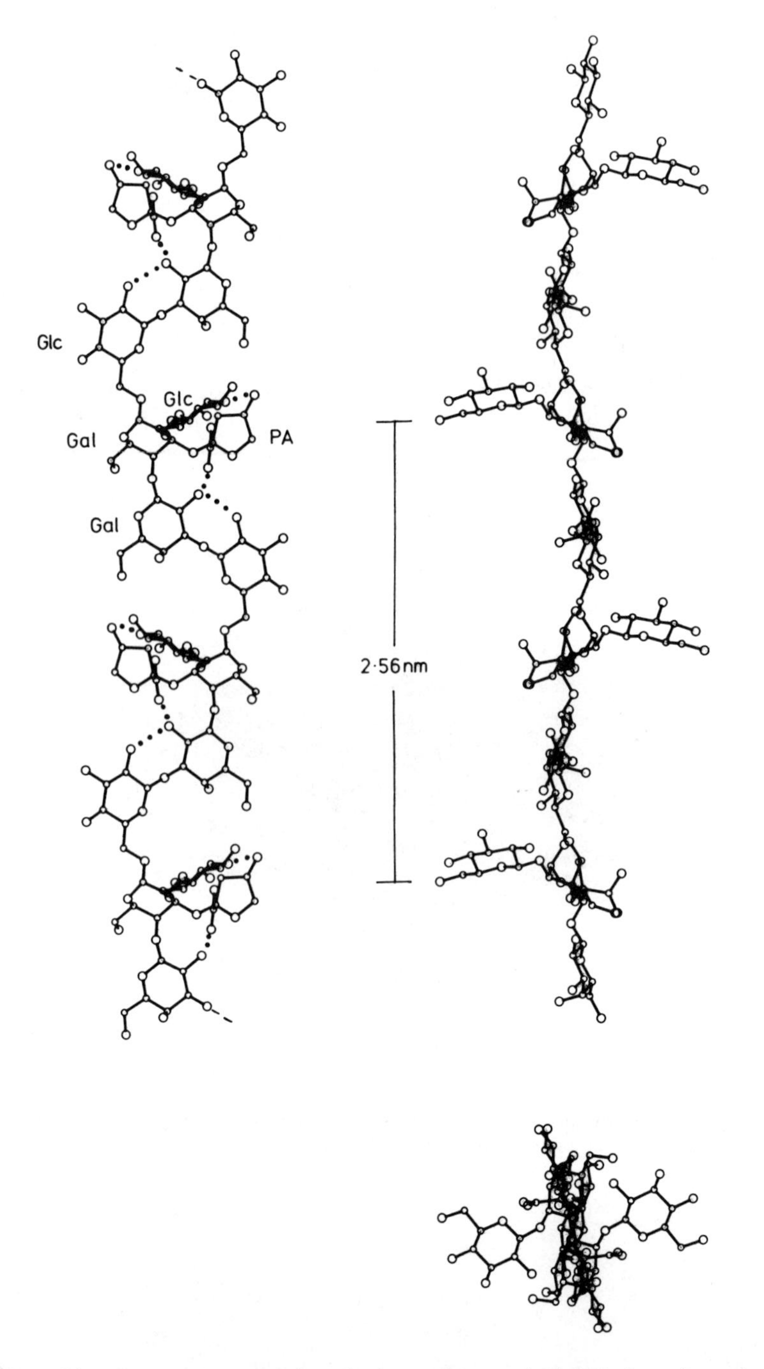

Figure 14. Computer-generated projections of a two-fold helical conformation for Klebsiella *K38*

obtained from the material shown in Figure 15 (14). The chemical of this polysaccharide was determined by Sutherland (17), Figure 16. In model building calculations left-handed helices were preferred as they allowed the incorporation of hydrogen bonds across two of the backbone linkages. Three projections of one such model are shown in Figure 17. Model building could not determine the position of the uronic acid side chain - there was no possibility of forming a hydrogen bond across the glycosidic linkage that defines its orientation. However refinement of the model against the observed X-ray data indicated a high sensitivity to the orientation of the uronic acid side chain and favoured the side chain being folded down along the backbone as shown in Figure 17 (14).

Klebsiella K83 and Klebsiella K55

Klebsiella K83 is the only 6-fold helical structure that has so far been found amongst this group of polysaccharides. In fact the X-ray diffraction patterns indicated that the molecule formed a 3-fold helical structure (D. Meader, Bristol, unpublished work). However the measured spacings are far too large for a repeat of three asymmetric units - in fact they are close to the theoretical maximum for six complete covalent repeats. This led to the conclusion that the structure is in fact a perturbed 6-fold helix - perhaps slight differences in the orientation of the side chain bringing up the additional meridional reflections. Both left- and right-handed helices were sterically acceptable, and three projections of a possible left-handed model which allows the incorporation of three stabilising intrachain hydrogen bonds are shown in Figure 18. (D. Meader, Bristol, unpublished work).

It is of interest to note here some of the similarities and differences between the chemical structures of Klebsiella K83, determined by Lindberg and Nimmich (18) and Klebsiella K55, determined by Bebault and Dutton (unpublished work), shown in Figures 19 and 20 respectively. In K55 the backbone has a glucose residue in place of the galactose residue in K83, but it is important to note that this does not affect the type of backbone linkage geometry which is identical in both polysaccharides. The other difference between the two chemical structures is the presence of an O-acetyl group attached to the backbone α-L-rhamnose residue in K55 that is absent in K83. From X-ray diffraction studies it is clear that these differences, and probably in particular the presence of the O-acetyl group, change the type of helix preferred in K55 from a perturbed 6- to a perturbed 4-fold. Two projections of a possible 4-fold helical model are shown in Figure 21. (D. Meader, Bristol, unpublished work). It is pertinent to recall here than in the case of Klebsiella K5, mentioned earlier in this paper, X-ray diffraction

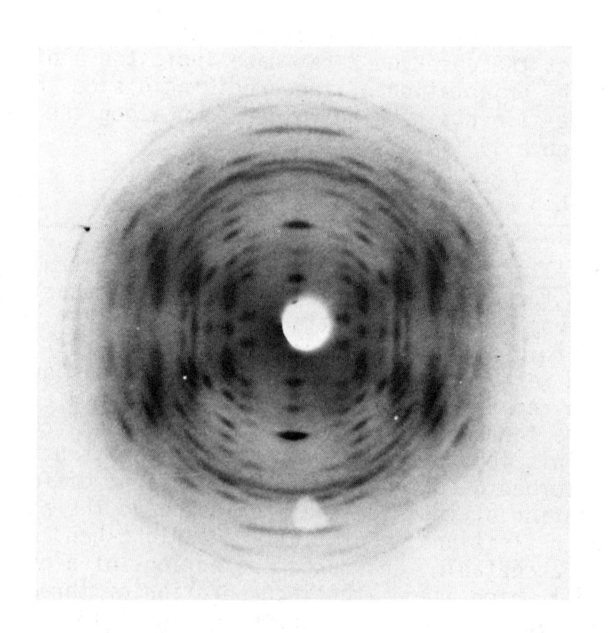

Figure 15. X-ray fiber diffraction pattern obtained from an oriented film of the sodium salt of Klebsiella *K8. It is evident that there is a slight perturbation from perfect four-fold symmetry giving rise to some weaker forbidden meridional reflections.*

Figure 16. Repeating chemical structure for Klebsiella *K8*

[1e→3e] [1a→3e] [1e→3e]

→3)-β-D-Gal-(1→3)-α-D-Gal-(1→3)-β-D-Glc-(1→
4
↑
1
α-D-GlcUA

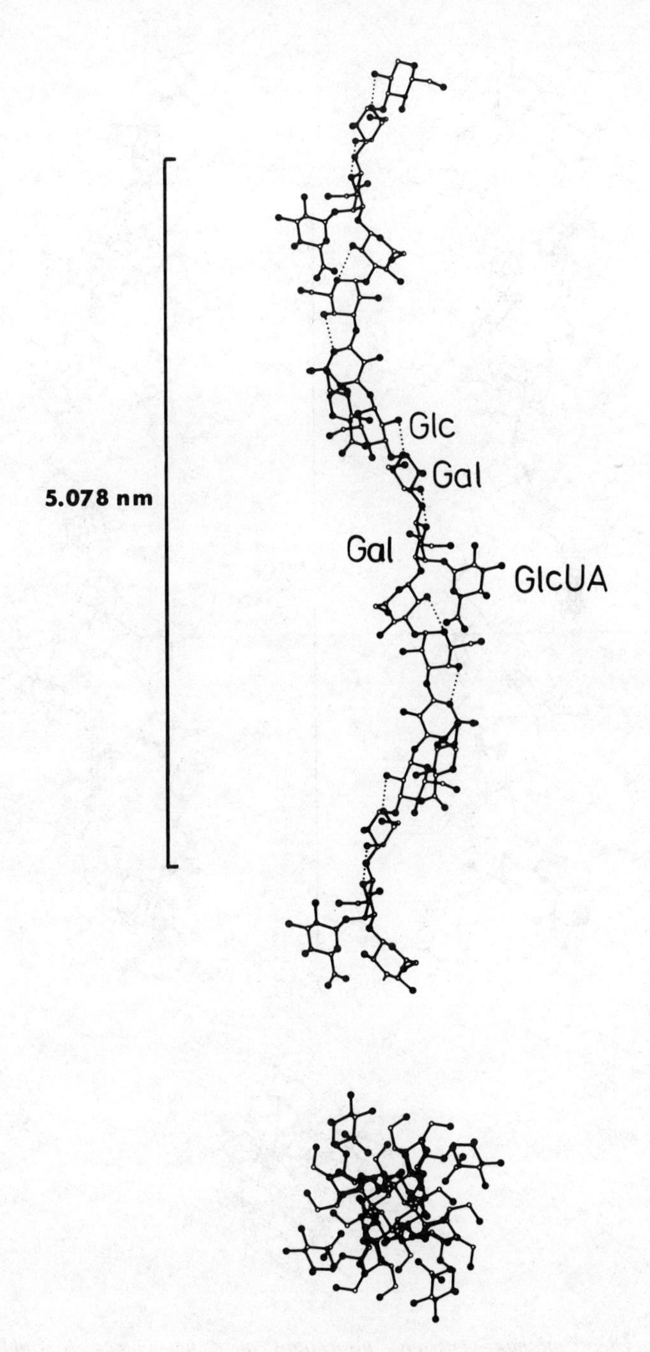

Figure 17. Computer-generated projections of a left-handed four-fold helical conformation for Klebsiella *K8*

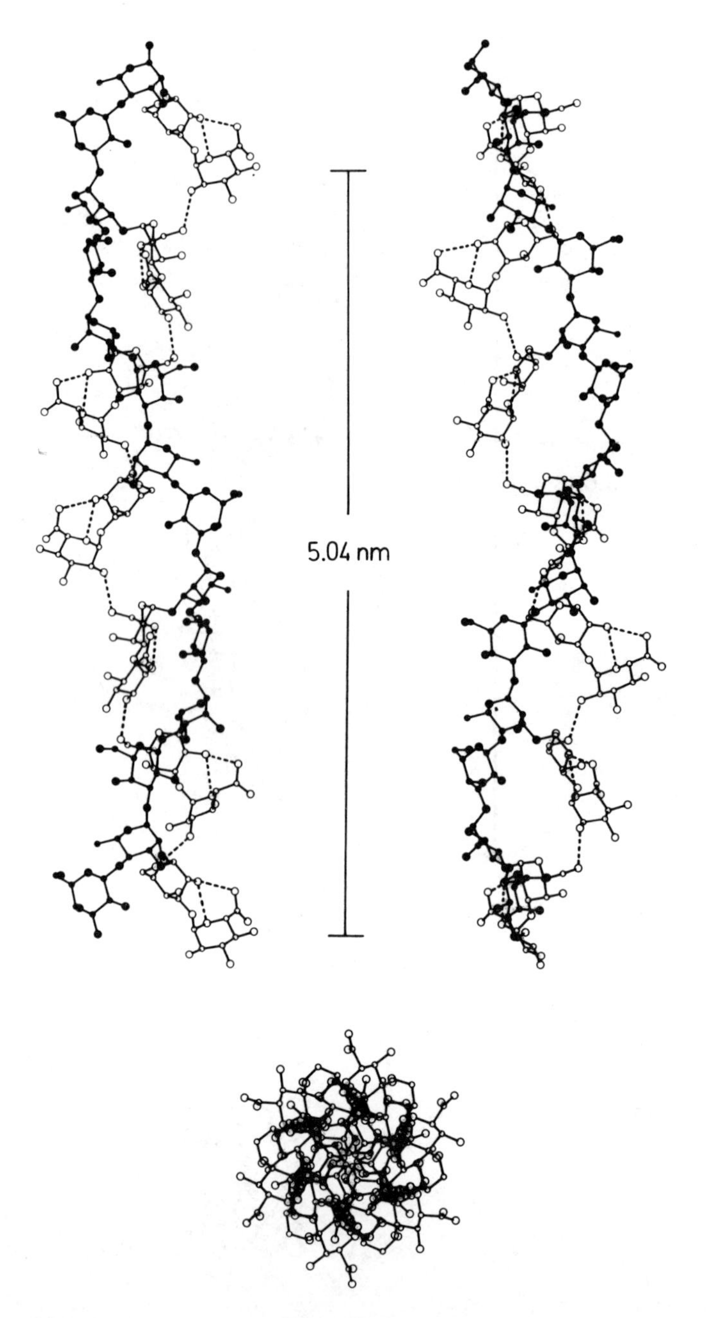

Figure 18. Computer-generated projections of a left-handed, six-fold helical conformation for Klebsiella *K83*

```
        [1e→4e]          [1a→3e]

→3)-β-D-Gal-(1→4)-α-L-Rha-(1→
                      3
                      ↑
                      1
                  α-D-Gal
                      3
                      ↑
                      1
                  α-D-GlcUA
```

Figure 19. Repeating chemical structure for Klebsiella *K83*

```
        [1e→4e]          [1a→3e]
                        OAc
                         ↑
                        2
→3)-β-D-Glc-(1→4)-α-L-Rha-(1→
                      3
                      ↑
                      1
                  α-D-Gal
                      3
                      ↑
                      1
                  α-D-GlcUA
```

Figure 20. Repeating chemical structure for Klebsiella *K55*

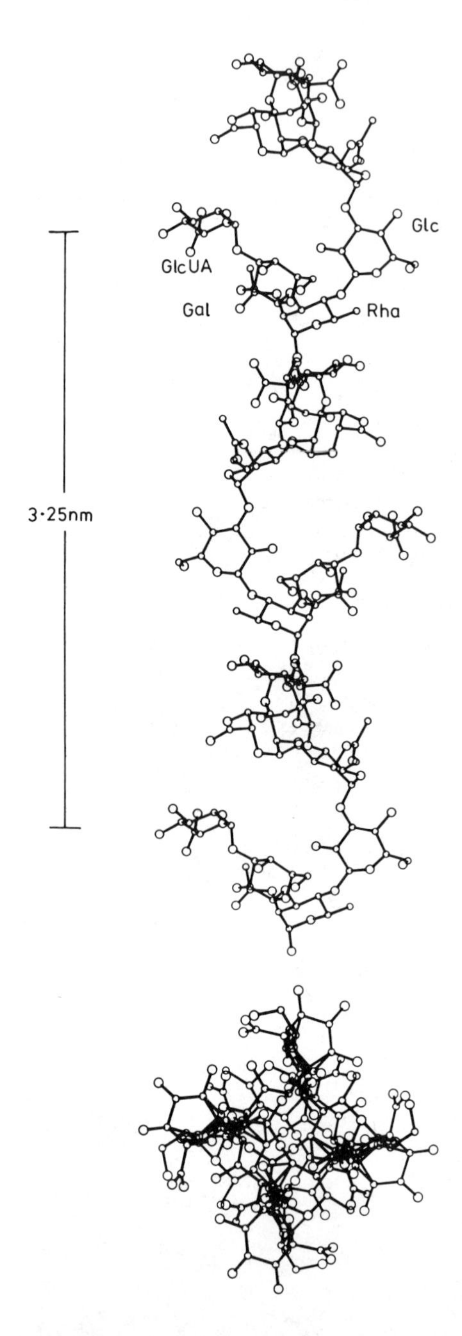

Figure 21. Computer-generated projections of a left-handed, four-fold helical conformation of Klebsiella *K55*

studies indicated that the presence or absence of O-acetyl groups did not alter the type of helix formed.

Klebsiella K18 and Klebsiella K30

Two of the most complex of this group of polysaccharides to have so far been investigated are Klebsiellas K18 and K30. The chemical structure of K18 was determined by Bebault et al (unpublished work) who found it consisted of a trisaccharide backbone with a trisaccharide side group which included a glucuronic acid residue (Figure 22). Figure 23 shows three projections of a stereochemically possible molecular model for this polysaccharide which, it was deduced from X-ray diffraction studies, formed an extended 2-fold helix. Again it was not possible to form a hydrogen bond across the glycosidic linkage that joins the side chain to the backbone. This means that the side chain has a fair amount of flexibility which might be of biological importance.

Klebsiella K30, whose chemical structure was elucidated by Lindberg et al (19) (Figure 24), is included in this review since it is the sole doubly-branched Klebsiella polysaccharide that has so far been investigated to any great extent. Although the quality of the X-ray diffraction patterns obtained were poorer than those obtained from other Klebsiella serotypes considered in this review, it was possible, from a comparison of other polysaccharides with similar backbone linkage geometries, to deduce a likely helical conformation for this molecule and three projections of an idealised two fold helical model for K30 are shown in Figure 25 (20).

Klebsiella K9 and Klebsiella K57

Klebsiella K9 has a particularly intriguing backbone linkage geometry. Its chemical structure, established by Lindberg et al (21), is given in Figure 26. There are three α-L-rhamnose residues in its backbone, two linked 1, 3 and one linked 1, 2. Up till now rhamnose residues have not been widely found in polysaccharides, though it is pertinent to note that both α 1, 2 and β 1, 2 linked rhamnose residues have been considered to have a kinking function in pectic substances where they occasionally interrupt a sequence of α 1, 4 linked galacturonic acid residues (22), X-ray diffraction patterns obtained from this material, of which a typical pattern is shown in Figure 27, showed that the polysaccharide formed an extended three fold helical structure and three projections of the most favourable conformation found from model building calculations are shown in Figure 28 (14). These projections show no evidence for any kinking occurring in this particular polysaccharide and indeed,

```
[1e→4e]              [1a→3e]              [1a→3e]

→3)-β-D-Gal-(1→4)-α-D-Glc-(1→3)-α-L-Rha-(1→
                       3
                       ↑
                       1
                    α-L-Rha
                       2
                       ↑
                       1
                    β-D-GlcUA
                       4
                       ↑
                       1
                    α-D-Glc
```

Figure 22. Repeating chemical structure for Klebsiella *K18*

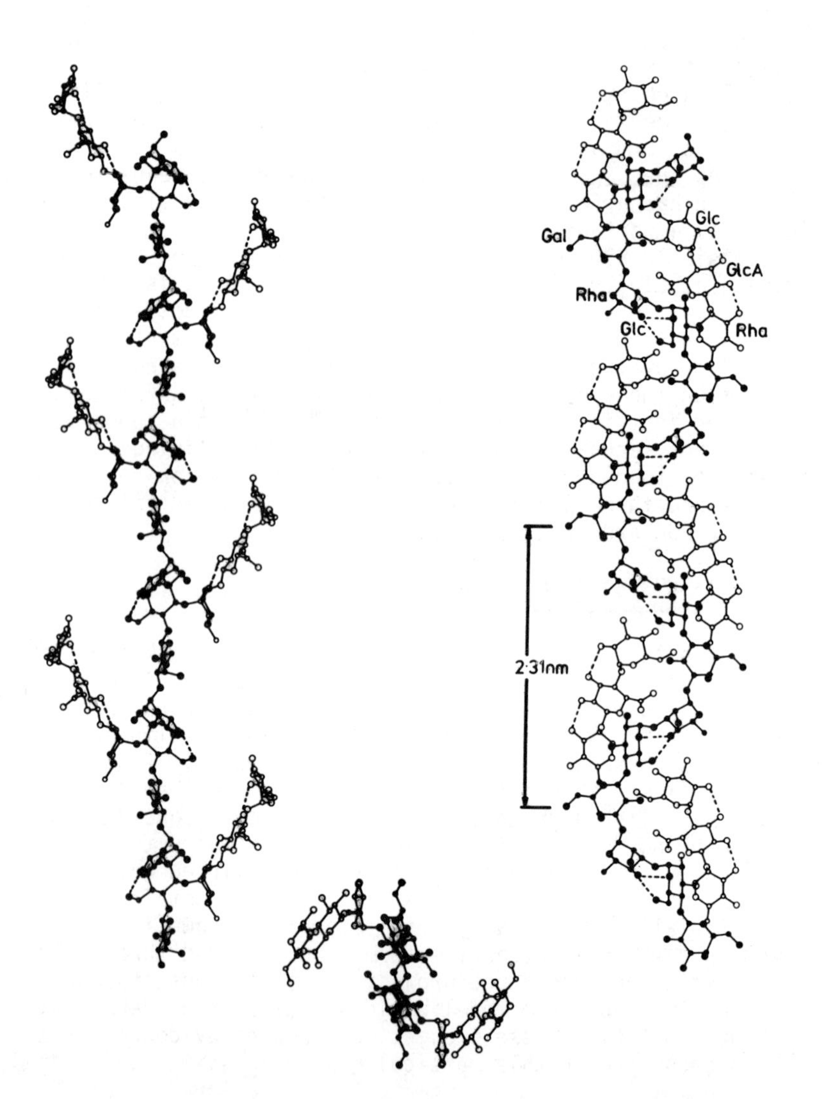

Figure 23. Computer-generated projections of a two-fold helical conformation for Klebsiella *K18*

[1e→4e] [1e→4e] [1e→4e]

α-D-GlcUA
1↓3
→4)-β-D-Glc-(1→4)-β-D-Man-(1→4)-β-D-Man-(1→
6↑1
β-D-Gal 3,4 C (COOH, CH_3)

Figure 24. Repeating chemical structure for Klebsiella *K30*

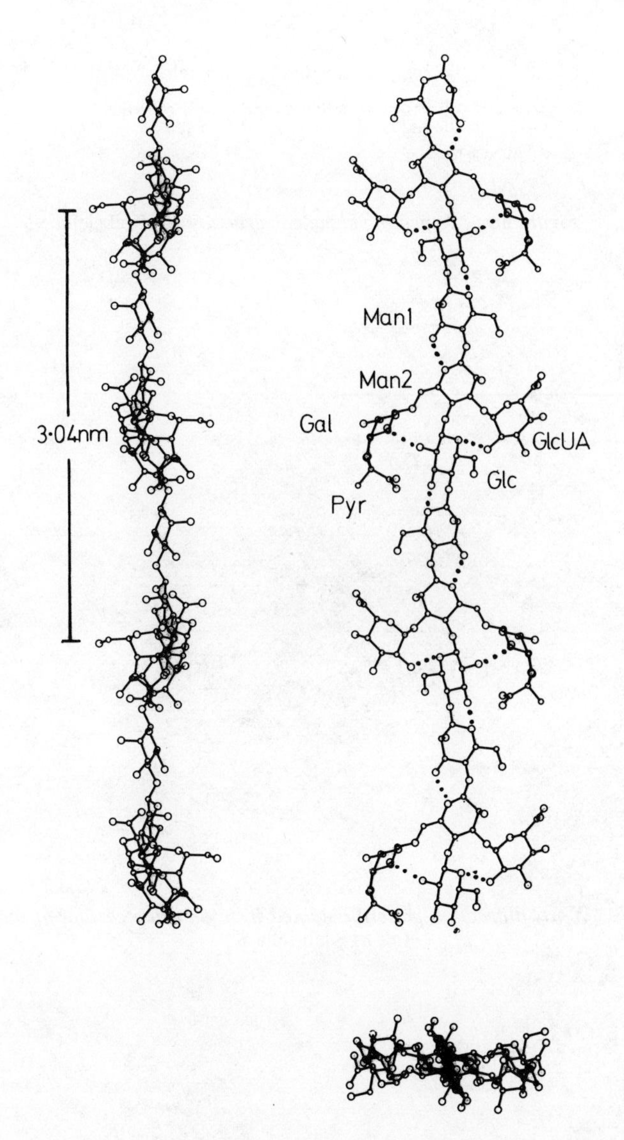

Figure 25. Computer-generated projections of an idealized two-fold helical conformation for Klebsiella *K30*

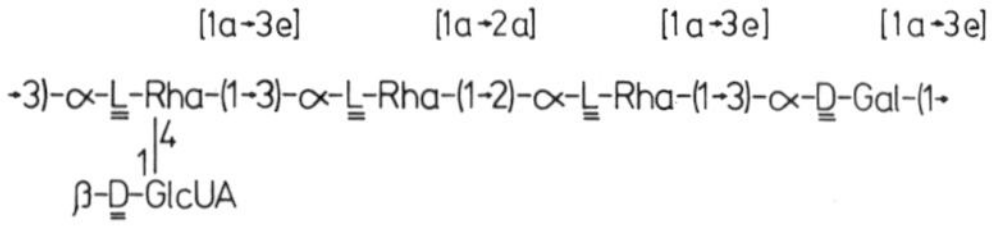

```
            [1a→3e]         [1a→2a]         [1a→3e]        [1a→3e]
→3)-α-L-Rha-(1→3)-α-L-Rha-(1→2)-α-L-Rha-(1→3)-α-D-Gal-(1→
        |4
       1|
   β-D-GlcUA
```

Figure 26. Repeating chemical structure for Klebsiella *K9*

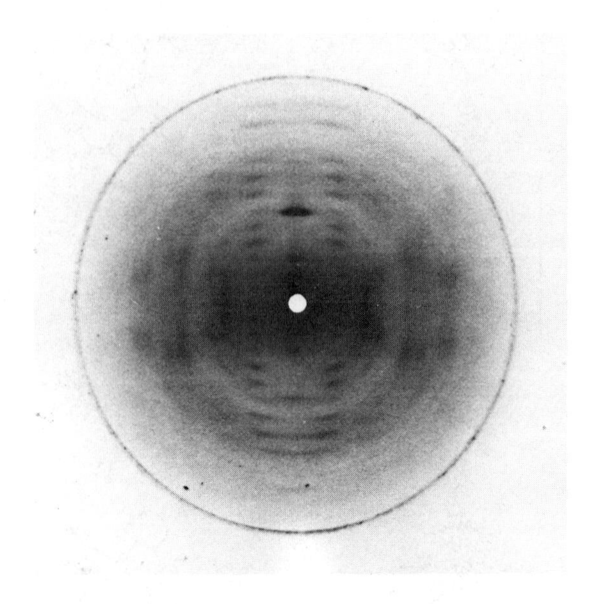

Figure 27. X-ray diffraction pattern obtained from an oriented film of the sodium salt of Klebsiella *K9*

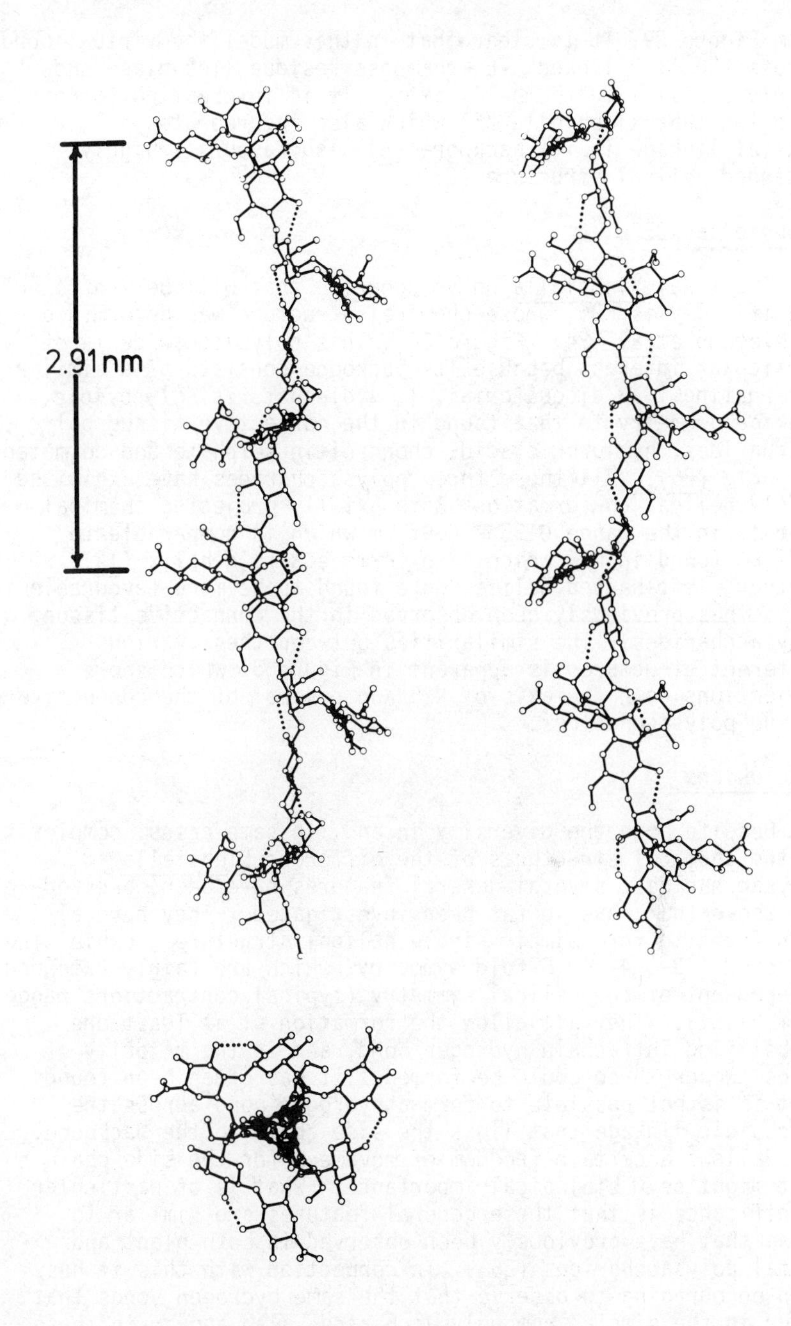

Figure 28. Computer-generated projections of a left-handed, three-fold helical conformation for Klebsiella *K9*

from Figure 29, it is clear that in this model the virtual bond across the 1, 2 linked α-L-rhamnose residue lies close and nearly parallel to the helix axis. It is interesting to note here too that *Klebsiella* K57 which also incorporates a 1, 2 diaxial linkage in its backbone (8) also favours a highly extended helical structure (23)

Klebsiella K25

The final *Klebsiella* polysaccharide that will be mentioned specifically is K25, whose chemical structure was determined by Niemann *et al* (24) (Figure 30). This polysaccharide is of particular interest because its backbone consists of a similar alternating 1, 3 diequatorial, 1, 4 diequatorial glycosidic linkage geometry to that found in the connective tissue polysaccharides, hyaluronic acid, chondroitin sulphate and dermatan sulphate (25). All these three polysaccharides have exhibited 3-fold helical conformations with axially projected chemical repeats in the range 0.95 - 0.97 nm which is comparable to 0.97 nm found in K25 which also forms a 3-fold helix (14). Further, left-handed helices were found to be more favourable in K25 as has previously been observed in the connective tissue polysaccharides. The similarities between these various different structures is apparent in Figure 31 which shows projections down the axis of K25 and several of the connective tissue polysaccharides.

Conclusions

Despite both the diversity in and, in some cases, complexity of the chemical structures of the different *Klebsiella* polysaccharides, several general features have been observed in all those that have so far been investigated. They have all been found to form simple single helical structures, exhibiting either 2-, 3-, 4- or 6-fold symmetry, which are fairly extended, independent of the helical symmetry (typical contractions range from 6-15%). They all allow the formation of at least one stabilising intrachain hydrogen bond, and in the majority of cases two or three could be formed. It has often been found that it is not possible to form a hydrogen bond across the glycosidic linkage that links the side chain to the backbone. This allows a certain freedom of movement for the side chain that might be of biological importance. What is of particular significance is that these general features are similar to those that have previously been observed in both plant and animal polysaccharides (26). In connection with this it has been encouraging to observe that the same hydrogen bonds that occur in the simpler homopolysaccharides also appear in these complex heteropolysaccharides. One example of this, concerning

helix
axis
1
Rha
2

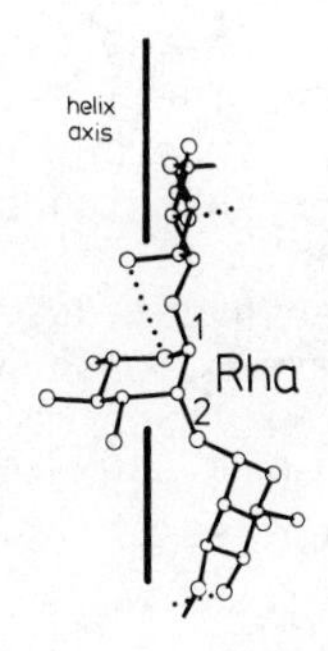

Figure 29. An enlargement of part of the conformation of Klebsiella *K9 shown in Figure 28 showing the orientation of the 1, 2-linked rhamnose residue with respect to the helix axis*

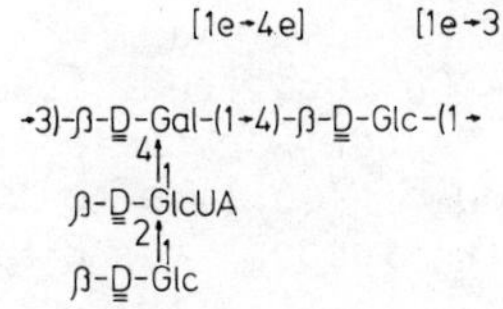

[1e→4e]
[1e→3e]
→3)-β-D-Gal-(1→4)-β-D-Glc-(1→
4↑1
β-D-GlcUA
2↑1
β-D-Glc

Figure 30. Repeating chemical structure for Klebsiella *K25*

HA
C4-S
C6-S
1nm
K25

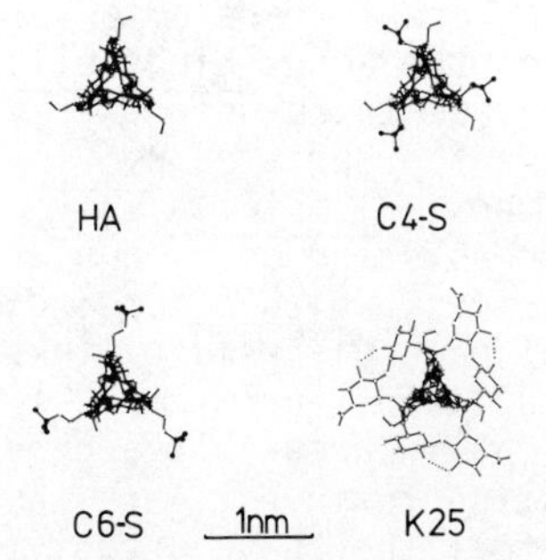

Figure 31. Computer-generated projections down the helix axis of the three-fold left-handed helical conformations of Klebsiella *K25, Hyaluronic acid, HA, Chondroitin 4-sulphate, C4-S, and Chondroitin 6-sulphate, C6-S*

Klebsiella K63 and the α 1, 3 glucans, has already been mentioned in the text, and a second example that can be cited concerns the familiar O(5)••H-O(3) hydrogen bond formed across the β 1, 4 linkage in cellulose (27) which can also occur in a large number of these *Klebsiella* polysaccharides.

There are many *Klebsiella* polysaccharides still to be investigated. In several instances very good X-ray diffraction patterns have been obtained and further investigation awaits the full elucidation of the chemical structure. On the other hand, several have been disappointing in showing no inclination to crystallise and/or orientate. We can see no inherent reason why they should not, and can only hope that a fresh batch of material may be more productive.

It is hoped that as more structures are investigated, further insight will be gained both concerning the relationship between the conformations and biological functions of these materials and also concerning the behaviour of individual glycosidic linkage geometries and a variety of different saccharide residues.

Acknowledgements

We wish to thank the following people who have most kindly supplied us with various *Klebsiella* serotypes: Professor G.G.S. Dutton, Professor S. Stirm, Dr. J.P. Joseleau, Dr. W. Nimmich and Dr. I.W. Sutherland. Our gratitude is also due to both Dr. K.H. Gardner and Mr. D. Meader for their help in computational investigations. We should also like to thank the Science Research Council for financial support for D.H.I. and H.F.E.

Abstract

The isolation and purification in recent years of a large number of microbial polysaccharides of the Genus *Klebsiella* has enabled conformational studies to be undertaken using X-ray diffraction techniques followed, where appropriate, by computational analysis. In general these *Klebsiella* polysaccharides are extracellular, each consisting of a regular repeating unit of up to seven saccharides, which often includes one or more side appendages. Several have glycosidic linkages which have not previously been crystallised, for example 1, 2 diaxial linkages (K9 and K57) and 1, 6 linkages (K38 and K54), and it has been particularly interesting to investigate the geometry of such linkages and the effect they have on polymer shape. These conformational investigations have indicated several characteristics of the structures of these *Klebsiella* polysaccharides, which are notably similar to those previously observed in the simpler plant or animal polysaccharides. The examination of the possible molecular conformations adopted by these polysaccharides is particularly pertinent and interesting in view of the fact that some of their biological functions are believed to be highly dependent on their individual conformations.

Literature Cited

1. Toenniessen, E. Zent. f. Backteriol, 1914, 75, 329.
2. Julianelle, L.A. J. Exp. Med., 1926, 44, 113.
3. Nimmich, W. Z. Med. Mikrobiol. u. Immunol., 1968, 154, 117.
4. Nimmich, W. Acta Biol. Med. Germ., 1971, 26, 397.
5. Lindberg, B., Samuelsson, K. and Nimmich, W. Carb. Res., 1973, 30, 63.
6. Dudman, W.F. and Wilkinson, J.F. Biochem. J., 1965, 62, 289.
7. Niemann, H., Beilharz, H. and Stirm, S. Carb. Res., 1978, 60, 353.
8. Isaac, D.H., Gardner, K.H., and Atkins, E.D.T. Carb. Res., 1978, 66, 43.
9. Dutton, G.G.S. and Mo-Tai Yan Can. J. Chem., 1973, 51, 1826.
10. Isaac, D.H., Gardner, K.H., Wolf-Ullish, C., Atkins, E.D.T. and Dutton, G.G.S. Int. J. Biol. Macromols., 1979, 1, 107.
11. Elloway, H.F. Ph.D. Thesis, Univ. of Bristol, 1977.
12. Jelsma, J. This volume.
13. Conrad, H.E., Bamburg, J.R., Epley, J.D. and Kindt, T.J. Biochemistry 1966, 5, 2808.
14. Atkins, E.D.T., Isaac, D.H. and Elloway, H.F. in "Microbial Polysaccharides and Polysaccharases" eds. Berkeley R.C.W., Gooday, G.W. and Ellwood, D.C., Academic Press, London, 1979, 161.
15. Kanters, J.A., Roelofsen, G., Doesburg, H.M., and Koops, T., Acta.Cryst., 1976, B32, 2830.
16. Berman, H.M., Acta. Cryst., 1970, B26, 290.
17. Sutherland, I.W., Biochemistry, 1970, 9, 2180.
18. Lindberg, B. and Nimmich, W. Carb. Res., 1976, 48, 81.
19. Lindberg, B., Lindh, F., Lönngren, J. and Sutherland, I.W. Carb. Res., 1979, 76, 281.
20. Elloway, H.F., Atkins, E.D.T., and Sutherland, I.W. Carb. Res., 1979, 76, 285.
21. Lindberg, B., Lönngren, J., Thompson, L. and Nimmich, W. Carb. Res., 1972, 25, 49.
22. Rees, D.A. and Wight, A.W., J. Chem. Soc. B, 1971, 1366.
23. Kammerling, J.P., Lindberg, B., Lönngren, J. and Nimmich, W. Acta. Chem. Scand., 1975, B29, 593.
24. Niemann, H., Kwiatkowski, B., Westphal, U, and Stirm, S. J. of Bacteriol 1977, 130, 366.
25. Atkins, E.D.T., Isaac, D.H., Nieduszynski, I.A., Phelps, C.F. and Sheehan, J.K., Polymer, 1974, 15, 263.

26. Atkins, E.D.T. in "Applied Fibre Science", ed. Happey, F., Academic Press, 1979, 3, 311.
27. Liang, G.Y. and Marchessault, R.H., J. Polymer Sci., 1959, 37, 385.

RECEIVED May 21, 1980.

28

Crystal Structures of Amylose and Its Derivatives

A Review

ANATOLE SARKO

Department of Chemistry, State University of New York,
College of Environmental Science and Forestry, Syracuse, NY 13210

PETER ZUGENMAIER

Institute of Macromolecular Chemistry, University of Freiburg,
D-7800 Freiburg i.Br., West Germany

Amylose, a linear, high molecular weight (1→4)-α-D-glucan, is one of the principal polysaccharides of starch. Because of the longstanding utility of starch as a raw material, and its widespread botanical availability, its structure and properties have been studied for centuries. Since the more recent realization that almost all varieties of starch are composed of two polysaccharides - the linear amylose and the branched amylopectin - a significant share of interest has shifted to the study of these components. Of particular interest has been the observation that both components occur naturally in crystalline form in the starch granule.

Structural studies of amylose have, in turn, revealed a wide range of crystalline polymorphy, both in chain conformation and in crystalline packing. An example is the group of V-amyloses that exist as complexes with small organic molecules, water, or iodine. The latter complex is particularly interesting because it displays an intense blue color. The V-amyloses can be prepared by precipitation or drying from solution, and they crystallize readily. Consequently, their crystal structures are of interest in connection with any regenerated form of starch material.

Another group of crystalline amyloses consists of complexes with ionic substances, for example, alkali or salts such as KBr. As will be shown later, these crystal structures differ considerably from the V-amyloses.

The amylose found in the native starch granule is, in many respects, the most fascinating polymorph. It is double-helical in structure, which raises a question how this molecule is synthesized and deposited in such a complex crystalline form into a layered, radially organized spherulitic granule, a morphology that has not been duplicated in the laboratory. The complexity does not stop with amylose, as the branched amylopectin almost certainly possesses the same double-helical structure. Also, there are two different types of crystalline starch granules: the A-starch of the cereals and the B-starch of the tubers. The two granules possess the same chain conformation, but differ in crystalline packing.

0-8412-0589-2/80/47-141-459$06.00/0

Finally, because the amylose molecule is a polyalcohol, chemical derivatives of it can be easily prepared. Included in this are the acetate, methyl, ethyl and similar derivatives, all of which belong to a class of polymers with completely different properties from the parent substance. Many of these derivatives demonstrate useful film and fiber properties, but they have not reached significant commercial utilization. All amylose derivatives crystallize easily, and many show interesting features in their crystalline state.

The relationships between the main classes of amylose polymorphs are schematically illustrated in Fig. 1. As shown by a selection of x-ray diffraction patterns in Fig. 2, all classes of amyloses generally yield good fiber diffraction diagrams. In addition, single crystals of a number of amylose polymorphs can be grown from dilute solutions. These crystals, shown in Fig. 3, also give good electron diffraction patterns. The analysis of the resulting x-ray and electron diffraction intensities, with the help of the modern modeling techniques such as described in this volume (1,2), has produced structural information of previously unavailable detail. In the following, we review the structures of amyloses determined in this fashion, including, however, only those results that have been obtained with the help of the more recent methods.

Classification of Amylose Structures and Their General Features

The unit cell dimensions of all crystalline amyloses that have been determined in some detail, are listed in Table I. Also included are some intermediate forms between the $V_{\bar{a}}$ and $V_{\bar{h}}$ amyloses (5) and some V-amylose complexes with *n*-butanol, which, although not yet completely determined, have been added to illustrate the range of variability in unit cell dimensions. In the case of the V_a-BuOH complex, a doubling of one unit cell axis was detected after a careful study of electron diffraction diagrams of single crystals (10). A consequence of the doubling is that the unit cell now contains four chains, instead of the two normally found in amylose structures. (In a strict sense, the A- and B-amyloses should also be considered as four-chain unit cells, but their double-helical structure still results in only two helices per cell) (13,14).

Almost all unit cells shown in Table I are either orthorhombic or pseudo-orthorhombic, with a majority of space groups $P2_12_12_1$ and $P2_1$. Only a few structures exhibit higher symmetry and none shows lower symmetry. All structures have an antiparallel packing of chains (however, see A- and B-amyloses). On the other hand, a large variety of helix characteristics are evident, in addition to the variability in the unit cell dimensions. Some of the features useful for classifying amylose structures are shown in Table II. The distance between the two nearest antiparallel-

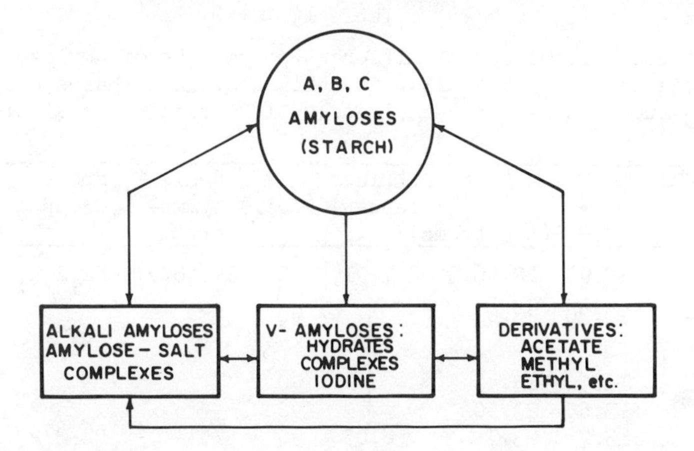

Figure 1. Relationships and conversion paths between different classes of amylose structures

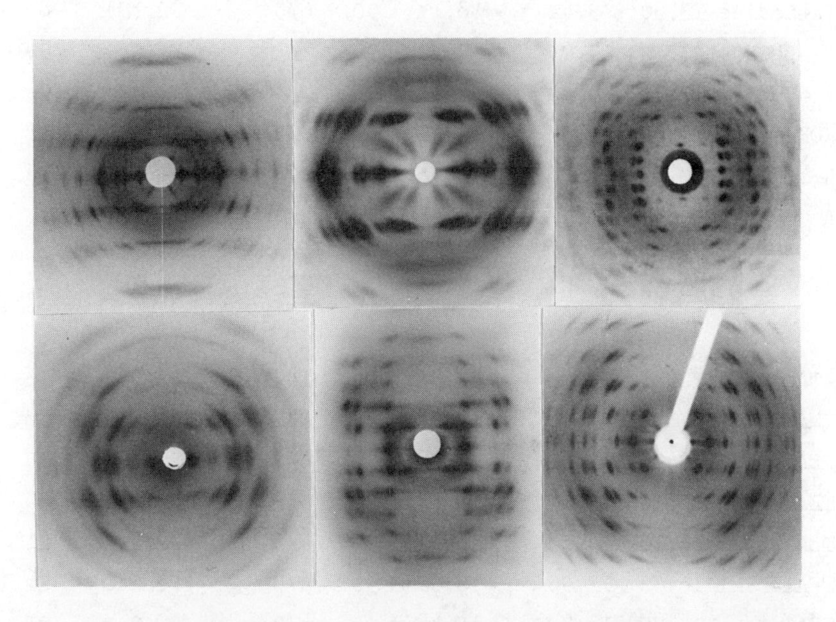

*Figure 2. X-ray fiber diffraction patterns for (*top, left to right*):* V_a*-amylose;* V_{DMSO}*-amylose; KOH-amylose; (*bottom, left to right*): B-amylose, amylose triacetate I, triethylamylose I-nitromethane complex*

Table I

Unit cell dimensions of different polymorphs of amylose and amylose derivatives. All unit cells contain 2 chains except V_a-BuOH and A- and B-amyloses, which contain 4 chains.

Structure	*a*	*b*	*c* (fiber repeat) (Ångstroms)	γ (deg.)	Helix symmetry	Space Group	Ref.
V_a	12.97	22.46	7.91	90	2_1(~6/5)	$P2_12_12_1$	3,4
Intermediate forms	13.30	23.0	↓	90			5
	13.50	23.45	↓	90			
	13.55	23.50	↓	90			
V_h	13.7	23.7	8.05	90	6/5	$P2_12_12_1$	6
V_{DMSO}	19.17	19.17	24.39	90	6/5 in 8.13 Å repeat (per turn)	$P2_12_12_1$	7
V_h-iodine	13.60	23.42	8.17	90	6/5	$P2_1$(S)	8
V_a-BUOH	26.4	27.0	7.92	90		$P2_12_12_1$	9,10
V_h-BUOH	13.7	25.8	8.10	90			9
KOH	8.84	12.31	22.41	90	6/5	$P2_12_12_1$	11
KBr	10.88	10.88	16.52	90	4/3	$P4_32_12$	12
A	11.90	17.70	10.52	90	2x6/1 in 21.04 Å repeat	$P2_1$(S)	13
B	18.50	18.50	10.40	120	2x6/1 in 20.8 Å repeat	$P3_121$	14
ATA I	10.87	18.83	52.53	90	14/11	$P2_1$(S)	16
TMA	17.24	8.70	15.64	90	2/1(~4/3)	$P2_12_12_1$	18
TEA1	16.13	11.66	15.48	90	4/3	$P2_12_12_1$	19
TEA3	15.36	12.18	15.48	90	4/3	$P2_12_12_1$	20
TEA1-C1	16.76	14.28	16.02	90	4/3	$P2_12_12_1$	21
TEA1-DCM1	16.52	13.95	16.02	90	4/3	$P2_12_12_1$	21
TEA1-N, -C2,-DCM2	14.70	14.70	15.48	90	4/3	$P2_12_12_1$	22

Table II

Interchain and intersheet spacings and rise per residue for different amyloses and derivatives.

Structure	$d_{\uparrow\downarrow}$ (Å)	d_{110} (Å)	h (Å)
V_a	12.97	11.23	1.32
V_h	13.69	11.86	1.34
V_{DMSO}	13.56	13.56	1.36
V_h-iodine	13.54	11.76	1.36
KOH	7.58	7.18	3.74
KBr	7.21	7.21	4.10
A	10.66	9.87	3.51
B	10.68	9.25	3.47
ATAI	10.87	9.41	3.75
TMA	9.66	7.77	3.91
TEA1	9.95	9.45	3.87
TEA3	9.80	9.54	3.87
TEA1-C1	11.01	10.87	4.01
TEA1-DCM1	10.81	10.66	4.01
TEA1-N, C2, DCM2	10.39	10.39	3.87

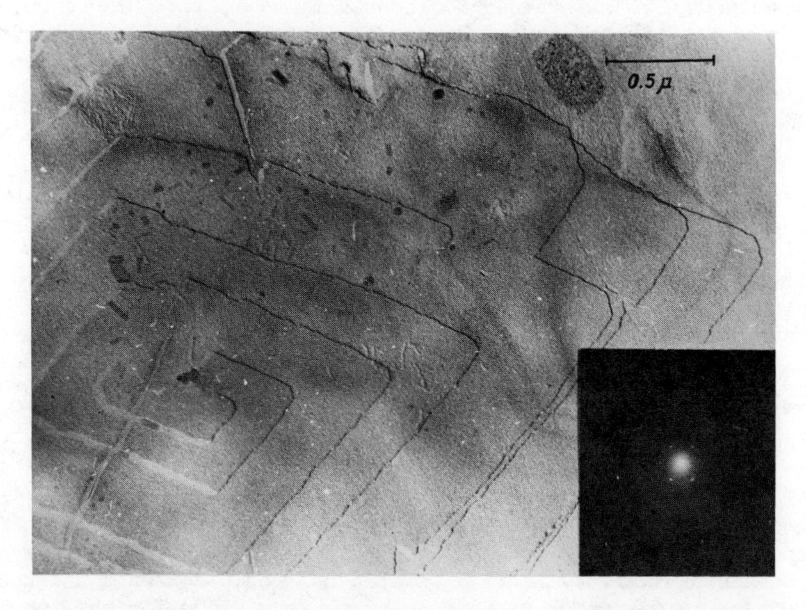

Figure 3. Electron micrograph and electron diffraction diagram of a single crystal of triethylamylose. (The diffraction rings are due to TlCl used for calibration.)

packed chains ($d_{\uparrow\downarrow}$) is listed in the first column. This usually represents the distance between a corner and a center in an orthorhombic, two-chain unit cell, and it invariably determines the closest contact distance between the chains in the crystal structure.

A similar feature, the Bragg d_{110} spacing, listed in the second column, usually represents the spacing between the sheets of closest-packing chains. It may be an important measure when adjacent sheets are separated by solvent molecules occupying the interstitial spaces in the unit cell. Finally, the rise per residue, h, listed in the last column of Table II, reflects the extension of the helix along its axis.

A clear dividing line separates the amylose structures with respect to h: all V-type structures fall on one side with h ranging from 1.32 to 1.36 Å, leaving all other structures in another group with values of h from 3.5 to 4.1 Å. The interchain distance $d_{\uparrow\downarrow}$ and the d_{110} spacing can be used to further classify the amylose structures into four groups. The V-type structures, forming one group, exhibit nearly identical values of $d_{\uparrow\downarrow}$, with only V_a somewhat out of line by ~0.7 Å, which is attributable to the absence of water molecules in its interstitial spaces. Such water molecules are found in the hydrated forms of V_h- and V_{DMSO}-amyloses. The d_{110} spacing also increases upon hydration by about the same amount, but jumps by 1.7 Å when large DMSO molecules locate between the sheets.

In the second group of structures, the KOH- and KBr-amyloses possess the lowest values for both $d_{\uparrow\downarrow}$ and d_{110}. This is not surprising because these amylose helices are much more extended than the V-type structures, with h at 3.74 and 4.10 Å, respectively. Considering that the KOH-amylose is a 6/5 helix and KBr-amylose is a 4/3 helix, the two values of h are surprisingly close.

The amylose derivatives, forming another group of structures, are characterized by h in the same range as in KOH- and KBr-amyloses, but with both $d_{\uparrow\downarrow}$ and d_{110} increased, because of the space required for the substituent groups. However, that $d_{\uparrow\downarrow}$ stays within ~1.2 Å for all derivatives, including the solvent-complexed structures, is unexpected. A larger range in the d_{110} spacing is more in line with the presumed effect of the solvent.

The double-helical structures of native A- and B-amyloses are found in the fourth group. It is interesting that in both h as well as the $d_{\uparrow\downarrow}$ and d_{110} spacings, they are comparable with the structure of amylose triacetate I (ATAI). In part, this may arise because the packing of the bulky acetate substituents in ATAI is similar to the close-packing of two amylose chains into a double helix. In the latter, one chain may act as the "substituent" for the other chain. At any rate, all three structures contain similar, cylindrical-shaped helices. Somewhat unexpectedly, the distances $d_{\uparrow\downarrow}$ and d_{110} are very close for the two native polymorphs, even though their unit cells and packing are different. As shown below, in the structure for B-amylose the sheets of antiparallel-packed double helices are interrupted by channels of water, while in A-amylose the sheets are intact and the water is located in the interstitial spaces between the double helices.

Originally, space groups were not assigned to ATAI and the A- and B-amyloses, and their structures were solved in the triclinic space group $P1$. It was later observed (P. Zugenmaier, unpublished work) that the same chain positions resulted when both ATAI and A-amylose were assigned the space group $P2_1$, but with the 2_1 screw axis positioned perpendicular to the helix axis (denoted here as $P2_1(S)$). Likewise, B-amylose could be assigned the trigonal space group $P3_121$ (P. Zugenmaier, unpublished work). The consequence of assigning space group $P2_1(S)$ to ATAI is that the complete 14/11 helix must be considered an asymmetric unit. Similarly, three glucose residues, or one-half turn of the helix, now form the asymmetric unit of A-amylose. Space group $P3_121$ for B-amylose requires an asymmetric unit of two glucose residues, or one-third of the helix turn.

In the light of these observations, it is clear that the amylose chain can adopt a variety of conformations. Predictions of the most probable conformations based on isolated chains, whether using hardsphere contact criteria (15) or potential energy functions (17), are not capable of producing the whole range of observed conformations. Furthermore, it is apparent that the molecular symmetry - such as the four- or six-fold helices - does not necessarily coincide with the symmetry of the crystalline packing or the space group. The lattice forces are thus probably as important as the intramolecular forces in determining the conformations of crystalline amyloses.

The four groups of amyloses are described individually in more detail in the following sections.

V-Amyloses

The structures of the V-amyloses were among the first polysaccharides on which the modern methods of conformation and packing refinement were tested. The sophistication of the computer programs has improved significantly over the years, as have some of the early ideas, particularly those concerning hydrogen bonding. In the early stages, conformational analysis was generally separated from the packing refinement, and both, in turn, were conducted separately from the x-ray analysis. In such modeling approaches, the conformation of the glucose ring remained invariant, and the packing refinement was sometimes completely omitted. For this reason, the earlier work has not been included in this review.

All V-amylose structures shown in Table I have in common a left-handed, six-residue helix, with h in a very narrow range from 1.32 to 1.36 Å, and an O-2...O-3(2) intramolecular hydrogen bond (cf. Fig. 5). Although the six-fold, right-handed helix appears to be equally probable from conformational analysis (17, 23), it has not been seen in any V-amylose. The same hydrogen bond also occurs in the maltose crystal structure (24,25,26) and in cyclohexaamylose (27,28). It limits the variability of the O-4...O-4(2) distance (the *virtual bond*) to a range of about 4.05 to 4.4 Å. When this hydrogen bond is not present, as in a variety of small-molecule crystal structures containing α-D-glucose, the length of the *virtual bond* extends as high as 4.6 Å (29). Other intrachain hydrogen bonds also form, primarily involving the O-6 hydroxyl and occurring between the turns of the helix. For example, in V_{DMSO}-amylose, there is an O-6(*i*+7) *gt*...O-2(*i*+1) hydrogen bond (*i*=0,1,2...), and possibly an additional O-6(*i*+7) *gt*... O-3(*i*+2) hydrogen bond through either DMSO or water. (For a description of the *gt*, *gg*, *tg* rotamer terminology, consult Ref. 41). The intramolecular hydrogen-bonding involving O-6 is slightly different in V_h- (6) and V_h-iodine-amyloses (8), because all O-6 are in the *gg* position in these structures. Nonetheless, hydrogen bonds still form between O-6(*i*+7) and O-2(*i*+1). In V_a-amylose, all three rotational sites for O-6 are found in successive residues (3), giving rise to all hydrogen bonds possible for O-6 in *gt*, *gg* and *tg* positions (cf. Fig. 4).

Largely because the O-6 hydroxyls of all six residues of one helix turn are in equivalent positions in the V_h, V_h-iodine and V_{DMSO} structures (*gg* in V_h and V_h-iodine, *gt* in V_{DMSO}), thus forming symmetric intramolecular hydrogen bonds, all residues in these structures are equivalent. However, in V_a-amylose, with its mixture of O-6 positions, molecular sixfold symmetry is not present in the helix and instead, a 2_1 screw axis along the helix axis exists, thus combining three residues of one half-turn into the asymmetric unit. Nonetheless, the helix backbone still resembles a six-fold helix.

Because *h* is small for the V-amyloses, a wide-diameter helix is characteristic of these structures. Complexing agents, such as DMSO, iodine, or water, are found inside the helix channel. For example, in V_{DMSO}-amylose, six DMSO molecules are accommodated inside the channel within one crystallographic repeat, which consists of three helix turns (*c*=24.39 Å). This fiber repeat is not the result of the intrachannel DMSO but is caused solely by the packing of the interstitial DMSO. A non-commensurable fiber repeat for the amylose helix and the intrahelical iodine is observed in V_h-iodine: approximately three iodines occupy the helix channel within one fiber repeat, but the iodines form an almost linear polyiodide chain of an undetermined length. In this respect, the structures of the V_h-iodine complex and the α-cyclodextrin-iodine complex (30) are similar.

Even though the inside of the helix channel of V-amyloses is primarily hydrophobic in character, intrahelical water has been found in all of the structures of complexes studied to date. The same was found to be the case in single crystals of hydrated cy-

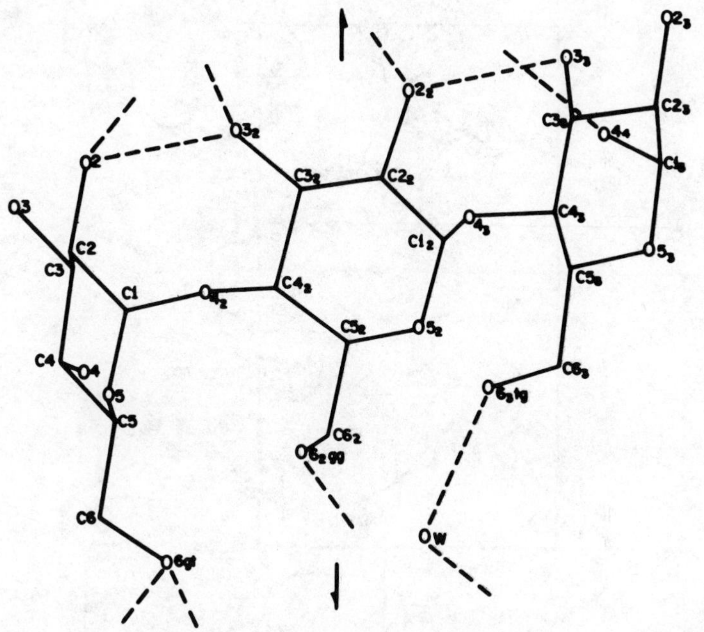

Biopolymers

Figure 4. The asymmetric unit of V_a-amylose. Intramolecular hydrogen bonds are shown by dashed lines (3).

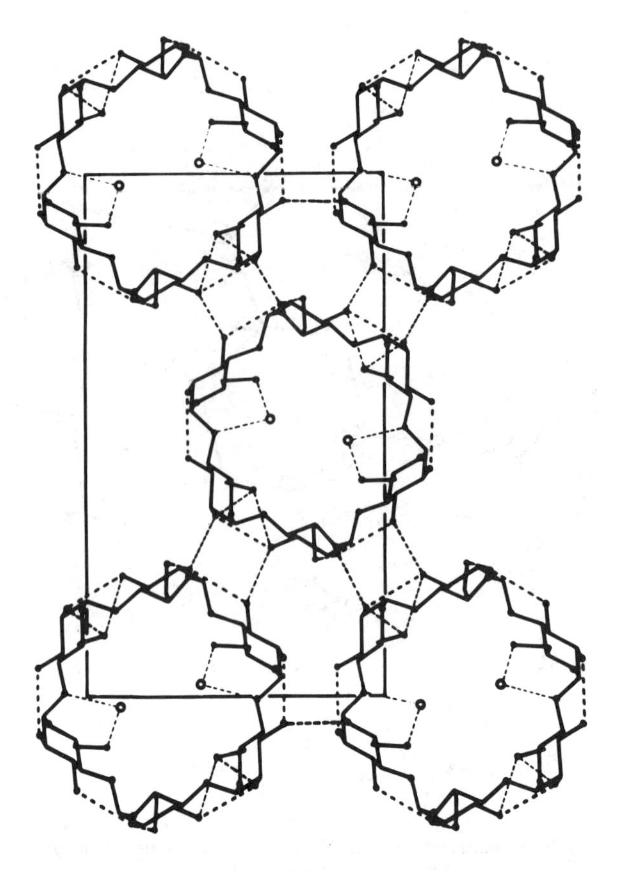

Figure 5. V_a-amylose in ab *projection. Positions of water molecules are shown by circles and hydrogen bonds are shown by dashed lines.*

clohexaamylose (31), in which the water, situated within the cyclic structure, is hydrogen-bonded to the 0-6 *gt* hydroxyl. In the V_h-amylose, all 0-6 are in the *gg* position and cannot hydrogen bond to the water in the channel, but water is still found in it, along with the iodine. Water is present even in the "anhydrous" V_a-amylose, although it is situated at a peripheral site in the helix channel and is hydrogen-bonded to an 0-6 *tg* hydroxyl (cf. Fig. 4). (The *tg* position has not yet been observed in monomer or oligomer carbohydrate single crystal structures, but has been found in the helices of several amylose derivatives, as shown later in this review. It has also been found in the polymorphic structures of cellulose). It is possible that the relatively empty helix of the V_a-amylose, in comparison with other V-complexes, can lower its energy only by distortion of the 6/5 helix into a 2/1 helix. The same apparently occurs in the cyclohexaamylose hexahydrate, also a relatively empty structure in comparison with other cyclohexaamylose complexes (31).

The V-amylose structures generally exhibit similar features of crystalline packing. All are orthorhombic, antiparallel, two-chain unit cells, with one chain at the corner and the other at the center of the cell (cf. Fig. 5), and all except one are in space group $P2_12_12_1$. The exception is V_h-iodine where the polyiodide chain disturbs the 2_1 screw axis along the chain, and the space group $P2_1(S)$ appears to fit the structure. The V_a-, V_h- and V_h-iodine-amyloses are also very close to hexagonal packing, as indicated by the ratio of the unit cell axes a/b being almost exactly $1/\sqrt{3}$. Some intermolecular hydrogen-bonding stabilizes the packing of the chains in all V-structures. In V_a-amylose, two hydrogen bonds are formed between the corner and center chains, and one hydrogen bond forms between two corner chains along the a axis. In V_h- and V_h-iodine-amyloses, additional hydrogen bonds form through interstitial water molecules. As each water molecule can participate in up to four hydrogen bonds, an extensive hydrogen bond network may be present in hydrated structures. Because the hydrogen atoms cannot be detected in polymer x-ray crystal studies, some indeterminacy in the hydrogen-bonding scheme cannot be avoided. The presence of a hydrogen bond is based solely on the oxygen-oxygen distance and, in some cases, on the bond angles about the presumed hydrogen bond.

The packing of the V_{DMSO}-structure is somewhat exceptional in that it shows features not found in other V-amyloses included in Table I. The interstitial DMSO molecules prevent any interaction between the corner chains of the unit cell, but allow hydrogen bonds to form between the corner and center chains. The location of the DMSO molecules between the parallel-packing corner chains results in a pseudotetragonal unit cell, but the space group is still $P2_12_12_1$. The interstitial DMSO molecules are also clearly responsible for the three-turn fiber repeat.

Alkali and Salt Complexes

A series of alkali-amylose complexes can be obtained during the solid-state deacetylation of amylose triacetate, as first described by Senti and Witnauer (32). The unit cells of the individual members of the series of LiOH-, KOH-, NH_3OH-, CsOH and guanidinium hydroxide-adducts of amylose appeared to fit an isomorphous series based on the space group $P2_12_12_1$, and a tentative crystal structure was proposed (32). The detailed structure of the KOH-amylose complex has now been determined (11) and the overall structure is similar to that proposed earlier. It is, therefore, likely that all members of the series are isomorphous.

The structure, as shown in Fig. 6, is based on a left-handed, 6/5 helix. The left-handed conformation is consistent with that of amylose triacetate from which it is derived by a solid state transformation, as well as with that of V-amylose into which it can be easily converted, likewise in the solid state. As shown in Table II, in extension the helix is very similar to that of amylose triacetate I (h = 3.74 Å for KOH-amylose and 3.75 Å for ATA I). Intramolecular hydrogen bonds are not possible in this extended conformation, which suggests that the forces responsible for the conformation of V-amyloses are not dominating here. As indicated by the ϕ,ψ angles of the KOH-amylose chain (cf. Fig. 7), its conformation does not coincide with the conformational minimum of the ϕ,ψ map. The intermolecular hydrogen bonds to water and the coordination of the K^+ ion apparently determine the energy minimum for this structure. Because there is one KOH and three water molecules for each half-turn of the helix, the asymmetric unit of the crystal structure consists of three glucose residues which differ slightly in conformation, particularly in glycosidic bond angles and the rotational positions of the hydroxymethyl groups. Therefore, six-fold molecular symmetry is not present, which is in agreement with the $P2_12_12_1$ space group. In this respect, the KOH- and V_a-amyloses are similar.

Because of the good x-ray data (a total of 99 intensities were available for refinement), difference Fourier techniques, such as described by Winter in this volume (33), could be used to locate the KOH and water molecules in this crystal structure. As shown in Fig. 6, the K^+ ion coordinates with four oxygens of the amylose chain and two water molecules. All three water molecules participate in hydrogen bonds, but the intermolecular hydrogen-bonding pattern is not extensive. This probably accounts for the water-solubility of the complex.

A similar series of salt complexes of amylose were also described earlier by Senti and Witnauer (34). A salt complex, such as KBr-amylose, is obtained from KOH-amylose by neutralization of the alkali. Although KBr-amylose has been studied since the initial description of the series, a definitive crystal structure determination by Miller and Brannon appears in this volume (12). It is clear that in its left-handed conformation and in the ex-

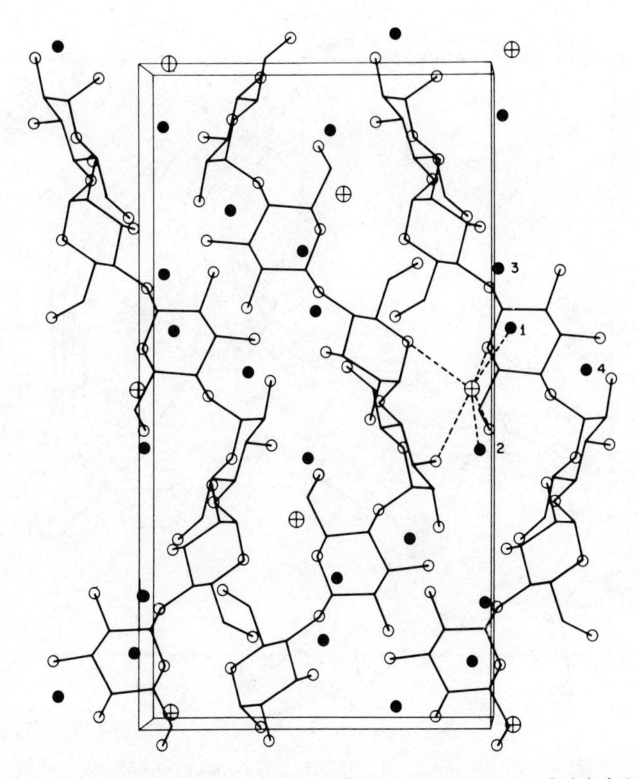

Carbohydrate Research

Figure 6. KOH-amylose in ac *projection. The coordination of the K^+ ion (shown as ⊕) is indicated by dashed lines. Water and OH^- oxygens are denoted by filled, numbered circles (*11*).*

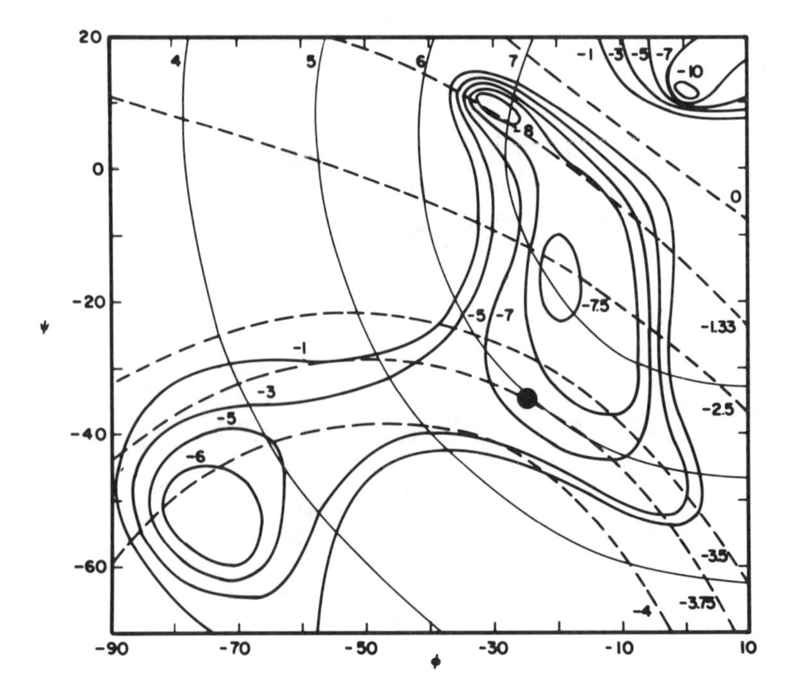

Carbohydrate Research

Figure 7. A section of the ϕ, ψ map for left-handed conformations of amylose, calculated with glycosidic bridge angle of 115° and a virtual bond length of 4.50 Å. The bold contours indicate nonbonded conformational energy in kcal/mol, the thin lines indicate the number of residues per turn (n), *and the dashed lines indicate the axial rise per residue* (h), *in Ångstroms. The position of the conformation of the KOH–amylose is shown in the map by a filled circle* (11).

tension of the helix, KBr-amylose resembles both amylose triacetate I and KOH-amylose.

The Native A, B and C Polymorphs

As is well known, starch granules are naturally crystalline and most starch varieties exhibit either the A or the B diffraction pattern, first described in 1930 by Katz and VanItallie (35). The A pattern is generally seen in cereal starches, whereas the starches of tubers and roots exhibit the B pattern. The much less common C pattern is given by the starches of some beans and banana. Despite the fact that these patterns had been known for nearly 50 years, a detailed analysis of native starch polymorphs was not completed until 1978 (13,14). Undoubtedly, the fact that only powder x-ray patterns are obtainable from ordinary starch granules has been, in part, responsible for this long delay, although Kreger, in 1951, by use of an ingenious microcamera and very large starch granules, was able to record a "fiber" pattern of a B-starch (36). Well-resolved fiber patterns duplicating the native A-, B- and C-starch patterns were finally obtained (17,38), using the method of Senti and Witnauer (32), and the structure analysis was successfully completed (13,14) following the suggestion by Kainuma and French (37) that the native polymorphs were double-stranded helices. These structure analyses showed that both A- and B-amyloses were nearly identical in molecular conformation, while differing considerably in the crystalline packing of the duplex helices. The C-polymorph was shown to be simply a mixture of the A- and B-polymorphs (38).

The main difficulty in all prior attempts to devise molecular models consistent with the A or B x-ray data stemmed from the fact that such models were sought in a ϕ,ψ conformational map for a single-stranded structure. Although suitable models residing in energy minima appeared to exist, none could be packed into the indicated unit cells. Not until double-stranded ϕ,ψ maps were calculated for amylose with different *virtual bond* lengths, were suitable molecular models found. A comparison of such single- and double-stranded maps is shown in Fig. 8. Once again, this illustrates that reliance on only ϕ,ψ maps in attempting crystal structure analysis may sometimes be misleading.

The molecular conformation and interstrand hydrogen bonding of the A- and B-amyloses are shown in Fig. 9. Both strands of the duplex are relatively extended (h = 3.47 Å for B and 3.51 Å for A), although not nearly as extended as the alkali or salt amyloses, or some of the derivative structures. The extension of the helix prevents the formation of the intramolecular O-2...O-3(2) hydrogen bond that occurs in V-amyloses. Conversely, the extended conformation permits the formation of the interstrand hydrogen bonds that appear instrumental in the stabilization of the structure. The two most interesting features of the duplex helix are that the strands are right-handed, as opposed to the left-handed conforma-

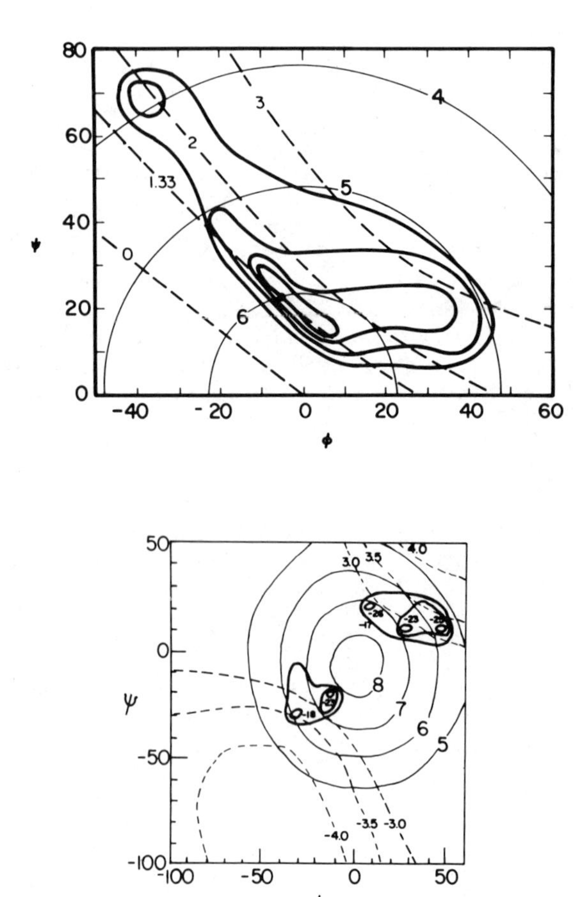

Carbohydrate Research

Figure 8. A comparison of the single-stranded (top) *and double-stranded* (bottom) *φ, ψ maps for the right-handed conformations of amylose* (see *caption of Figure 7 for details)* (14)

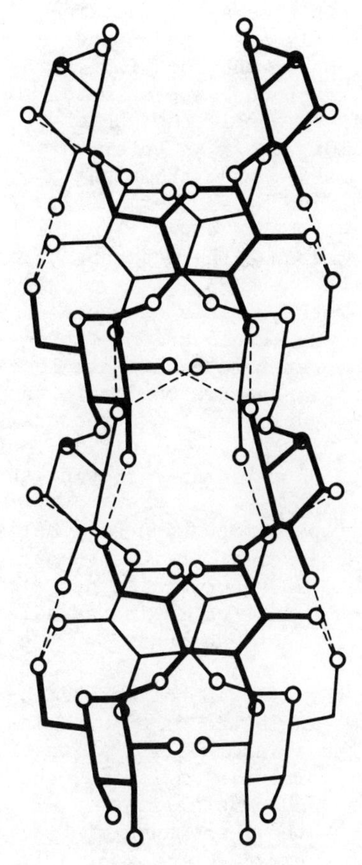

Figure 9. Side view of the double helix of A- and B-amyloses. Interstrand hydrogen bonds are shown by dashed lines (38).

tions of all V- and ionic complex amyloses, and that the packing of the two strands is parallel. Furthermore, the parallel strands of the duplex are packed in phase along their respective helical paths, which causes all odd-order layer lines to be absent in the diffraction pattern. This circumstance presented an added obstacle in the way of an earlier recognition of the true structure of these polymorphs. However, the packing of the duplexes into the crystal lattice is antiparallel in both structures.

The different packing of the A- and B-amylose helices is shown in Fig. 10. Both structures exhibit hexagonal packing (although for the A-amylose it is slightly distorted); however, the main difference between the two lies in the location of the water molecules. In the B-structure, a channel approximately of the same diameter as that of the helix exists in the center of the hexagon of helices. As much as ~30% water can enter this channel. In the A-structure, on the other hand, the hexagon is slightly larger and its center is occupied by another helix, while a smaller amount of water is distributed equally in the interstial spaces between the helices. Because there is no "hole" present in the center of the duplex, as there is in the single helix of V-amylose, water cannot enter the helices of the A- and B-amyloses. Further, in the B-structure, the location of the water in the channel suggests that the water molecules are loosely held and are non-crystalline. This is in agreement with the facile and reversible dehydration-rehydration of the structure upon vacuum-drying and exposure to high relative humidity. The expected changes in the x-ray diffraction patterns following such treatment are observed.

It is also probable that under proper circumstances a helix could displace the water in the open channel of the B-structure, thus converting it to the A-structure. This conversion has also been observed. A mixture of the A and B unit cells in the same structure is also possible and this, in fact, accounts for the C-polymorph.

The double helical structure of the A and B polymorphs is in agreement with the physical properties exhibited by crystalline starch and amylose, particularly in their insolubility in water, lack of complexing ability with small molecules or iodine, and their gelling behavior. The hardening and crystallization of gels of amylose and starch upon aging - the phenomenon of "retrogradation" familiar to starch chemists - may occur as a result of the formation of double helical "junction zones" in the solution of amylose, followed by aggregation and crystallization of the double helices, as shown schematically in Fig. 11.

Because the varieties of starch that contain only amylopectin are also crystalline, exhibiting the same diffraction patterns as starches containing amylose, there is a strong likelihood that the extensively branched amylopectin molecule also crystallizes in a double-helical form. In turn, this implies that linear sequences in amylopectin remain sufficiently long to

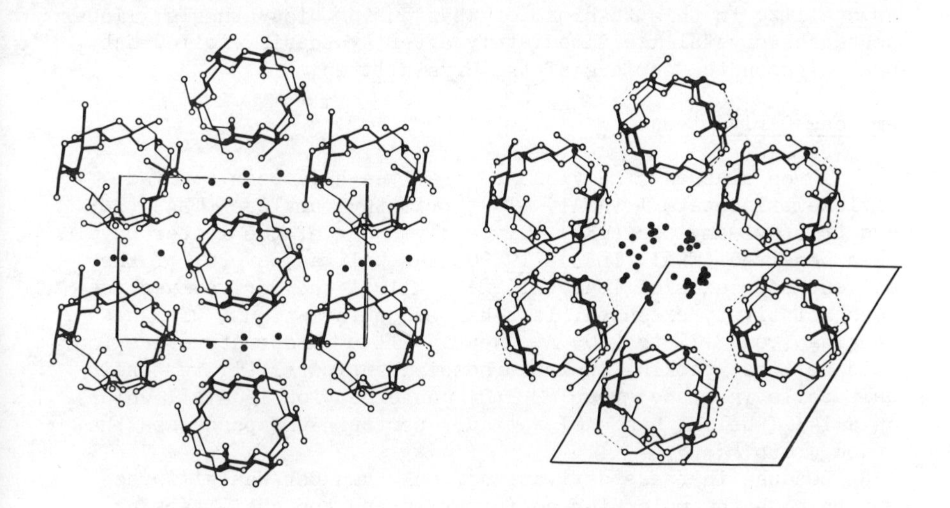

Carbohydrate Research

Figure 10. Comparison of the unit cells and helix packing of (left) *A- and* (right) *B-amylose* (13)

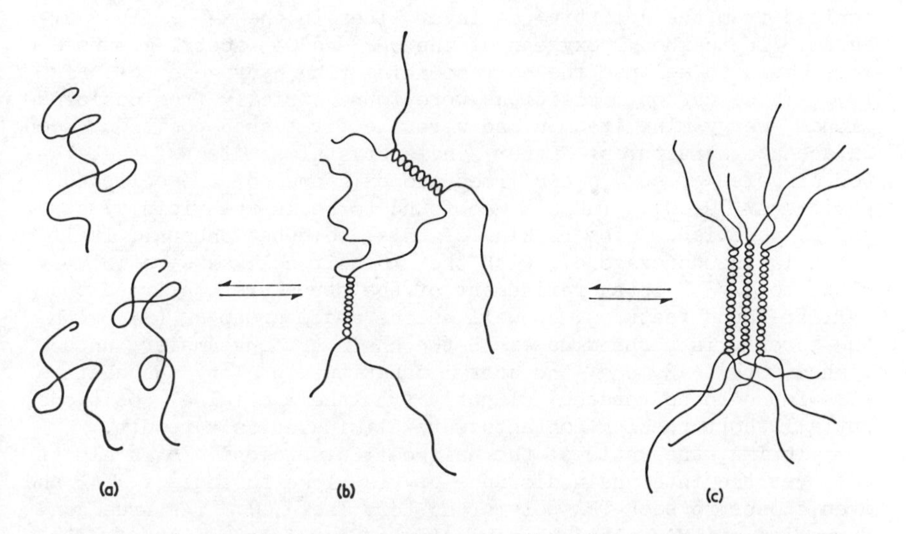

Journal of Molecular Biology

Figure 11. Possible mechanism of crystalline gel formation from aqueous amylose solutions (left: *solution;* middle: *gel;* right: *crystalline gel)* (42)

crystallize in this fashion, or that during biosynthesis, linear sequences crystallize immediately after synthesis, followed by branching on the surface of the crystallites.

Amylose Derivatives

Three classes of amylose derivatives have been studied: amylose triacetate I (ATAI) (16), trimethyl-amylose (TMA) (18) and triethyl-amylose (TEA). Two polymorphs of the latter have been observed (TEA1, TEA3) (19,20), as well as several of the solvent complexes of TEA1 (21,22). All of the derivatives possess relatively extended, left-handed conformations, with h ranging from 3.75 to 4.05 Å. Nearly all of the structure are four-fold helices, packing in orthorhombic, antiparallel, two-chain unit cells in space group $P2_12_12_1$. An exception is ATAI, which as a 14/11 helix, has 4.67 residues per turn and possesses the space group $P2_1(S)$.

Because in these derivatives, only van der Waals' forces govern both the molecular conformation and the chain packing, it is imperative that both are refined simultaneously during the structure analysis. Although this was originally not done with ATAI, its structure was successfully solved by stereochemical methods, more than 10 years ago, the first polysaccharide structure solved by these means. Its lefthanded, non-integral helix, with its very long fiber repeat of 52.53 Å, was essentially recognized from the splitting of layer lines in the x-ray fiber diagram. The carbonyl oxygens of the O-2 and O-3 acetyl groups were found to eclipse the corresponding ring hydrogens (cf. Fig. 12). These carbonyl positions were found strictly from conformational energy minimization and were the first such positions seen in acetate structures. Later single crystal studies of acetylated oligomes of sugars confirmed the existence of the eclipsed positions (39,40). The O-6 was found to be in the vicinity of the tg position. The packing of this, somewhat unusual, 14/11 helix is pseudohexagonal, with the a/b unit cell axis ratio very close to $1/\sqrt{3}$. Later refinement of the structure confirmed the earlier found features, as well as the space group $P2_1(S)$, with the screw axis perpendicular to the chain (P. Zugenmaier, unpublished work). Some of the acetyl substituents along the chain were found to be rotated slightly from their original positions, but all short packing contacts were eliminated as a result.

In TMA, the ratio of the unit cell dimensions a/b is almost 2/1, yet the interchain distance $d_{\uparrow\downarrow}$ is close to that of ATAI and even closer to both TEA polymorphs (cf. Table II). The true conformation of TMA, although nominally a four-fold helix (cf. Fig. 13), is governed by a 2_1 screw axis, as evidenced by a strong second order meridional reflection. The reason for the lower symmetry is that the O-6 methyl groups on two successive residues are in different rotational positions, one close to tg and the other close to gt. The dimer residue is, therefore, the true asymmetric unit in accordance with the $P2_12_12_1$ space group. When

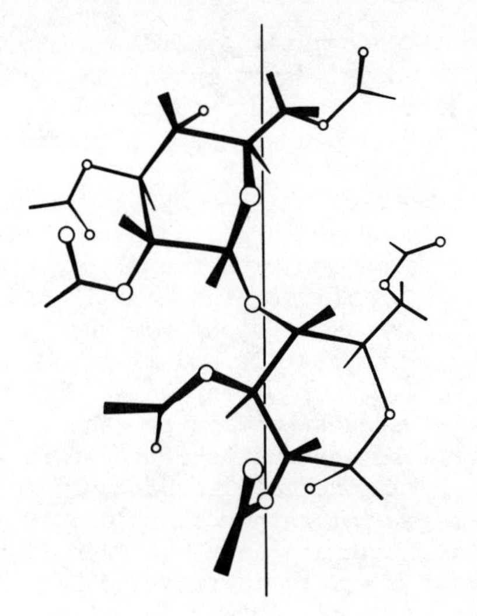

Figure 12. A section of the amylose triacetate helix showing acetate positions

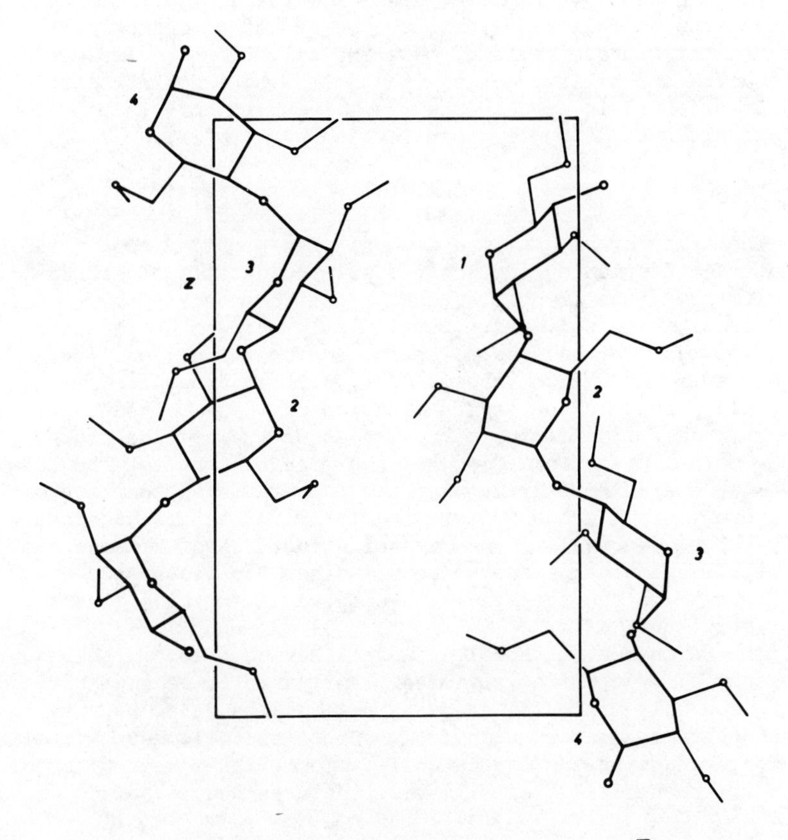

Figure 13. Chain conformation of trimethylamylose, shown in $\bar{1}10$ projection

complexing agents are inserted into the crystal lattice of TMA, solid state transformations occur. These result in all of the methyl O-6 groups rotating into one position and a true four-fold helix is established.

The TEA derivative yields an abundant number of easily obtainable crystalline polymorphs, most of which complex with solvent molecules. The complexes that have been studied exhibit a four-fold, lefthanded helical conformation with the rotation of the ethyl O-6 group again close to *tg*. Introducing a small number of solvent molecules (*e.g.*, chloroform or dichloromethane, one molecule per two glucose units), the TEA1-Cl and TEA1-DCM1 polymorphs are obtained, respectively. The solvent molecules are located both in the interstitial spaces between the chains of the same polarity as well as between the sheets of antiparallel-packed chains. This results in an increase in both $d_{\uparrow\downarrow}$ and d_{110} distances (cf. Table II). As several rather short Cℓ...O distances occur between the complexing agents and the TEA helices, it was concluded that strong dipole interactions force the solvent molecule into the observed positions (21).

Another series of polymorphs is obtained when more complexing agent, such as chloroform, dichloromethane or nitromethane, is added to the TEA1 structure, to the extent of one solvent molecule per glucose residue. Now the solvent molecules are found along the helix grooves with their dipole moments statistically oriented in space. The resulting unit cell is pseudo-tetragonal, but the space group remains $P2_12_12_1$ (22).

Conclusions

The study of crystalline amyloses has provided considerable structural information, not only about the details of structure, but also about the principles that appear to govern the crystallization of these polymers. Particularly noteworthy is the observation that for diverse molecular conformations, only two space groups account for the majority of the structures. It seems clear that in the crystallization of these polysaccharides, energy minimization of the structure is not solely the property of the molecular conformation, as had been thought not too long ago. It is also noteworthy that the *tg* rotational position of the hydroxymethyl group appears regularly. This position has not yet been observed in the crystal structures of glucose dimers or oligomers, thus one wonders whether the occurrence of this rotamer in the polymer is in some way a function of the chain structure.

The refinement methodology has, likewise, been furthered by the study of crystalline amyloses. For example, we now know that given only minimally adequate diffraction data, but having on hand detailed information from monomer and oligomer crystal structures, and having the capability to refine models with complete stereochemical flexibility, structures can be solved with high degrees of reliability. With improvements in intensity measurements on the horizon, as a number of papers in this volume indicate, the reliability of structure analysis should soon approach that of single-crystal analysis. It is already impressively demonstrated by the results of the KBr-amylose study (12). Similarly, it is becoming clear that electron diffraction will play an increasingly important role in future structural work.

And as one final observation, the structures of the native amylose polymorphs suggest once more that synthetic polymer chemists have much to learn from natural polymer processes. At this time, only nature seems to be able to assemble such an elegant supermolecular structure as the starch granule, perfectly suited to its needs.

Acknowledgements

This work has been supported by National Science Foundation grant CHE7727749 (to A. S.) and a grant from Deutsche Forschungsgemeinschaft (to P. Z.). Cooperative efforts of this work have also been supported by a NATO Research Grant No. 1386, to both authors.

Literature Cited

1. Smith, P.J.C., This symposium.
2. Zugenmaier, P.; Sarko, A., This symposium.
3. Zugenmaier, P.; Sarko, A., Biopolymers, 1976, 15, 2121-2136.
4. Winter, W.T.; Sarko, A., Biopolymers, 1974, 13, 1447-1460.
5. Zobel, H.; French, A.D.; Hinkle, M.E., Biopolymers, 1967, 5, 837-845.
6. Rappenecker, G.; Zugenmaier, P., Carbohydr. Res., in press.
7. Winter, W.T.; Sarko, A. Biopolymers, 1974, 13, 1461-1482.
8. Bluhm, T.L.; Zugenmaier, P., Carbohydr. Res., in the press.
9. Hinkle, M.E.; Zobel, H., Biopolymers, 1968, 6, 1119-1128.
10. Booy, F.P.; Chanzy, H.; Sarko, A., Biopolymers, 1979, 18, 2261-2266.
11. Sarko, A.; Biloski, A., Carbohydr. Res., in the press.
12. Miller, D.P.; Brannon, R.C., This symposium.
13. Wu, H.C.; Sarko, A., Carbohydr. Res., 1978, 61, 27-40.
14. Wu, H.C.; Sarko, A., Carbohydr. Res., 1978, 61, 7-25.
15. Rees, D.A.; Scott, W.E., J. Chem. Soc. B, 1971, 469-479.
16. Sarko, A.; Marchessault, R.H., J. Amer. Chem. Soc., 1970, 89, 6454-6462.
17. Blackwell, J.; Sarko, A.; Marchessault, R.H., J. Mol. Biol., 1969, 42, 379-383.
18. Zugenmaier, P.; Kuppel, A.; Husemann, E., ACS Symposium Series No. 48, American Chemical Society: Washington, D.C., 1977, pp. 115-132.
19. Bluhm, T.L.; Rappenecker, G.; Zugenmaier, P., Carbohydr. Res., 1978, 60, 241-250.

20. Bluhm, T.L.; Zugenmaier, P., Carbohydr. Res., 1979, 68, 15-21.
21. Bluhm, T.L.; Zugenmaier, P., Polymer, 1979, 20, 23-30.
22. Bluhm, T.L.; Zugenmaier, P., Progr. Colloid and Polymer Sci., 1979, 64, 132-138.
23. Zugenmaier, P.; Sarko, A., Biopolymers, 1973, 12, 435-444.
24. Quigley, G.J.; Sarko, A.; Marchessault, R.H., J. Amer. Chem. Soc., 1970, 92, 5834-5839.
25. Gress, M.E.; Jeffrey, G.A., Acta Cryst., 1977, B33, 2490-2495.
26. Takusagawa, F.; Jacobson, R.A., Acta Cryst., 1978, B34, 213-218.
27. Hingerty, B.; Saenger, W., J. Amer. Chem. Soc., 1976, 98, 3357-3365.
28. Hybl, A.; Rundle, R.E.; Williams, D.E., J. Amer. Chem. Soc., 1965, 87, 2779-2788.
29. French, A.D.; Murphy, V.G., Carbohydr. Res., 1973, 27, 391-406.
30. Noltemeyer, M.; Saenger, W., Nature, 1976, 259, 629-632.
31. Manor, P.C.; Saenger, W., J. Amer. Chem. Soc., 1974, 96, 3630-3639.
32. Senti, F.R.; Witnauer, L.P., J. Amer. Chem. Soc., 1948, 70, 1438-1444.
33. Winter, W.T., This symposium.
34. Senti, F.R.; Witnauer, L.P., J. Polym. Sci., 1952, 9, 115-132.
35. Katz, J.R.; VanItallie, T.B., Z. Physik. Chem., 1930, A150, 90-99.
36. Kreger, D.R., Biochem. Biophys. Acta, 1951, 6, 406-425.
37. Kainuma, K,; French, D., Biopolymers, 1972, 11, 2241-2250.
38. Sarko, A.; Wu, H.C., Stärke, 1978, 30, 73-78.
39. Leung, F.; Marchessault, R.H., Can. J. Chem., 1973, 51, 1215-1222.
40. Leung, F.; Chanzy, H.D., Perez, S.; Marchessault, R.H., Can. J. Chem., 1976, 54, 1365-1371.
41. Sarko, A.; Marchessault, R.H., J. Polym. Sci., 1969, C28, 317-331.
42. Anderson, N.S.; Campbell, J.W.; Harding, M.M.; Rees, D.A., Samuel, J.W.B., J. Mol. Biol., 1969, 45, 85-99.

RECEIVED May 21, 1980.

29

Some New Polynucleotide Structures and Some New Thoughts About Old Structures

R. CHANDRASEKARAN, STRUTHER ARNOTT, A. BANERJEE, S. CAMPBELL–SMITH, A. G. W. LESLIE, and L. PUIGJANER

Department of Biological Sciences, Purdue University, West Lafayette, IN 47907

For more than a quarter of a century, X-ray fibre diffraction analysis has been the most direct experimental technique to elucidate the structures of nucleic acids. The paucity of diffraction data from these fibres does not permit us to deduce atomic positions from electron density maps as in single crystal structure determination. The linked-atom treatment (1) using bond lengths and bond angles fixed at their respective average values, as obtained from a survey of nucleoside and nucleotide crystal structures enables to redefine the problem to one of dealing with the rotations about single bonds, $\alpha,\beta,\gamma,\delta,\varepsilon,\zeta$ and χ (Figure 1) as the major variable independent parameters. For simplicity of computation, the five-membered sugar ring is quite often frozen at one of the more commonly observed puckered shapes: C3'-*endo* ($\zeta \simeq 85^{0}$) and C2'-*endo* ($\zeta \simeq 156^{0}$) or a minor variant. When appropriate, the endocyclic conformation angles ν_0, ν_1, ν_2, ν_3 (equivalent to ζ) and ν_4, and of necessity, the endocyclic bond angles can be treated as variables tied elastically to standard values. This procedure in effect allows flexibility of the sugar ring shape similar to that observed in single crystals of nucleotides.

Molecular models of polynucleotide helices consistent with the observed axial rise per nucleotide (h) and its turn angle (t), which are accurately measurable from the fibre

0-8412-0589-2/80/47-141-483$05.00/0

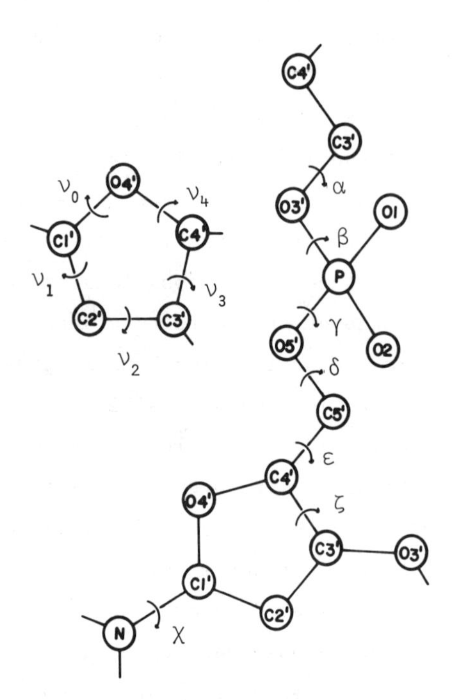

Figure 1. The conformation angles of a nucleotide unit

diffractograms, are refined against the available X-ray data so as to provide satisfactory agreement using the linked-atom least-squares (LALS) procedure (2,3). In the course of the refinement, there are also provisions for the minimization of steric compression and optimisation of hydrogen bond lengths.

Classification of Polymononucleotide Secondary Structures

Usually the asymmetric unit of a helical polynucleotide chain is effectively one nucleotide residue. To date, over two dozen nucleic acid structures of this kind--both native and synthetic and of varying base composition and sequence have been defined accurately and in most cases refined extensively. In addition to establishing the fine details of the individual structures, we are now able to describe a classification of nucleic acids from a structural view point (Table I). These studies bring to light several important observations. For example, all the nucleic acid structures currently known can be classified under one of the two families: *A* with C3'-*endo* sugar pucker and *B* with C2'-*endo* sugar pucker. Those in the *A*-family are always characterized by a positive tilt of the bases when viewed normal to the helix axis (Figure 2), but the large tilt of the eponymous *A*-DNA conformation is not necessary. Duplexes with C3'-*endo* sugars invariably have their base-pairs some distance in front of the helix axis. In contrast, the base-pairs in the *B*-family are generally tilted negatively and positioned either astride the helix axis as in *B*-DNA or behind the helix axis as in *C* and *D*-DNA (see Figure 7b). The location of the bases relative to the helix axis determines the widths and depths of the minor and major grooves of double and triple helices of nucleic acids. Defining the minor groove to be the one in which the two glycosidic bonds of the Watson-Crick base-pair make an obtuse angle, the minor and major grooves are

Table I

CLASSIFICATION OF HELICAL SECONDARY STRUCTURES OF POLYMONONUCLEOTIDES

Family	A		B	
	C3'-*endo*	(Low *anti* χ)	C2'-*endo*	(High *anti* χ)
Genera	tg-g-tg^+	tg-ttt	ttg-tg^+	$ttttt$
Species	19	1	4	1
hÅ	2.6 - 3.4	3.1	3.0 - 3.4	3.25
t°	30.0 - 32.7†	36.0	36.0 - 45.0	48.0
	Large Δh &	Small Δt	Small Δh &	Large Δt
Examples:				
1. Duplex	DNA-DNA	A*-RNA	B-DNA	E-DNA
	RNA-RNA		B'-DNA	
	DNA-RNA		C-DNA	
			D-DNA	
2. Triplex	U.A.U.,I.A.I.			
	dT.dA.dT			
3. Quadruplex	I.I.I.I.			

†Exceptions to this are:

1. Poly(C) h = 3.0Å, t = 60.0°
2. Polymer-Monomer Complexes

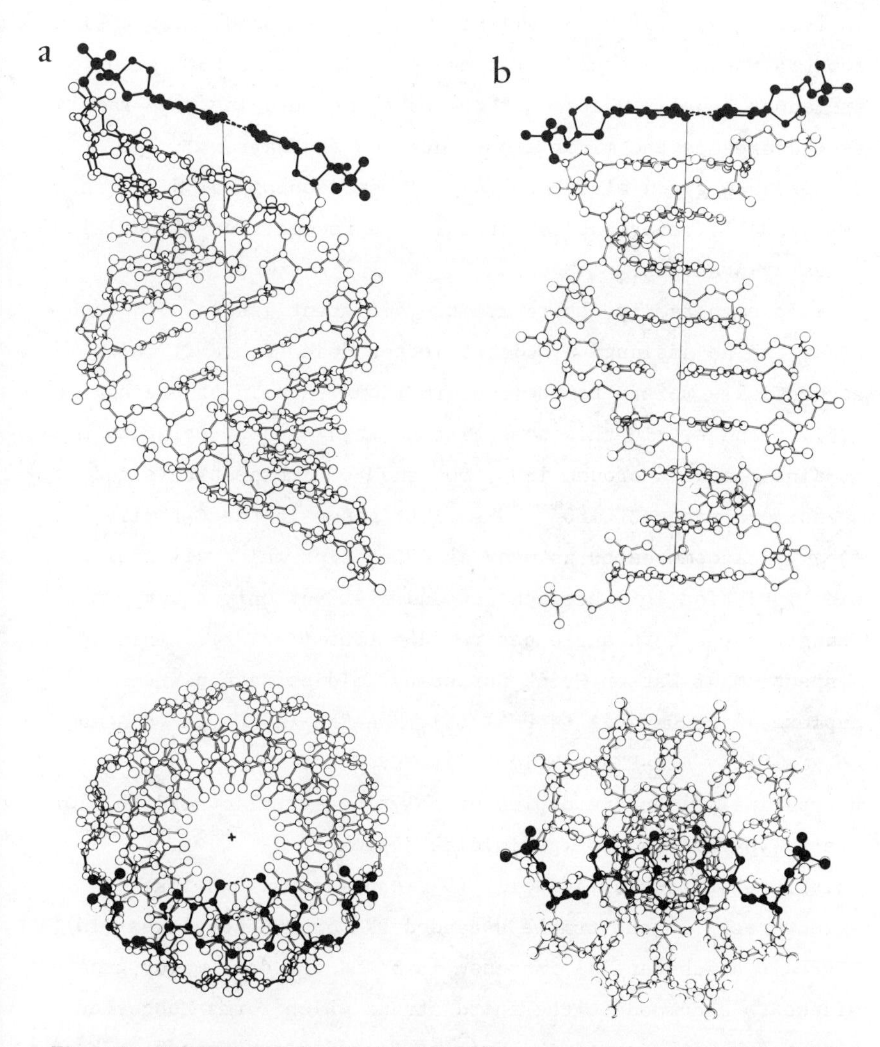

Figure 2. (a) One pitch length of A-*DNA viewed normal to (*top*) and along the helix axis (*bottom*). The top base-pair (thick line) shows the positive tilt (*top*) and the forward positioning relative to the helix axis (*bottom*). (b) One pitch length of* B-*DNA showing that the base pairs are nearly at right angles to (*top*) and astride the helix axis (*bottom*). The helix axis is marked by a vertical line (*top*) or a cross (*bottom*). The hydrogen bonds (dashed line) are shown for the top base pair.*

usually shallow and deep respectively in the *A*-family and it is this deep major groove that becomes wider as a consequence of increasing *h*. In the *B*-family, on the other hand, groove sizes show the opposite trend. For example, in *B*-DNA, both grooves are about equally deep. For the other members in the *B*-family, as *h* decreases the minor groove deepens and narrows. The details are given elsewhere (4). These topological differences are probably important features in the recognition of nucleic acids by proteins.

In each family, there are two different genera. To each genus can be assigned a quintet letter code (Table I) which sequentially refers to the conformational domain of the angles $\alpha, \beta, \gamma, \delta$ and ε. In this nomenclature, t, g^+ and g^- refer to the domains centered around 180°, 60° and -60° respectively and spanning a range of $\pm 60^\circ$. The first genus in the *A*-family, $tg^-g^-tg^+$ accommodates as many as 19 species which all show a wide variation in *h* between 2.6 and 3.4Å but only a very small change in the turn angle per residue ($30^\circ \leq t \leq 32.7^\circ$). This covers a spectrum of Watson-Crick purine-pyrimidine base paired duplexes like DNA-DNA (*A*-DNA) (5), RNA-RNA (*A*-RNA, *A'*-RNA and *A''*-RNA) (6), DNA-RNA hybrids (7). Even the polypurine-polypurine duplex of polyriboxanthylic acid (8), or the polypyrimidine-polypyrimidine duplex of polyribo-2-thiouridylic acid (9), can adopt *A*-DNA-like structures. In the triple stranded DNA or RNA complexes (10), there is a substantial increase in *h* and the deep major groove widens to accommodate the third strand which forms Hoogsteen base-pairs with one of the Watson-Crick duplex strands. There are also isomorphous four-stranded structures with all parallel strands for poly (I) and poly (G) (11,12,13). The three single-stranded 6-fold structures of poly (C) (14), poly (Cm) (15) and poly (Ce) (16) ($t = 60^\circ$ and $h = 3$Å) also belong to this genus of the *A*-family and are notable in being stabilized by base-stacking interactions alone. Quite recently, we have found that polymer-monomer complexes (17) such as poly (C)·pG and poly (U)·A·poly (U), are virutally isomorphous with their polymer-polymer analogues.

The second genus in the *A*-family, *tg-ttt*, is represented by one DNA-RNA hybrid, poly d(I)·poly (C) ($t = 36^{\circ}$ and $h = 3.1$Å) (18).

So far, only DNA duplexes with Watson-Crick base-pairing have been observed in the *B*-family. There are four distinct species in the $ttg^{-}tg^{+}$ genus: *B*-DNA (19), *B'*-DNA, *C*-DNA and *D*-DNA (20) which are characterized by $3.0 \leq h \leq 3.4$Å and $36^{\circ} \leq t \leq 45^{\circ}$. The structure of poly d(IIT)·poly d(CCA) or *E*-DNA (20) is the sole member of the *ttttt* genus with $t = 48^{\circ}$ and $h = 3.25$Å.

Thus poly d(I)·poly (C) in the *A*-family and *E*-DNA of the *B*-family are extremely important as they have led to the diversification of the two families.

A-DNA-like Structures of DNA-RNA and RNA-RNA Duplexes

The great significance of the orthodox *A*-DNA structure cannot be underestimated as the intensity distribution of the *A*-DNA diffraction pattern is repeatedly observed on three different occasions viz., for the RNA-DNA hybrid poly (A)·poly d(T), the polypurine duplex poly (X)·poly (X) and, finally, the polypyrimidine duplex poly (s^2U)·poly (s^2U) (Figure 3). The close resemblance of the 'fingerprints' between all of them does indeed suggest that the *A*-DNA sugar-phosphate chain is almost unaltered in all these structures. It is achieved almost immediately for the hybrid, as it involves just the conventional purine-pyrimidine base pairing (Figure 4a) and hence the *A*-DNA structure can be substituted for this model with only the addition of the 2'-OH for the RNA strand. This addition does not produce steric anomalies.

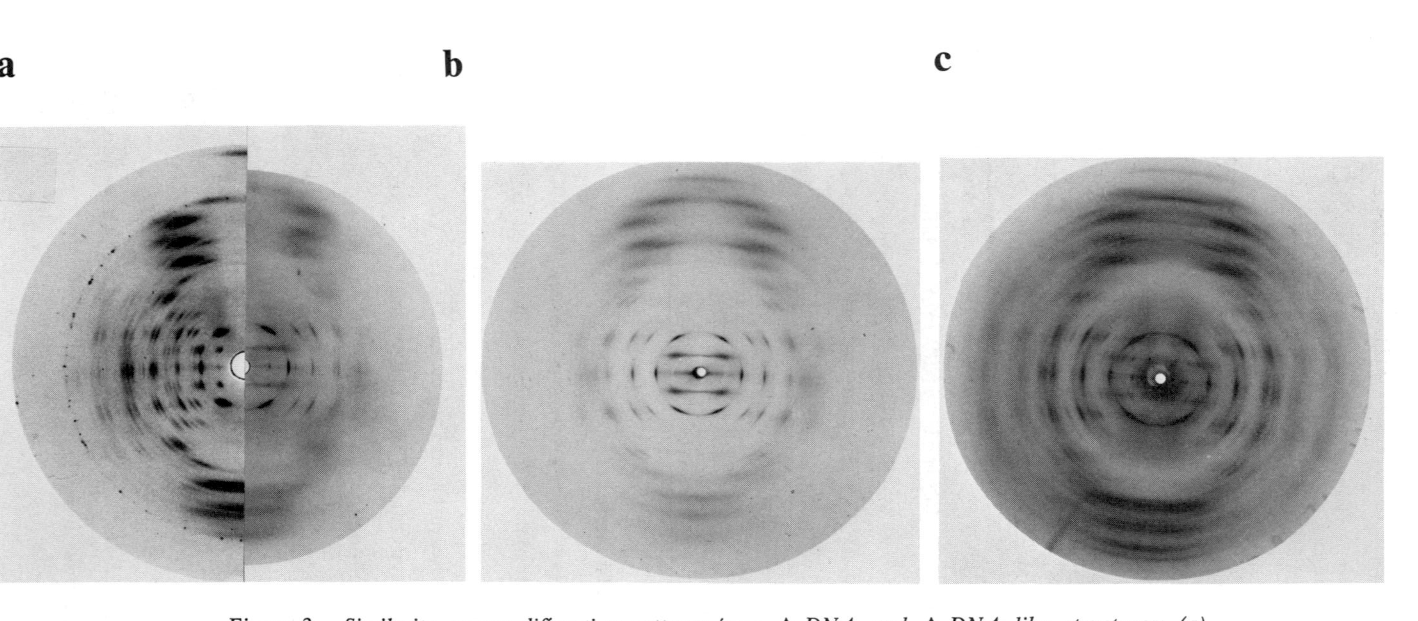

Figure 3. Similarity among diffraction patterns from A-*DNA and* A-*DNA–like structures: (a) composite of* A-*DNA* (left) *and poly(A) · poly d(T)* (right)*; (b) poly(X) · poly(X); and (c) poly-(s^2U) · poly(s^2U)*

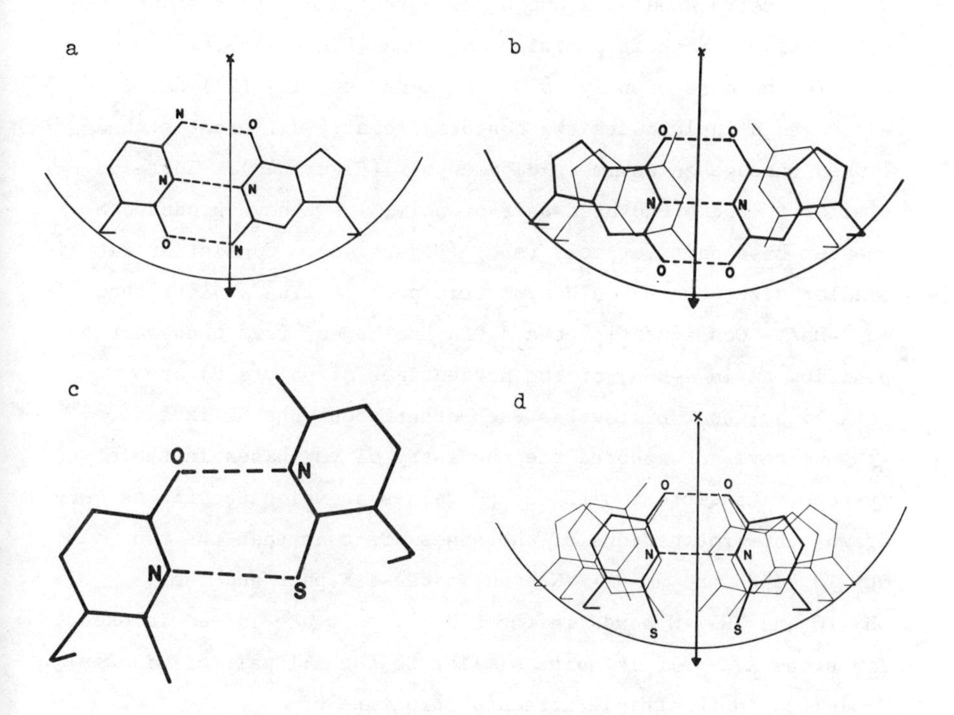

Figure 4. Proposed base-pairing schemes that accommodate the A-*DNA-like structures.*

(a) Watson–Crick purine:pyrimidine; (b) the broad X:X; and (d) the narrow s^2U:s^2U base pairs. The glycosidic bonds in the pairs are related by dyads indicated by the arrows. The C1′ . . . C1′ distances are 10.8, 12.8, and 9.2 Å, respectively. The helix axis is marked by a cross. The circular arc is the locus of the phosphate group of A-*DNA in this projection which remains virtually the same in all three structures (also are marked in thin line in (b) and (d) the pu:py pair and in (d) the X:X pair for mutual comparison). (c) The asymmetric s^2U:s^2U base pair of Mazumdar et al.* (21).

There exists a symmetric purine-purine base-pair for poly (X) which is nearly 2Å broader than the Watson-Crick base-pair (Figure 4b). To accommodate this broad base-pair and yet retain the molecular diameter nearly the same as in *A*-DNA, the *A*-DNA backbones of the nucleotides in the two antiparallel strands are rotated by about 6° away from each other about the helix axis from their position in *A*-DNA (Figure 4a).

In the case of poly ($s^{2}U$), Mazumdar et al., (21) had envisaged a duplex with two conformationally different strands linked through an asymmetric base-pair (Figure 4c). Instead, similar to the broad base-pair of poly(X), we have produced a shorter base-pair for poly ($s^{2}U$) (Figure 4d). Consistent with its smaller size, the $s^{2}U:s^{2}U$ symmetric pair is 1.6Å smaller than in *A*-DNA. Consequently, the *A*-DNA backbones, from the starting position as in *A*-DNA, of the nucleotides of poly($s^{2}U$) are rotated by about 6° towards each other about the helix axis.

We have not ignored the chemistry of the bases in these systems. Consistent with the pH values at which the fibres were drawn, the protonations of the bases are such that the two OH···O bonds and an NH···N bond in the X:X pair and the OH···O and NH···N bonds in the $s^{2}U:s^{2}U$ pair are indeed in order. The broad X:X pair is quite similar to the A:I pair originally exploited in the triple stranded structure of poly(I)·poly(A)·poly(I) (10).

Remarkable Behavior of Polymer-Monomer Complexes in the *A*-family

As mentioned earlier, the $tg^{-}g^{-}tg^{+}$ genus is thickly populated with a wide variety of duplexes, triplexes and a quadruplex which all have t in a very narrow range of 30° to 32.7^{o}. But the occurrence of the single-stranded poly(C) (14) and its alkylated derivatives (15,16), all having the largest t of 60°, in this genus is exceptional. This in fact is accomplished by a reduction in the values of α to about -123°, the monomer standard value (22).

Poly(C), when complexes with pG (guanosine-5'-monophosphate) in the ratio 1:1, changes its conformation so as to resemble that of the polymer-polymer duplex poly(A)·poly(U) (6). This is strikingly obvious from the composite diffraction shown in Figure 5a which suggests strong similarity in their intensity distributions and hence structural similarities. Although the pitch of the molecule is nearly the same in both cases (c = 28.3Å), poly(C)·pG forms a 10_1 helix in contrast with the 11_1 helix for the other.

Further comparison of the X-ray diffractograms (Figure 5b) from the polymer-monomer complex of 2 poly(U)·A with the corresponding polymer analogue of 2 poly(U)·poly(A) (23) leads to a similar conclusion that in this case also stacks of monomers are able to substitute the role of the polymer strand in the triplex.

Thus, structurally similar to their respective $tg^{-}g^{-}tg^{+}$ polymer analogues, the monomers in the polymer-monomer complexes, are all interlinked through O3'···O (phosphate) hydrogen bonds in the duplex and through O3'...O5' hydrogen bonds in the triplex (17).

Discovery of the Original Crick-Watson Double Helix

Although all the RNA-RNA and DNA-RNA duplexes we have examined so far belong to the same $tg^{-}g^{-}tg^{+}$ genus in the A-family, there is the exception of poly d(I)·poly(C) whose diffraction pattern is shown in Figure 6a. The strong meridional reflection occurs on the 10th layer line (h = 3.1Å) which assigns the largest turn angle of $t = 36^{\circ}$ for a duplex in the A-family. These features lead us to a new set of conformational

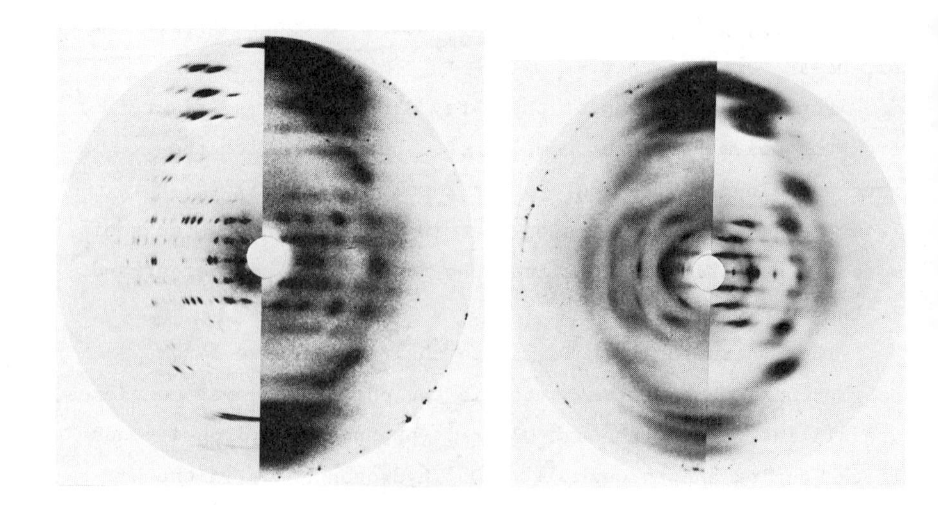

Figure 5. Composite diffraction diagrams of: (a) the 11_1 form of poly(A) · poly(U) (left) *and the 10_1 form of poly(C) · pG* (right)*; and (b) the 11_1 form of 2 poly(U) · A* (left) *and the screw disordered 11_1 form of 2 poly(U) · poly(A)* (right)

parameters characterized by the genus *tg⁻ttt* for this *A**-RNA. This is essentially the same structure published by Crick and Watson (24) and supports their general proposal for the structure of DNA. Although they intended to model *B*-DNA, once they had assigned the wrong sugar conformation, the other conformationl peculiarities followed.

Topologically, however, the positive tilt and a forward displacement of the bases (Figure 6b) as well as the gross groove dimensions of *A**-RNA are quite reminiscent of the other duplexes in the *A*-family.

Old Structures Revised and New Structures in the *B*-family

Until a short time ago, there was just a single genus $tg^-g^-tg^+$ which accommodated all the known DNA duplexes such as *B*,*C* and *D*-DNA in the *B*-family. Recently, however, in the course of the detailed re-refinement of the structure of *B*-DNA employing newly measured intensity data set and allowing for the flexibility of the sugar rings (19), we have discovered that the conformation angles β and γ , defining the internucleotide linkage, are tg^- $(-157^o,-41^o)$ rather than g^-g^- $(-99^o,-39^o)$ (5) used in the past. Consequently, we have replaced the previous genus ($tg^-g^-tg^+$) by the current one denoted by ttg^-tg^+.

The structure of *C*-DNA has hitherto been not refined since its X-ray diffraction (25) corresponded to a screw-disordered (26) fiber. We now have *C*-DNA in a polycrystalline system containing 9_1 helices of the synthetic DNA poly d(GGT)·poly d(CCA) (Figure 7a). Detailed X-ray analysis shows that *C*-DNA belongs to the same ttg^-tg^+ genus as both $B(10_1)$ and *D*-DNA (8_1). Figure 7b shows the negative tilt for the bases and the backward positioning of the base-pairs relative to the helix axis.

Substitution of I for G produces an entirely different, complex diffraction pattern (Figure 8a) which demonstrates the

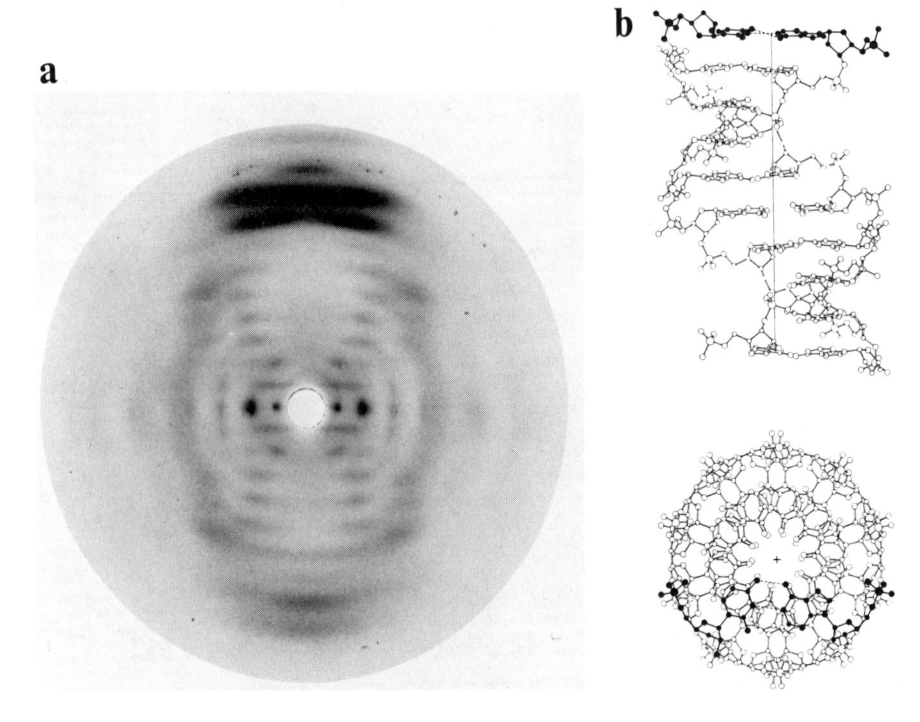

*Figure 6. (a) X-ray diffraction pattern of the 10_1 form of poly d(I) · poly(C); and (b) a pitch length of this structure viewed normal to (*top*) and along the helix axis (*bottom*). Note the small positive tilt of the bases and the forward positioning of the base-pairs characteristic of the A-family.*

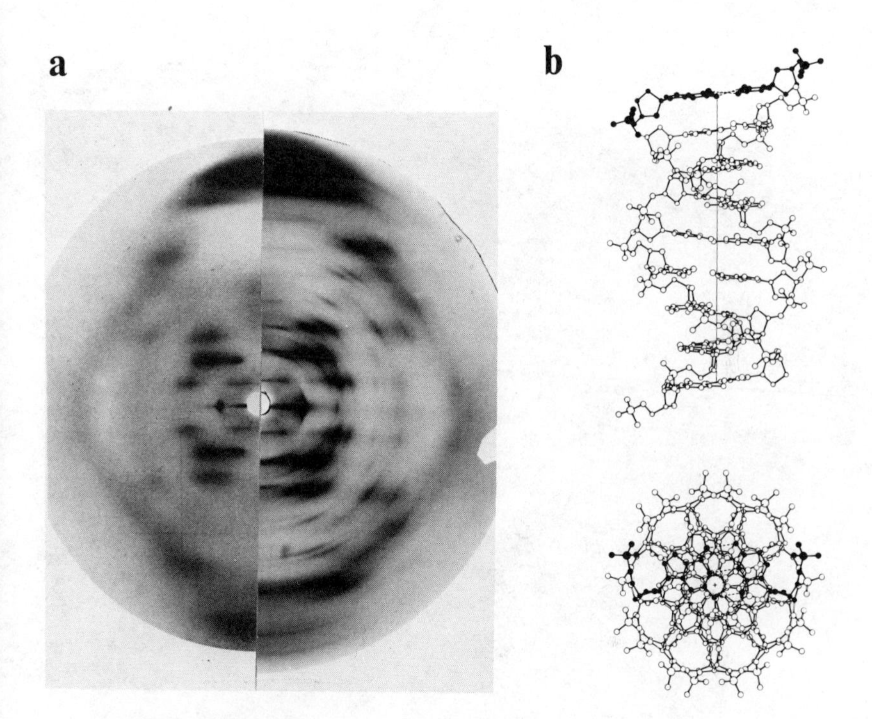

Figure 7. (a) Composite diffraction pattern from C-*DNA. Screw disordered,* 28_3 *form of native DNA* (left) *and crystalline* 9_1 *form of poly d(GGT) · poly d(CCA)* (right)*; (b) views of the* 9_1 *duplex helix normal to* (top) *and along the helix axis* (bottom)*, showing the negative tilt and backward displacement of the bases.*

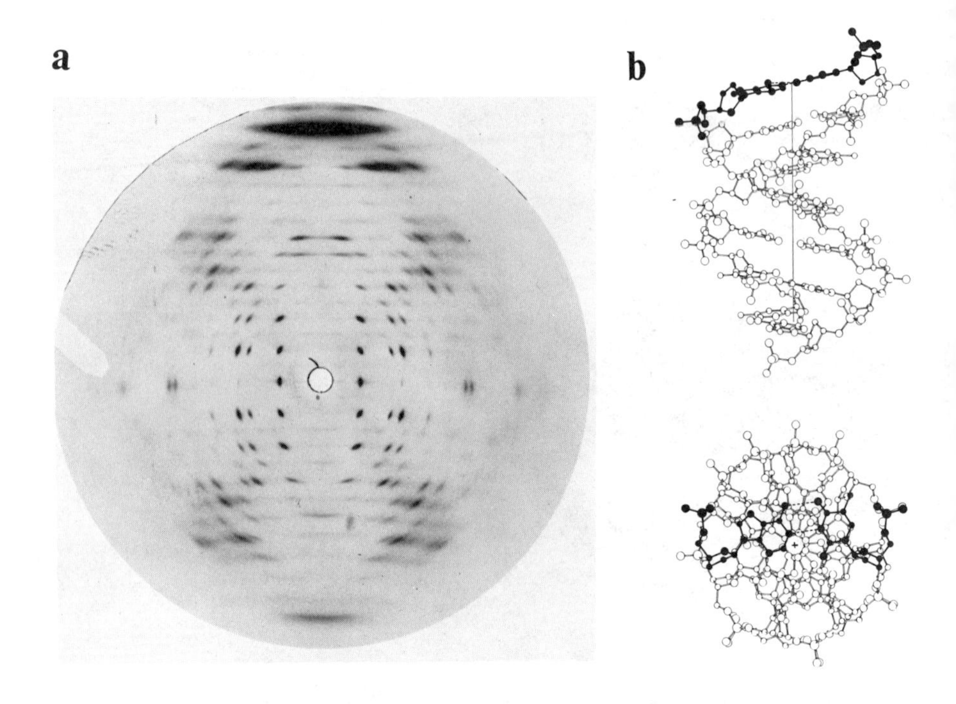

Figure 8. (a) Diffraction from poly d(IIT) · poly d(CCA) and (b) views of E-*DNA normal to (*top*) and along the helix axis (*bottom*)*

influence of base sequence on the secondary structure. This is all the more appealing since one strand is the same in both cases. Although an I:C pair is known to behave like an A:T pair, this duplex of poly d(IIT)·poly d(CCA) is not similar to that of the *D*-form of poly d(AAT)·poly d(TTA) (27). Instead, retaining *h* (=3.25Å) the same as in *C*-DNA, *t* has increased to 48°, 8° more than in *C*-DNA, so as to produce the first 15_2 helix, *E*-DNA. This new DNA structure has attained such a large turn angle per residue mainly by adopting the all *trans* conformational angles for the backbone. Thus, as with *A**-RNA, we have the first member of a new genus *ttttt* for *E*-DNA in the *B*-family.

Two views of the 15_2 helix of *E*-DNA shown in Figure 8b clearly demonstrate the morphological resemblance with *C*-DNA. An interesting feature of *E*-DNA diffractogram in Figure 8a is the presence of additional meridional reflections on every third layer line which makes the 'true' *E*-DNA double helix to have a turn angle of 240° for each pentanucleotide as the real repeating unit.

Left-handed Polydinucleotide Helices

Increasing the asymmetric unit from a mono- to a dinucleotide obviously increases the conformational flexibility of the nucleic acid structure. When the two nucleotides are only minor variants of one another, such as the differences between them are one of degree rather than kind, the resulting structure would still be of the same kind as that of the parent polymononucleotide helix. On the other hand, if the two nucleotides appear in quite different conformational domains, this could eventually lead to unusual secondary structures. For exampls, we have recently obtained from the synthetic DNA poly d(GC)·poly d(GC) diffraction patterns which can be interpreted in terms of a left-handed helix with sixfold screw symmetry and h = 7.25Å for a dinucleotide (28). The polydinucleotide model incorporates C3-*endo syn*-G and C2-*endo anti*-C nucleoside conformations. Similar structures are observed with polymers containing other alternating purine-pyrimidine base sequences.

Conclusion

It is extremely remarkable that there is as much diversity in the small number of species in the *B*-family as the overwhelming conformational similarity in the more populated *A*-family. Obviously, such a multitude of secondary structures should stem from the flexibility of the nucleotide units in both families. Our new findings demonstrate the conservation of the orthodox *A*-DNA conformation even in the novel RNA duplexes involving only polypurine strands (poly(X)) and those involving only polypyrimidine strands (poly(s^2U)) and, what is more, have shown that some polymer-monomer complexes can very readily resemble their respective polymer analogues. The discoveries of *A**-RNA and *E*-DNA and hence the additions of two new genera to the taxonomic tree of nucleic acids are noteworthy.

Finally, the recent discovery of the unusual left-handed polydinucleotide helices for the alternating purine-pyrimidine polymers opens up a whole new field of nucleic acid secondary structures. The relevance and importance of these new structures in the visualization of overwound, as well as underwound, and supercoiled DNA molecules in biological systems need hardly be emphasized.

This work was generously supported by the U.S. Public Health Service (Grant #GM17371).

Literature Cited

1. Arnott, S.; Wonacott, A. J. Polymer, 1966, 7, 157.
2. Arnott, S.; Dover, S. D.; Wonacott, A. J. Acta Cryst., 1969, B25, 2192.
3. Smith, P.J.C.; Arnott, S. Acta Cryst., 1978, A34, 3.
4. Arnott, S. "Organization and Expression of Chromosomes"; Allfrey, V. G.; Bautz, E.K.F.; McCarthy, B. J.; Schimke, R. T.; Tissieres, A., Eds. Dahlem Konferenzen: Berlin, 1976, p. 209.
5. Arnott, S.; Campbell Smith, P. J.; Chandrasekaran, R. Nucl. Acids, 1976, 2, 411.
6. Arnott, S.; Hukins, D.W.L.; Dover, S. D.; Fuller, W.; Hodgson, A. R. J. Mol. Biol., 1973, 81, 107.
7. O'Brien, E. J.; MacEwan, A. W. J. Mol. Biol., 1970, 48, 243.
8. Arnott, S.; Chandrasekaran, R.; Day, W. A.; Puigjaner, L; Watts, L. J. Mol. Biol., in preparation.
9. Arnott, S.; Chandrasekaran, R.; Leslie, A.G.W.; Puigjaner, L.; Saenger, W. J. Mol. Biol., in preparation.
10. Arnott, S.; Bond, P. J.; Selsing, E.; Smith, P.J.C. Nucl. Acids Res., 1976, 3, 2459.
11. Arnott, S.; Chandrasekaran, R.; Martilla, C. M. Biochem. J., 1974, 141, 537.
12. Chou, C. H.; Thomas, G. J.; Arnott, S.; Campbell Smith, P. J. Nucl. Acids. Res., 1977, 4, 2407.
13. Zimmerman, S. B.; Cohen, G. H.; Davies, D. R. J. Mol. Biol., 1975, 92, 181.
14. Arnott, S.; Chandrasekaran, R.; Leslie, A.G.W. J. Mol. Biol., 1976, 106, 735.
15. Leslie, A.G.W.; Arnott, S. J. Mol. Biol., 1978, 119, 399.
16. Banerjee, A.; Chandrasekaran, R.; Leslie, A.G.W.; Arnott, S.; Shugar, D. J. Mol. Biol., in preparation.

17. Smith, S. C. M.S. Thesis, Purdue University, 1978.
18. Banerjee, A.; Arnott, S.; Leslie, A.G.W.; Selsing, E. J. Mol. Biol., in preparation.
19. Arnott, S; Chandrasekaran, R. J. Mol. Biol., in preparation.
20. Chandrasekaran, R.; Arnott, S.; Banerjee, A.; Leslie, A.G.W.; Selsing, E. J. Mol. Biol., in preparation.
21. Mazumdar, S. K.; Saenger, W.; Scheit, K. H. J. Mol. Biol., 1974, 85, 213.
22. Arnott, S.; Hukins, D.W.L. Biochem. J., 1972, 130, 453.
23. Arnott, S.; Bond, P. J. Nature New Biol., 1973, 244, 99.
24. Crick, F.H.C.; Watson, J. D. Proc. Roy. Soc., 1954, 223A, 80.
25. Marvin, D. A.; Spencer, M.; Wilkins, M.H.F.; Hamilton, L. D. J. Mol. Biol., 1961, 3, 547.
26. Arnott, S. Trans. Am. Crystallogr. Assoc., 1973, 9, 31.
27. Arnott, S.; Selsing, E. J. Mol. Biol., 1974, 88, 509.
28. Arnott, S.; Chandrasekaran, R.; Birdsall, D. L.; Leslie, A.G.W.; Ratliff, R. L. Nature, 1980, in press.

RECEIVED February 19, 1980.

INDEX

R

S